D0202775

Elementary Functions

Elementary Functions

Adil Yaqub

University of California
Santa Barbara

Houghton Mifflin Company Boston

Atlanta · Dallas · Geneva, Illinois
Hopewell, New Jersey · Palo Alto · London

To Nancy

Printed in the United States of America

Library of Congress Catalog Card Number: 73–11772

ISBN: 0–395–17093–1

Contents

Preface

This book is primarily designed to provide the student with the subject matter needed for the study of calculus. The text has been written with the recommendations of the Committee on the Undergraduate Program in Mathematics (CUPM) in mind. Since this book contains *more* precalculus topics than recommended by the CUPM, it may also be used as a text for an integrated algebra and trigonometry course. Some familiarity with elementary algebra is assumed.

Throughout the book, the author has tried to present the subject matter in a mathematically honest fashion, with particular emphasis on clarity, logical development, and sound pedagogy. It is the author's belief that students at all levels, and especially at this level, learn best by means of examples. For this reason, many fully worked out examples, illustrating the various concepts involved, are found in every section of the text. Also, a large number of problems, ranging from routine to challenging, appear in every chapter. These exercises form an integral part of the text and should be seriously attempted in order to gain a thorough understanding of the subject matter. Graphs and answers to odd-numbered problems appear at the end of the text. Also, a brief summary which presents the highlights of each chapter appears at the end of the chapter. We now give a brief description of each chapter.

Chapter 1 After an introductory, but thorough, section on sets, this chapter deals with functions and their properties. In particular, the notion of inverse of a function is treated with care, and several examples are given to clarify this concept. Graphic representation of functions is also considered. Other basic concepts, such as domain, codomain, range, one-to-one and onto functions, and composition are given a thorough and clear treatment. These concepts permeate the entire text.

Chapter 2 In this chapter, we deal with the simplest types of functions, namely, constant, linear, and quadratic functions, with particular emphasis on the graphs of these functions. In attempting to find the x intercept points of the graph of a quadratic function, we are led in a natural way to a study of the quadratic formula. Also, a method for finding the maximum or minimum point of the graph of a quadratic function is discussed, and many examples are given.

Chapter 3 Absolute values and inequalities are discussed in this chapter with extreme care, and frequent references to geometric interpretations are given. It is the author's experience that students at this level, and even later, have difficulty with the concept of absolute value and with solving inequalities. For this reason, we give a detailed discussion, accompanied by numerous examples, of the absolute value function, together with its graph and properties. We then turn our attention to solving inequalities that involve certain polynomial or rational expressions, or absolute values of such expressions. We also sketch the solution sets of these inequalities in order to give the student geometric interpretations of the results.

Chapter 4 In this chapter, we study complex numbers, together with their properties and the operations associated with them. We also give geometric interpretations throughout. Moreover, important concepts, such as complex conjugate and absolute value, included with equalities and inequalities relating to these concepts, are discussed in detail. Again, geometric interpretations of the results are stressed.

Chapters 5 and 6 We give in these two chapters a thorough exposition of polynomial functions and their properties and an especially thorough discussion of their zeros. Thus, the Remainder theorem and its important corollary, known as the Factor theorem, are proved and illustrated by numerous examples. We then apply the Factor theorem to find the zeros of polynomials. In order to determine the zeros of polynomials with *real* coefficients, we discuss theorems such as the Location theorem, the Upper Bound and Lower Bound theorems, the Rational and Integral Zeros theorem, and the Complex Conjugates theorem. In order to determine the zeros of polynomials with *any complex* (i.e., real or nonreal) coefficients, we discuss the important theorem, known as the Fundamental Theorem of Algebra, and the general form of this theorem. We do not give a proof of the Fundamental Theorem of Algebra becase a proof of this theorem is well beyond the scope of this text. We also discuss a method for finding rational approximations for irrational zeros of polynomials with real coefficients. Finally, we discuss briefly graphs of polynomial functions.

Chapter 7 Here, we discuss rational functions and manipulations involving rational functions. Linear rational functions are treated in detail, and their graphs give rise in a natural way to the study of horizontal and vertical asymptotes. Graphs of arbitrary rational functions are also considered in detail, and numerous examples are given. The important concepts of symmetry and asymptotes are emphasized throughout.

Chapter 8 In this chapter, we give an intuitive account of the laws of exponents, in which the exponent may be *any* real number. We also consider manipulations involving rational exponents as well as radicals. Logarithms are then introduced as inverses of exponential functions, and the graphs of both exponential and logarithmic functions are studied and sketched. The laws of logarithms are also established, and the use of logarithms in computations is discussed. Finally, certain applications of exponents to problems in growth, decay, and compound interest are given.

Chapters 9 and 10 These two chapters are devoted to the study of trigonometric functions, their properties, and their graphs. Trigonometric functions may be viewed as functions of real numbers or as functions of angles, and both points of view are carefully discussed. Also, the trigonometric functions are all periodic functions—an important consideration in sketching their graphs. Numerous trigonometric identities are established, and some of these identities are used to shed some light on the graphs of the trigonometric functions. These graphs are discussed in detail and then sketched. Moreover, solutions of triangles, along with the important laws of sines and cosines, are considered. Also, the polar form of a complex number, together with an important theorem, known as De Moivre's theorem, is introduced. The application of De Moivre's theorem to finding the *n*th roots of complex numbers is then discussed, and the results are interpreted geometrically. Finally, inverse trigonometric functions are introduced, and a brief discussion of their graphs and properties is given.

Chapter 11 This chapter is concerned with analytic geometry in the plane. The distance formula and the slope formula are first derived. Next, an extensive discussion of straight lines, together with their equations and properties, is given. Then circles are considered, followed by a detailed discussion of other conic sections (i.e., parabolas, ellipses, and hyperbolas). Numerous equations of conic sections are discussed, and their graphs are carefully sketched.

Chapter 12 In this final chapter, we give a brief account of three important topics: mathematical induction, progressions, and the Binomial theorem. We give numerous illustrations of proofs by induction. We also indicate situations in which induction does not work. As further illustrations of proofs by mathematical induction, we use induction to prove the formulas for the sum of an arithmetic progression and for the sum of a geometric progression. We also use mathematical induction to establish the Binomial theorem.

This book may be used as a text for the following courses:

1. One-semester precalculus course meeting 5 times per week. The entire text can be covered in such a course.
2. One-semester precalculus course meeting 4 times per week, or a two-quarter course meeting 3 times per week. In such a course, the entire text, with the possible exception of Secs. 3.2, 3.3, and 8.4, can be covered.
3. One-quarter precalculus course meeting 5 times per week. In such a course, the entire text, with the possible exception of Secs. 3.2, 3.3, 8.4, 12.1, 12.2, and 12.3, can be covered.
4. One-semester course in algebra and trigonometry meeting 4 times per week, or a two-quarter course meeting 3 times per week. In such a course, the entire text, with the possible exception of Chap. 11, can be covered.
5. One-year course in the final year of high school. In such a course, the entire text can be covered *at a leisurely pace.*

The author wishes to express his sincere thanks and deep appreciation to Professors V. C. Cateforis of SUNY, Potsdam; L. J. Chatterley of Brigham Young University; R. A. Fritz of Moraine Valley Community College, Palos Hills, Illinois; and R. A. Little of Kent State University, Stark Regional Campus. Their thorough reviews and constructive criticisms have been of great help in improving the quality of the text. Professor Little has kindly reread the entire manuscript and has made some further helpful suggestions for which the author is very grateful.

The author also wishes to express his great appreciation to Professors J. Chunn of Fullerton Junior College, Fullerton, California, and S. Cotter of Foothill College, Los Altos Hills, California, for reading the early part of the manuscript. Their valuable suggestions and comments, too, have resulted in improving the quality of this text. It is also a pleasure to acknowledge the superb work of Louise Kraus, Sonia Ospina, and, above all, Delores Brannon, for the fine job they did in typing the manuscript. My deep thanks go to Barbara Federman for the splendid job she did in drawing the figures. The author also wishes to express his sincere appreciation to the staff of Houghton Mifflin for their excellent professional work in producing this book.

All answers and solutions have been checked for accuracy, and it is hoped that the text is relatively free of errors. The author will be very grateful for any corrections, criticisms, or suggestions.

Adil Yaqub

1 Functions

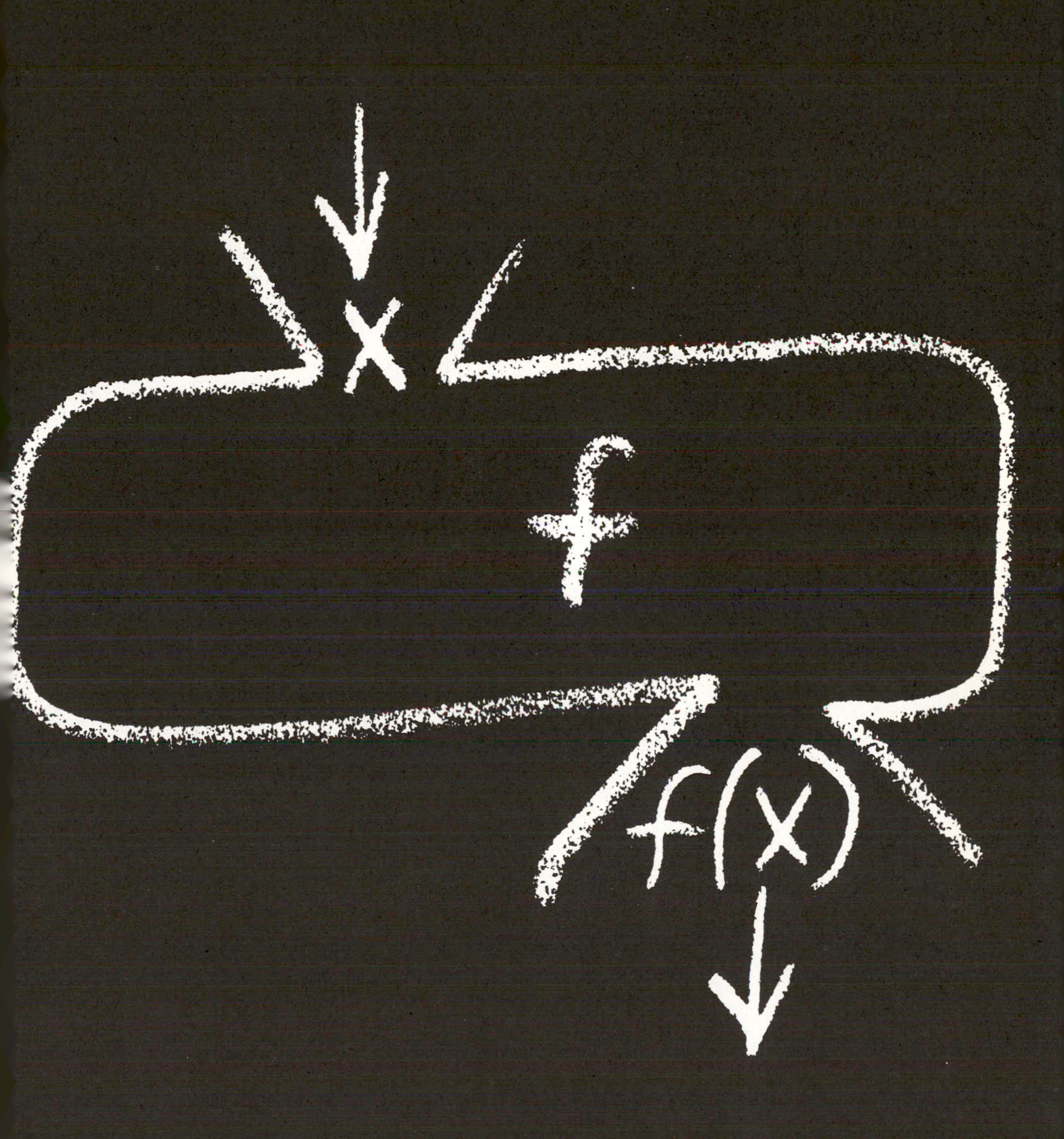

1.1 Sets

A *set* is a collection of objects. These objects need not be numbers. Thus, one may speak of the set of students in a college or the set of chairs in a room. In order to describe a set, a rule must be given by which we can tell whether a given object is in our collection. The objects which belong to the set are called *elements* of the set.

It is extremely convenient and customary to use the following notation: If S is a set in which the elements are given by a certain rule, we write $S = \{x|x$ satisfies the given rule$\}$, read as "S is the set of all x such that x satisfies the given rule." Thus, $\{x|x$ is a positive integer$\}$ is the set of all positive integers, while $\{x|x$ is a real number and $-3 < x < 1\}$ is the set of all real numbers strictly between -3 and 1. A convenient notation to describe a set with a few elements is to write within braces all the elements of the set. Thus $\{-2, 0, 1\}$ is the set whose elements are -2, 0, and 1. If x is an element of a set S, we write $x \in S$, while if x is not an element of S, we write $x \notin S$. We read $x \in S$ as "x belongs to S" or "x is a member of S." Moreover, we read $x \notin S$ as "x does not belong to S" or "x is not a member of S." Thus $1 \in \{x|x$ is a positive integer$\}$ while $-1 \notin \{x|x$ is a positive integer$\}$.

Now, suppose S and T are sets. We say that

> *S is a **subset** of T or S is **contained** in T, in symbols $S \subseteq T$, if every element of S is also an element of T. In this case, we may also write $T \supseteq S$, read as "T **contains** S."*

Thus, for example, $\{1, 2\} \subseteq \{1, 2, 5\}$ and $\{1, 3\} \supseteq \{3\}$. If $S \subseteq T$ and $T \subseteq S$, we say that S and T are *equal*, and write $S = T$. In other words, $S = T$ simply means that S and T have precisely the same elements. For example, if $S = \{-3, 3\}$ and $T = \{x|x^2 = 9\}$, then $S = T$.

It may well happen that S is a subset of T but S and T are not equal. In this case, we say S is a *proper subset* of T and denote this by $S \subset T$ or by $T \supset S$. We can illustrate these facts by drawing a diagram, called a *Venn diagram.* Thus, in Fig. 1.1 S is understood to consist of all points of the region within the smaller oval, while T is to consist of all points of the region within the larger oval. This figure illustrates a situation in which $S \subset T$.

The concept of the empty set is a convenient one to have. We define the *empty set* to be a set which has no elements at all. Thus, the set of all numbers which are less than 1 *and* larger than 2 is clearly the empty set. Often the empty set (also called the *void,* or *null,* set) is denoted by $\varnothing$. The empty set $\varnothing$ is a subset of *every* set S. This is because there is no element in the empty set $\varnothing$ which is not in S. Note, moreover, that $S \subseteq S$, for any set S. Thus, every set S has the subsets $\varnothing$ and S.

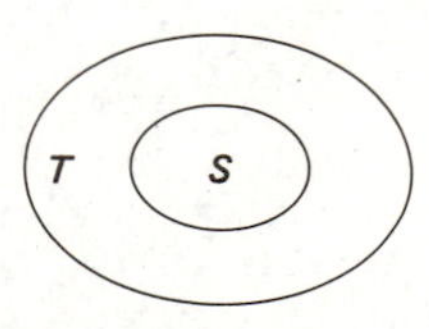

Fig. 1.1 A Venn diagram for $S \subset T$

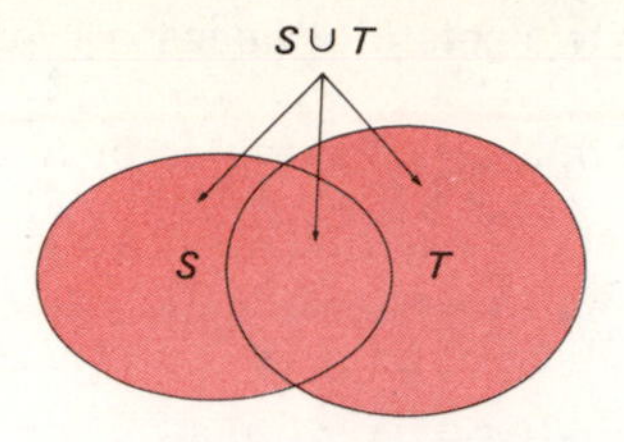

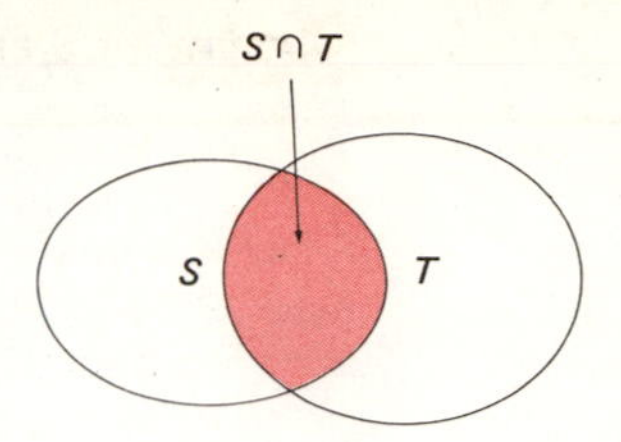

Fig. 1.2 A Venn diagram for $S \cup T$ and $S \cap T$

We now introduce the concepts of union and intersection of sets. Suppose S and T are sets.

*The **union** of S and T, denoted by $S \cup T$, is the set of all elements which are in S or in T (or in both). Thus,*

$$S \cup T = \{x | x \in S \text{ or } x \in T\}$$

*The **intersection** of S and T, denoted by $S \cap T$, is the set of all elements which are common to both S and T. Thus,*

$$S \cap T = \{x | x \in S \text{ and } x \in T\}$$

Figure 1.2 illustrates these concepts. Thus, $S \cup T$ consists of all the points which are in at least one of the two ovals, while $S \cap T$ consists of all points which are common to both ovals. For example, if $S = \{1, 2, 3, 4\}$ and $T = \{1, 4, 5, 10\}$, then $S \cup T = \{1, 2, 3, 4, 5, 10\}$ and $S \cap T = \{1, 4\}$.

Generally speaking, all the sets that we consider are viewed as subsets of a fixed set U, called a *universal set.* Let S be a subset of the universal set U. By definition,

*The **complement** of S relative to U, denoted by S', is the set of all elements in U which are **not** in S; that is,*

$$S' = \{x | x \in U \text{ and } x \notin S\}$$

Observe that S' depends on what universal set is being used. In Fig. 1.3, the shaded region denotes S'. For example, if $U = \{1, 2, 3, 4, 5\}$ and $S = \{1, 4, 5\}$, then $S' = \{2, 3\}$.

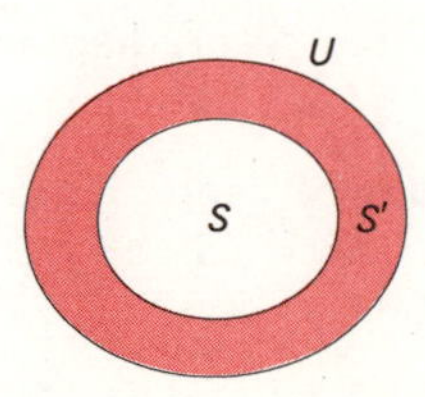

Fig. 1.3 A Venn diagram for S'

Finally, let S and T be any two sets (not necessarily distinct). By definition,

*The **Cartesian product** of S and T, denoted by $S \times T$, is the set of all ordered pairs (a,b), where $a \in S$ and $b \in T$. Thus,*

$$S \times T = \{(a,b) \mid a \in S \text{ and } b \in T\}$$

For example, if $S = \{a, b, c\}$ and $T = \{1, 2\}$, then $S \times T = \{(a,1), (a,2), (b,1), (b,2), (c,1), (c,2)\}$. Two ordered pairs (a,b) and (a',b') in $S \times T$ are said to be equal if and only if $a = a'$ and $b = b'$.

The student is probably familiar with the fact that if $S = T =$ set of all real numbers, then $S \times T$ is simply the set of all points in the xy (Cartesian) plane since a point in the xy plane may be identified with an ordered pair (x,y) of real numbers x and y. (See Sec. 1.3 for a detailed discussion of this topic.) In the next section, we shall see how the concept of Cartesian product is useful in studying functions.

Problem Set 1.1

In these problems, all sets are supposed to be subsets of a fixed universal set U.

1.1.1 Let $S = \{1, 2\}$ and $T = \{1, 4\}$. Find $S \cup T$, $S \cap T$, $S \times T$, $T \times S$, $S \times S$, and $T \times T$. If U is the set of all positive integers smaller than 10, find S' and T'.

1.1.2 Let $S = \{x \mid x$ is a positive integer less than $10\}$ and $T = \{2, 4, 6, 8\}$. Find $S \cup T$, $S \cap T$, and $S \times T$. Is it true that $T \subseteq S$? Is it true that $S \cup T = S$ and $S \cap T = T$?

1.1.3 Let $S = \{x \mid x$ is a positive integer$\}$ and $T = \{x \mid x$ is a negative integer$\}$. Is it true that $S \cap T$ is an empty set? Is it true that $S \cup T$ is the set of *all* integers?

1.1.4 Suppose that S is a set consisting of exactly m elements and T is a set consisting of exactly n elements. How many elements does the Cartesian product $S \times T$ have? Explain.

1.1.5 Suppose $S_1 = \{a\}$, $S_2 = \{b, c\}$, and $S_3 = \{d, e, f\}$. Write all subsets of S_1, S_2, and S_3 (including both $\varnothing$ and the set itself). How many subsets does each of S_1, S_2, and S_3 have?

1.1.6 Suppose S is a set consisting of n elements. Show that the number of subsets of S (including both $\varnothing$ and S) is equal to 2^n. (*Hint*: For each of the n elements of S, we have two choices, namely, to include it or to not include it in the subset under consideration.)

1.1.7 Show that $(S')' = S$.

1.1.8 Verify by means of Venn diagrams that $(S \cup T)' = S' \cap T'$ and $(S \cap T)' = S' \cup T'$.

1.1.9 Verify by means of Venn diagrams that $S \cup (T \cap W) = (S \cup T) \cap (S \cup W)$ and $S \cap (T \cup W) = (S \cap T) \cup (S \cap W)$. Also, verify that $S \cup (S \cap T) = S$ and $S \cap (S \cup T) = S$.

1.1.10 Suppose that $S \subseteq T$ and $T \subseteq W$. Show that $S \subseteq W$.

1.1.11 Suppose that $S \cup T = S \cap T$. What can you say about S and T?

1.1.12 Determine the elements of the following sets:
(*a*) The set of all integers x such that $x^2 = 900$
(*b*) The set of all integers x such that $2x + 1 = 7$
(*c*) The set of all living former Presidents of the United States

1.1.13 Show that $\{x|x \text{ is real and } x \geq 5\} \cap \{x|x \text{ is real and } x \leq 5\} = \{5\}$.

1.1.14 Show that $\{x|x \text{ is real and } \sqrt{x} = 10\} = \{100\}$ (recall $\sqrt{x}$ denotes the *nonnegative* square root of x).

1.1.15 Show that $\{x|x \text{ is an integer and } 2x = 1\} = \varnothing$ (the empty set).

1.1.16 Show that $\{x|x \text{ is real and } x^2 = 1\} = \{x|x \text{ is real and } x^4 = 1\} = \{1, -1\}$.

1.1.17 Show that $\{x|x \text{ is real and } \sqrt{x^2} = 4\} = \{4, -4\}$.

1.1.18 Suppose that S_1 = set of all rectangles, S_2 = set of all squares, S_3 = set of all parallelograms, and S_4 = set of all quadrilaterals. Which of these sets are proper subsets of the others?

1.2 Functions

One of the most important and universal concepts in mathematics is that of a *function.* In this section, we give a brief account of this important concept. Let us first of all give a precise definition of a function.

> *Let S and T be sets. A* ***function*** *from set S into set T is a subset f of the Cartesian product $S \times T$ such that for any given $x \in S$ there is* ***exactly one*** *element $y \in T$ for which $(x,y) \in f$.*

This unique element y is called the *functional value* of f at x or the *image* of x under f, and is usually denoted by $f(x)$ (read as "f of x"). There are several other ways in which we can describe a function. Thus, we may say that

> *A* ***function*** *$f: S \to T$ (read as "f is a function from S into T") is a rule which assigns to every element x in S exactly one element, denoted by $f(x)$, in T.*

Another extremely useful way to define a function is the following:

> *A* ***function,*** *$f: S \to T$, from set S into set T is a correspondence which associates with each element x in S exactly one element, denoted by $f(x)$, in T.*

This last formulation of a function is a very convenient one, and we shall use it often throughout this text. First, we introduce some terminology and notation.

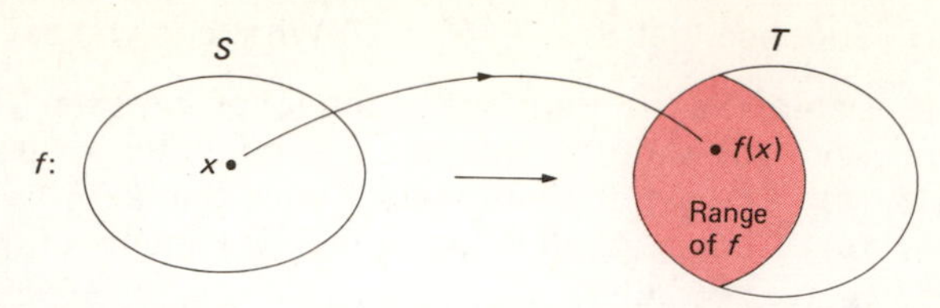

Fig. 1.4 Range of f

Suppose $f: S \to T$ is a function from S into T. The set S is called the ***domain*** *of f, while the set T is called the* ***codomain*** *of f. The set of all functional values $f(x)$, where x is any element in S, is called the* ***range*** *of f. In other words,*

Range of $f = \{y | y = f(x) \text{ for some } x \in S\}$

Clearly, the range of f is contained in the codomain T of f (see Fig. 1.4). However, it is *not* necessarily true that the range of f equals the codomain of f.

A function $f: S \to T$ is often called a *mapping* from (or of) S into T. It is also sometimes written as $S \xrightarrow{f} T$, $f: x \to y$, or $f: x \to f(x)$.

In cases where the domain of a function $f: S \to T$ is a finite, small set, we may represent the function by drawing an arrow from each element of the domain to its functional value in the codomain (see the examples below).

Remark Figure 1.5 does *not* represent a function since the element x in the domain goes under the action of f to *two distinct* elements (namely, y and z) in the codomain T. That is, the correspondence f does not associate a unique element for each element in the domain, and hence f is not a function. We can thus say that whenever f gives rise to the above picture, f is not a function.

Clearly, a function f is completely determined when we know its domain S and when we know the image $f(x)$ for each $x \in S$. In fact, we often describe a function f by prescribing $f(x)$ for each x in the domain of f. Thus, the function f given by

$$f(x) = x^2 \qquad x \text{ is a real number}$$

is the set of all ordered pairs (x,x^2), where x is a real number.

We may also describe this function as the correspondence (or rule) which associates with each real number x the number x^2. We may also denote this function by

$$f: x \to x^2 \qquad x \text{ is a real number}$$

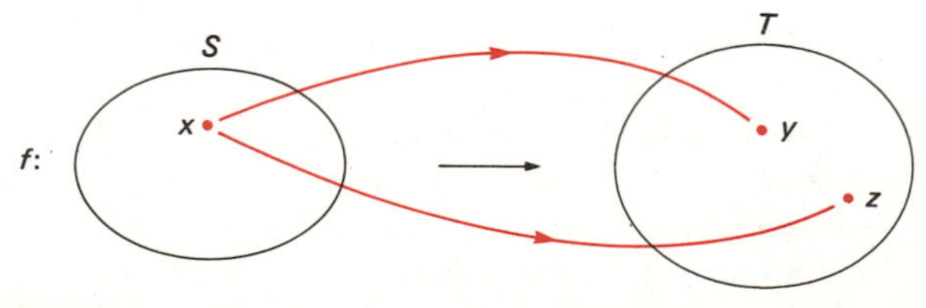

Fig. 1.5 f *not* a function

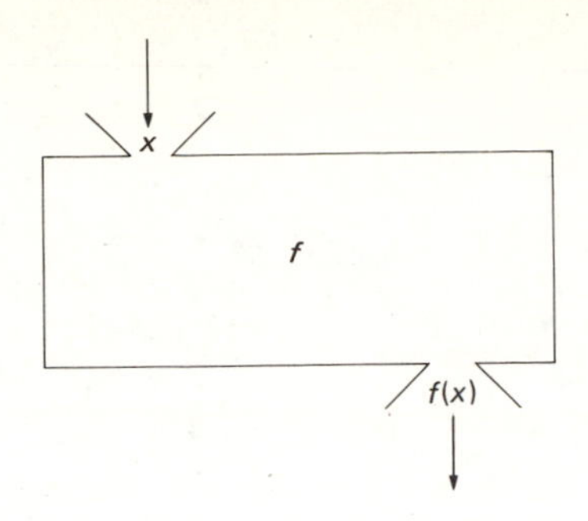

Fig. 1.6 Function as a machine

Another way to describe a function f is to think of it as a machine. Each x in the domain goes into the machine f and after being operated on comes out as the element $f(x)$ in the range of f (see Fig. 1.6). x represents the input, while $f(x)$ represents the output.

Let S be a subset of the set R of real numbers. A function $f:S \to R$ is often called a *real-valued function* of a real variable. We now make the following convention: When the domain of a real-valued function f of a real variable is not given, we shall understand the domain of f to be the set of all those real numbers x for which $f(x)$ is real. Thus, the function f given by $f(x) = x^2$ is understood to have a domain consisting of the set of *all* real numbers (since x^2 is a real number for all real numbers x). However, the function g given by $g(x) = \sqrt{x}$ is understood to have a domain consisting of the set of all *non-negative* real numbers (since $\sqrt{x}$ is a real number if and only if $x \geq 0$).

Example 1.1 Let $S = \{1, 2, 3\}$ and $T = \{a, b, c\}$, where a, b, and c are distinct. Let

$$f_1 = \{(1,a),\ (2,a),\ (3,b)\}$$

$$f_2 = \{(1,a),\ (1,b),\ (2,a),\ (3,c)\}$$

Clearly, f_1 *is* a function from S into T since there are no two ordered pairs in f_1 with the same first element but with different second elements. However, f_2 is *not* a function from S into T since both $(1,a)$ and $(1,b)$ are in f_2 and these are two distinct ordered pairs with the same first element. Observe that $f_1(1) = a$, $f_1(2) = a$, and $f_1(3) = b$.

Example 1.2 Let $S = \{1, 2, 3\}$ and $T = \{a, b\}$, where a and b are distinct. Let f_1 and f_2 be the following correspondences:

$$f_1{:}1 \to a \quad 2 \to b \quad 3 \to b$$

$$f_2{:}1 \to b \quad 1 \to a \quad 2 \to b \quad 3 \to a$$

Clearly, f_1 *is* a function from S into T. However, f_2 is *not* a function from S into T since f_2 associates with the element 1 in S *two* distinct elements in T (namely, b and a).

Now, suppose that $f:S \to T$ is a function from set S into set T. We say that

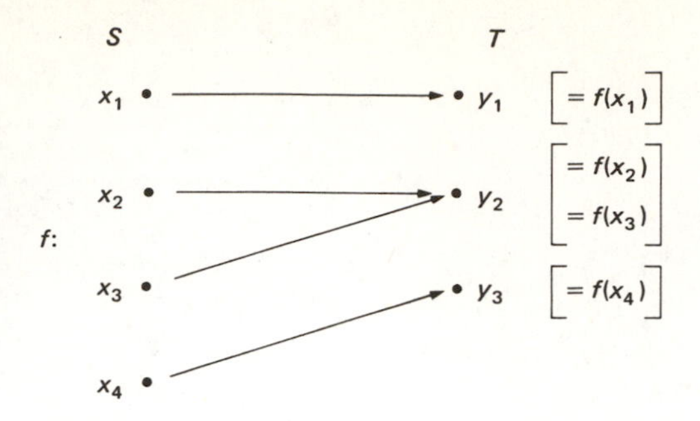

Fig. 1.7 f is onto, or surjective

> *f is **onto,** or **surjective,** if the range of f equals the codomain of f. In other words, f is onto if for every $y \in T$, there exists at least one element $x \in S$ such that $f(x) = y$ (see Fig. 1.7).*

For example, suppose, as indicated in Fig. 1.7, that

$$S = \{x_1, x_2, x_3, x_4\} \qquad \text{and} \qquad T = \{y_1, y_2, y_3\}$$

Define $f: S \to T$ as follows (see Fig. 1.7):

$$f(x_1) = y_1 \qquad f(x_2) = y_2 \qquad f(x_3) = y_2 \qquad f(x_4) = y_3$$

Then, recalling the definition of range of a function, we have

$$\text{Range of } f = \{f(x_1), f(x_2), f(x_3), f(x_4)\} = \{y_1, y_2, y_3\} = \text{codomain of } f$$

Thus,

$$\text{Range of } f = \text{codomain of } f$$

and hence, by definition, the function $f: S \to T$ is onto, or surjective.

As a further illustration of this concept, let $g: R \to R$ (R is the set of real numbers) be given by $g(x) = x - 5$. If t is any real number, then $g(t+5) = (t+5) - 5 = t$. Thus, for each t in R there exists an element x_0 in R (namely, $t+5$) such that $g(x_0) = t$; therefore the function $g: R \to R$ is onto, or surjective.

Next, suppose that $f: S \to T$ is a function from set S into set T. We say that

> *f is **one-to-one,** or **injective,** if distinct elements x and x' in S have distinct functional values $f(x)$ and $f(x')$ in T; that is, $x \neq x'$ always implies $f(x) \neq f(x')$ (see Fig. 1.8).*

Another equivalent definition of a one-to-one function, which is often useful, is the following:

> *A function $f: S \to T$ is **one-to-one,** or **injective,** if for all x and x' in S $f(x) = f(x')$ always implies $x = x'$.*

In general, in verifying that a function is one-to-one it is usually more convenient to apply this second definition.

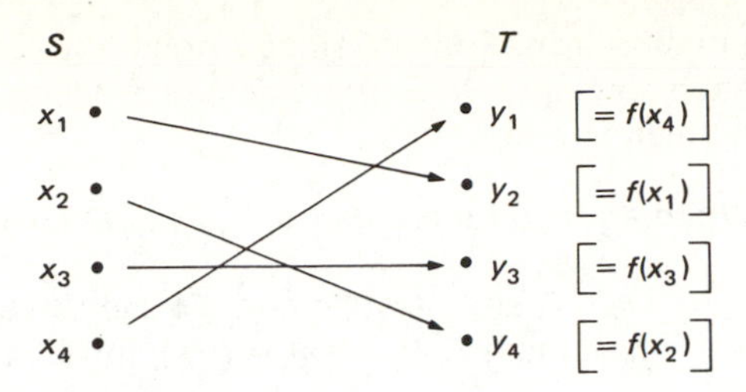

Fig. 1.8 f is one-to-one, or injective

As an illustration of this concept, suppose, as indicated in Fig. 1.8, $S = \{x_1, x_2, x_3, x_4\}$ and $T = \{y_1, y_2, y_3, y_4\}$. Now, define $f:S \to T$ as follows (see Fig. 1.8):

$$f(x_1) = y_2 \qquad f(x_2) = y_4 \qquad f(x_3) = y_3 \qquad f(x_4) = y_1$$

Since distinct elements in the domain S have distinct functional values in the codomain T, f is a one-to-one (or injective) function. It is easy to see that f is also an onto (or surjective) function.

As a further illustration, suppose that $g:R \to R$ (R is the set of real numbers) is given by $g(x) = 2x - 3$. Also, suppose x and x' are any real numbers such that $g(x) = g(x')$; that is, $2x - 3 = 2x' - 3$. Then, clearly, $x = x'$. In other words, we have shown that $g(x) = g(x')$ always implies that $x = x'$; hence $g:R \to R$ is a one-to-one, or injective, function.

A function which is both one-to-one (injective) and onto (surjective) is called a *bijection.* As we shall soon see, a bijection always possesses a very desirable property.

Example 1.3 Let $S = T =$ the set of real numbers, and let $f:S \to T$ be given by $f(x) = x^2$. Clearly, f is a function from S into T (since, for every $x \in S$, f associates exactly one element in T, namely, x^2). However, f is not onto since there is *no* real number x such that $f(x) = -1$. Moreover, f is not one-to-one since, for example, $f(1) = f(-1)$ even though $1 \neq -1$.

Example 1.4 Let $S =$ the set of real numbers, $T =$ the set of nonnegative real numbers, and $f:S \to T$ be given by $f(x) = x^2$. Once more, we easily see that f is a function from S into T. This time, however, f is onto (or surjective) because for any *nonnegative* real number y (in T) there is indeed a solution x (in S) to the equation $f(x) = y$. In fact, we may take $x = \sqrt{y}$ since $f(\sqrt{y}) = (\sqrt{y})^2 = y$; and, of course $\sqrt{y} \in S$ because $y \geq 0$. Finally, f is still not one-to-one since again $f(1) = f(-1)$ even though $1 \neq -1$.

Example 1.5 Let $S = T =$ the set of nonnegative real numbers, and once again let $f:S \to T$ be given by $f(x) = x^2$. As in the above example, we easily verify that f is a function from S onto T. This time, however, this function is also one-to-one (or injective), for suppose $f(x_1) = f(x_2)$. Then $x_1{}^2 = x_2{}^2$, and hence $x_1 = \pm x_2$. However, since both x_1 and x_2 are *nonnegative,* we must have $x_1 = x_2$. Thus, $f(x_1) = f(x_2)$ always implies $x_1 = x_2$; hence f is one-to-one (or injective). Therefore, f is a bijection in this case.

Example 1.6 Let S be the set of all nonnegative real numbers, and let T be the set of all real numbers. Suppose that for each x in S f associates $\pm\sqrt{x}$. In this case, f is not a function from S into T since f does not associate exactly one element in T for each element x in S. For example, f associates with 1 two distinct numbers (namely, 1 and -1).

When are two functions f and g equal? The answer is

> $f = g$ if and only if f and g have the same domain S and the same codomain T, and $f(x) = g(x)$ for all x in S (their common domain).

For example, suppose $S = T =$ the set of all nonnegative real numbers. Suppose also that $f: S \to T$ and $g: S \to T$ are defined by

$$f(x) = \sqrt{x^2} \qquad \text{for all } x \text{ in } S$$

$$g(x) = x \qquad \text{for all } x \text{ in } S$$

Observe that f and g have the same domain (namely, S) and the same codomain (namely, T). Moreover, $f(x) = g(x)$ for all x in S (the common domain) since $\sqrt{x^2} = x$ for all nonnegative real numbers x. Hence, by definition, the two functions f and g are equal.

Warning If x is a *negative* real number, then $\sqrt{x^2} = -x$ (recall that $\sqrt{\ }$ always denotes the *nonnegative* square root). Therefore, if now S and T denote the set of *all* real numbers, then f and g will *not* be equal.

Example 1.7 We consider below a function g, which may be called a "grading" function. Suppose $S =$ the set of students enrolled in Mathematics 1 in college X, $T = \{A, B, C, D, F\}$ (that is, T is the set of grades being used), and $g(x) =$ the grade which student x gets in Mathematics 1. Then, $g: S \to T$ is a function from S into T. This follows since each student x (in S) gets in Mathematics 1 exactly one grade $g(x)$. Incidentally, the range of g need not be equal to the codomain of g since it is conceivable that every student might have passed Mathematics 1. Thus, $g: S \to T$ is not necessarily an onto function. Observe also that g is *not* a *one-to-one* function if the class has more than five students. (Why?)

Problem Set 1.2

In the exercises below, R denotes the set of all real numbers.

1.2.1 Let $f: R \to R$ be given by $f(x) = -2x + 9$. Is f a function from R into R? Is f onto? Is f one-to-one? Is f a bijection?

1.2.2 Let $f: R \to R$ be given by $f(x) = x^2 + 5$. Is f onto? Is f one-to-one? Is f a bijection?

In Probs. 1.2.3 to 1.2.6, S and T are given sets, and $f: S \to T$ is also given. Which of these functions are onto? Which are one-to-one? Which are bijections?

1.2.3 $S = \{-1, 0, 1\}$, $T = \{-1, 0, 1\}$, and $f: x \to x^4$

1.2.4 $S = \{0, 1\}$, $T = \{-1, 0, 1\}$, and $f: x \to x^4$

1.2.5 $S = \{0, 1\}$, $T = \{0, 1\}$, and $f: x \to x^4$

1.2.6 $S = \{-1, 0, 1\}$, $T = \{0, 1\}$, and $f: x \to x^4$

1.2.7 Let $S = \{x_1, x_2, \ldots, x_m\}$ and $T = \{y_1, y_2, \ldots, y_n\}$ be sets consisting of exactly m elements and n elements, respectively. Let $f: S \to T$ be a function from S into T. Show that:

(a) If f is onto, then $n \leq m$.
(b) If f is one-to-one, then $n \geq m$.
(c) If f is a bijection, then $n = m$.

1.2.8 Show that if S and T are sets with exactly n elements each, then there exists a bijection $f: S \to T$.

1.2.9 Prove that the function $f: R \to R$ defined by $f(x) = 5x - 8$ is a bijection.

1.2.10 Suppose that $f: R \to R$ is defined by

$$f(x) = \begin{cases} x^2 - 1 & \text{if } x \geq 0 \\ x^2 + 1 & \text{if } x \leq 0 \end{cases}$$

Show that f is not a function.

1.2.11 Suppose that $f: R \to R$ is defined by

$$f(x) = \begin{cases} 1 - x^2 & \text{if } x \geq 0 \\ 1 + x^2 & \text{if } x \leq 0 \end{cases}$$

Is f a function? Explain.

1.2.12 Suppose $f: S \to S$ is a function from S into S, where S consists of exactly n elements (n is a positive integer). Show that f is one-to-one if and only if f is onto.

1.2.13 Suppose S is a set consisting of exactly n elements, say, $S = \{1, 2, \ldots, n\}$. How many bijections does S have? [*Hint:* If $f: S \to S$ is a bijection, then $f: 1 \to a_1$, $2 \to a_2, \ldots, n \to a_n$, where $a_1, a_2, \ldots, a_n$ are just the integers $1, 2, \ldots, n$ in some order (all the a_i's are distinct). How many choices does a_1 have? Having chosen a_1, how many choices remain for a_2, etc.?]

1.2.14 Let S and T be defined by

$S =$ the set of all states in the United States

$T =$ the set of all congressmen in the United States

Suppose that f is the correspondence which associates with each x in S the congressmen from state x. Is f a function from S into T? Explain.

1.2.15 Express the perimeter of a square as a function of the length of one of its sides.

1.2.16 Express the area of a circle as a function of the length of its radius.

1.2.17 Express the volume of a cube as a function of the length of one of its sides.

1.2.18 Suppose that $f: R \to R$ and $g: R \to R$ are defined by

$$f(x) = \sqrt{x^2} \qquad \text{for all } x \in R \text{ (that is, } x \text{ real)}$$

$$g(x) = \begin{cases} x & \text{if } x \geq 0 \\ -x & \text{if } x < 0 \end{cases}$$

Show that the functions f and g are equal.

1.2.19 Let $S = \{x, y, z\}$ and $T = \{1, 2, 3\}$. Construct two one-to-one functions: f of S onto T and g of T onto S.

1.3 Graphs of Functions

An illuminating way to view a function is to represent it graphically. Indeed, graphic representation of data is an indispensable device, used in almost all disciplines. In order to achieve this graphic representation, we first construct two perpendicular lines and agree on a unit of measurement to be used along both of these lines (see Fig. 1.9). It is customary to call the horizontal line the *x axis* and the vertical line the *y axis.* The point of intersection of these axes $\mathcal{O}$ is called the *origin.* By common agreement, the positive directions along the x axis and the y axis are as indicated by the arrows in Fig. 1.9. Thus, when we measure from the origin $\mathcal{O}$ to any point on the x axis to the right of $\mathcal{O}$, we assign a positive x coordinate. A negative x coordinate is assigned if the measurement along the x axis is from the origin $\mathcal{O}$ to a point to the left of $\mathcal{O}$. Similarly, a positive y coordinate is assigned if the measurement along the y axis is in the upward direction, while a negative y coordinate is assigned if the measurement along the y axis is in the downward direction.

We now proceed to show the important fact that there is a one-to-one correspondence between the set of all ordered pairs (x,y) of real numbers x and y (i.e., the Cartesian product $R \times R$, where R is the set of real numbers) and the set of all points in the plane. This is easily achieved by the simple device of drawing through any given point P in the plane two lines which are perpendicular to the x axis and y axis, respectively, and then associating with the point P the ordered pair of numbers (x_1,y_1), where x_1 and y_1 are as indicated in Fig. 1.10. The number x_1 is called the *x coordinate*, or *abscissa*, of P, while the number y_1 is called the *y coordinate*, or *ordinate*, of P. We usually write $P:(x_1,y_1)$, or $P(x_1,y_1)$, to indicate that the point P has an x coordinate equal to x_1 and a y coordinate equal to y_1. Conversely, if x_1 and y_1 are *any* real numbers (positive, negative, or zero), we associate with the ordered pair (x_1,y_1) exactly one point, namely, the point P with x coordinate equal to x_1 and with y coordinate equal to y_1 (see Fig. 1.10). Thus in Fig. 1.9 the point P_0 and the ordered pair (3,1) are associated with each other. Note that the point P_0:(3,1) is quite different from

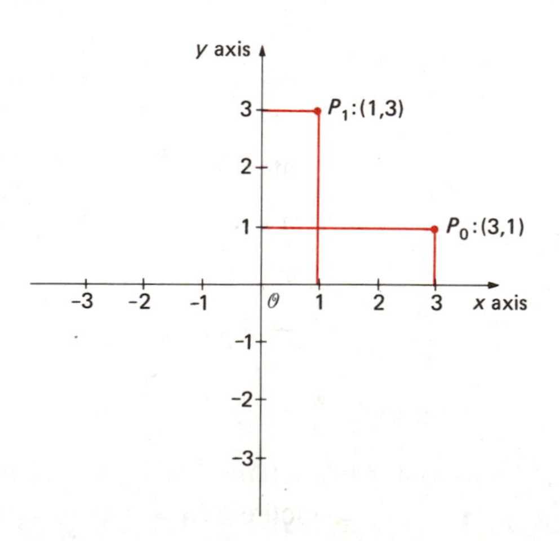

Fig. 1.9 Coordinates of points in the plane

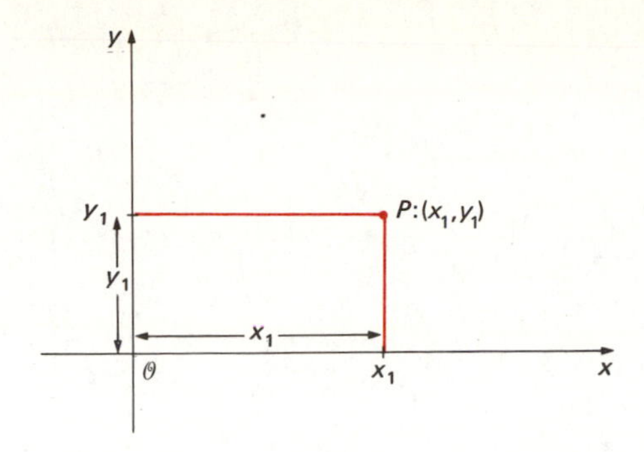

Fig. 1.10 Identification of points and ordered pairs of real numbers

the point P_1: (1,3). It should now be clear that every point on the x axis has coordinates of the form $(x,0)$, while every point on the y axis has coordinates of the form $(0,y)$. In particular, the origin $\mathcal{O}$ has coordinates (0,0). We follow the usual practice of calling the four quadrants into which the x axis and y axis divide the plane quadrants I, II, III, and IV, as indicated in Fig. 1.11. It is quite easy to see that the algebraic values of the coordinates of points in the plane are also as indicated in Fig. 1.11. For example, points in quadrant II have their x coordinates negative and their y coordinates positive.

Example 1.8 Plot the following points: P_1: (1,4), P_2: (4,1), P_3: (−1,2), P_4: (2,−1), P_5: (−1,−3), P_6: (−4,0), and P_7: (0,7/2).

Solution Recall the discussion at the beginning of this section; it is readily verified that the given points are as indicated in Fig. 1.12.

In Sec. 1.2., we saw that a function $f:S \to T$ is the set of all ordered pairs $(x,f(x))$, where x is any element in the domain S. Suppose that $f:S \to R$ is a real-valued function of a real variable. That is, the domain S of f is now assumed to be a subset of the set R of real numbers while the codomain of f is the set R itself (see Sec. 1.2). In this case, the ordered pair $(x,f(x))$ is simply an ordered pair of *real numbers*, and thus $(x,f(x))$ can be identified with a unique point in the xy (Cartesian) plane, as explained at the beginning of this section.

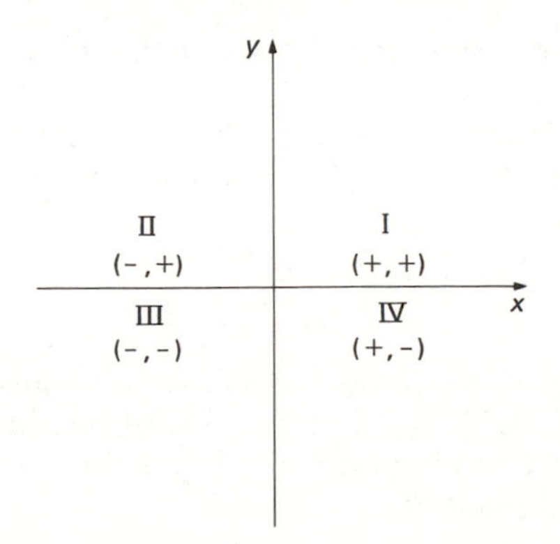

Fig. 1.11 Coordinates and quadrants

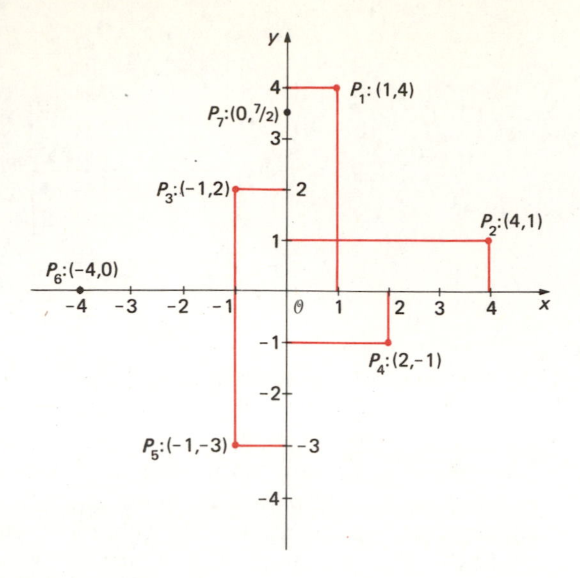

Fig. 1.12 Coordinates of points

This motivates the following definition:

> *Suppose R is the set of real numbers and $S \subseteq R$. Suppose $f:S \to R$ is a real-valued function of a real variable. The graph of the function f [or equation $y = f(x)$] is the set of all ordered pairs (of real numbers) $(x, f(x))$, where $x \in S$.*

In other words, we are now identifying the *real-valued function* $f:S \to R$ with its graph. To sketch the graph of a real-valued function $f:S \to R$ of a real variable, we first find a "reasonable" number of ordered pairs (x,y) of real numbers x and y which satisfy the equation $y = f(x)$. Then, we locate the points in the xy (Cartesian) plane which are associated with these ordered pairs $(x,f(x))$. Finally, we join these points together by means of a curve, as illustrated in the examples below.

Example 1.9 Sketch the graph of the function f given by $f(x) = 2x + 4$.

Solution The equation $y = f(x) = 2x + 4$ is certainly satisfied by an infinite number of ordered pairs (x,y). We tabulate below a few of these ordered pairs.

x	0	1	−1	2	−2
$f(x)$	4	6	2	8	0

We find the points associated with these ordered pairs (x,y) [namely, $(0,4)$, $(1,6)$, $(-1,2)$, $(2,8)$, and $(-2,0)$] and join them together (see Fig. 1.13). It seems clear that the graph of the function f [or the equation $y = f(x)$] is a straight line in this case.

Example 1.10 Sketch the graph of the function f given by $f(x) = x^2$.

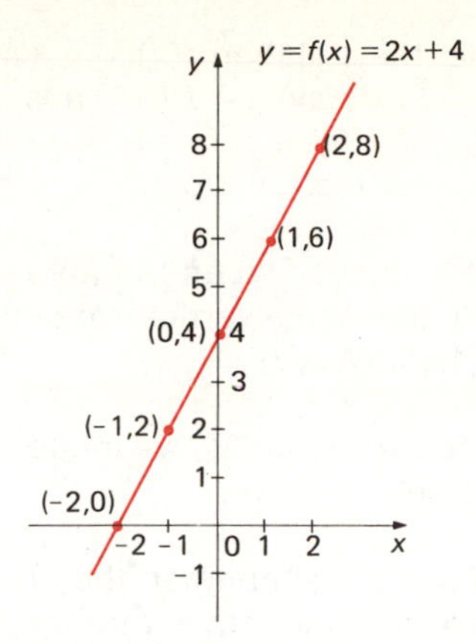

Fig. 1.13 Graph of $y = f(x) = 2x + 4$

Solution Once again, we tabulate some of the ordered pairs (x,y) which satisfy the equation $y = f(x) = x^2$.

x	0	1	2	−1	−2
$f(x)$	0	1	4	1	4

Next, we plot the points whose coordinates are (x,y) as given in the above table. We then join these points together by means of a curve, as shown in Fig. 1.14.

Of special interest are those points (if any) on the graph of a function where the graph crosses the coordinate axes. The points of intersection of a graph with the x axis are called the x *intercept points*, while the points of intersection of a graph with the y axis are called the y *intercept points*. Thus, in Example 1.9 the x intercept point is $(-2,0)$, while the y intercept point is $(0,4)$, as indicated in Fig. 1.13. In Example 1.10, both x and y intercept points are $(0,0)$. In general, to find the x intercept points of the graph of $y = f(x)$ set $y = 0$, and solve (if

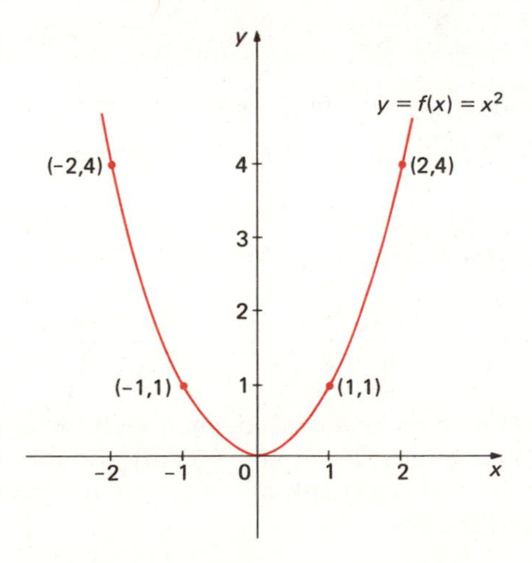

Fig. 1.14 Graph of $y = f(x) = x^2$

possible) the equation $f(x) = 0$. (Why?) Similarly, to find the y intercept point set $x = 0$ to obtain $y = f(0)$. Thus, *assuming that zero is in the domain of f* [and hence $f(0)$ is defined] the point $(0, f(0))$ is the *only* y intercept point of the graph of the function f.

Remark 1

f is *not* a function when the following situation arises: The graph of f has the property that some vertical line intersects it at *more* than one point, as indicated in Fig. 1.15. (Why?)

Remark 2

In the above examples, we made use of the familiar so-called *functional notation,* namely

> If a is a real number, then $f(a)$ is simply the number obtained by replacing x by a throughout the expression giving $f(x)$.

For example, if $f(x) = x^3 - 2x + 1$ for any real number x, then

$$f(0) = 0^3 - 2(0) + 1 = 1$$

$$f(1) = 1^3 - 2(1) + 1 = 0$$

$$f(a) = a^3 - 2a + 1$$

$$f(x^2) = (x^2)^3 - 2(x^2) + 1 = x^6 - 2x^2 + 1$$

$$f(-x) = (-x)^3 - 2(-x) + 1 = -x^3 + 2x + 1$$

and so on.

Also, observe that the functions defined below are all *equal* to the function f defined above:

$f(t) = t^3 - 2t + 1$ for any real number t

$f(y) = y^3 - 2y + 1$ for any real number y

$g(x) = x^3 - 2x + 1$ for any real number x

$g(z) = z^3 - 2z + 1$ for any real number z

and so on. This follows at once from the definition of equality of functions stated earlier.

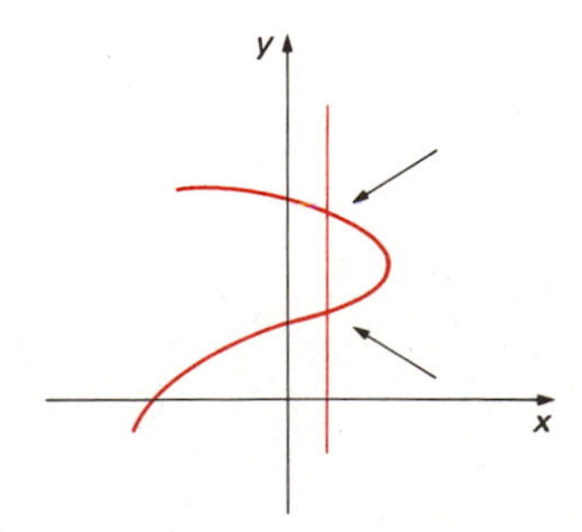

Fig. 1.15 f is not a function

Problem Set 1.3

1.3.1 Plot the following points: (1,2), (2,1), (3,0), (–1,0), (0,4), (0,–1), (–1,–2), (–1,3), and (1,–3).

1.3.2 Describe the coordinates of the points on a horizontal line which passes through the point (–1,–4).

1.3.3 Describe the coordinates of the points on a vertical line which passes through the point (1,–2).

1.3.4 Suppose that the functions f and g are given by

$$f(x) = x^2 + 5 \qquad g(x) = x - 1$$

Find $f(0)$, $g(0)$, $f(g(0))$, $g(f(0))$, $f(4)$, $f(-4)$, $f(-x)$, $g(-x)$, $f(x^2)$, $g(x^2)$, $f(g(x))$, $g(f(x))$, $f(f(x))$, *and* $g(g(x))$.

1.3.5 Sketch the graphs of the functions f given by the following equations. In each case, give the domain of the function.

(*a*) $f(x) = -3x + 2$
(*b*) $f(x) = -x^2$
(*c*) $f(x) = x^2 + 2$
(*d*) $f(x) = 0$
(*e*) $f(x) = -5$
(*f*) $f(x) = 1/x^2$
(*g*) $f(x) = 4x$
(*h*) $f(x) = \sqrt{x-1}$

1.3.6 Find the x intercept points (if any) and the y intercept points (if any) of the graph of each of the functions in Prob. 1.3.5.

1.3.7 Sketch the graph of $y = f(x) = 1/x$. What is the domain of the function f? Are there any x intercept points and y intercept points for this graph?

1.3.8 Let S and T be any sets (not necessarily distinct). A *relation* is defined to be any subset of the Cartesian product $S \times T$. In other words, a relation is simply a set of ordered pairs (x,y), where $x \in S$ and $y \in T$. Thus, in particular, every function is a relation, but the converse is not necessarily true. In fact, a function is a relation f such that if (x,y) and (x,y') are both in f, then $y = y'$. Sketch the graph of each of the following relations. (Here both x and y are assumed to be real numbers.) Which of these relations are functions?

(*a*) $\{(x,y) \mid x^2 + y^2 = 1\}$
(*b*) $\{(x,y) \mid y = \sqrt{1 - x^2}\}$
(*c*) $\{(x,y) \mid y = -\sqrt{1 - x^2}\}$
(*d*) $\{(x,y) \mid x = y^2\}$
(*e*) $\{(x,y) \mid y = \sqrt{x}\}$
(*f*) $\{(x,y) \mid y = -\sqrt{x}\}$
(*g*) $\{(x,y) \mid 2x - 3y = 6\}$
(*h*) $\{(x,y) \mid x = 0\}$
(*i*) $\{(x,y) \mid y = 0\}$
(*j*) $\{(x,y) \mid x = -3\}$
(*k*) $\{(x,y) \mid y = 2\}$

1.3.9 Do the following graphs represent functions? Give reasons. (Use the vertical-line test.)

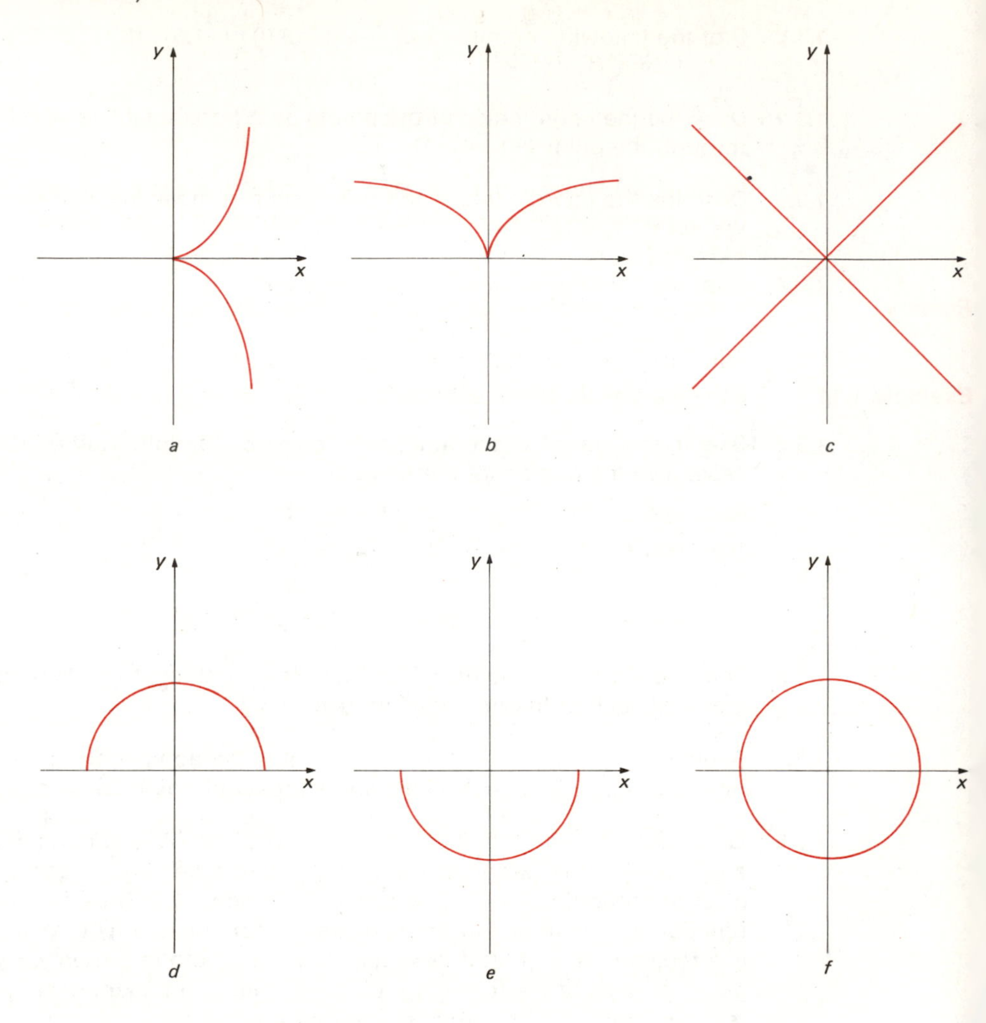

1.4 Operations on Functions; Inverses

In this section, we consider the question of forming new functions from old ones. First, we introduce algebraic operations on functions; that is, we define the sum, difference, product, and quotient of certain functions. Later, we discuss the notion of *composition* of functions and the concept of *inverse* of a function.

Suppose that S is any set, and suppose R is the set of real numbers. Suppose $f: S \to R$ and $g: S \to R$. We define the sum $f + g$ of the functions f and g by

$$f + g : x \to f(x) + g(x) \qquad x \in S$$

Similarly, we define the difference $f - g$, the product fg, and the quotient f/g by

$$f - g: x \to f(x) - g(x)$$
$$fg: x \to f(x)g(x)$$
$$\frac{f}{g}: x \to \frac{f(x)}{g(x)} \qquad \text{provided } g(x) \neq 0$$

We thus see that $f + g$, $f - g$, and fg all have the same common domain which f and g have (namely, S) while the domain of f/g is that subset of S for which $g(x) \neq 0$.

Remark It should be emphasized that in the above definitions we do *not* require that S be a subset of R.

Example 1.11 Let R be the set of real numbers, and let $f: R \to R$ and $g: R \to R$ be defined by

$$f(x) = x^2 \qquad g(x) = 5x - 1 \qquad (x \text{ is any real number})$$

In this case,

$$(f + g)(x) = f(x) + g(x) = x^2 + 5x - 1$$
$$(f - g)(x) = f(x) - g(x) = x^2 - (5x - 1) = x^2 - 5x + 1$$
$$(fg)(x) = f(x)g(x) = x^2(5x - 1) = 5x^3 - x^2$$
$$\left(\frac{f}{g}\right)(x) = \frac{f(x)}{g(x)} = \frac{x^2}{5x - 1} \qquad \left(x \neq \frac{1}{5}\right)$$

The domain of each of $f + g$, $f - g$, and fg is the set R of all real numbers, while the domain of f/g is the set R of all real numbers with the exception of $1/5$.

Next, we consider the question of composition of functions. Thus, suppose that S, T, and W, are sets, and suppose that $f: S \to T$ and $g: T \to W$ are functions. For every $x \in S$, $f(x) \in T$. Hence, $g(f(x)) \in W$ since g associates with elements in T [such as $f(x)$] elements in W. Keeping this fact in mind, we define the *composition* of g and f, denoted by $g \circ f$, as follows: If $f: S \to T$ and $g: T \to W$, then $g \circ f: S \to W$ is defined by (see Fig. 1.16)

$$(g \circ f)(x) = g(f(x)) \qquad \text{for all } x \in S$$

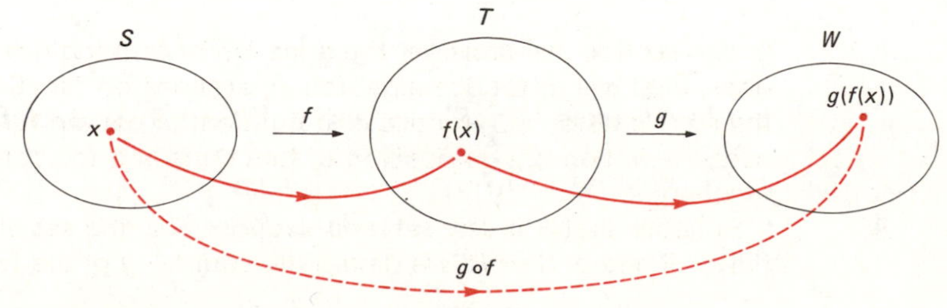

Fig. 1.16 Composition of functions

Example 1.12 Let $S = T = W =$ set of all real numbers, and define $f: S \to T$ and $g: T \to W$ by

$$f(x) = x^2 \qquad \text{and} \qquad g(x) = 3x + 1$$

Then (see Remark 2 at the end of Sec. 1.3)

$$(g \circ f)(x) = g(f(x)) = g(x^2) = 3x^2 + 1$$

Hence, $g \circ f: S \to W$ is the function given by

$$(g \circ f)(x) = 3x^2 + 1 \qquad x \text{ is any real number (in } S)$$

Example 1.13 Let S, T, W, f, and g be as in Example 1.12. Can one form the composition $f \circ g$? The answer is "yes." For, by definition of $f \circ g$ we have

$$(f \circ g)(x) = f(g(x)) = f(3x + 1) = (3x + 1)^2 = 9x^2 + 6x + 1$$

In other words, $f \circ g$ is the function given by

$$(f \circ g)(x) = 9x^2 + 6x + 1 \qquad x \text{ is any real number}$$

These two examples show that even if both $f \circ g$ and $g \circ f$ are defined, $g \circ f$ is in general *not* equal to $f \circ g$. (Clearly, $3x^2 + 1 \neq 9x^2 + 6x + 1$ except when $x = 0$ or $x = -1$.)

Example 1.14 Let $i: S \to S$ be the function given by $i(x) = x$ for all $x \in S$. Let $f: S \to S$ be any function from S into S. Let us compute $f \circ i: S \to S$ and $i \circ f: S \to S$. By definition,

$$(f \circ i)(x) = f(i(x)) = f(x)$$

and

$$(i \circ f)(x) = i(f(x)) = f(x)$$

Hence $f \circ i = i \circ f = f$. This function $i: S \to S$ is called the *identity function* (on S). It has the interesting property that $i \circ f = f \circ i = f$, for all functions $f: S \to S$.

Example 1.15 Let S be the set of all real numbers, and define $f: S \to S$ and $g: S \to S$ by

$$f(x) = 2x + 1 \qquad g(x) = \tfrac{1}{2}(x - 1)$$

Let us calculate $f \circ g: S \to S$, and $g \circ f: S \to S$. By definition,

$$(f \circ g)(x) = f(g(x)) = f(\tfrac{1}{2}(x - 1)) = 2[\tfrac{1}{2}(x - 1)] + 1 = x$$

and

$$(g \circ f)(x) = g(f(x)) = g(2x + 1) = \tfrac{1}{2}[(2x + 1) - 1] = x$$

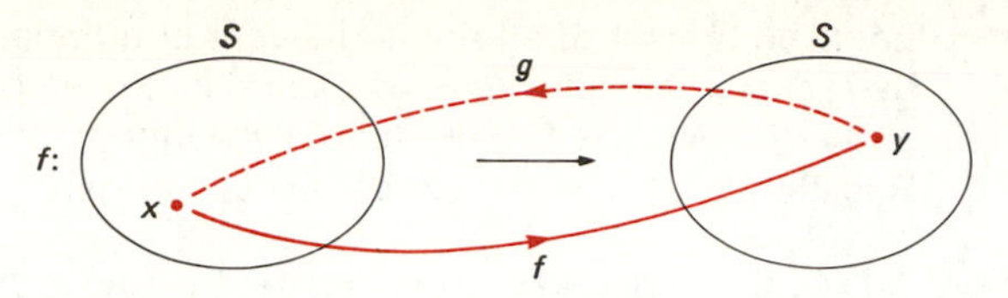

Fig. 1.17 g is an inverse of f

We thus see that $(f \circ g)(x) = (g \circ f)(x) = x$ for all $x \in S$, and hence $f \circ g = g \circ f = i$ (identity function on S).

With an eye on this last example, we now introduce the concept of *inverse* of a function. For simplicity's sake, we shall assume that $f: S \to S$; that is, the domain and codomain of f are the same. (In Chaps. 7 and 8, we extend this definition to the general case in which $f: S \to T$ for *any* sets S and T.) We say that a function $g: S \to S$ is an inverse for (or an *inverse* of) the function $f: S \to S$ if

$$f \circ g = g \circ f = i \qquad \text{(identity function on } S\text{)}$$

(see Examples 1.14 and 1.15). In this case, we write $g = f^{-1}$ (read as "f inverse"). Thus, in Example 1.15 we see that the function $g: S \to S$ is an inverse for the function $f: S \to S$.

We shall now show that

If $f: S \to S$ is both one-to-one and onto (*i.e., a bijection*), *then f has an* ***inverse*** $f^{-1}: S \to S$.

Thus, suppose $y \in S$ (see Fig. 1.17). Since f is onto, there exists at least one element $x \in S$ such that $f(x) = y$. Moreover, because f is one-to-one, this element x is *unique*. Thus, for *any* $y \in S$, there exists exactly one element x in S such that $f(x) = y$. Now define $g(y) = x$ (see Fig. 1.17). In other words, $g(y) = x$ holds if and only if $f(x) = y$. We claim that $g = f^{-1}$. For, suppose $x \in S$, and suppose $f(x) = y$. Then by definition of g, $g(y) = x$; therefore,

$$(g \circ f)(x) = g(f(x)) = g(y) = x \tag{1.1}$$

Since $x \in S$, $g(x)$ is defined (see Fig. 1.18). Let $g(x) = z$. Then by definition of g again, we have $f(z) = x$ [since $g(x) = z$], which yields

$$(f \circ g)(x) = f(g(x)) = f(z) = x \tag{1.2}$$

Thus, in view of (1.1) and (1.2), $g \circ f = f \circ g = i$, and hence $g = f^{-1}$.

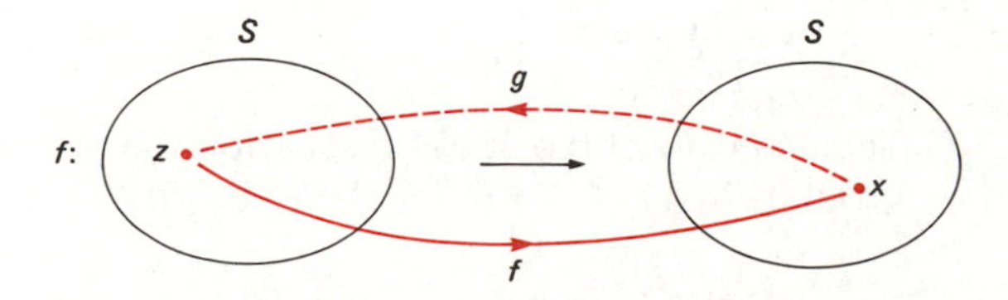

Fig. 1.18 g is an inverse of f

Example 1.16 Let S be the set of all nonnegative real numbers, and let $f: S \to S$ be defined by $f(x) = x^2$ for all $x \in S$. In Example 1.5, we proved that f is both one-to-one and onto. What is f^{-1}? First, let us define $g: S \to S$ by $g(x) = \sqrt{x}$, where $x \in S$. Recalling that $\sqrt{x^2} = x$ (since $x \geq 0$), we get

$$(f \circ g)(x) = f(g(x)) = f(\sqrt{x}) = (\sqrt{x})^2 = x \tag{1.3}$$

and

$$(g \circ f)(x) = g(f(x)) = g(x^2) = \sqrt{x^2} = x \qquad (\text{since } x \geq 0) \tag{1.4}$$

Hence, (by (1.3) and (1.4),) $f \circ g = g \circ f = i$ (identity function on S); thus $f^{-1} = g$, that is, $f^{-1}(x) = \sqrt{x}$, where $x \in S$.

Remark In order to obtain the inverse of a bijection $f: S \to S$, we may do the following: First, set $y = f(x)$. Then try to solve this equation for x (in terms of y) to get $x = g(y)$, say. Now if we should happen to get more than one such solution $x = g(y)$, we choose that solution $g(y)$ which lies in S [since $g(y) = x \in S$]. This g so chosen is then the desired inverse of f. This is so since we defined the inverse $g: S \to S$ of $f: S \to S$ by the condition

$$g(y) = x \quad \text{if and only if} \quad y = f(x) \qquad (x, y \in S)$$

Thus, to find $g(y)$, that is, x, we need only put the equation $y = f(x)$ in the form $x = g(y)$; i.e., find x in terms of y from the original equation $y = f(x)$. Finally, if in solving for x we obtain more than one answer $g(y)$, we certainly must choose that answer $x = g(y)$ which lies in S (since $x \in S$).

We illustrate this procedure by finding the inverse of the bijection given in Example 1.16. In that example, $f: S \to S$ is given by $f(x) = x^2$. The equation $y = x^2$ formally has two solutions for x, namely, $x = \sqrt{y}$ and $x = -\sqrt{y}$. Since $x \in S$ (the set of all *nonnegative* real numbers), we must have $x \geq 0$, and hence $x = \sqrt{y}$ is the only acceptable solution in our present situation. Thus the inverse of f is the function g given by $g(y) = \sqrt{y}$ for all nonnegative real numbers y, or equivalently by

$$g(x) = \sqrt{x} \qquad \text{for all nonnegative real numbers } x$$

As a further illustration, let $f_1: R \to R$ be given by $f_1(x) = -3x + 7$ (here R denotes the set of real numbers). Solving the equation $y = f_1(x) = -3x + 7$ for x, we obtain

$$x = -\tfrac{1}{3}y + \tfrac{7}{3}$$

Thus, in view of the above discussion the inverse of f_1 is the function $g_1: R \to R$ given by $g_1(y) = -\tfrac{1}{3}y + \tfrac{7}{3}$. The function g_1 can also be described, of course, by

$$g_1(x) = -\tfrac{1}{3}x + \tfrac{7}{3} \qquad (x \in R)$$

Problem Set 1.4

In the problems below, R denotes the set of real numbers.

1.4.1 Let $f:R \to R$ and $g:R \to R$ be given by $f(x) = x^2 + 1$ and $g(x) = 2x + 3$. Find $f+g$, $f-g$, $g-f$, fg, f/g, and g/f. What is the domain of each of these functions?

1.4.2 Let f and g be as in Prob. 1.4.1. Find $f \circ g$, $g \circ f$, $f \circ f$, and $g \circ g$. What is the domain of each of these functions?

1.4.3 Let f be as in Prob. 1.4.1. Is f onto? Is f one-to-one? Is f a bijection? If f is a bijection, find f^{-1}.

1.4.4 Let g be as in Prob. 1.4.1. Is g onto? Is g one-to-one? Is g a bijection? If g is a bijection, find g^{-1}.

1.4.5 Let f and g be as in Prob. 1.4.1. Find $f(0)$, $g(0)$, $f(-x)$, $g(x^2)$, $(f \circ g)(4)$, $(g \circ f)(4)$, $(f \circ f)(3)$, and $(g \circ g)(5)$.

1.4.6 Let f and g be defined by $f(x) = x^2$ and $g(x) = 1/(x+1)$. Find the domain of f and the domain of g.

1.4.7 Let f and g be as in Prob. 1.4.6. Find $f \circ g$, $g \circ f$, $f \circ f$, and $g \circ g$. What are the domains of $f \circ f$, $g \circ g$, $f \circ g$, and $g \circ f$? Is it true that $f \circ g \neq g \circ f$?

1.4.8 Let f and g be as in Prob. 1.4.6. Find $f+g$, $f-g$, fg, ff, gg, and f/g. What are the domains of these functions? Does fg equal $f \circ g$?

1.4.9 Are the functions f and g in Prob. 1.4.6 one-to-one?

1.4.10 For the functions f and g in Prob. 1.4.6, find fg and gf, and verify that $fg = gf$. Do the same for $f+g$ and $g+f$.

1.4.11 Let f and g be as in Prob. 1.4.6; find $f(0)$, $f(1)$, $f(-1)$, $f(-x)$, $f(g(2))$, $g(f(2))$, $f(f(5))$, $g(g(0))$, $g(1/x)$, and $g(5x-1)$.

1.4.12 Let f and g be defined by $f(x) = x^2$ and $g(x) = x^3$. Show that $f(x) = f(-x)$ and $g(-x) = -g(x)$.

1.4.13 Verify that the following functions $f:R \to R$ are bijections, and find the inverse of each function:

(a) $f(x) = x$ for every x in R

(b) $f(x) = -x$ for every x in R

(c) $f(x) = 1 + 3x$ for every x in R

(d) $f(x) = -2 - 7x$ for every x in R

(e) $f(x) = x^3 + 5$ for every x in R

(f) $f(x) = -x^3 + 4$ for every x in R

1.4.14 Sketch, using the same axes, the graphs of f and f^{-1} for each function f in Prob. 1.4.13.

1.4.15 For every real number x, x°F is equivalent to $f(x)$°C where $f(x) = \frac{5}{9}(x - 32)$. Let $g:R \to R$ be the function which converts degrees Celsius into degrees Fahrenheit. Show that $g = f^{-1}$, and find g.

1.5 Summary

Suppose S and T are sets (or collections) of objects. The union of S and T is the set of all elements which are in S or in T (or in both). The intersection of S and T is the set of all elements which are common to both S and T. The Cartesian product of S and T is the set of all ordered pairs (x,y), where x is in S and y is in T. S is a subset of T if every element of S is also in T.

A function f from S into T (written as $f:S \to T$) is a correspondence which associates with each element x in S *exactly one* element, denoted by $f(x)$, in T. If $f:S \to T$, then S is called the domain of f, T is called the codomain of f, and the set of all functional values $f(x)$ (where x is in S) is called the range of f. If the range of f equals the codomain of f, the function f is called onto (or surjective). If distinct elements x and x' in the domain of a function f have distinct functional values $f(x)$ and $f(x')$, the function f is called one-to-one (or injective). If a function is both onto and one-to-one, it is called a bijection. Two functions f and g are equal if they have the same domain, if they have the same codomain, and if $f(x) = g(x)$ for all x in their common domain.

A real-valued function $f:S \to R$ of a real variable is a function whose domain S is a subset of the set R of real numbers and whose codomain is the set R itself. Real-valued functions of a real variable can conveniently be represented by means of graphs. The algebraic operations of addition, subtraction, multiplication, and division for *any* two functions $f:S \to R$ and $g:S \to R$ (S need *not* be a subset of R here) are exactly as we learned in high school algebra. For example, the product fg is the function $fg:S \to R$ given by $(fg)(x) = f(x)g(x)$ for all x in S.

Suppose that $f:S \to T$ and $g:T \to W$ are functions, not necessarily real-valued functions. The composition of g and f, denoted by $g \circ f$, is the function $g \circ f:S \to W$, defined by $(g \circ f)(x) = g(f(x))$ for all x in S. The identity function (on S) is the function $i:S \to S$ with the property that $i(x) = x$ for all x in S. If $f:S \to S$ and $g:S \to S$ are such that $f \circ g = g \circ f = i$, then g is called an inverse of f. We have seen that if $f:S \to S$ is a bijection, then f has an inverse $g:S \to S$. In fact, g is defined by $g(y) = x$ if and only if $f(x) = y$, where y is any element in S.

2 Linear and Quadratic Functions

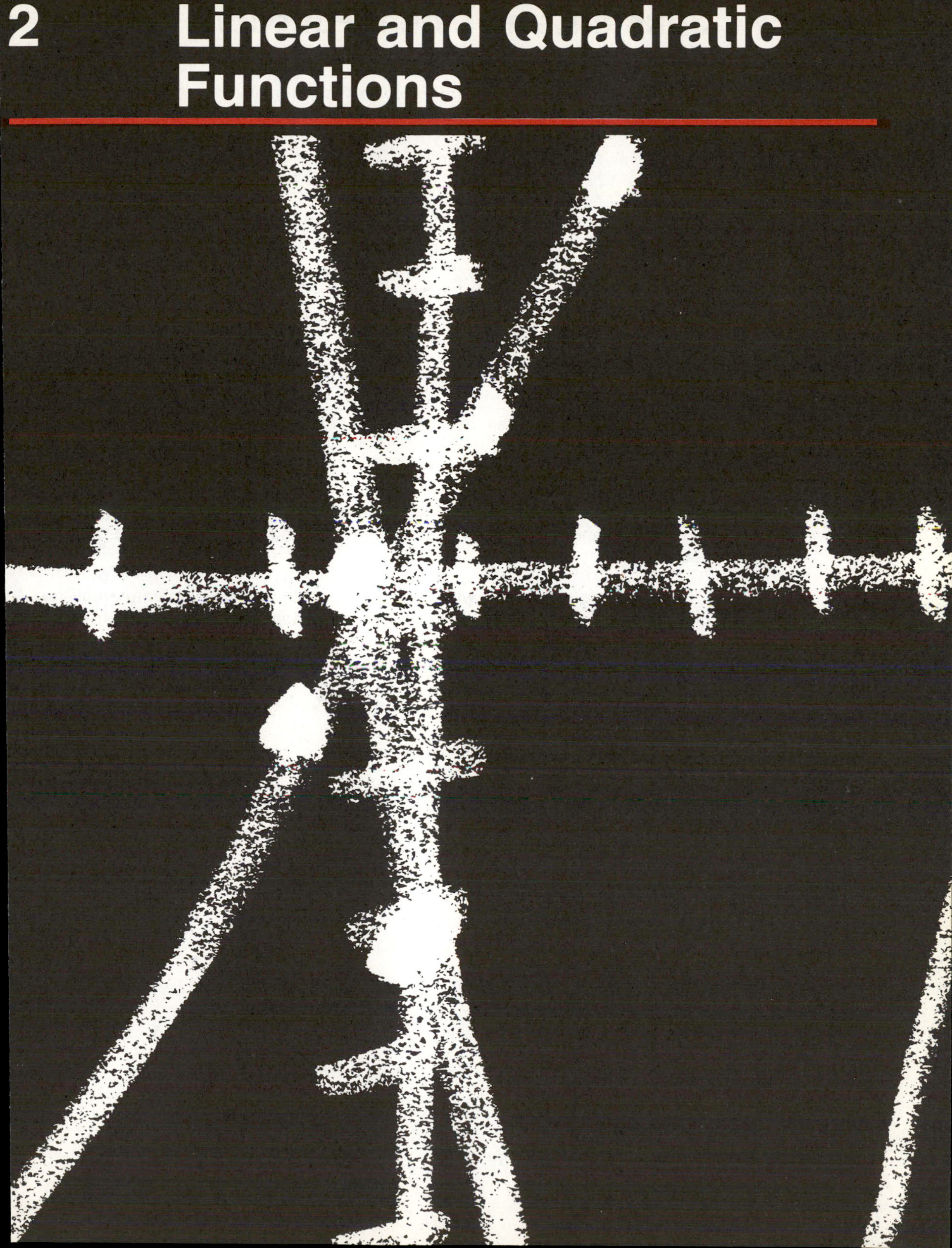

2.1 Constant and Linear Functions

A function f is called a *constant function* if it assumes only one functional value c, say. That is,

$$f(x) = c \qquad \text{for all real numbers } x$$

Thus, the functions f, g, and h, defined by

$$f(x) = 1 \qquad g(x) = 0 \qquad h(x) = -1$$

are all examples of constant functions. Clearly, the graph of every constant function is a horizontal line (parallel to the x axis), as shown in Fig. 2.1.

A function f is called a *linear function* if

$$f(x) = ax + b \qquad a \neq 0$$

where a and b are any real numbers, except that we require, of course, that $a \neq 0$.

Example 2.1 The function f defined by $f(x) = x$ is also a linear function with $a = 1$ and $b = 0$. It is called the *identity function,* and its graph is a straight line passing through the origin $\mathcal{O}$ and bisecting the right angles forming the first and third quadrants (see Fig. 2.2).

The last example, together with some of the examples we encountered in the last chapter, suggests that the graph of *every* linear function is a straight line. This is indeed the case, as we now proceed to show. Thus, suppose $P_1: (x_1, y_1)$, $P_2: (x_2, y_2)$, and $P_3: (x_3, y_3)$ all lie on the graph of the linear function f given by

$$f(x) = ax + b \qquad a, b \text{ fixed real numbers}$$

and suppose x_1, x_2, and x_3 are all distinct. Then,

$$y_1 = ax_1 + b \qquad y_2 = ax_2 + b \qquad y_3 = ax_3 + b$$

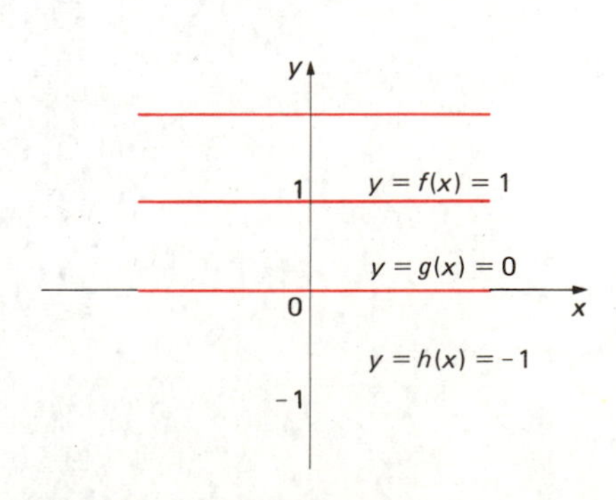

Fig. 2.1 Constant functions

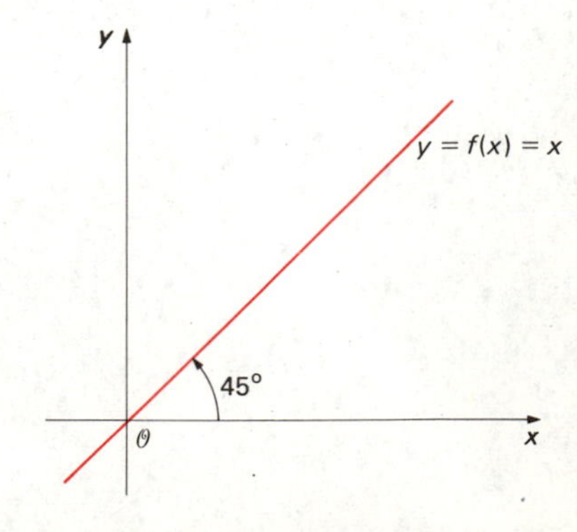

Fig. 2.2 The function $y = f(x) = x$

Hence, $y_1 - y_2 = a(x_1 - x_2)$, and $y_2 - y_3 = a(x_2 - x_3)$. Therefore, since $x_1 \neq x_2$ and $x_2 \neq x_3$, by dividing we get

$$\frac{y_1 - y_2}{x_1 - x_2} = \frac{y_2 - y_3}{x_2 - x_3} = a \tag{2.1}$$

Comparing the two triangles in Fig. 2.3, we see that they are right triangles with proportional sides. Hence, by high school geometry these two triangles are similar. In particular, $\sphericalangle \alpha = \sphericalangle \beta$ (see Fig. 2.3), and hence the three points P_1, P_2, and P_3 are on the same straight line. In other words, any three distinct points on the graph of the function f all lie on the same straight line, and thus the graph of f is a straight line. In view of this and high school geometry, any two distinct points on the graph of $y = f(x) = ax + b$ are sufficient to determine the entire graph. Moreover, keeping Eq. (2.1) in mind, the number a in the equation $y = f(x) = ax + b$ is called the *slope* of the line. It is also very easy to check that $(0,b)$ is the y intercept point of this line.

Now, let functions f and g be given by

$$f(x) = ax + b \qquad g(x) = cx + d \tag{2.2}$$

Clearly, the graphs of these two functions (being straight lines, as we have just shown) are parallel, coincident (i.e., the same line), or they intersect at exactly one point. Recalling (2.1) and some well-known facts from high school geometry, it is easy to show that the distinct lines corresponding to (2.2) are parallel if and only if $a = c$ (that is, their slopes are equal). Furthermore, the lines corresponding to (2.2) are coincident if and only if $a = c$ *and* $b = d$. (Why?) In this case, the two equations in (2.2) are identical. Finally, the two lines corresponding to (2.2) intersect at exactly one point if and only if $a \neq c$. For in this case these two lines are not parallel, and their unique point of intersection is obtained by solving simultaneously the equations (why?)

$$y = ax + b \qquad y = cx + d \qquad (a \neq c)$$

as indicated in the example below.

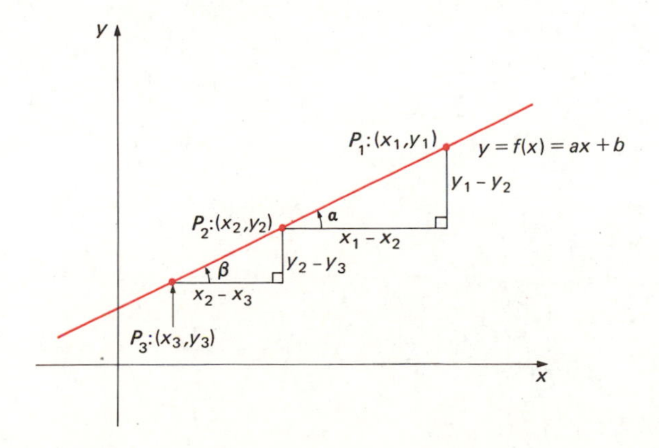

Fig. 2.3 Linear functions

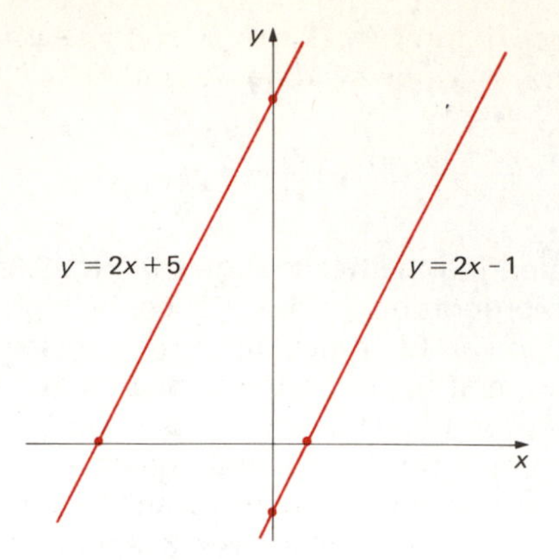

Fig. 2.4 Graphs of $y = 2x + 5$ and $y = 2x - 1$

Example 2.2 Are the following lines (*a*) $y = 2x + 5$, $y = 2x - 1$ and (*b*) $y = 3x + 1$, $y = -3x + 2$ parallel or intersecting at a unique point? If the lines intersect at a unique point, find it. Also, sketch the graphs in each case.

Solution In (*a*), the two distinct lines have the same slope (namely, 2) and are thus parallel (see Fig. 2.4). In (*b*) the two lines intersect at a unique point. To obtain this point, we solve simultaneously the equations $y = 3x + 1$ and $y = -3x + 2$. Eliminating y, we get $3x + 1 = -3x + 2$, and hence $x = 1/6$. Substituting this value of x in the equation $y = 3x + 1$ (say) gives

$$y = 3(1/6) + 1 = 3/2$$

Thus the unique point of intersection is $(1/6, 3/2)$. The graphs of these lines appear in Fig. 2.5.

Further results on linear functions will appear in Chap. 11 (on analytic geometry).

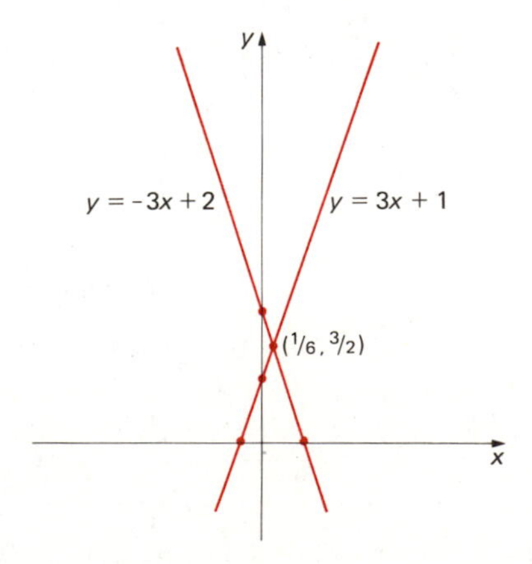

Fig. 2.5 Graphs of $y = -3x + 2$ and $y = 3x + 1$

Problem Set 2.1

2.1.1 Sketch the graphs of the following functions:

(*a*) $f(x) = -2x - 4$ $\quad g(x) = -3x + 1$

(*b*) $f(x) = 5x + 1$ $\quad g(x) = 5x - 1$

(*c*) $f(x) = 0$ $\quad g(x) = 4x - 1$

2.1.2 Which pairs of the lines in Prob. 2.1.1 are parallel, and which have a unique point of intersection? In the latter case, find this point.

2.1.3 Find all possible intersections of $y = 1 - x$, $y = 2x - 1/2$, and $y = 4x + 1$.

2.1.4 Are the following lines *concurrent* (i.e., do all pass through the same point)? If so, find this common point.

$$y = 2x + 1 \qquad y = 5x - 1 \qquad y = -x + 3$$

2.1.5 Are the following points on the line $y = 3x - 5$? Why?

$$(0,5) \qquad (1,-2) \qquad (2,11) \qquad (5/3,0) \qquad (-1,-8)$$

2.1.6 Find a linear function f such that the points (1,2) and (3,4) lie on the graph of $y = f(x)$.

2.1.7 The lines $y = ax + 1$ and $y = 2x + 3$ are parallel. Find a. Also, sketch the graphs of both lines.

2.1.8 Find the point of intersection of the lines $y = -2x - 4$ and $y = 3/2x + 7/2$.

2.1.9 Find a linear function f such that the graph of $y = f(x)$ has an x intercept $(a,0)$ and a y intercept $(0,b)$.

2.1.10 Find the constant function whose graph passes through (–3,4).

2.1.11 For what values of a, b, and c will the function f given by $f(x) = ax^2 + bx + c$ for any real number x be (*a*) a constant function or (*b*) a linear function? Explain.

2.1.12 Find the constant function whose graph passes through the point of intersection of $y = 2x + 1$ and $y = 3x - 2$.

2.1.13 Let f be the linear function given by $f(x) = ax + b$, where $a \neq 0$. Find f^{-1}.

2.1.14 Let f and g be the functions given by $f(x) = 2x + 1$ and $g(x) = 3x + b$. Find the value of b for which $f \circ g = g \circ f$.

2.1.15 Let R be the set of all real numbers, and let $f: R \to R$ be a bijection. Let $g: R \to R$ be the inverse of f, so that $g = f^{-1}$. (That f^{-1} exists was shown in Chap. 1.) The graph of f is, by definition, the set of all points (x,y) such that $y = f(x)$. Now, draw the line $y = x$. Show that the graph of g $[=f^{-1}]$ is simply the mirror reflection with respect to the line $y = x$ of the graph of f. [*Hint:* If (a,b) is any point on the graph of f, show that (b,a) is on the graph of g and, moreover, the line $y = x$ is the perpendicular bisector of the line segment joining the points (a,b) and (b,a).] See Fig. 2.6.

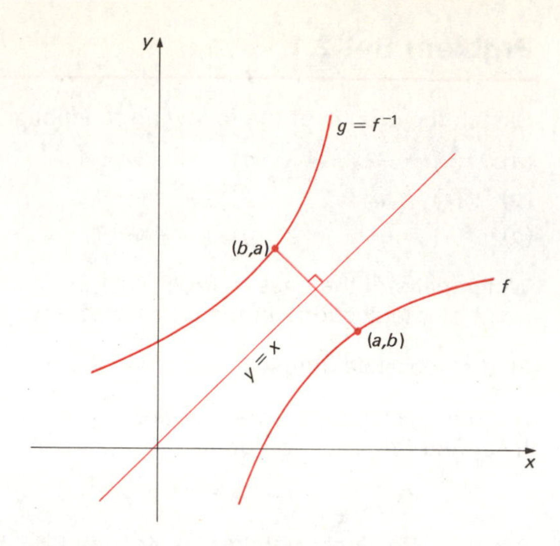

Fig. 2.6 Graphs of f and f^{-1}

2.1.16 Show that the lines $y = ax + b$ and $y = cx + d$ $[a \neq c]$ intersect at a unique point P and that the coordinates of P are

$$\left(\frac{d-b}{a-c}, \frac{ad-bc}{a-c}\right)$$

2.2 Quadratic Functions

A function f is called a *quadratic function* if for all real numbers x

$$f(x) = ax^2 + bx + c \qquad a \neq 0$$

where a, b, and c are some fixed real numbers. These fixed real numbers a, b, and c may be positive, negative, or zero, except that we naturally require that $a \neq 0$ (if $a = 0$, we have either a linear or a constant function). We have already encountered some examples of quadratic functions. Unlike the situation with linear functions, the graph of a quadratic function is *not* a straight line. The graph of $y = f(x) = ax^2 + bx + c$ $(a \neq 0)$ is called a *parabola* and can, in general, be sketched by preparing first a table of values. Some of the most interesting points on a graph are the x and y intercept points, and we now direct our attention to these points. To begin with, the y intercept point of the graph of $y = f(x) = ax^2 + bx + c$ is simply the point $(0, f(0))$, that is, $(0,c)$. What are the x intercept points of this graph? This is, of course, essentially the same question as asking for the *solutions*, or *roots*, of the quadratic equation

$$ax^2 + bx + c = 0 \qquad (a \neq 0) \tag{2.3}$$

We thus turn our attention now to the problem of determining the real-number solutions (if any) of Eq. (2.3). The method we are about to introduce for solving (2.3) is known as "completing the square." We describe below this method, together with the reasons for each step.

$a^2 + bx + c = 0$	Given
$x^2 + \frac{b}{a}x + \frac{c}{a} = 0$	Divide by a ($a \neq 0$)
$x^2 + \frac{b}{a}x = \frac{-c}{a}$	Add $\frac{-c}{a}$ to both sides
$x^2 + \frac{b}{a}x + \left(\frac{b}{2a}\right)^2 = \frac{-c}{a} + \left(\frac{b}{2a}\right)^2$	Add $\left(\frac{b}{2a}\right)^2$ to both sides
$\left(x + \frac{b}{2a}\right)^2 = \frac{-c}{a} + \frac{b^2}{4a^2}$	Algebra
$\left(x + \frac{b}{2a}\right)^2 = \frac{b^2 - 4ac}{4a^2}$	Algebra
$x + \frac{b}{2a} = \pm\frac{\sqrt{b^2 - 4ac}}{2a}$	Take square roots of both sides
$x = \frac{-b}{2a} \pm \frac{\sqrt{b^2 - 4ac}}{2a}$	Add $\frac{-b}{2a}$ to both sides
$x = \frac{-b \pm \sqrt{b^2 - 4ac}}{2a}$	Algebra

We are thus led to the following so-called *quadratic formula*: The solutions to the quadratic equation $ax^2 + bx + c = 0$, where $a \neq 0$, are given by

$$x = \frac{-b \pm \sqrt{b^2 - 4ac}}{2a}$$

Example 2.3 Solve $2x^2 - 5x + 2 = 0$.

Solution Apply the quadratic formula with $a = 2$, $b = -5$, and $c = 2$; we get

$$x = \frac{-(-5) \pm \sqrt{(-5)^2 - 4(2)(2)}}{2(2)}$$

$$= \frac{5 \pm 3}{4}$$

Hence, $x = (5 - 3)/4 = 1/2$, or $x = (5 + 3)/4 = 2$. Thus the solutions are $1/2$ and 2.

Alternative Solution By factoring the quadratic expression, our quadratic equation becomes $(2x - 1)(x - 2) = 0$. For a product to be equal to zero, one of the factors must equal zero; this leads to

$$2x - 1 = 0 \quad \text{or} \quad x - 2 = 0$$

Hence, $x = 1/2$ or $x = 2$.

Example 2.4 Solve $x^2 - 2x - 1 = 0$.

Solution Again, apply the quadratic formula, with $a = 1$, $b = -2$, and $c = -1$; we get

$$x = \frac{-(-2) \pm \sqrt{(-2)^2 - 4(1)(-1)}}{2(1)}$$

$$= \frac{2 \pm \sqrt{8}}{2}$$

$$= \frac{2 \pm 2\sqrt{2}}{2}$$

$$= 1 \pm \sqrt{2}$$

Thus the solutions are $1 - \sqrt{2}$ and $1 + \sqrt{2}$. Observe that factoring here in order to determine the solutions would have been hard.

Example 2.5 Solve $4x^2 - 4x + 1 = 0$.

Solution In the quadratic formula, set $a = 4$, $b = -4$, and $c = 1$; we get

$$x = \frac{-(-4) \pm \sqrt{(-4)^2 - 4(4)(1)}}{2(4)}$$

$$= 4/8 = 1/2$$

Thus, the two solutions happen to be equal in this case: $1/2$ and $1/2$.

Example 2.6 Solve $x^2 - 2x + 3 = 0$.

Solution In the quadratic formula, set $a = 1$, $b = -2$, and $c = 3$; thus

$$x = \frac{-(-2) \pm \sqrt{(-2)^2 - 4(1)(3)}}{2(1)}$$

$$= \frac{2 \pm \sqrt{-8}}{2} = 1 \pm \sqrt{-2}$$

Thus, the two solutions are *not* real numbers since $\sqrt{-2}$ is not a real number. This is so because if $\sqrt{-2} = z$, where z is a real number, then $z^2 = -2$. But this is absurd since the square of *any* real number is always positive or zero (and hence cannot be equal to -2). The quadratic equation $x^2 - 2x + 3 = 0$ thus has *no real solutions*.

Returning to the graph of

$$y = f(x) = ax^2 + bx + c \qquad a \neq 0 \tag{2.4}$$

we can now say the y intercept point is $(0,c)$ while the x intercept points are

$$\left(\frac{-b+\sqrt{b^2-4ac}}{2a}, 0\right) \qquad \left(\frac{-b-\sqrt{b^2-4ac}}{2a}, 0\right)$$

assuming that $b^2 - 4ac \geq 0$. Observe that if $b^2 - 4ac < 0$, then $\sqrt{b^2 - 4ac}$ is *not* a real number. Geometrically speaking, this says that if $b^2 - 4ac < 0$, then the graph of (2.4) *does not intersect* the x axis (and thus there are *no* x intercept points).

Example 2.7 Find the x and y intercept points for each of the quadratic functions in Examples 2.3 to 2.6.

Solution In view of the results of these examples, we can now tabulate these intercepts:

Function	x Intercept Points	y Intercept Points
$y = 2x^2 - 5x + 2$	$(\frac{1}{2},0)$, $(2,0)$	$(0,2)$
$y = x^2 - 2x - 1$	$(1 - \sqrt{2}, 0)$, $(1 + \sqrt{2}, 0)$	$(0,-1)$
$y = 4x^2 - 4x + 1$	$(\frac{1}{2},0)$	$(0,1)$
$y = x^2 - 2x + 3$	none	$(0,3)$

It is interesting to note what the above table tells us about the graphs of these four quadratic functions. The graph of the first function $y = 2x^2 - 5x + 2$ crosses the x axis at two distinct points [namely, $(\frac{1}{2},0)$ and $(2,0)$] and crosses the y axis at $(0,2)$. This graph appears in Fig. 2.7 (see Example 2.9). The graph of the second function $y = x^2 - 2x - 1$ again crosses the x axis at two distinct points [namely, $(1 - \sqrt{2}, 0)$ and $(1 + \sqrt{2}, 0)$] and crosses the y axis at $(0,-1)$. This graph appears in Fig. 2.8 (see Example 2.9). Next, the graph of the third function crosses the x axis at just one point (or, as we sometimes say, two coincident

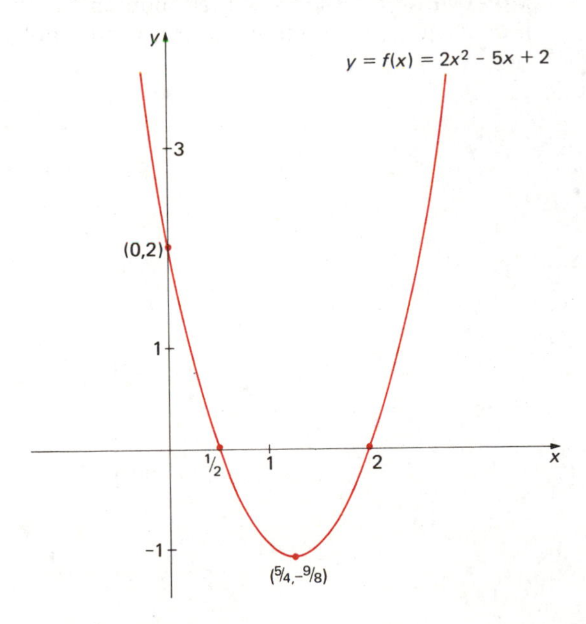

Fig. 2.7 Graph of $y = f(x) = 2x^2 - 5x + 2$

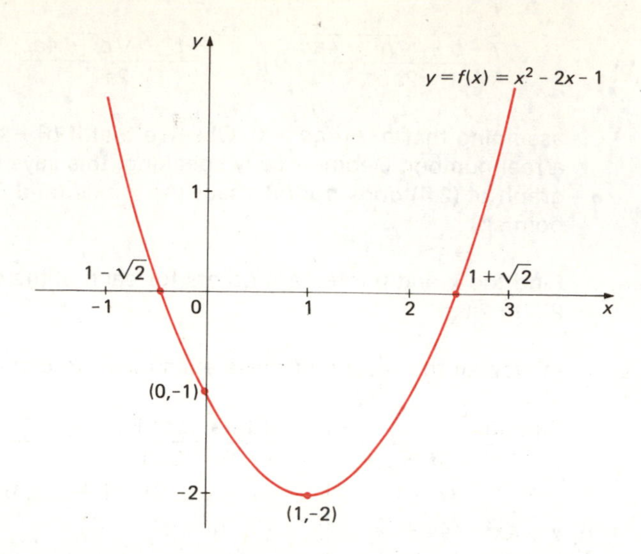

Fig. 2.8 Graph of $y = f(x) = x^2 - 2x - 1$

points). In other words, this graph just "touches" the x axis at $(^1/_2,0)$. This graph crosses the y axis at (0,1). A graph of this function appears in Fig. 2.9 (see Example 2.9). Finally, the graph of the fourth function does not have any point in common with the x axis, but it does cross the y axis at (0,3). Note that the solutions of $x^2 - 2x + 3 = 0$ (namely, $x = 1 \pm \sqrt{-2}$) cannot appear anywhere on the x axis! (Why?) The graph of this function appears in Fig. 2.10 (see Example 2.9). Observe that if $b^2 - 4ac > 0$, then there are *two distinct* x intercept points. Moreover, if $b^2 - 4ac = 0$, then there is *exactly* one x intercept point. Finally, if $b^2 - 4ac < 0$, then there are *no* x intercept points.

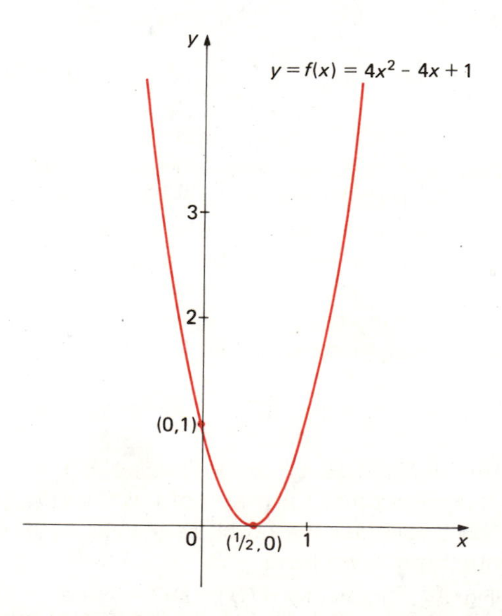

Fig. 2.9 Graph of $y = f(x) = 4x^2 - 4x + 1$

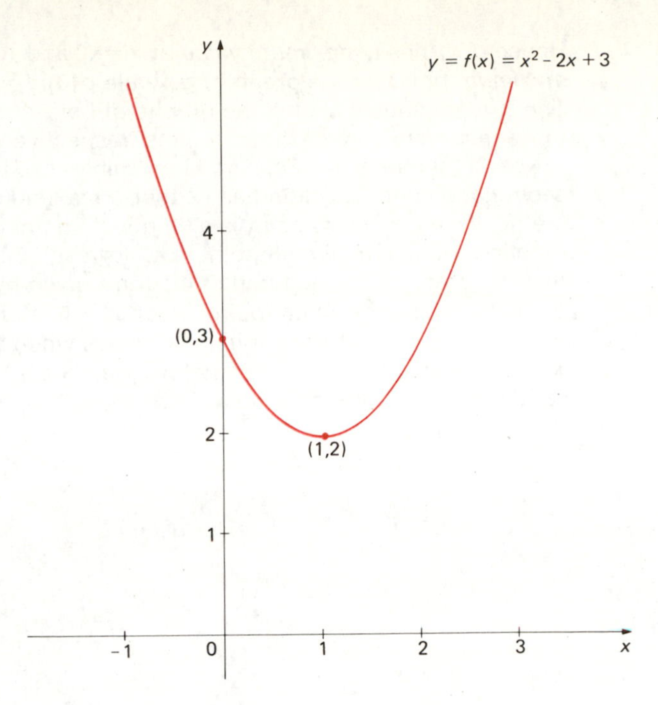

Fig. 2.10 Graph of $y = f(x) = x^2 - 2x + 3$

In studying functions and their graphs it is often helpful to determine whether a given function has a maximum or a minimum value. To study this problem for general functions requires a knowledge of calculus, which is beyond the scope of this text. However, if we confine our attention to quadratic functions, we are able to give a solution to this problem. Thus, suppose that $f: R \rightarrow R$, where R is the set of all real numbers, is a quadratic function defined by

$$f(x) = ax^2 + bx + c \qquad a \neq 0;\ a,\ b,\ c \text{ fixed real numbers} \tag{2.5}$$

We say that $f(x)$ has a *maximum value* at $x = x_0$ if $f(x) \leq f(x_0)$ for *all* real numbers x. In this case, we also say that the point $(x_0, f(x_0))$ is a *maximum point* for the graph of f and $f(x_0)$ is the *maximum value* of f. Similarly, we say that $f(x)$ has a *minimum value* at $x = x_1$ if $f(x) \geq f(x_1)$ for *all* real numbers x. In this case, we also say that the point $(x_1, f(x_1))$ is a *minimum point* for the graph of f and $f(x_1)$ is the *minimum value* of f.

As an illustration of these concepts, suppose that $f(x) = x^2 + 4$. Since $x^2 \geq 0$ for all real numbers x, we have

$$f(x) \geq 4 \ [= f(0)] \qquad \text{for all real numbers } x$$

Hence $f(x)$ has a *minimum value* at $x = 0$, and $(0, f(0))$ [that is, $(0,4)$] is a *minimum point* for graph of f, while $f(0)$ $[=4]$ is the *minimum value* of f.
Next, consider the function g given by $g(x) = 9 - x^2$. Since $-x^2 \leq 0$ for all real numbers x, we have

$$g(x) \leq 9 \ [= g(0)] \qquad \text{for all real numbers } x$$

Hence $g(x)$ has a *maximum value* at $x = 0$, and $(0, g(0))$ [that is, (0,9)] is a *maximum point* for the graph of g, while $g(0)$ [=9] is the *maximum value* of g. The student should sketch the graphs of both of these functions and identify these minimum and maximum points, respectively.

Not all functions have maximum or minimum values. For example, it is easily seen that a *linear* function has neither a maximum nor a minimum value. Geometrically, a maximum point is a "highest" point on the graph of the given function, while a minimum point is a "lowest" point on the graph. It turns out that the graph of the quadratic function f given by (2.5) always has a minimum point (but not a maximum point) when $a > 0$. Moreover, the graph of f has a maximum point (but not a minimum point) when $a < 0$. To see this, once again we use the technique of "completing the square," which we used in the derivation of the quadratic formula. Thus,

$f(x) = ax^2 + bx + c$	Given
$= a\left(x^2 + \frac{b}{a}x\right) + c$	Algebra
$= a\left[x^2 + \frac{b}{a}x + \left(\frac{b}{2a}\right)^2\right] + c - a\left(\frac{b}{2a}\right)^2$	Add and subtract $a\left(\frac{b}{2a}\right)^2$
$= a\left(x + \frac{b}{2a}\right)^2 + c - \frac{b^2}{4a}$	Algebra
$= a\left(x + \frac{b}{2a}\right)^2 + \frac{4ac - b^2}{4a}$	Algebra

We have thus shown that $f(x) = ax^2 + bx + c$ is equivalent to

$$f(x) = a\left(x + \frac{b}{2a}\right)^2 + \frac{4ac - b^2}{4a} \tag{2.6}$$

Now we distinguish two cases.

Case 1 $a > 0$

In this case, $a[x + b/(2a)]^2 \geq 0$ since the square $[x + b/(2a)]^2$ is always positive (or zero) and a is positive. Hence (2.6) now implies that $f(x) \geq (4ac - b^2)/(4a)$ for *all* real numbers x. Moreover, by (2.6) again, $f(x) = (4ac - b^2)/(4a)$ if and only if $x + b/(2a) = 0$, that is, if and only if $x = -b/(2a)$. Thus, $f[-b/(2a)] = (4ac - b^2)/(4a)$; using our above inequality, we get $f(x) \geq f[-b/(2a)]$ for *all* real numbers x. Hence $f[-b/(2a)] = (4ac - b^2)/(4a)$ is a *minimum value* for the function f, and the point $[-b/(2a), (4ac - b^2)/(4a)]$ is a *minimum point* for the graph of f. Finally, $f(x)$ has *no* maximum value in our present case since $a[x + b/(2a)]^2$ can be made as large as we please by choosing x large enough [see (2.6)].

Case 2 $a < 0$

In this case, $a[x + b/(2a)]^2 \leq 0$ since the square $[x + b/(2a)]^2$ is always positive (or zero) and a is negative. If we argue as above, (2.6) now implies that $f(x) \leq (4ac - b^2)/(4a) = f[-b/(2a)]$ for all real numbers x. Thus, $f[-b/(2a)] = (4ac - b^2)/(4a)$ is a *maximum value* for the function f, and the point

$[-b/(2a), (4ac - b^2)/(4a)]$ is a *maximum point* for the graph of f. Finally, $f(x)$ has *no* minimum value in our present case because by choosing x large enough we can make the *negative* number $a[x + b/(2a)]^2$ less than any negative number we can think of [see (2.6)].

To sum up, we have the following facts:

1. If $a > 0$, then the graph of the quadratic function f given by $f(x) = ax^2 + bx + c$ has a minimum point at $[-b/(2a), (4ac - b^2)/(4a)]$ but has no maximum points.
2. If $a < 0$, then the graph of the quadratic function f given by $f(x) = ax^2 + bx + c$ has a maximum point at $[-b/(2a), (4ac - b^2)/(4a)]$ but has no minimum points.

Example 2.8 Use the above method of "completing the square" to find the minimum point of the graph of the function f given by $f(x) = 2x^2 - 6x + 1$.

Solution As in the derivation of the quadratic formula, we start with "completing the square":

$2x^2 - 6x + 1 = 2(x^2 - 3x) + 1$	Algebra
$= 2[x^2 - 3x + (-3/2)^2] + 1 - 2(-3/2)^2$	Add and subtract $2(-3/2)^2$
$= 2(x - 3/2)^2 + 1 - 9/2$	Algebra
$= 2(x - 3/2)^2 - 7/2$	Algebra

Thus [compare with (2.6)]

$$f(x) = 2(x - 3/2)^2 - 7/2$$

Hence, as we have seen in the above discussion, the minimum point of the graph of the above quadratic function is $(3/2, -7/2)$. Observe that we could have also applied the results obtained above directly to conclude that the minimum point is

$$\left(\frac{-b}{2a}, \frac{4ac - b^2}{4a}\right) = \left(\frac{-(-6)}{2(2)}, \frac{4(2)(1) - (-6)^2}{4(2)}\right) = \left(\frac{3}{2}, \frac{-7}{2}\right)$$

The student should sketch the graph of this function and verify that $(3/2, -7/2)$ is the minimum point.

Example 2.9 Find the minimum point for each of the graphs of the quadratic functions in Examples 2.3 to 2.6.

Solution First, consider the quadratic function $y = f(x) = 2x^2 - 5x + 2$. Here $a > 0$, and hence the graph has a *minimum point* at

$$\left(-\frac{b}{2a}, \frac{4ac - b^2}{4a}\right) = \left(-\frac{(-5)}{2(2)}, \frac{4(2)(2) - (-5)^2}{4(2)}\right) = \left(\frac{5}{4}, \frac{-9}{8}\right)$$

Next consider the quadratic function $y = f(x) = x^2 - 2x - 1$. Here $a > 0$ again, and the graph has a *minimum point* at

$$\left(-\frac{b}{2a}, \frac{4ac-b^2}{4a}\right) = \left(-\frac{(-2)}{2(1)}, \frac{4(1)(-1)-(-2)^2}{4(1)}\right) = (1,-2)$$

Similarly, we find that the graph of $y = f(x) = 4x^2 - 4x + 1$ has a *minimum point* at $(1/2, 0)$, and the graph of $y = f(x) = x^2 - 2x + 3$ has a *minimum point* at $(1,2)$. The graphs of these four quadratic functions appear in Figs. 2.7 to 2.10 and are discussed in Example 2.7.

Example 2.10 Discuss the graphs of the four quadratic functions in Examples 2.3 to 2.6.

Solution Combining the facts from Examples 2.7 and 2.9, we can easily check that the desired graphs are as indicated in Figs. 2.7 to 2.9. In each graph, we labeled the intercept points and the minimum point. Observe that the graph of $y = f(x) = x^2 - 2x + 3$ does *not* intersect the x axis (see Fig. 2.10).

Example 2.11 Find the intercept points and the maximum point, and sketch the graph of the quadratic function f given by $y = f(x) = -x^2 - 2x + 3$.

Solution The solutions to the quadratic equation $-x^2 - 2x + 3 = 0$ are easily seen to be -3 and 1 (apply the quadratic formula). Thus, $(-3,0)$ and $(1,0)$ are the x intercept points. Moreover, since $a = -1 < 0$, we know that the graph has a *maximum point* at

$$\left(-\frac{b}{2a}, \frac{4ac-b^2}{4a}\right) = \left(-\frac{(-2)}{2(-1)}, \frac{4(-1)(3)-(-2)^2}{4(-1)}\right) = (-1,4)$$

Finally, setting $x = 0$ gives $f(0) = 3$; hence $(0,3)$ is the y intercept point. Combining these facts, we easily see that the desired graph is as indicated in Fig. 2.11.

Example 2.12 Sketch, using the same coordinate axes, the graphs of the linear function $y = f(x) = 2x + 1$ and the quadratic function $y = g(x) = x^2 - 4x - 5$. Find the points of intersection.

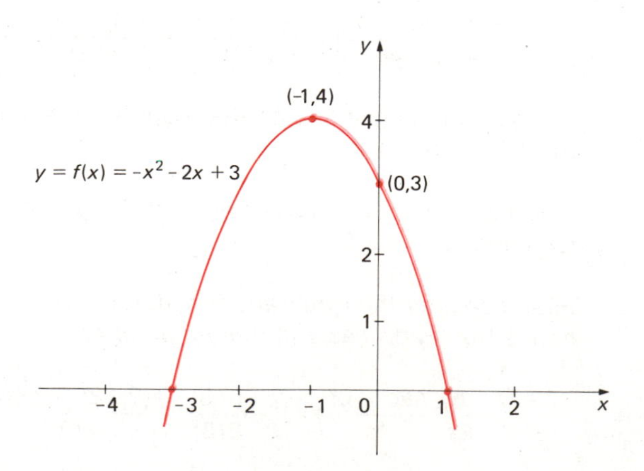

Fig. 2.11 Graph of $y = f(x) = -x^2 - 2x + 3$

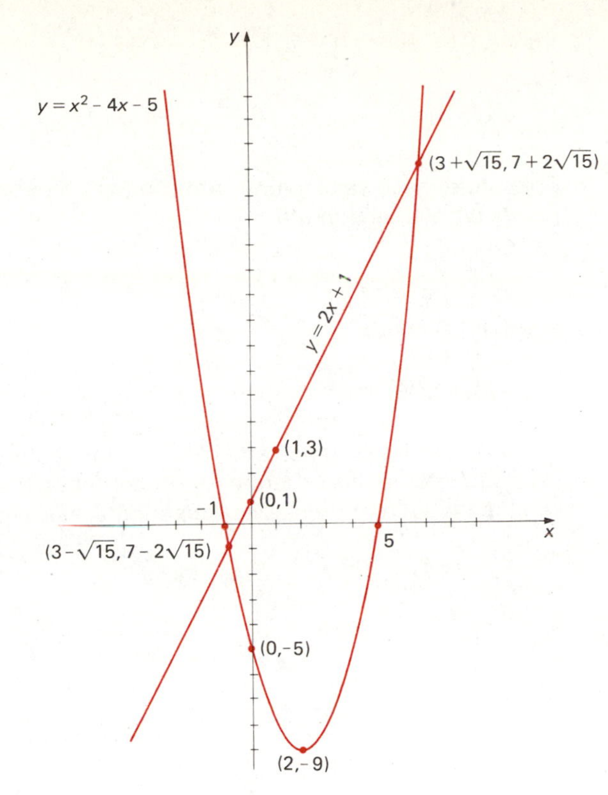

Fig. 2.12 Graphs of $y = x^2 - 4x - 5$ and $y = 2x + 1$

Solution

First, observe that the points (0,1) and (1,3) both lie on the line $y = 2x + 1$. (Why?) Now, by the quadratic formula the solutions of the equation $x^2 - 4x - 5 = 0$ are easily seen to be -1 and 5. Also with $a > 0$ the graph of $y = x^2 - 4x - 5$ has a minimum point at

$$\left(-\frac{b}{2a}, \frac{4ac - b^2}{4a}\right) = \left(-\frac{(-4)}{2(1)}, \frac{4(1)(-5) - (-4)^2}{4(1)}\right)$$

that is, the point $(2,-9)$ is the minimum point on the graph of the quadratic function. Also, the y intercept point for the graph of the quadratic function is $(0,-5)$. This leads to the graph sketched in Fig. 2.12.

To find the points of intersection of these two graphs, we solve simultaneously the equations

$$y = 2x + 1 \qquad y = x^2 - 4x - 5$$

Eliminating y, we get $2x + 1 = x^2 - 4x - 5$, or $x^2 - 6x - 6 = 0$. By the quadratic formula, the solutions of this quadratic equation are

$$x = \frac{-(-6) \pm \sqrt{(-6)^2 - 4(1)(-6)}}{2(1)}$$

that is,

$$x = \frac{6 \pm \sqrt{60}}{2} = \frac{6 \pm 2\sqrt{15}}{2} = 3 \pm \sqrt{15}$$

Substitute these values for x into $y = 2x + 1$ to find the y coordinates. Thus, the points of intersection are

$$(3 + \sqrt{15}, 2(3 + \sqrt{15}) + 1) \qquad (3 - \sqrt{15}, 2(3 - \sqrt{15}) + 1)$$

that is,

$$(3 + \sqrt{15}, 7 + 2\sqrt{15}) \qquad (3 - \sqrt{15}, 7 - 2\sqrt{15})$$

Example 2.13 A farmer desires to fence a field bordering a straight stream with 600 yd of fencing material. If this farmer does not fence the side bordering the stream, what is the area of the largest rectangular field he can fence?

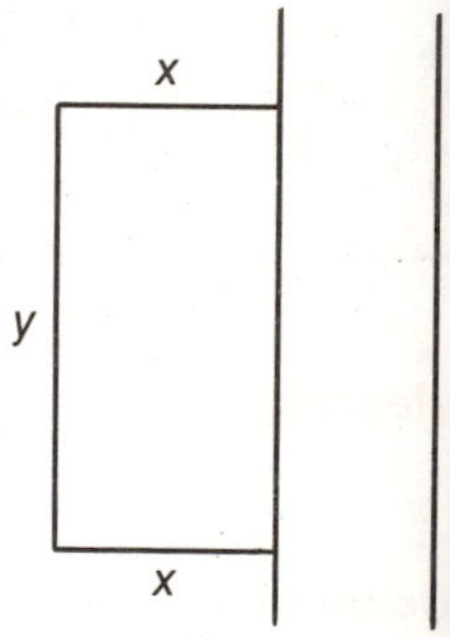

Solution Suppose that the dimensions of the desired rectangle are x by y yd, as indicated in the present figure. Then the total amount of fencing material is $2x + y$. Hence, by hypothesis

$$2x + y = 600 \tag{2.7}$$

Now let A be the area of the rectangle in the above figure. Then

$$A = xy \tag{2.8}$$

By (2.7), $y = 600 - 2x$. Substituting this into (2.8), we get

$$A = x(600 - 2x) = -2x^2 + 600x \tag{2.9}$$

Thus,

$$A = -2x^2 + 600x \tag{2.10}$$

Our problem now is to find the value of x which makes the area A assume its maximum value. By the results of this section, we know that the quadratic function A given by (2.10) has a maximum value (since $a = -2 < 0$). Moreover this maximum value is obtained when $x = -b/(2a) = -(600)/[2(-2)] = 150$; that is, $x = 150$. Substituting this value of x into (2.7) we obtain $y = 300$. Thus, the area A of the largest rectangular field is given by $A = (150)(300) = 45{,}000$ sq yd.

Example 2.14 Find a quadratic function f such that the graph of $y = f(x)$ passes through the points (0,0), (1,3), and (2,–1).

Solution Let $y = f(x) = ax^2 + bx + c$, where $a \neq 0$, be the desired function. Since (0,0), (1,3), and (2,–1) all lie on the graph of $y = f(x)$, we get (by substituting these coordinates in this equation)

$$0 = a(0)^2 + b(0) + c = c$$

$$3 = a(1)^2 + b(1) + c = a + b + c$$

$$-1 = a(2^2) + b(2) + c = 4a + 2b + c$$

The first of these equations gives $c = 0$; the last two equations now reduce to

$$a + b = 3$$

$$4a + 2b = -1$$

Multiplying the first of these equations by 2 gives

$$2a + 2b = 6$$

Subtracting the last two equations, we get $2a = -7$; hence, $a = -7/2$. Substituting this in the equation $a + b = 3$ we get $b = 13/2$. Therefore, the desired quadratic function is $y = f(x) = -7/2x^2 + 13/2x$.

Remark Strictly speaking, what we have shown is that *if* there is a quadratic function f passing through the above given points, *then* f is as given in the above equation. Accordingly, we still have to verify that the points (0,0), (1,3), and (2,–1) do indeed lie on the graph of $y = f(x) = -7/2x^2 + 13/2x$. We leave this easy verification to the student.

Problem Set 2.2

2.2.1 Solve the following quadratic equations:

(*a*) $2x^2 - x - 1 = 0$ (*c*) $x^2 - 4x - 1 = 0$

(*b*) $9x^2 - 6x + 1 = 0$ (*d*) $x^2 - 2x + 4 = 0$

2.2.2 Sketch the graph of each of the quadratic functions corresponding to the left side of each equation in Prob. 2.2.1. Label the x intercept points, the y intercept point, and the minimum point for each graph.

2.2.3 Find the points of intersection of the graphs of $y = 3x - 1$ and $y = -x^2 + 2x + 5$, and sketch the graphs.

2.2.4 Find the x intercepts (if any), the y intercept, and the maximum point of the graph of $y = -x^2 + 6x - 10$. Sketch the graph.

2.2.5 Does the point (1,–1) lie on the graph of $y = x^2 + x - 1$? Why?

2.2.6 Find a quadratic function f such that the graph of $y = f(x)$ passes through (1,–1), (–1,1), and (2,3). Sketch the graph.

2.2.7 Do the following graphs have maximum points or minimum points? Find these points. Do this by "completing the square," as explained in Example 2.8.

(*a*) $y = -2x^2 - x + 1$ (*c*) $y = -x^2 - 2x - 1$

(*b*) $y = x^2 - 4x + 10$

2.2.8 The graph of $y = x^2 + bx + c$ has a minimum point at (1,2). Find the values of b and c and sketch the graph. Does the graph have any x intercepts?

2.2.9 Let a, b, and c be any real numbers such that $a \neq 0$ and $b^2 - 4ac < 0$. Prove that for all real numbers x, $ax^2 + bx + c$ always has the same sign as a. (*Hint*: Complete the square.)

2.2.10 Find the points of intersection of $y = x^2 + x - 2$ and $y = -x^2 + 3x + 2$. Sketch the graphs.

2.2.11 When are the roots of the quadratic equation $ax^2 + bx + c = 0$ $[a \neq 0]$ equal? When are they real? When are they rational?

2.2.12 The equation $x^2 - 2x + c = 0$ has 4 as a solution. Find c.

2.2.13 The equation $x^2 + bx - 3 = 0$ has -1 as a solution. Find b.

2.2.14 The equation $x^2 + bx + 4 = 0$ has no real solutions. What can you say about b? (*Hint*: Recall the quadratic formula.)

2.2.15 The equation $x^2 + 3x + c = 0$ has both of its solutions real. What can you say about c?

2.2.16 The equation $x^2 + bx + c = 0$ has 1 and -2 for solutions. Find b and c.

2.2.17 The equation $x^2 + bx + 8 = 0$ has two equal solutions. Find b.

2.2.18 The sum of two integers is 100. What should these integers be so that their product is a maximum?

2.2.19 Find the x intercepts and the y intercept, and sketch the graph of the quadratic function given in Example 2.8.

2.2.20 The difference of the two solutions of the quadratic equation $x^2 + bx - 16 = 0$ is 10. Find b and the solutions. (There are two answers for b.)

2.2.21 Find the rectangle with the largest area which can be inscribed in a right triangle with sides of length 6 and 8 in. (*Hint*: Use the fact that the three triangles in the figure are similar and, hence, their corresponding sides are proportional. This will give y as a linear function of x.)

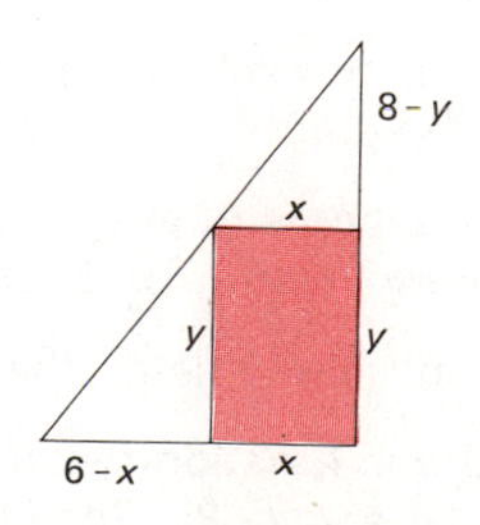

2.3 Summary

In this chapter, we studied properties and graphs of constant, linear, and quadratic functions, with heavy emphasis on quadratic functions. We have seen that the graph of every constant function, as well as every linear function, is a straight line. On the other hand, the graph of a quadratic function f, given by $f(x) = ax^2 + bx + c$, is a parabola which has a maximum point (but not a minimum point) if $a < 0$. However, such a graph has a minimum point (but not a maximum point) if $a > 0$. Both maximum and minimum points occur at $(-b/(2a), f[-b/(2a)])$. Furthermore, the quadratic formula was derived and used in finding the x intercepts of the graph of a quadratic function. Also, solutions of two linear, or one linear and one quadratic, equations were discussed. In the latter case, the method of substitution (or elimination) resulted in a quadratic equation of the form $ax^2 + bx + c = 0$ (for $a \neq 0$). The solutions of this quadratic equation are given by the *quadratic formula*:

$$x = \frac{-b \pm \sqrt{b^2 - 4ac}}{2a}$$

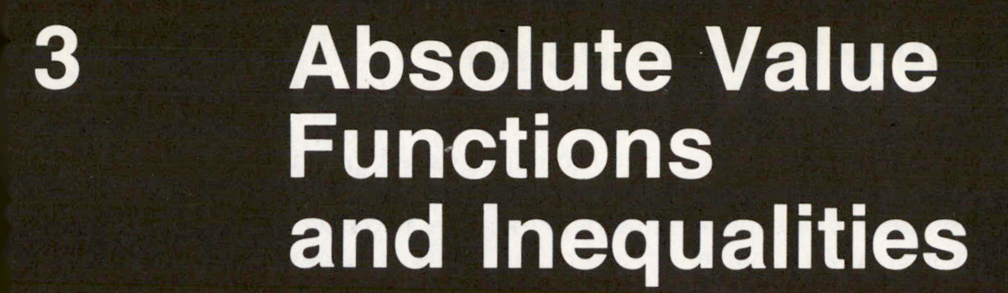

3 Absolute Value Functions and Inequalities

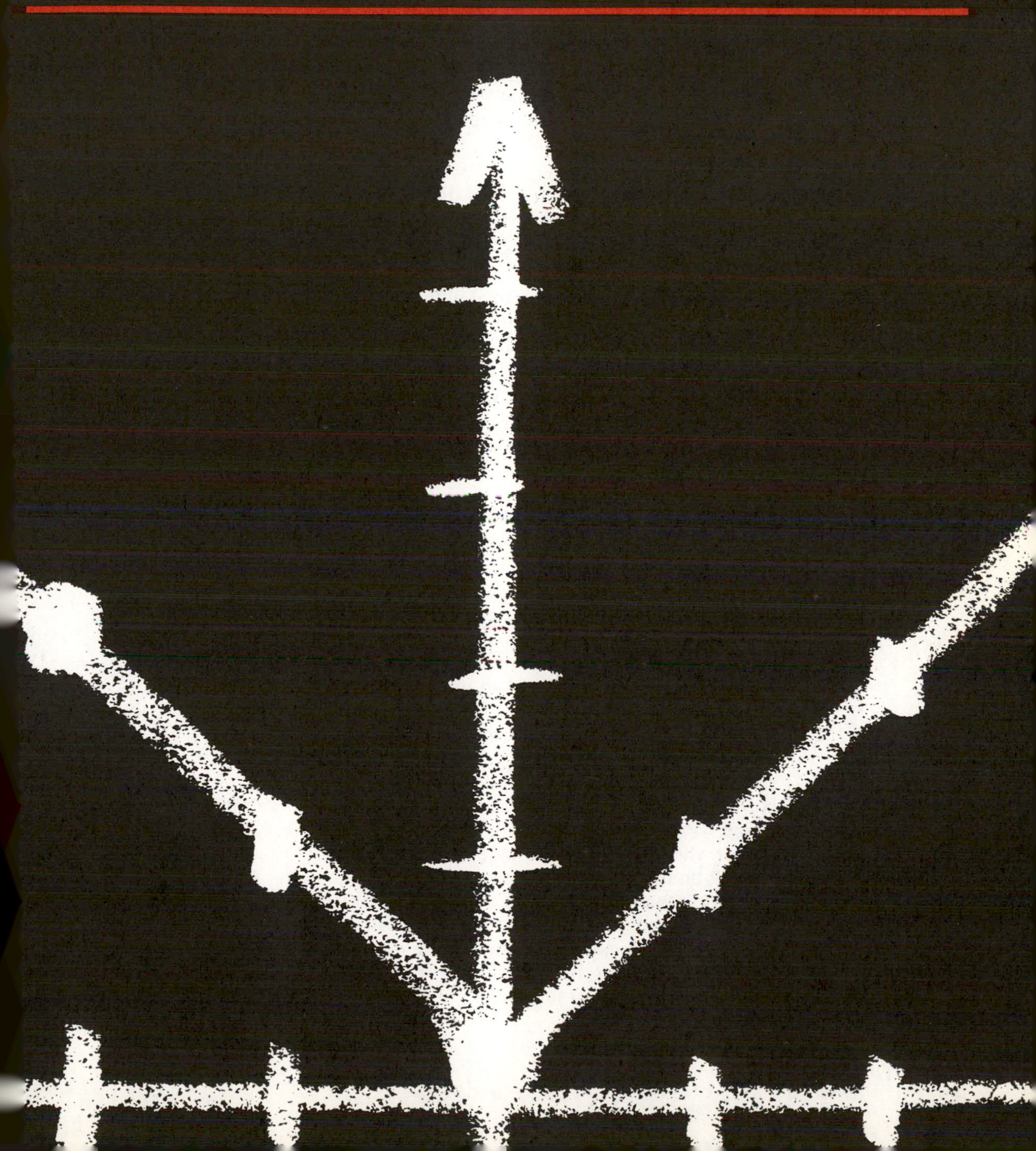

3.1 Absolute Value Function

In this section, we study a very useful function, called the *absolute value function.* We begin with a brief review of the properties of inequalities.

Suppose a and b are any real numbers (positive, negative, or zero). An *inequality* is simply a statement in any of the forms

$$a < b \qquad a \leq b \qquad a > b \qquad \text{or} \qquad a \geq b$$

We recall that $a < b$ means that $b - a$ is positive while $a \leq b$ means that $b - a$ is positive or zero. Similarly, $a > b$ means that $a - b$ is positive, while $a \geq b$ means that $a - b$ is positive or zero. We now recall some familiar and basic facts about inequalities.

Rule 1

An inequality is not affected by adding or subtracting the same number to both sides: $a < b$ is equivalent to $a + c < b + c$ and to $a - c < b - c$. For example, $1 < 2$ is equivalent to $1 + 3 < 2 + 3$ and to $1 - 3 < 2 - 3$.

Rule 2

An inequality is not affected by multiplying or dividing both sides by the same *positive* number: $a < b$ is equivalent to $ac < bc$ and to $a/c < b/c$ *provided c is positive.* For example, $4 < 8$ is equivalent to $4 \cdot 3 < 8 \cdot 3$ and to $4/3 < 8/3$.

Rule 3

The direction of an inequality must be *reversed* if both sides are multiplied or divided by a *negative* number. That is, $a < b$ is equivalent to $ac > bc$ and to $a/c > b/c$ *if c is negative.* For example, $4 < 8$ is equivalent to $4(-2) > 8(-2)$ and to $4/-2 > 8/-2$.

Now suppose that x is any real number. We define the *absolute value* of x, denoted by $|x|$, as follows:

$$|x| = \begin{cases} x & \text{if } x \geq 0 \\ -x & \text{if } x < 0 \end{cases} \tag{3.1}$$

Thus, according to this definition we have $|1| = 1$, $|5/2| = 5/2$, $|0| = 0$, $|-1| = -(-1) = 1$, $|-1/3| = -(-1/3) = 1/3$, and so on. These examples suggest that

$$|x| \geq 0 \qquad \text{for } all \text{ real numbers } x \tag{3.2}$$

This is indeed the case for the following reasons: First, if $x \geq 0$, then $|x| = x \geq 0$, and thus $|x| \geq 0$. On the other hand, if $x < 0$, then $|x| = -x > 0$ (since $x < 0$), and thus $|x| > 0$. Hence, in any case $|x| \geq 0$; in fact, $|x| > 0$ except when $x = 0$ (recall that $|0| = 0$, by definition).

Let R be the set of all real numbers, and define a function $f: R \to R$ by $f(x) = |x|$ for *all* real numbers x in R. In other words, each of the domain and codomain of the function f is the set R itself, and $f: x \to |x|$ for all x in R. This function f is called the *absolute value function,* and it is instructive to sketch the graph of this function f. For this purpose, we tabulate below a few of the points on the graph of the equation $y = f(x) = |x|$ and plot these points (see Fig. 3.1).

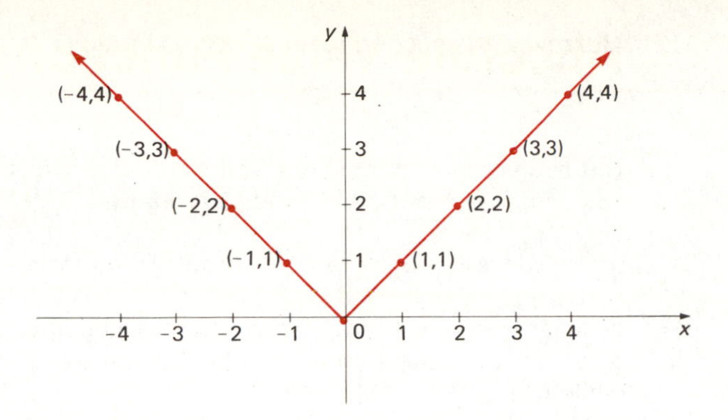

Fig. 3.1 Graph of $y = f(x) = |x|$

x	0	1	2	3	4	−1	−2	−3	−4
$y = \|x\|$	0	1	2	3	4	1	2	3	4

Observe that the graph of the absolute value function is a V-shaped curve. Indeed, this graph consists of two rays which begin at the origin; also each ray bisects one of the right angles formed by the axes. Observe also that, in view of the definition of absolute value, the ray in the first quadrant is a part of the graph of the line $y = x$ (since $|x| = x$ for all $x \geq 0$) while the ray in the second quadrant is a part of the graph of the line $y = -x$ (since $|x| = -x$ for all $x < 0$). Furthermore, the absolute value function is clearly *not* a one-to-one function since, for example, $f(1) = |1| = 1$ and $f(-1) = |-1| = 1$. Thus, $f(1) = f(-1)$ even though $1 \neq -1$. Moreover, since $|x| \geq 0$ for all real x, the range of the absolute value function is the set of all *nonnegative* real numbers.

Keeping in mind the definition of absolute value [as given in (3.1)], it is easy to see that $|x| \geq x$ and $|x| \geq -x$ for *all* real numbers x. The first inequality above is, of course, equivalent to $x \leq |x|$, while the second inequality is equivalent to $-|x| \leq x$ since multiplying both sides of an inequality by -1 reverses its direction (by Rule 3). The net result, then, is

$$-|x| \leq x \leq |x| \qquad \text{for } all \text{ real numbers } x \tag{3.3}$$

Next, we consider $|x_1 + x_2|$, where x_1 and x_2 are any real numbers (positive, negative, or zero). Now, by (3.3) we have $-|x_1| \leq x_1 \leq |x_1|$ and $-|x_2| \leq x_2 \leq |x_2|$. Hence, by adding these inequalities we obtain

$$-|x_1| + (-|x_2|) \leq x_1 + x_2 \leq |x_1| + |x_2| \tag{3.4}$$

Since $-|x_1| + (-|x_2|) = -(|x_1| + |x_2|)$, Eq. (3.4) becomes

$$-(|x_1| + |x_2|) \leq x_1 + x_2 \leq |x_1| + |x_2| \tag{3.5}$$

Multiplying the *first* inequality in (3.5) by -1 will have the effect of *reversing* the direction of this inequality. The result is $|x_1| + |x_2| \geq -(x_1 + x_2)$; therefore,

$$-(x_1 + x_2) \leq |x_1| + |x_2| \tag{3.6}$$

Moreover, the second inequality in (3.5) reads

$$(x_1 + x_2) \leq |x_1| + |x_2| \tag{3.7}$$

But by definition of absolute value $|x_1 + x_2|$ is always equal to $(x_1 + x_2)$ or to $-(x_1 + x_2)$; hence by (3.6) and (3.7) we get

$$|x_1 + x_2| \leq |x_1| + |x_2| \qquad \text{for all real numbers } x_1, x_2 \tag{3.8}$$

In other words, we have shown that the absolute value of a sum is always *equal to or less than* the sum of the absolute values. For example, $|3 + 7| = |3| + |7|$, while $|3 + (-7)| < |3| + |-7|$.

What about the absolute value of a product? The answer is

$$|x_1x_2| = |x_1||x_2| \qquad \text{for all real numbers } x_1, x_2 \tag{3.9}$$

This follows at once from the definition of absolute value because each side of (3.9) is equal to $\pm x_1x_2$ (whichever of these is positive or zero) and both sides of (3.9) certainly agree in sign (since $|x| \geq 0$ for all x). Similarly, we show that

$$\left|\frac{x_1}{x_2}\right| = \frac{|x_1|}{|x_2|} \qquad \text{for all real numbers } x_1, x_2, \text{ where } x_2 \neq 0 \tag{3.10}$$

Indeed, both sides of (3.10) are equal to $\pm x_1/x_2$ (whichever of these is positive or zero), and, moreover, both sides of (3.10) agree in sign since $|x| \geq 0$.

Let us summarize the properties of the absolute value function which we have encountered so far:

1. The absolute value function $f: R \to R$ is a function of the set R of real numbers into R given by $f: x \to |x|$ [that is, $f(x) = |x|$]. By definition, $|x| = x$, if $x \geq 0$, while $|x| = -x$ if $x < 0$. The absolute value function is *not* a one-to-one function, and its range is the set of all *nonnegative* real numbers. The graph of the absolute value function is V-shaped and lies entirely in the first and second quadrants.

2. Let x, x_1, and x_2 be any real numbers. Then
   (*a*) $|x| \geq 0$ (equality holds if and only if $x = 0$)
   (*b*) $-|x| \leq x \leq |x|$
   (*c*) $|x_1 + x_2| \leq |x_1| + |x_2|$
   (*d*) $|x_1x_2| = |x_1||x_2|$
   (*e*) $\left|\frac{x_1}{x_2}\right| = \frac{|x_1|}{|x_2|}$ provided that $x_2 \neq 0$

Example 3.1 Let f be the function defined by $f(x) = |x|$. Find $f(4)$, $f(-4)$, $f(5/2)$, $f(-5/2)$, and $f(0)$.

Solution Recalling the definition of absolute value, we have $f(4) = |4| = 4$; $f(-4) = |-4| = -(-4) = 4$; $f(5/2) = |5/2| = 5/2$; $f(-5/2) = |-5/2| = -(-5/2) = 5/2$; and $f(0) = |0| = 0$.

Example 3.2 Sketch the graph of the function f given by $f(x) = |x - 1|$.

Solution In view of the definition of absolute value, we naturally distinguish two cases depending on where $x-1$ changes sign.

Case 1 $x-1 \geq 0$; that is, $x \geq 1$

In this case, $|x-1| = x-1$ (by definition), and hence $f(x) = x-1$. Thus, we have shown that

$$f(x) = x - 1 \qquad \text{if } x \geq 1 \tag{3.11}$$

Case 2 $x-1 < 0$; that is, $x < 1$

In this case, $|x-1| = -(x-1)$ (by definition), and hence $f(x) = -(x-1) = -x+1$. Thus, we have shown that

$$f(x) = -x + 1 \qquad \text{if } x < 1 \tag{3.12}$$

In other words, our function f is graphed in two stages (depending on whether $x \geq 1$ or $x < 1$) as given in (3.11) and (3.12). Thus, the graph of the function f [or the equation $y = f(x) = |x-1|$] is simply that part of the line $y = x-1$ which lies to the *right* of $x = 1$ together with that part of the line $y = -x+1$ which lies to the *left* of $x = 1$. The graph of this function f appears in Fig. 3.2. Observe that, once again, the graph of the present function f is V-shaped and that the two rays described above now meet at (1,0). Note, also, the great similarity between the graphs of $y = |x|$ and $y = |x-1|$, sketched in Figs. 3.1 and 3.2. In fact, it is easily seen that one obtains the graph of $y = |x-1|$ from the graph of $y = |x|$ by the simple devise of moving the y axis in Fig. 3.1 one unit to the *left* (keeping the direction of this y axis unchanged).

Example 3.3 Sketch the graph of the function f given by $f(x) = |x+1|$.

Solution Arguing as in Example 3.2, we easily see that

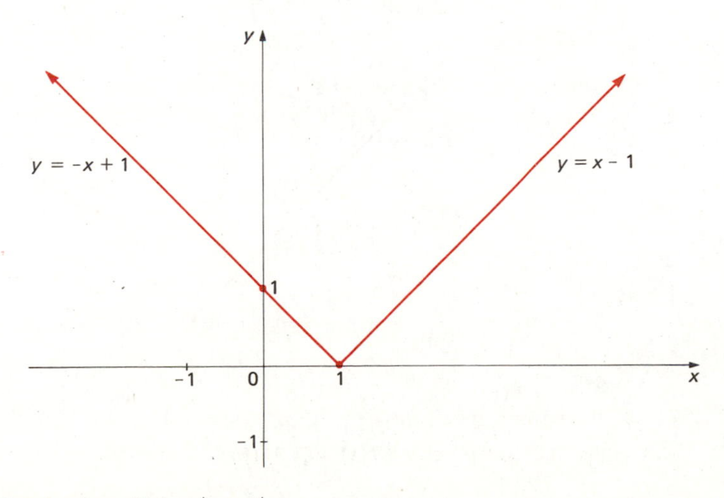

Fig. 3.2 Graph of $y = f(x) = |x-1|$

$$f(x) = \begin{cases} x+1 & \text{if } x \geq -1 \\ -x-1 & \text{if } x < -1 \end{cases}$$

Thus, the graph of the function f is simply that part of the line $y = x + 1$ which lies to the *right* of $x = -1$ together with that part of the line $y = -x - 1$ which lies to the *left* of $x = -1$. The graph of f appears in Fig. 3.3. Again, observe that the graph of $y = |x + 1|$ can be obtained from the graph of $y = |x|$ by simply moving the y axis in Fig. 3.1 one unit to the *right* (keeping the direction of this y axis unchanged).

Example 3.4 Sketch the graph of the function f given by $y = f(x) = |2x|$.

Solution Arguing as in the previous examples, we easily verify that f is given by

$$f(x) = \begin{cases} 2x & \text{if } x \geq 0 \\ -2x & \text{if } x < 0 \end{cases}$$

The graph of f is sketched in Fig. 3.4. Once again, the graph is V-shaped. This time, however, the two rays $y = 2x$ and $y = -2x$, in the first and second quadrants, respectively, are steeper than those rays we encountered in the preceding examples.

Example 3.5 Sketch the graph of the function f given by $f(x) = |x + 1| + |x - 1|$.

Solution We tabulate below the values of x at which $x + 1$ or $x - 1$ changes sign together with the corresponding values of $f(x)$.

	$x < -1$	$-1 \leq x < 1$	$x \geq 1$
$\|x + 1\|$	$-(x + 1)$	$x + 1$	$x + 1$
$\|x - 1\|$	$-(x - 1)$	$-(x - 1)$	$x - 1$
$y = \|x + 1\| + \|x - 1\|$	$-(x + 1) - (x - 1)$	$(x + 1) - (x - 1)$	$(x + 1) + (x - 1)$

In verifying the above table, observe that if, for example, $x < -1$, then $x + 1 < 0$,

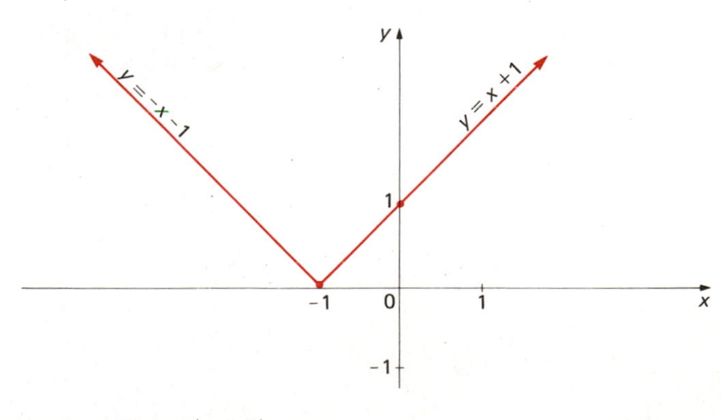

Fig. 3.3 Graph of $y = f(x) = |x + 1|$

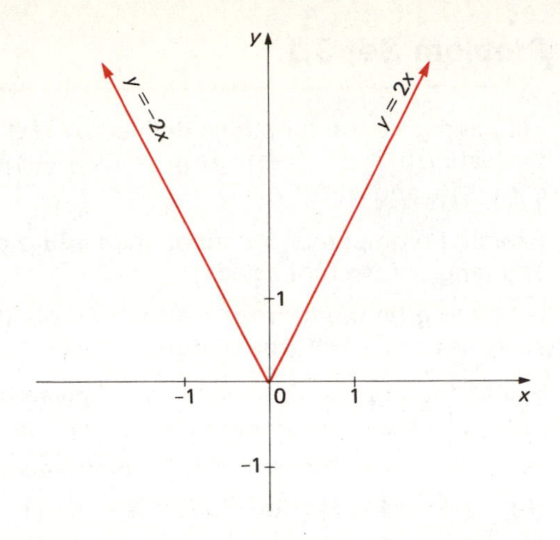

Fig. 3.4 Graph of $y = f(x) = |2x|$

and hence $|x+1| = -(x+1)$. This is the entry in our table at the intersection of the row headed by $|x+1|$ and the column headed by $x < -1$. Similarly, we verify the rest of the entries. Thus, the function f is graphed in three stages as follows:

$$f(x) = \begin{cases} -(x+1) - (x-1) = -2x & \text{if } x < -1 \\ (x+1) - (x-1) = 2 & \text{if } -1 \le x < 1 \\ (x+1) + (x-1) = 2x & \text{if } x \ge 1 \end{cases}$$

The graph of the function f appears in Fig. 3.5.

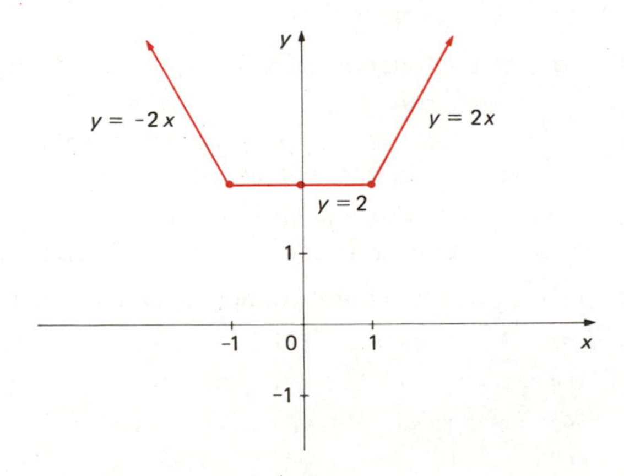

Fig. 3.5 Graph of $y = |x+1| + |x-1|$

Problem Set 3.1

3.1.1 Let f and g be the functions defined by $f(x) = |x|$ and $g(x) = |2x - 3|$. Evaluate $f(0)$, $f(1)$, $f(-1)$, $g(0)$, $g(1)$, $g(-1)$, $f(g(-1))$, $g(f(-1))$, $f(f(-4))$, $g(g(-4))$, and $g(3/2)$.

3.1.2 Sketch the graphs of the functions f and g given in Prob. 3.1.1. Find the domain and range of each of f and g.

3.1.3 Let f and g be the functions given in Prob. 3.1.1. Is it true that $f(x) = f(-x)$ for all values of x? Is it true that $g(x) = g(-x)$ for all values of x? Explain.

3.1.4 Sketch the graphs of the functions f given by the following equations. In each case, give the domain and range of the function.

(*a*) $f(x) = |-x|$ (How does this compare with the graph in Fig. 3.1?)

(*b*) $f(x) = |1 - x|$ (How does this compare with the graph in Fig. 3.2?)

(*c*) $f(x) = |2x + 3|$

(*d*) $f(x) = |3x - 1|$

(*e*) $f(x) = |3x| - 1$

(*f*) $f(x) = |1 + 2x| + |1 - 2x|$

(*g*) $f(x) = |2x + 1| + |2x - 1|$ (How does this graph compare with the graph in Prob. 3.1.4*f*?)

(*h*) $f(x) = x - |x|$

(*i*) $f(x) = x + |x|$

(*j*) $f(x) = \dfrac{|x|}{x}$

3.1.5 Suppose $f: R \to R$ (R is the set of real numbers) is defined by

$$f(x) = \begin{cases} |x| & \text{if } |x| \leq 5 \\ 5 & \text{if } |x| \geq 5 \end{cases}$$

Show that f is indeed a function, and sketch the graph of f. What are the domain and range of f?

3.1.6 Let f be the function given in Prob. 3.1.5. Find $f(0)$, $f(1)$, $f(-1)$, $f(5)$, $f(-5)$, $f(-7)$, and $f(8)$.

3.1.7 Suppose that $f: R \to R$ (R is the set of real numbers) is defined by $f(x) = \sqrt{x^2}$ ($\sqrt{\ }$ denotes the *nonnegative* square root). Show that $f(x) = |x|$.

3.1.8 Let $f: R \to R$ (R is the set of real numbers) be the absolute value function [that is, $f(x) = |x|$ for all x in R]. Is the function f onto (or surjective)? Explain.

3.1.9 Find all real numbers x (if any) such that the following are true:

(*a*)	$\sqrt{x^2} = x$	(*f*)	$\|x\| = 0$	(*k*)	$\|x\| = 1$
(*b*)	$\sqrt{x^2} = -x$	(*g*)	$\|x\| \leq 0$	(*l*)	$\|x - 5\| = 1$
(*c*)	$\sqrt{x^2} = \|x\|$	(*h*)	$\|x\| = \|-x\|$	(*m*)	$\|x + 5\| = 0$
(*d*)	$\|x\| = x$	(*i*)	$\|-x\| = -\|x\|$	(*n*)	$\|-x - 4\| = -1$
(*e*)	$\|x\| = -x$	(*j*)	$\|-x\| < 0$	(*o*)	$\|x\| + \|-x\| = 1$

3.2 Inequalities Involving Polynomial and Rational Expressions

In this section, we discuss the solution sets of certain inequalities. We shall be primarily concerned with inequalities which involve linear expressions, quadratic expressions, and rational expressions composed of linear and quadratic expressions. The following examples illustrate the techniques involved.

Example 3.6 Find all real numbers x such that

$$\frac{1}{1-x} > \frac{1}{2} \tag{3.13}$$

Solution Observe that the above inequality is equivalent to $1/(1-x) - \frac{1}{2} > 0$, $[1(2) - 1(1-x)]/[2(1-x)] > 0$, and finally $(x+1)/[2(1-x)] > 0$. Hence, both $x+1$ and $1-x$ must have the *same* sign; that is,

$$x+1>0 \text{ and } 1-x>0 \qquad \text{or} \qquad x+1<0 \text{ and } 1-x<0$$

These inequalities, in turn, are equivalent to

$$x>-1 \text{ and } x<1 \qquad \text{or} \qquad x<-1 \text{ and } x>1$$

But both $x<-1$ and $x>1$ cannot be true simultaneously (since no number x can possibly be smaller than -1 and larger than 1 at the same time), and hence the only alternative is ($x>-1$ and $x<1$). In other words, the solution of the inequality $1/(1-x) > \frac{1}{2}$ is simply the set of all real numbers x such that $x>-1$ and $x<1$; that is, $-1<x<1$.

We now proceed to study linear inequalities. A *linear inequality* is any inequality which is equivalent to any one of the following forms:

$$ax+b<0 \qquad ax+b\le 0 \qquad ax+b>0 \qquad ax+b\ge 0$$

where a and b are given real numbers and $a \neq 0$. The method for solving a linear inequality is similar to that of solving a linear equation, as illustrated in the examples below. The set of all real numbers which satisfy (or are solutions of) a given inequality is called the *solution set* of the inequality.

Example 3.7 Find the solution set of the inequality $2x-3<0$.

Solution

$2x-3<0$	Given
$2x<3$	Add 3 to both sides
$x<\frac{3}{2}$	Divide both sides by the *positive* number 2

Thus, the solution set of $2x - 3 < 0$ is the set of all real numbers x such that $x < 3/2$; that is, $\{x|x \text{ real and } x < 3/2\}$.

Example 3.8 Find the solution set of the inequality $-3x + 4 \leq 5x - 8$.

Solution As in a linear equation, the general idea is to *collect* all the terms involving *x on one side* and all the remaining constant terms *on the other side*. We indicate below the various steps involved, together with the reasons.

$-3x + 4 \leq 5x - 8$	Given
$-3x + 4 - 5x \leq 5x - 8 - 5x$	Subtract $5x$ from both sides
$-8x + 4 \leq -8$	Algebra
$-8x + 4 - 4 \leq -8 - 4$	Subtract 4 from both sides
$-8x \leq -12$	Algebra
$x \geq {}^{-12}/_{-8}$	Divide both sides by the *negative* number -8. This *reverses* the direction of the inequality (see Rule 3).
$x \geq 3/2$	Algebra

Thus, the solution set of the given inequality is the set of all real numbers x such that $x \geq 3/2$; that is, $\{x|x \text{ real and } x \geq 3/2\}$.

Next, we consider quadratic inequalities. A *quadratic inequality* is any inequality which is equivalent to any one of the following forms:

$$ax^2 + bx + c < 0 \qquad ax^2 + bx + c > 0$$

$$ax^2 + bx + c \leq 0 \qquad ax^2 + bx + c \geq 0$$

where a, b, and c are given real numbers and $a \neq 0$. In studying quadratic inequalities, the following fact (stated in Prob. 2.2.9) is sometimes useful:

> If $b^2 - 4ac < 0$, then for all real numbers x the quadratic expression $ax^2 + bx + c$ always has the same sign as a. (Here a, b, and c are real numbers, and $a \neq 0$.) (3.14)

To prove (3.14), we use the method of "completing the square" to obtain

$$ax^2 + bx + c = a\left(x + \frac{b}{2a}\right)^2 + \frac{4ac - b^2}{4a} \tag{3.15}$$

[This identity was established in Eq. (2.6).] Since by hypothesis $b^2 - 4ac < 0$, we have $-(b^2 - 4ac) > 0$, and thus $4ac - b^2 > 0$. Hence, if a is *positive,* then $a[x + b/(2a)]^2 \geq 0$ (recall that a square of any real number is always positive or zero), $(4ac - b^2)/(4a) > 0$, and thus the right side of (3.15) is positive. On the other hand, if a is *negative,* then $a[x + b/(2a)]^2 \leq 0$, $(4ac - b^2)/(4a) < 0$, and thus the right side of (3.15) is negative. This proves (3.14). The following two examples illustrate the use of (3.14) in regard to finding the solution sets of certain types of quadratic inequalities.

Example 3.9 Find the solution set of the inequality $4x^2 - 2x + 1 > 0$.

Solution Here $a = 4$, $b = -2$, and $c = 1$; hence $b^2 - 4ac = -12 < 0$. Thus (3.14) applies, and by (3.14) we know that for *all* real numbers x the quadratic expression $4x^2 - 2x + 1$ always has the same sign as a [$=4$]. Since $a > 0$, it follows that $4x^2 - 2x + 1 > 0$ for *all* real numbers x. Thus the solution set of the inequality $4x^2 - 2x + 1 > 0$ is the set of *all* real numbers.

Example 3.10 Find the solution set of the inequality $3x^2 - 2x + 1 \leq 0$.

Solution Here $a = 3$, $b = -2$, and $c = 1$, and hence $b^2 - 4ac = -8 < 0$. Thus (3.14) applies, and by (3.14) we know that for *all* real numbers x the quadratic expression $3x^2 - 2x + 1$ always has the same sign as a [$=3$]. Since $a > 0$, it follows that $3x^2 - 2x + 1 > 0$ for *all* real numbers x; therefore our inequality does *not* have a single real number as a solution. In other words, the solution set of our inequality is the empty set.

In view of (3.14), we have thus disposed completely of the case in which the discriminant $b^2 - 4ac < 0$, and we now direct our attention to the case in which $b^2 - 4ac \geq 0$. This case is of considerable interest as far as inequalities are concerned. To begin with, we recall the quadratic formula, which asserts that the solutions of the quadratic equation $ax^2 + bx + c = 0$, where a, b, and c are real numbers ($a \neq 0$), are r and s (say), where

$$r = \frac{-b + \sqrt{b^2 - 4ac}}{2a} \qquad s = \frac{-b - \sqrt{b^2 - 4ac}}{2a} \tag{3.16}$$

If we take the sum and the product of these two solutions, we get (after simplifying)

$$r + s = \frac{-b}{a} \qquad rs = \frac{c}{a} \tag{3.17}$$

Thus, by the usual rules of high school algebra we get

$$\begin{aligned} a(x-r)(x-s) &= a[x^2 - (r+s)x + rs] \\ &= a\left[x^2 - \left(\frac{-b}{a}\right)x + \frac{c}{a}\right] \qquad \text{by (3.17)} \\ &= ax^2 + bx + c \end{aligned}$$

In other words, we have established the following important identity, which tells us how to factor a quadratic expression:

> $ax^2 + bx + c = a(x-r)(x-s)$, where r and s are the solutions of the quadratic equation $ax^2 + bx + c = 0$. Moreover, both r and s are real numbers if $b^2 - 4ac \geq 0$ [see (3.16)]. (3.18)

In view of (3.18), the quadratic inequality $ax^2 + bx + c < 0$, say, is equivalent to the inequality $a(x-r)(x-s) < 0$. However, the latter inequality is a lot easier to solve. In fact, the inequality $a(x-r)(x-s) < 0$ can easily be reduced to a couple of *linear* inequalities, as we shall see in the following examples. But let us first summarize the results we have obtained thus far: To find the solution set of a given quadratic inequality involving the quadratic expression $ax^2 + bx + c$, first compute the discriminant $b^2 - 4ac$. If $b^2 - 4ac < 0$, apply (3.14), as illustrated in Examples 3.9 and 3.10. If, on the other hand, $b^2 - 4ac \geq 0$, then factor $ax^2 + bx + c$ as indicated in (3.18). This factorization will reduce at once the original quadratic inequality to a couple of linear inequalities (involving real numbers only), as shown in the examples below. Finally, solve these linear inequalities as shown earlier.

Example 3.11 Find the solution set of

$$x^2 - x - 2 < 0 \tag{3.19}$$

Solution Here $a = 1$, $b = -1$, and $c = -2$, which means $b^2 - 4ac = 9 > 0$. In view of this, we know that $x^2 - x - 2$ can be factored into a product of linear factors, using real numbers only as coefficients. In fact, we can factor $x^2 - x - 2$ directly [without using (3.18)]. We get $x^2 - x - 2 = (x - 2)(x + 1)$. Hence, the inequality (3.19) is equivalent to

$$(x - 2)(x + 1) < 0 \tag{3.20}$$

The only way a product of two real numbers can be negative is for one of them to be positive and the other negative. Hence (3.20) is equivalent to

$$x - 2 > 0 \text{ and } x + 1 < 0 \quad \text{or} \quad x - 2 < 0 \text{ and } x + 1 > 0 \tag{3.21}$$

But $(x - 2 > 0$ and $x + 1 < 0)$ together is equivalent to $(x > 2$ and $x < -1)$, which is impossible (since no number x can possibly be greater than 2 and less than -1 at the same time). Hence, only the second alternative in (3.21) is feasible, and (3.21) is thus equivalent to

$$x - 2 < 0 \quad \text{and} \quad x + 1 > 0$$

The last inequalities are, of course, equivalent to $x < 2$ and $x > -1$; that is, $-1 < x < 2$. Hence, the solution set of $x^2 - x - 2 < 0$ is the set of all real numbers x such that $-1 < x < 2$, that is, the set of all real numbers x strictly between -1 and 2.

The following procedure can also be used. First, find the solutions of the quadratic equation $x^2 - x - 2 = 0$. These solutions are given by $x = -1$ and $x = 2$ [recall that $x^2 - x - 2 = (x - 2)(x + 1)$]. We now represent these solutions on the x axis at the points A and B, say. Observe that the two points A and B divide the x axis into *three* parts, namely, the ray to the left of A, AB, and the ray to the right of B (Fig. 3.6). These three parts actually correspond to the set of all real numbers x such that $x < -1$, $-1 < x < 2$, and $x > 2$, respec-

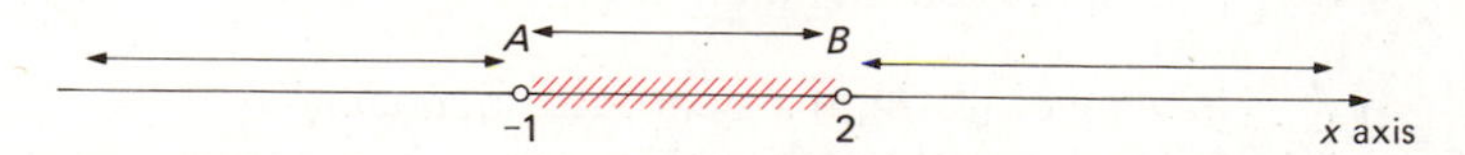

Fig. 3.6 Solution set of $x^2 - x - 2 < 0$. The numbers -1 and 2 are excluded.

tively. [Observe that we excluded -1 and 2 since they are *not* solutions of (3.19).]

Next, we take any sample point (or number) in each one of these three parts and substitute this sample number in the inequality (3.20), which of course is equivalent to (3.19) but has the advantage of being in factored form. If, after this substitution is made in (3.20), we get a true statement, we then keep that *entire* part of the x axis as a part of our solution set. If, on the other hand, this substitution results in a false statement, we do *not* keep that part of the x axis at all as a part of our solution set. We illustrate this procedure for our present inequality (3.20) (see Fig. 3.7).

1. A sample in part 1 is $x = -2$. Substituting $x = -2$ in (3.20), we get $(-2 - 2)(-2 + 1) < 0$, which is certainly false. Thus, we *exclude* this entire part 1 from our solution set.
2. A sample in part 2 is $x = 0$. Substituting $x = 0$ in (3.20), we get $(0 - 2)(0 + 1) < 0$, which is certainly true. Thus, we *include* this entire part 2 in our solution set.
3. A sample in part 3 is $x = 3$. Substituting $x = 3$ in (3.20), we get $(3 - 2)(3 + 1) < 0$, which is certainly false. Thus, we *exclude* this entire part 3 from our solution set.

We have thus seen that the solution set of our given inequality (3.20) [or (3.19)] is simply the set of all real numbers in *part 2 only,* that is, the set of all real numbers x such that $-1 < x < 2$.

The above procedure can easily be justified by considering the signs of each factor in (3.20) in the various parts 1, 2, and 3. We illustrate this in the following table (see Fig. 3.7).

	Part 1 $x < -1$	Part 2 $-1 < x < 2$	Part 3 $x > 2$
$x + 1$	$-$	$+$	$+$
$x - 2$	$-$	$-$	$+$
$(x - 2)(x + 1)$	$+$	$-$	$+$

For example, *any* number x in part 1 satisfies $x < -1$, and therefore $x + 1 < 0$ for *all* numbers x in part 1. For this reason, we put $-$ in the upper left corner of the above table. Similarly, we verify all the remaining signs in the table above.

The last row in the above table tells us precisely when $(x - 2)(x + 1) < 0$. Indeed, a glance at this last row shows that $(x - 2)(x + 1) < 0$ if and only if x is in part 2, that is, $-1 < x < 2$. Thus, the solution set of (3.20) [and (3.19)] is simply the set of all real numbers x such that $-1 < x < 2$.

We wish to make two further remarks. First, the above method is *quite general* and is indeed *valid* for any *inequality* involving *any product (or quotient) of any number of linear polynomials* on one side and zero on the

Part 1 Part 2 Part 3

$x < -1$ -1 $-1 < x < 2$ 2 $x > 2$ x axis

Fig. 3.7 Geometric illustration for inequality (3.20)

other side (see the following examples for illustration). Secondly, the reason we did not include the points $x = -1$ and $x = 2$ in the above discussion (or in the solution set) is that the inequality (3.19) is a *strict* inequality (i.e., it involves $<$ instead of $\leq$). Indeed, had we been asked instead to find the solution set of the inequality $x^2 - x - 2 \leq 0$ [which, of course, is equivalent to $(x-2)(x+1) \leq 0$], the solution set would have to include both $x = -1$ and $x = 2$ [since $(x-2)(x+1) = 0$ for $x = -1$ and $x = 2$].

Example 3.12 Find the solution set of Eq. (3.22).

$$2x^2 + 3x - 2 \geq 0 \tag{3.22}$$

Solution Here the discriminant $b^2 - 4ac > 0$; hence, the above quadratic polynomial can be factored as a product of two linear polynomials, involving *real-number coefficients only.* In fact, we easily see that

$$(2x^2 + 3x - 2) = (x+2)(2x-1) = 2(x+2)(x - 1/2)$$

hence (3.22) is equivalent to $2(x+2)(x - 1/2) \geq 0$. Thus, (3.22) is also equivalent to

$$(x+2)(x - 1/2) \geq 0 \tag{3.23}$$

Following the procedure given in the preceding example, we first solve the quadratic equation $(x+2)(x - 1/2) = 0$ to obtain the solutions -2 and $1/2$. Now -2 and $1/2$ clearly divide the x axis into three parts, namely, part 1: $\{x | x \leq -2\}$; part 2: $\{x | -2 < x < 1/2\}$; and part 3: $\{x | x \geq 1/2\}$. Proceeding exactly as in the preceding example, we now tabulate the signs of each factor in (3.23) in the various parts 1, 2, and 3.

	Part 1 $x \leq -2$	Part 2 $-2 < x < 1/2$	Part 3 $x \geq 1/2$
$x + 2$	$-$ (or 0)	$+$	$+$
$x - 1/2$	$-$	$-$	$+$ (or 0)
$(x+2)(x - 1/2)$	$+$ (or 0)	$-$	$+$ (or 0)

The last row in this table tells us that $(x+2)(x - 1/2) \geq 0$ if and only if x lies in part 1 or 3. Thus, the solution set of (3.23) [and (3.22)] is the set of all real numbers x such that $x \leq -2$ or $x \geq 1/2$. This solution set is indicated in Fig. 3.8.

Remark Since (3.22) involves $\geq$, it is necessary to *include* both $x = -2$ and $x = 1/2$ in the solution set. Indeed, these values of x make (3.22) become an equality.

Example 3.13 Find the solution set of

$$\frac{(x+1)(x-3)}{x^2 - 4} < 0 \tag{3.24}$$

Fig. 3.8 Solution set of $2x^2 + 3x - 2 \geq 0$. The numbers -2 and $1/2$ are included.

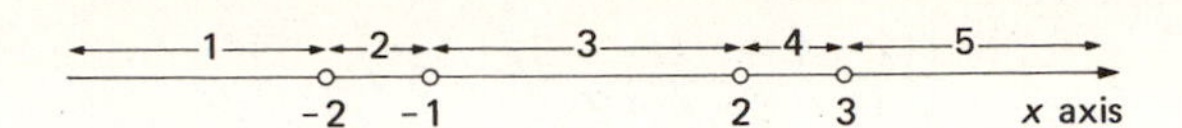

Fig. 3.9 Points arising from (3.25)

Solution

First, we factor $x^2 - 4$ and get $x^2 - 4 = (x+2)(x-2)$. Hence, (3.24) is equivalent to

$$\frac{(x+1)(x-3)}{(x+2)(x-2)} < 0 \tag{3.25}$$

Next, equating each of the four linear factors in (3.25) to zero gives the boundary points, which divide the x axis into *five* parts (see Fig. 3.9). These points on the x axis, when ordered from left to right, are given by $x = -2$, $x = -1$, $x = 2$, and $x = 3$. The five parts are tabulated below (see Fig. 3.9) together with the signs of each of the four linear factors in (3.25) in these various parts. [Observe that we exclude -2, -1, 2, and 3 since they are *not* solutions of (3.25).]

	Part 1 $x < -2$	Part 2 $-2 < x < -1$	Part 3 $-1 < x < 2$	Part 4 $2 < x < 3$	Part 5 $x > 3$
$x+1$	$-$	$-$	$+$	$+$	$+$
$x-3$	$-$	$-$	$-$	$-$	$+$
$x+2$	$-$	$+$	$+$	$+$	$+$
$x-2$	$-$	$-$	$-$	$+$	$+$
$\frac{(x+1)(x-3)}{(x+2)(x-2)}$	$+$	$-$	$+$	$-$	$+$

A glance at the last row shows that the solution set to (3.25) [and (3.24)] is $\{x \mid -2 < x < -1\} \cup \{x \mid 2 < x < 3\}$. This solution set is sketched in Fig. 3.10.

Example 3.14

Find the solution set of

$$\frac{1}{x-3} > \frac{1}{x-2} \tag{3.26}$$

Solution

Before we can apply the method used thus far, we must first make one side of (3.26) zero. This is easily achieved by subtracting $1/(x-2)$ from both sides of (3.26). The result is

$$\frac{1}{x-3} - \frac{1}{x-2} > 0 \tag{3.27}$$

and, of course, (3.27) is equivalent to (3.26). Next, we simplify the left side of (3.27) by the usual rules of high school algebra; we get

Fig. 3.10 Solution set of $(x+1)(x-3)/(x^2-4) < 0$. The numbers -2, -1, 2, and 3 are excluded.

$$\frac{1(x-2)-1(x-3)}{(x-3)(x-2)} > 0$$

which, after simplifying, becomes

$$\frac{1}{(x-3)(x-2)} > 0 \tag{3.28}$$

Since the numerator (namely, 1) is positive, the denominator $(x-3)(x-2)$ must be positive also. Thus, (3.28) is equivalent to

$$(x-3)(x-2) > 0 \tag{3.29}$$

Hence, both $x-3$ and $x-2$ are positive together, or both $x-3$ and $x-2$ are negative together. That is,

$$x > 3 \text{ and } x > 2 \qquad \text{or} \qquad x < 3 \text{ and } x < 2$$

These inequalities, in turn, are easily seen to be equivalent to

$$x > 3 \qquad \text{or} \qquad x < 2$$

Thus, the solution set of (3.26) is the set of all real numbers $x > 3$ together with the set of all real numbers $x < 2$, as indicated in Fig. 3.11.

Example 3.15 Find the solution set of (3.30):

$$x^2 + x - 17 > -2x^2 + 4x - 2 \tag{3.30}$$

Solution Once again, in order to use the method of this section, we must first make one side of (3.30) zero. Thus, we subtract $-2x^2 + 4x - 2$ from both sides of (3.30) to get

$$(x^2 + x - 17) - (-2x^2 + 4x - 2) > 0 \tag{3.31}$$

Clearly (3.31) is equivalent to (3.30). Simplifying (3.31) by the usual rules of high school algebra, we get

$$3x^2 - 3x - 15 > 0 \tag{3.32}$$

By Rule 2 we still get an equivalent inequality if we divide both sides of (3.32) by the *positive* number 3; the result is

$$x^2 - x - 5 > 0 \tag{3.33}$$

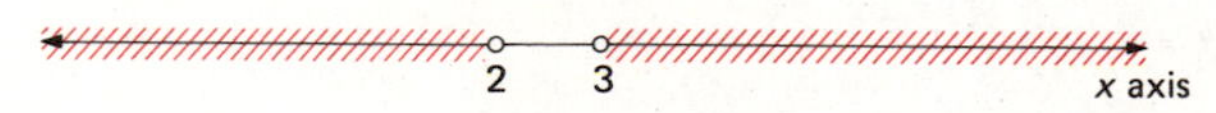

Fig. 3.11 Solution set of $1/(x-3) > 1/(x-2)$. The points 2 and 3 are excluded.

Thus, (3.30) is equivalent to (3.33), and we now proceed to find the solution set of (3.33). It is not obvious how to factor the quadratic expression in (3.33). However, computing the discriminant of $x^2 - x - 5$, we get $b^2 - 4ac = (-1)^2 - 4(1)(-5) = 21$, which is *positive*. Hence, recalling (3.18), we know that

$$x^2 - x - 5 = (x - r)(x - s) \tag{3.34}$$

where r and s are the solutions of the quadratic *equation* $x^2 - x - 5 = 0$. By the quadratic formula, we find that

$$r = \frac{-(-1) + \sqrt{21}}{2(1)} \qquad s = \frac{-(-1) - \sqrt{21}}{2(1)}$$

that is,

$$r = \frac{1 + \sqrt{21}}{2} \qquad s = \frac{1 - \sqrt{21}}{2}$$

Hence, by (3.34)

$$x^2 - x - 5 = \left(x - \frac{1 + \sqrt{21}}{2}\right)\left(x - \frac{1 - \sqrt{21}}{2}\right) \tag{3.35}$$

and (3.33) now becomes

$$\left(x - \frac{1 + \sqrt{21}}{2}\right)\left(x - \frac{1 - \sqrt{21}}{2}\right) > 0 \tag{3.36}$$

Arguing as we did in the previous examples, we easily check that the solution set of (3.36) [as well as (3.30)] is

$$\left\{x | x > \frac{1 + \sqrt{21}}{2}\right\} \cup \left\{x | x < \frac{1 - \sqrt{21}}{2}\right\}$$

This solution set is sketched in Fig. 3.12.

Let us summarize the general procedure for finding the solution set of an inequality. It is well to keep in mind Example 3.13 throughout.

Step 1 Make sure that one side of the inequality is zero. If need be, subtract one side from the other.

Step 2 Factor (if possible) the expression you now have into a product of *linear factors* using only *real numbers as coefficients* [in the case of a fractional

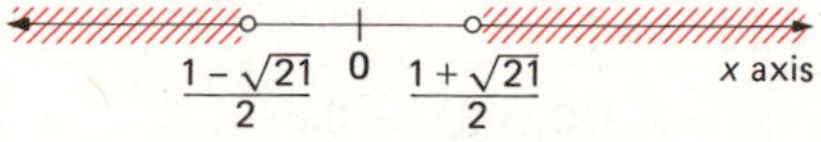

$\frac{1-\sqrt{21}}{2}$ 0 $\frac{1+\sqrt{21}}{2}$ x axis

Fig. 3.12 Solution set of (3.30). The numbers indicated are excluded.

expression factor *both* numerator and denominator into such a product of *linear* factors (if possible)].

Step 3 Set every single linear factor to zero, and solve. Sketch these solutions on the x axis, and describe (in inequality form) each one of the parts of the x axis which these solutions give rise to.

Step 4 Tabulate the signs of each linear factor in these various parts. This could easily be achieved by taking any sample point in each part and substituting it in each one of the linear expressions. Determine, eventually, those (and only those) parts on the x axis for which the original inequality (now in factored form, see step 2) is satisfied. These parts will give rise to the desired solution set of the original inequality.

Problem Set 3.2

3.2.1 Find the solution sets of the following inequalities. In each case, sketch the solution set on the x axis.

(*a*) $x - 2 \leq 1$
(*b*) $2x + 3 \geq 5 - x$
(*c*) $-4x + 2 < x + 5$
(*d*) $3x + 4 > x - 5$
(*e*) $-3x < 9$
(*f*) $-2x \geq 5$

3.2.2 Find the solution sets of the following inequalities. In each case, sketch the solution set on the x axis.

(*a*) $x^2 \leq 4$
(*b*) $x^2 \geq 5$
(*c*) $x^2 - 5x + 6 < 0$
(*d*) $-x^2 + 2x + 3 \leq 0$
(*e*) $x^2 - x + 2 \leq 0$
(*f*) $-x^2 + 2x - 3 < 0$
(*g*) $-x^2 + 3x + 2 \geq 0$

3.2.3 Find the solution sets of the following inequalities. In each case, sketch the solution set on the x axis.

(*a*) $\frac{1}{x} \leq 2$

(*b*) $\frac{-1}{x} < -5$

(*c*) $\frac{1}{x+1} < 0$

(*d*) $\frac{2}{x-3} \geq \frac{1}{x-1}$

(*e*) $(x-3)(x-2)(x-1) \leq 0$

(*f*) $\frac{(x+3)(x+2)}{x(x-1)} \geq 0$

(*g*) $\frac{x^2-4}{4x^2-25} < 0$ (*Hint*: Factor both numerator and denominator.)

(*h*) $x^2 - x + 3 \geq 3x^2 - 3x + 1$

(*i*) $\frac{2}{x-1} - \frac{3}{2x-5} < 1$

3.2.4 Prove that for all real numbers x, (*a*) $x^2 + x + 1 > 0$ and (*b*) $x^2 - x + 1 > 0$.

3.2.5 Prove that for all real numbers x, (*a*) $-x^2 + 2x - 1 \leq 0$ (when does equality hold?), and, (*b*) $-2x^2 + 3x - 4 < 0$.

3.2.6 Show that if $0 < a < 1$, then $a^2 < a$.

3.2.7 Show that if $1 < a$, then $a < a^2$.

3.2.8 Show that if $0 < a < b$, then $a^2 < b^2$, $a^3 < b^3$, etc.

3.2.9 If $-1 < x - 2 < 1$, what can you say about $3x - 6$ and about $4 - 2x$?

3.3 Inequalities Involving Absolute Values

In this section, we consider the problem of finding the solution sets of inequalities involving the *absolute values* of certain polynomial or rational expressions. With the aid of some important properties of absolute values, which we establish first, we show that such inequalities reduce, essentially, to certain combinations of inequalities of the types we discussed in Sec. 3.2. The first important and useful fact is the following:

Suppose x is any real number. Then $|x|$ is simply the positive length of the line segment joining the origin $\mathcal{O}$ and the point P on the x axis with x coordinate x. (3.37)

This follows at once from the definition of absolute value. Keeping (3.37) in mind, the following can be shown:

Suppose $a > 0$. Then $|x| < a$ is equivalent to $-a < x < a$. (3.38)

To see (3.38) intuitively, observe that by (3.37), the assertion in (3.38) is equivalent to saying that the points on the x axis whose positive distances from the origin are less than a are precisely those (and only those) points with x coordinates strictly *between* $-a$ and a. This, of course, is clear from a geometric point of view.

In view of (3.38), we also have the following:

Suppose $a > 0$. Then $|x| \geq a$ is equivalent to $(x \geq a \text{ or } x \leq -a)$. (3.39)

To see this, observe that if $|x| \geq a$, then we *cannot* possibly have $-a < x < a$ [since if we did, we would have $|x| < a$ by (3.38)]; hence we must have either $x \geq a$ or $x \leq -a$. Conversely, if $x \geq a$ or $x \leq -a$, then we *cannot* possibly have $|x| < a$ [since if we did, we would have $-a < x < a$ by (3.38)] and hence $|x| \geq a$. This proves (3.39).

It should be pointed out that with only a few minor and obvious modifications in the proofs of (3.38) and (3.39) we can also show that

$$|x| \le a \quad \text{is equivalent to} \quad -a \le x \le a \qquad (a > 0) \tag{3.38a}$$

$$|x| > a \quad \text{is equivalent to} \quad x > a \text{ or } x < -a \qquad (a > 0) \tag{3.39a}$$

We wish to further remark that all of (3.38), (3.39), (3.38*a*), and (3.39*a*) remain valid if we replace x by $f(x)$ throughout, where $f(x)$ is any real number. Thus

$$|f(x)| < a \quad \text{is equivalent to} \quad -a < f(x) < a \qquad (a > 0) \tag{3.38b}$$

$$|f(x)| \le a \quad \text{is equivalent to} \quad -a \le f(x) \le a \qquad (a > 0) \tag{3.38c}$$

$$|f(x)| \ge a \quad \text{is equivalent to} \quad f(x) \ge a \text{ or } f(x) \le -a \qquad (a > 0) \tag{3.39b}$$

$$|f(x)| > a \quad \text{is equivalent to} \quad f(x) > a \text{ or } f(x) < -a \qquad (a > 0) \tag{3.39c}$$

The reason for this is the following: Since (3.38), (3.38*a*), (3.39), and (3.39*a*) are true for *all* real numbers x, they certainly remain true if throughout these inequalities we replace x by *any* real number whatsoever. In particular, since $f(x)$ is a real number, (3.38), (3.38*a*), (3.39), and (3.39*a*) remain true if we replace x by $f(x)$ throughout all these inequalities. When we perform such a replacement, we obtain (3.38*b*), (3.38*c*), (3.39*b*), and (3.39*c*).

In view of the above inequalities, we see that the study of inequalities involving absolute values is essentially reduced (or is equivalent) to the study of certain inequalities *not* involving absolute values. This fact is extremely useful in the applications, as the following examples will show. Before we give these examples, however, it is instructive to give some geometric interpretations of some of the above inequalities. First, consider (3.38). The set of all real numbers x such that $-a < x < a$ $[a > 0]$ is represented in Fig. 3.13. Observe that both points with x coordinates $-a$ and a are *excluded.* Similarly, we represent in Figs. 3.14 to 3.16 the sets of real numbers corresponding to the inequalities given in (3.38*a*), (3.39*a*), and (3.39), respectively.

Next, we illustrate geometrically the inequalities in (3.38*b*), (3.38*c*), (3.39*b*), and (3.39*c*) in the all-important case in which $f(x)$ is the linear expression given by $f(x) = x - c$, where c is a positive real number, say. The case in which c is negative can be illustrated in a similar way. First, consider (3.38*b*).

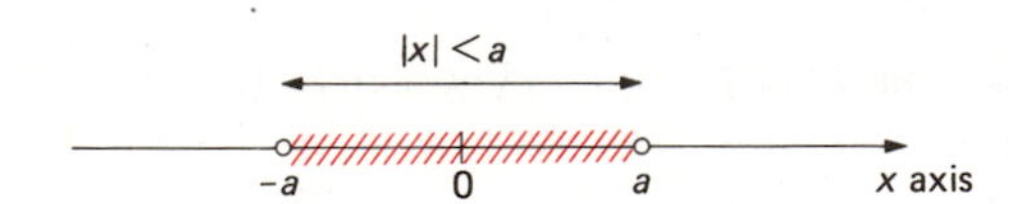

Fig. 3.13 $\{x | -a < x < a\}$. The points $-a$ and a are excluded.

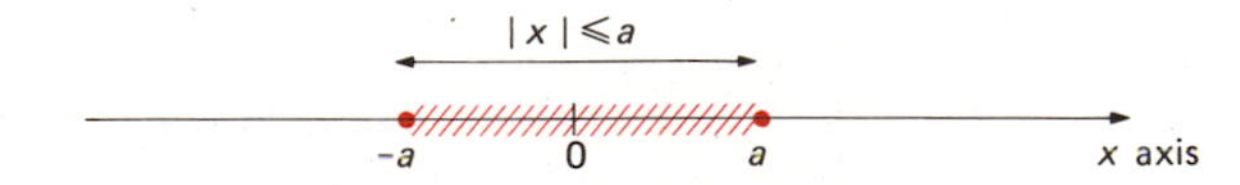

Fig. 3.14 $\{x | -a \le x \le a\}$. The points $-a$ and a are included.

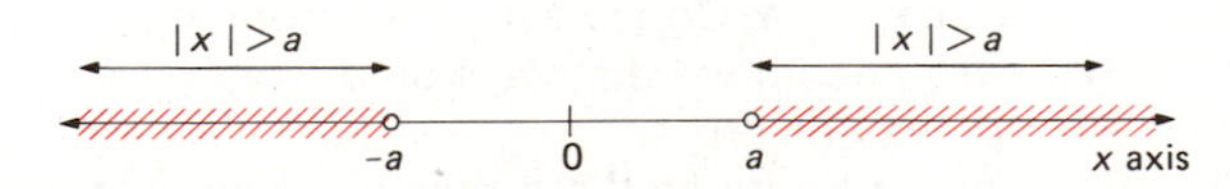

Fig. 3.15 $\{x | x > a \text{ or } x < -a\}$. The points $-a$ and a are excluded.

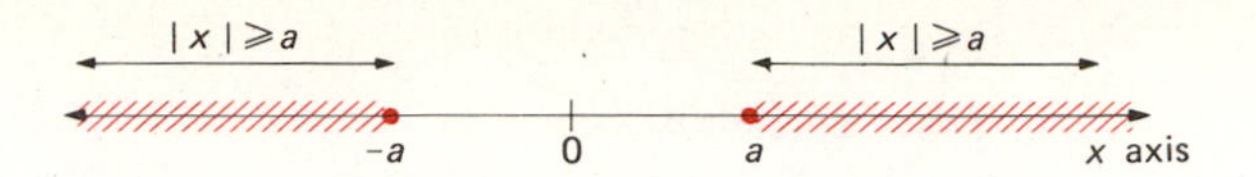

Fig. 3.16 $\{x | x \geq a \text{ or } x \leq -a\}$. The points $-a$ and a are included.

Observe that if we substitute $f(x) = x - c$ in (3.38*b*), we obtain

$$|x - c| < a \quad \text{is equivalent to} \quad -a < x - c < a \qquad (a > 0) \tag{3.40}$$

and this, in turn, is equivalent to (by adding c to the latter inequalities)

$$|x - c| < a \quad \text{is equivalent to} \quad c - a < x < c + a \qquad (a > 0) \tag{3.41}$$

We illustrate this in Fig. 3.17. Observe that in view of (3.40) and (3.41), we may always interpret $|x - c|$ geometrically as the *positive* distance of the line segment on the x axis whose endpoints have x coordinates x and c. This geometric interpretation of absolute values is extremely helpful. Thus, one may interpret (3.40) and (3.41) geometrically as follows: The locus of all points on the x axis whose positive distances from $(c,0)$ are strictly less than a is simply the *open* line segment AB in Fig. 3.17 (with both A and B excluded because the inequalities are strict). In a similar way, we easily verify, taking $f(x) = x - c$ in (3.38*c*), (3.39*b*), and (3.39*c*) that

$$|x - c| \leq a \quad \text{is equivalent to} \quad c - a \leq x \leq c + a \qquad (a > 0) \tag{3.42}$$

$$|x - c| \geq a \quad \text{is equivalent to} \quad x \geq c + a \text{ or } x \leq c - a \qquad (a > 0) \tag{3.43}$$

$$|x - c| > a \quad \text{is equivalent to} \quad x > c + a \text{ or } x < c - a \qquad (a > 0) \tag{3.44}$$

These inequalities are illustrated in Figs. 3.18, 3.19, and 3.20, respectively, and can be given a geometric interpretation similar to that given for (3.40) and (3.41).

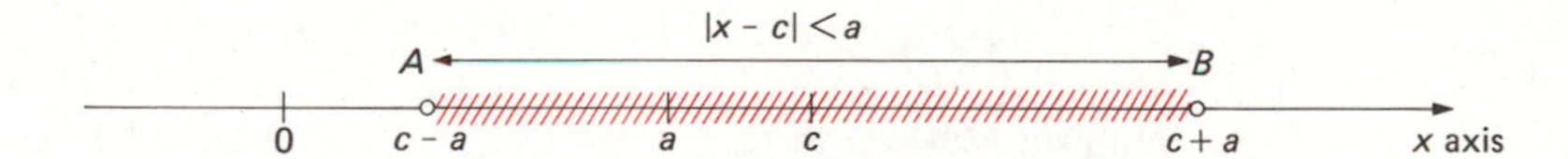

Fig. 3.17 $\{x | c - a < x < c + a\}$. The points $c - a$ and $c + a$ are excluded.

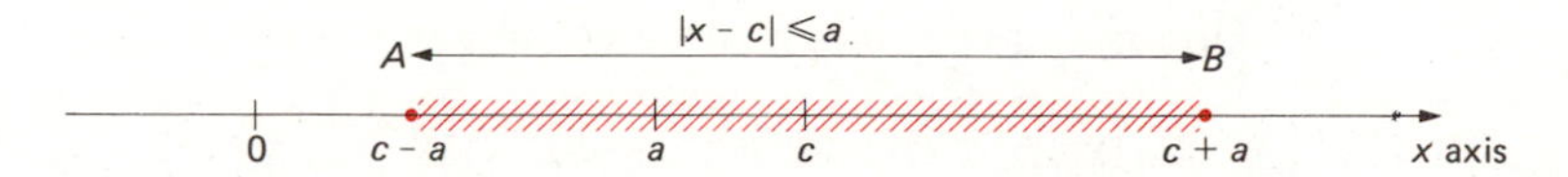

Fig. 3.18 $\{x | c - a \leq x \leq c + a\}$. The points $c - a$ and $c + a$ are included.

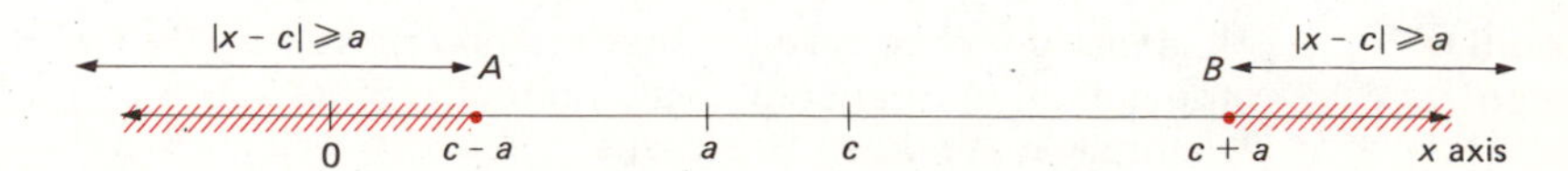

Fig. 3.19 $\{x | x \geq c + a \text{ or } x \leq c - a\}$. The points $c - a$ and $c + a$ are included.

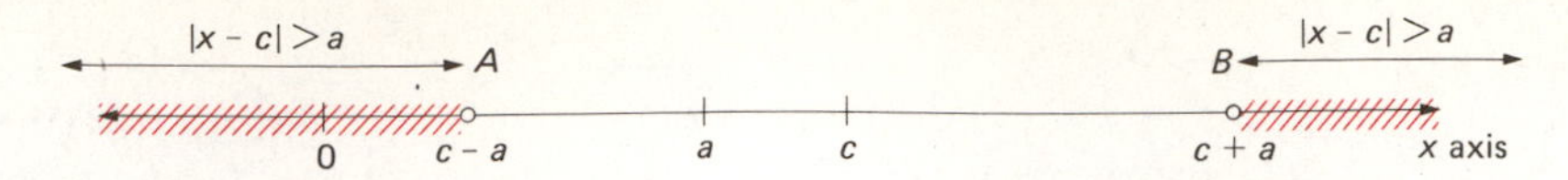

Fig. 3.20 $\{x|x > c + a \text{ or } x < c - a\}$. The points $c - a$ and $c + a$ are excluded.

Example 3.16 Find the solution set of (3.45):

$$|x - 1| < 4 \tag{3.45}$$

Solution By (3.40) we know that (3.45) is equivalent to $-4 < x - 1 < 4$, and this, in turn, is equivalent to (upon adding 1 to all sides) $1 - 4 < x < 1 + 4$, or $-3 < x < 5$. Thus the solution set of $|x - 1| < 4$ is the set of all real numbers x such that $-3 < x < 5$. This solution set is sketched in Fig. 3.21.

Example 3.17 Find the solution set of (3.46):

$$0 < |x - 1| < 4 \tag{3.46}$$

Solution Clearly (3.46) is equivalent to a combination of two inequalities, namely,

$$0 < |x - 1| \quad \text{and} \quad |x - 1| < 4 \tag{3.47}$$

The solution set of the second inequality in (3.47) was found in Example 3.16 to be

$$\{x|-3 < x < 5\} \tag{3.48}$$

and there only remains to determine the solution set of the first inequality in (3.47), namely, $0 < |x - 1|$, or, $|x - 1| > 0$. But by (3.2) $|x - 1| \geq 0$ for *all* real numbers x, and, moreover, $|x - 1| > 0$ *always except when* $|x - 1| = 0$. Furthermore, $|x - 1| = 0$ if and only if $x = 1$ (recall the definition of absolute value). The net result, then, is that

$$|x - 1| > 0 \quad \text{for } all \text{ real numbers } x \text{ } except \text{ } x = 1 \tag{3.49}$$

Combining (3.48) and (3.49), we get the following: The solution set of (3.46) is

$$\{x|-3 < x < 5\} \qquad except\ x = 1 \tag{3.50}$$

Inequality (3.50) can also be written as

$$\{x|-3 < x < 1\} \cup \{x|1 < x < 5\} \tag{3.51}$$

This solution set [given in (3.50) or (3.51)] is sketched in Fig. 3.22.

Alternative Solution It is instructive to give another method of solving $0 < |x - 1| < 4$ which uses the definition of absolute value directly. Since $x - 1$ changes sign at $x = 1$, we naturally distinguish two cases.

Case 1 $x - 1 \geq 0$

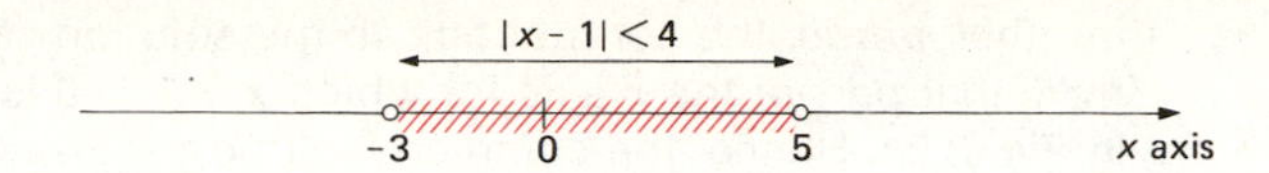

Fig. 3.21 Solution set of $|x-1|<4$. The points -3 and 5 are excluded.

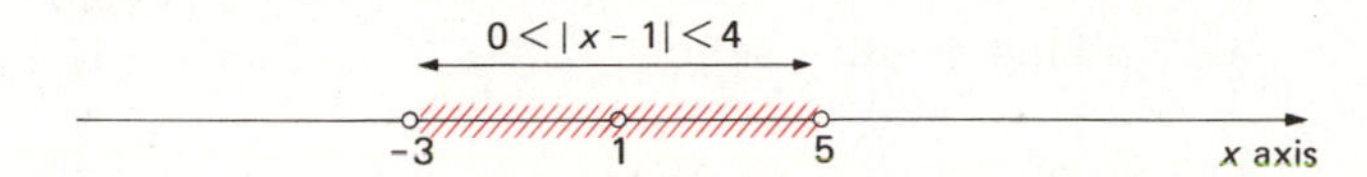

Fig. 3.22 Solution set of $0<|x-1|<4$. The points -3, 1, and 5 are excluded.

In this case, $|x-1|=x-1$ (by definition of absolute value), and hence (3.46) now becomes

$$0<x-1<4 \tag{3.52}$$

Adding 1 to all sides of (3.52), we get

$$1<x<5 \tag{3.53}$$

In other words, the contribution to the solution set of $0<|x-1|<4$ arising from that part of the x axis for which $x-1\geq 0$ is $\{x|1<x<5\}$, as indicated in Fig. 3.23.

Case 2 $x-1<0$

In this case, $|x-1|=-(x-1)$ (by definition of absolute value), and hence (3.46) now becomes

$$0<-(x-1)<4 \tag{3.54}$$

Multiplying (3.54) by the *negative* number -1 has the effect of *reversing* the directions of the inequalities in (3.54). The result is

$$0>x-1>-4 \tag{3.55}$$

or (by adding 1 throughout)

$$1>x>-3 \tag{3.56}$$

which, in turn, is equivalent to

$$-3<x<1 \tag{3.57}$$

Fig. 3.23 Contribution to solution set of (3.46) when $x-1\geq 0$. The points 1 and 5 are excluded.

In other words, the contribution to the solution set of $0 < |x - 1| < 4$ arising from that part of the x axis for which $x - 1 < 0$ is $\{x|-3 < x < 1\}$, as indicated in Fig. 3.24. Hence, the complete solution set of (3.46) is [see (3.57) and (3.53)] $\{x|-3 < x < 1\} \cup \{x|1 < x < 5\}$, which, of course, agrees with (3.51) and (3.50).

Example 3.18 Find the solution set of

$$\left|\frac{3x - 1}{x + 2}\right| \geq 4 \tag{3.58}$$

Solution By (3.39*b*), we know that inequality (3.58) is equivalent to

$$\frac{3x - 1}{x + 2} \geq 4 \quad \text{or} \quad \frac{3x - 1}{x + 2} \leq -4 \tag{3.59}$$

Let us now consider the first inequality in (3.59). Indeed,

$$\frac{3x - 1}{x + 2} \geq 4$$

is equivalent to

$$\frac{3x - 1}{x + 2} - 4 \geq 0 \qquad \frac{(3x - 1) - 4(x + 2)}{x + 2} \geq 0 \qquad \text{and} \qquad \frac{-x - 9}{x + 2} \geq 0$$

By multiplying both sides of the last inequality by the *negative* number -1, we can now conclude that

$$\frac{3x - 1}{x + 2} \geq 4 \quad \text{is equivalent to} \quad \frac{x + 9}{x + 2} \leq 0$$

Now, the last inequality is equivalent to saying that $x + 9$ (if not zero) and $x + 2$ have opposite signs, and this leads to: $(x + 9)/(x + 2) \leq 0$ is equivalent to $-9 \leq x < -2$. (Note that $x \neq -2$.) The net result is

$$\frac{3x - 1}{x + 2} \geq 4 \quad \text{is equivalent to} \quad -9 \leq x < -2 \tag{3.60}$$

Next, consider the second inequality in (3.59): $(3x - 1)/(x + 2) \leq -4$ is equivalent to $(3x - 1)/(x + 2) + 4 \leq 0$, which, after simplification, becomes $(7x + 7)/(x + 2) \leq 0$; this, in turn, is equivalent to $7(x + 1)/(x + 2) \leq 0$ and to $(x + 1)/(x + 2) \leq 0$. We have thus shown that $(3x - 1)/(x + 2) \leq -4$ is equivalent to $(x + 1)/(x + 2) \leq 0$. The last inequality is, of course, equivalent to saying that $x + 1$ (if not zero) and $x + 2$ have opposite signs, and this leads to

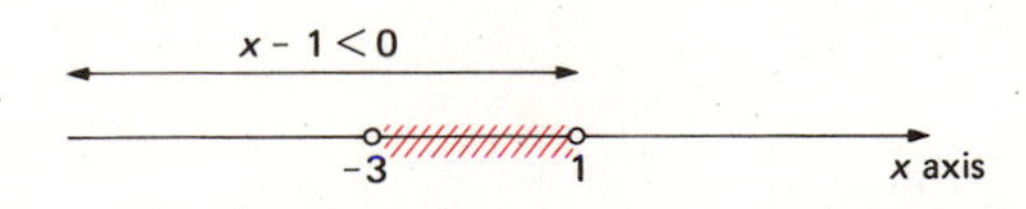

Fig. 3.24 Contribution to solution set of (3.46) when $x - 1 < 0$. The points -3 and 1 are excluded.

$$\frac{3x-1}{x+2} \leq -4 \quad \text{is equivalent to} \quad -2 < x \leq -1 \tag{3.61}$$

Combining (3.59) to (3.61), we conclude that the solution set of (3.58) is

$$\{x|-9 \leq x < -2\} \cup \{x|-2 < x \leq -1\}$$

This solution set is sketched in Fig. 3.25.

Example 3.19 Find the solution set of (3.62):

$$\left|\frac{2x-5}{x-1}\right| < 1 \tag{3.62}$$

Solution By (3.38*b*) we know that

$$\text{Inequality (3.62)} \quad \text{is equivalent to} \quad -1 < \frac{2x-5}{x-1} < 1 \tag{3.63}$$

We can now proceed, as in the previous example, by first finding the solution set of each of the two inequalities $-1 < (2x-5)/(x-1)$ and $(2x-5)/(x-1) < 1$ and then by finding the *intersection* of these two solution sets [since (3.62) is equivalent to the conjunction of *both* of these inequalities]. A somewhat different method, which is also a little shorter, involves taking a closer look at $x-1$ and considering two cases depending on whether $x-1$ is positive or negative. We proceed to discuss this method and these two cases.

Case 1 $x-1 > 0$

Since we are presently confining ourselves just to the case in which $x-1$ is positive, we may, by Rule 2, safely multiply by $x-1$ to conclude that

$$-1 < \frac{2x-5}{x-1} < 1 \quad \text{is equivalent to} \quad -1(x-1) < 2x-5 < 1(x-1) \tag{3.64}$$

The last inequalities in (3.64) are clearly equivalent to $-x+1 < 2x-5 < x-1$, which, in turn, is equivalent to ($-x+1 < 2x-5$ *and* $2x-5 < x-1$) and then to ($x > 2$ *and* $x < 4$). In view of this, (3.64) now becomes

$$-1 < \frac{2x-5}{x-1} < 1 \quad \text{is equivalent to} \quad x > 2 \textit{ and } x < 4 \tag{3.65}$$

In other words, the contribution to the solution set of (3.62) in that part of the x axis for which $x > 1$ is simply the set of all real numbers strictly between 2 and 4 (see Fig. 3.26).

Case 2 $x-1 < 0$

Fig. 3.25 Solution set of $|(3x-1)/(x+2)| \geq 4$. The points -9 and -1 are included, but the point -2 is not.

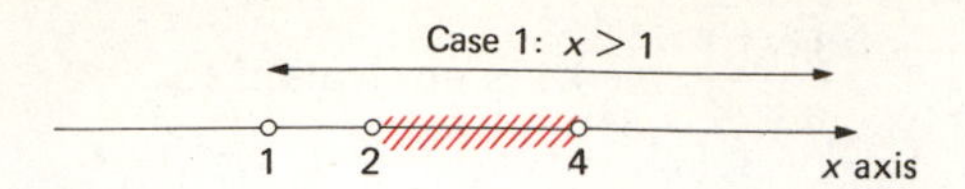

Fig. 3.26 Contribution to solution set of (3.62) when $x > 1$. The points 2 and 4 are excluded.

In this case, we argue exactly as in Case 1, except that the signs of all the inequalities involved are now *reversed* because in our present case we are multiplying by the *negative* expression $x - 1$ (recall Rule 3). The net result is [see (3.65)]:

$$-1 < \frac{2x-5}{x-1} < 1 \quad \text{is equivalent to} \quad x < 2 \text{ and } x > 4 \tag{3.66}$$

However, $x < 2$ and $x > 4$ is impossible (since there is no number x which is less than 2 and larger than 4 at the same time), and, hence, by (3.66), $-1 < (2x-5)/(x-1) < 1$ is also impossible to hold for any number x for which $x < 1$ (Case 2). In other words, Case 2 *does not contribute* anything at all to the solution set of (3.62) (we may also say that such a contribution is the empty set). The net result, then, is that the total solution set of (3.62) is $\{x|2 < x < 4\}$.

Example 3.20 Find the solution set of (3.67):

$$|x^2 - x - 1| \leq 1 \tag{3.67}$$

Solution By (3.38*c*) we know that

$$\text{Inequality (3.67)} \quad \text{is equivalent to} \quad -1 \leq x^2 - x - 1 \leq 1 \tag{3.68}$$

Clearly, $-1 \leq x^2 - x - 1$ is equivalent to $0 \leq x^2 - x$, that is, $x(x-1) \geq 0$. Moreover, it is easy to see that the inequality $x(x-1) \geq 0$ is equivalent to $(x \geq 1 \text{ or } x \leq 0)$. We have thus shown:

$$\text{The solution set of } -1 \leq x^2 - x - 1 \text{ is } \{x|x \geq 1 \text{ or } x \leq 0\}. \tag{3.69}$$

Next, consider the inequality $x^2 - x - 1 \leq 1$. Clearly, $x^2 - x - 1 \leq 1$ is equivalent to $x^2 - x - 2 \leq 0$ or to $(x+1)(x-2) \leq 0$, and it is very easy to see that this last inequality is equivalent to $-1 \leq x \leq 2$. We have thus shown:

$$\text{The solution set of } x^2 - x - 1 \leq 1 \text{ is } \{x|-1 \leq x \leq 2\}. \tag{3.70}$$

Hence, combining (3.68) to (3.70), we get:

$$\text{The solution set of (3.67) is } \{x|x \geq 1 \text{ or } x \leq 0\} \cap \{x|-1 \leq x \leq 2\}. \tag{3.71}$$

(Observe that we took the *intersection* because $-1 \leq x^2 - x - 1 \leq 1$ means $-1 \leq x^2 - x - 1$ *and* $x^2 - x - 1 \leq 1$.) In Fig. 3.27 we sketch each of the sets displayed in (3.71). Observe that (3.71) could be written, more simply, as follows

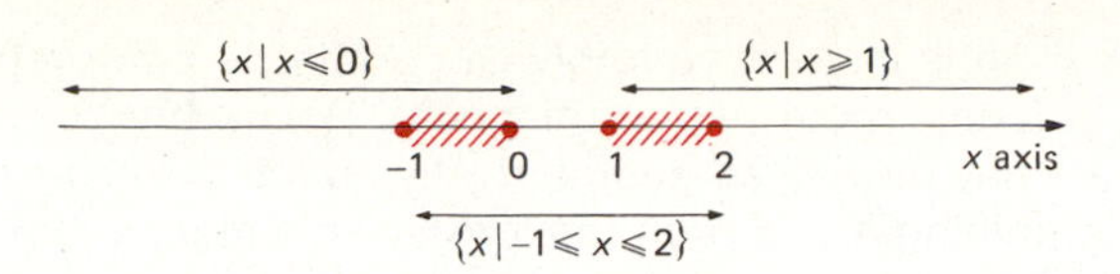

Fig. 3.27 Solution set of $|x^2 - x - 1| \leq 1$. The points −1, 0, 1, and 2 are all included.

$$\text{The solution set of (3.67) is } \{x|-1 \leq x \leq 0\} \cup \{x|1 \leq x \leq 2\}. \tag{3.72}$$

That is, the solution set of (3.67) is simply the set of all real numbers between −1 and 0 (inclusive), together with those between 1 and 2 (inclusive).

Example 3.21 Find the solution set of (3.73):

$$|x + 1| + |x - 1| > 2 \tag{3.73}$$

Solution Because $x + 1$ changes sign at $x = -1$ and $x - 1$ changes sign at $x = 1$, we distinguish three cases (see Fig. 3.28).

Case 1 $x < -1$

In this case, $x + 1 < 0$ and $x - 1 < 0$. Hence (3.73) now becomes (recall the definition of absolute value) $-(x + 1) - (x - 1) > 2$, or $-2x > 2$, which, of course, is equivalent to $x < -1$. In other words, we have shown that under Case 1 (3.73) is equivalent to $x < -1$. But $x < -1$ is certainly true (since we are in Case 1), and hence (3.73) is true for *all numbers satisfying Case 1* (that is, $x < -1$).

Case 2 $-1 \leq x \leq 1$

In this case, $x + 1 \geq 0$, but $x - 1 \leq 0$; hence (applying the definition of absolute value again), (3.73) now becomes $(x + 1) - (x - 1) > 2$, that is, $2 > 2$, which is certainly false. In other words, we have shown that there is not a single number x arising from Case 2 (that is, $-1 \leq x \leq 1$) which satisfies (3.73); thus Case 2 *does not contribute at all* to the solution set of (3.73).

Case 3 $x > 1$

In this case, $x + 1 > 0$ and $x - 1 > 0$; hence (by definition of absolute value), (3.73) now becomes $(x + 1) + (x - 1) > 2$, that is, $2x > 2$, or $x > 1$. In other words, we have shown that under Case 3 (3.73) is equivalent to $x > 1$. However, $x > 1$ is certainly true (we are in Case 3). Therefore (3.73) is *true* for *all* numbers satisfying Case 3 (that is, $x > 1$).

To sum up, the solution set of (3.73) is simply $\{x|x < -1\} \cup \{x|x > 1\}$. In other words, the solution set of (3.73) consists of all real numbers x except those which lie between −1 and 1, inclusive (that is, except $\{x|-1 \leq x \leq 1\}$).

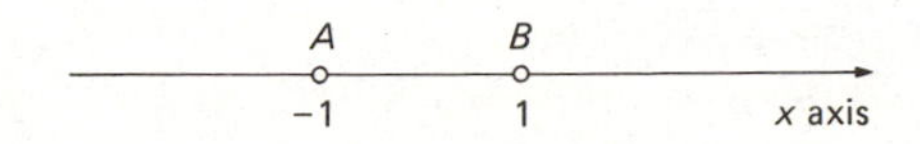

Fig. 3.28 Points at which $x + 1$ or $x - 1$ changes sign.

It is instructive to give a geometric interpretation of the fact just established (concerning this solution set). Recall that for any real numbers a and b, $|a - b|$ may be viewed geometrically as the *positive* (or zero) length of the line segment joining a and b on the x axis. This means that since $x + 1 = x - (-1)$, we have

$|x + 1|$ is the *positive* distance from x to -1

$|x - 1|$ is the *positive* distance from x to 1

Thus, we may interpret the above results geometrically as follows: The locus of all points $P{:}x$ on a number line such that the sum of the *positive* distances from P to $A{:}-1$ and $B{:}1$, respectively, is greater than 2 is simply the set of all points on the x axis except those which lie between A and B (inclusive). This result, of course, is obvious from a geometric point of view.

Finally, we wish to remind the reader that the graph of the function f given by $f(x) = |x + 1| + |x - 1|$ has been sketched in Example 3.5. The reader will easily see that the information obtained here is consistent with that graph.

In the next chapter, we shall see how to extend the set of real numbers to the set of complex numbers. Once this extension is done, we shall then see that many of the properties we established for absolute values of real numbers are still valid for absolute values of complex numbers.

Problem Set 3.3

3.3.1 Find the solution sets of the following inequalities. Also, sketch these solution sets on the x axis, and interpret the results geometrically.

(*a*) $|x| \leq 1$ (*c*) $0 < |x - 1| < 1$

(*b*) $|x| > 2$ (*d*) $|x + 2| \geq 1$

3.3.2 Find the solution sets of the following inequalities, and sketch these solution sets on the x axis.

(*a*) $0 < |2x + 3| < 1$ (*b*) $|1 - 4x| \geq 1$

3.3.3 Find the solution sets of the following inequalities, and sketch these solution sets on the x axis:

(*a*) $\left|\dfrac{1}{x + 1}\right| \leq 1$

(*b*) $\left|\dfrac{2}{3 - x}\right| \geq 1$

(*c*) $\left|\dfrac{1 - 4x}{2 + 3x}\right| < 1$

(*d*) $\left|\dfrac{3x + 1}{1 - 2x}\right| > 1$

(*e*) $|x^2 + x - 2| \leq 4$ [*Hint*: Recall (3.14).]

(*f*) $|x^2 + x - 1| \geq 1$

3.3.4 Find the solution sets of the following inequalities. Also, sketch these solution sets on the x axis, and interpret the results geometrically.

(a) $|x+1| - |x-1| > 0$

(b) $|x+3| + |x-2| \geq 5$

(c) $|x+1| + |x-1| \leq 4$ (Verify that the results are consistent with the graph of Example 3.5.)

3.3.5 Sketch the graphs of the functions f given below. In each case, find the domain and range of the function. Compare your answers with the results in Prob. 3.3.4*a* and *b*.

(a) $y = |x+1| - |x-1|$ (b) $y = |x+3| + |x-2|$

3.3.6 Suppose a and b are any nonzero real numbers. Prove that a/b always has the same sign as ab. [*Hint*: b^2 is positive, and $ab = (a/b)(b^2)$.] Accordingly, any fractional inequality can be changed to one without fractions. For example, $a/b > 0$ is equivalent to $ab > 0$. Also, $a/b < 1$ is equivalent to $(a-b)b < 0$. (Why?)

3.3.7 Suppose that a, b, and c are any real numbers such that the discriminant $b^2 - 4ac < 0$. Prove that for all real numbers x we have:

(a) $|ax^2 + bx + c| = ax^2 + bx + c$ if $a > 0$

(b) $|ax^2 + bx + c| = -(ax^2 + bx + c)$ if $a < 0$

[*Hint*: Recall (3.14).]

3.3.8 Let a be any positive real number. Prove that:

(a) $|x| < a$ is equivalent to $x^2 < a^2$.

(b) $|x| > a$ is equivalent to $x^2 > a^2$, and similarly for $|x| \leq a$ and $|x| \geq a$. [*Hint*: Recall that $x^2 - a^2 = (x+a)(x-a)$.]

3.3.9 Find the domain and range of the function f given in Example 3.5.

3.3.10 Prove that, for all real numbers x_1 and x_2, $|x_1 - x_2| = |x_2 - x_1|$. Deduce that $|-x_2| = |x_2|$. [*Hint*: $x_1 - x_2 = (-1)(x_2 - x_1)$. For the last part of the problem, take $x_1 = 0$.]

3.3.11 Prove that, for all real numbers x_1 and x_2, $|x_1 - x_2| \geq |x_1| - |x_2|$ and $|x_1 - x_2| \geq |x_2| - |x_1|$. Deduce that $|x_1 - x_2| \geq ||x_1| - |x_2||$. [*Hint*: Observe that $x_1 = (x_1 - x_2) + x_2$. Now, use the fact that the absolute value of a sum is less than or equal to the sum of the absolute values.]

3.3.12 Prove that $|x_1 + x_2| = |x_1| + |x_2|$ if and only if both x_1 and x_2 are nonnegative together or nonpositive together.

3.4 Summary

In this chapter, we introduced the concepts of absolute value and absolute value function. We also studied some of the properties of this function. In particular, we learned that the graph of the absolute value function is V-shaped, it lies entirely above the x axis, and it passes through the origin. Moreover, we saw that $|x_1 + x_2| \leq |x_1| + |x_2|$, $|x_1 - x_2| \geq ||x_1| - |x_2||$, $|x_1x_2| = |x_1||x_2|$, and $|x_1/x_2| = |x_1|/|x_2|$. We also discussed the solution sets of linear and quadratic inequalities, together with the solution sets of certain inequalities involving polynomial expressions or quotients of polynomial expressions. A general method for finding the solution sets of such inequalities was described for use if such polynomial, or rational, expressions can be factored fully into products (or quotients) of *linear* factors with *real* coefficients. Finally, we studied the solution sets of inequalities involving absolute values and saw that these, in turn, are essentially equivalent to certain combinations of inequalities of the above mentioned types. This follows from facts such as $|x| < a$ is equivalent to $-a < x < a$ and $|x| > a$ is equivalent to ($x > a$ or $x < -a$). In this connection, a very useful geometric interpretation of $|x_1 - x_2|$, where x_1 and x_2 are any real numbers, is the following: $|x_1 - x_2|$ is the nonnegative length of the line segment joining the two points on the x axis with x coordinates x_1 and x_2, respectively.

4 Complex Numbers

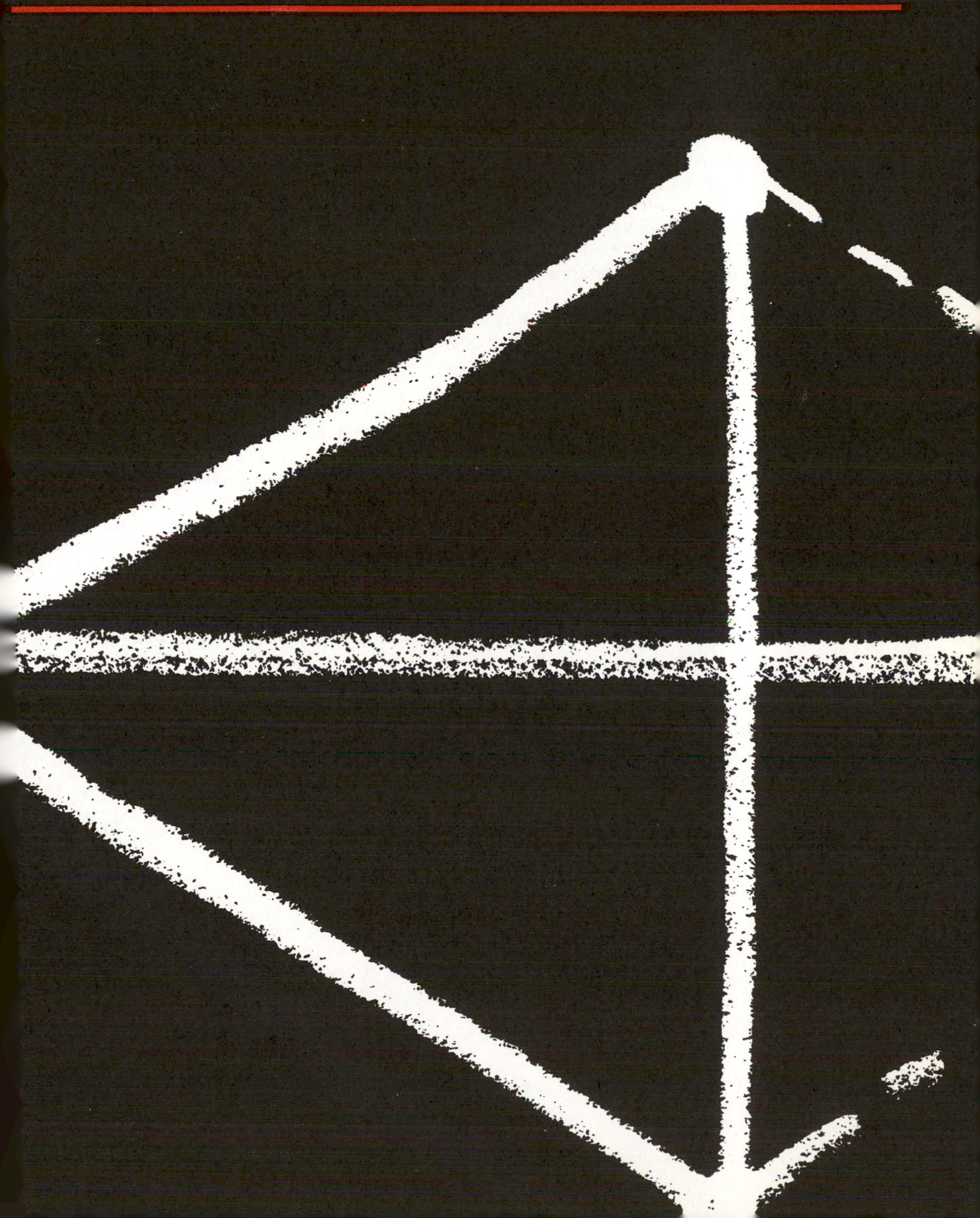

4.1 Definition of and Operations on Complex Numbers

Consider for a moment the equation $x^2 + 1 = 0$. Does this equation have any real solutions? The answer is "no" for the following reason: We know that $x^2 \geq 0$ for all real numbers x and, hence, $x^2 + 1 > 0$ for all real numbers x. In view of this, if we wish to have a solution of the equation $x^2 + 1 = 0$, we must then extend our number system beyond the set of real numbers, and this we do as follows: We agree to introduce a number i satisfying $i^2 = -1$. Observe that, in view of the above discussion, this new number i *cannot* possibly be a real number. Moreover, i is a solution of the equation $x^2 + 1 = 0$. In fact, the solutions of the equation $x^2 + 1 = 0$ are given by $x = \pm i = \pm\sqrt{-1}$. Now, armed with this new number i, we define a complex number as follows:

> *A* ***complex number*** *is any number of the form $a + bi$, where a and b are both real numbers and i satisfies the equation $i^2 = -1$. If, further, $b \neq 0$, the complex number $a + bi$ is called* ***imaginary.***

We further agree to identify every real number a with the complex number $a + 0i$. In view of this, *every real number may be viewed as a complex number.* We also agree to identify i with the complex number $0 + 1i$, and more generally, we identify any number of the form bi, where b is a real number, with the complex number $0 + bi$. Incidentally, the complex number bi $[= 0 + bi]$, where b is any real number *different from zero,* is called *pure-imaginary.* Furthermore, if $a + bi$ is any complex number (and thus a and b are real), then a is called the *real part,* while b (not bi) is called the *imaginary part* of this number. Thus, the so-called imaginary part of the complex number $a + bi$ (which is b by definition) is indeed a *real* number. (The term *imaginary* was historically introduced before complex numbers were properly understood, and this terminology still persists.)

There is an extremely convenient way to represent complex numbers geometrically. In brief, we agree to identify the complex number $a + bi$ (a and b are real numbers) with the point $P: (a,b)$ in the xy plane (see Fig. 4.1), and, in fact, we will sometimes find it convenient to write $(a,b) = a + bi$. This xy

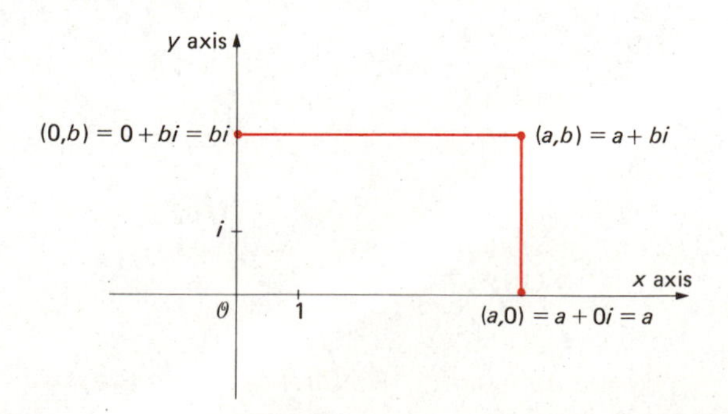

Fig. 4.1 Identification of the complex number $a + bi$ and the point (a,b)

plane is sometimes called the *complex plane.* In particular, the real number a $[= a + 0i]$ is identified with the point $(a,0)$, and this point, of course, lies on the x axis. For this reason, the x axis is called the *real axis.* Moreover, since the pure imaginary number bi $[= 0 + bi]$ is identified with the point $(0,b)$, which lies on the y axis, the y axis is called the *imaginary axis.* Furthermore, by definition, $a + bi = c + di$ (a, b, c, and d all real) if and only if $a = c$ and $b = d$. The set C of all complex numbers may thus be viewed as the Cartesian product $R \times R$ of R and R, where R denotes the set of all real numbers.

How do we operate on complex numbers? Certainly, whatever we do, we would like to preserve the basic rules of high school algebra. Keeping this in mind, we *define* addition, subtraction, and multiplication for complex numbers as follows:

$$(a + bi) + (c + di) = (a + c) + (b + d)i \tag{4.1}$$

$$(a + bi) - (c + di) = (a - c) + (b - d)i \tag{4.2}$$

$$(a + bi)(c + di) = (ac - bd) + (ad + bc)i \tag{4.3}$$

The definitions of addition and subtraction above are plausible enough. In order to make the definition of multiplication more transparent, let us recall that, *by definition,* $i^2 = -1$; this explains the reason for the term $-bd$ $[= (bi)(di)]$ on the right side of (4.3). In fact, the right side of (4.3) is obtained by simply multiplying $a + bi$ by $c + di$, as we learned in high school, and then substituting $i^2 = -1$.

So far, we have seen how to add, subtract, and multiply any two complex numbers, and it is now natural to ask how we divide a complex number $a + bi$ by a *nonzero* complex number $c + di$. In other words, what is $(a + bi)/(c + di)$ equal to? To answer this question, let us multiply both numerator and denominator of the above fraction by $c - di$.

$$\frac{a + bi}{c + di} = \frac{(a + bi)(c - di)}{(c + di)(c - di)} = \frac{(ac + bd) + (bc - ad)i}{c^2 + d^2}$$

Hence,

$$\frac{a + bi}{c + di} = \left(\frac{ac + bd}{c^2 + d^2}\right) + \left(\frac{bc - ad}{c^2 + d^2}\right)i \tag{4.4}$$

Observe that since $c + di \neq 0 + 0i$, we *cannot* possibly have $c^2 + d^2 = 0$ (since $c^2 + d^2 = 0$ implies $c = 0$ *and* $d = 0$); hence, the real numbers on the right side of (4.4) are well defined. Observe also that in view of Eqs. (4.1) to (4.4) we have the following fact: The *sum, difference, product,* and *quotient* (with nonzero denominator) of *any two complex numbers* is again a *complex number.* (Recall that $a + 0i$ is still a complex number.)

There is a very easy way to present the sum or difference of two complex numbers geometrically. Thus, suppose that

$$z_1 = a + bi \qquad z_2 = c + di \qquad (a, b, c, d \text{ real}) \tag{4.5}$$

By the definition of addition given in (4.1), we have

$$z_1 + z_2 = (a + c) + (b + d)i = (a + c, b + d) \tag{4.6}$$

We claim that the points $\mathcal{O}$, z_1, z_2, and $z_1 + z_2$ (see Fig. 4.2) form a parallelogram. This is easy to verify since the two right triangles in Fig. 4.2 are indeed congruent. To sum up, we have the following.

Parallelogram Rule

Let $\mathcal{O}$ be the origin (or complex number $0 + 0i$), and let z_1 and z_2 be any complex numbers. Then the sum $z_1 + z_2$ is simply the fourth vertex of the parallelogram shown in Fig. 4.2, where $\mathcal{O}$, z_1, and z_2 are the other vertices.

Next, we seek a geometric representation of the difference $z_1 - z_2$ of the complex numbers z_1 and z_2 given in (4.5). Since $z_2 = c + di$, $-z_2 = -c + (-d)i$, and thus $-z_2$ is simply the reflection of z_2 through the origin $\mathcal{O}$ (see Fig. 4.3). Because $z_1 - z_2 = z_1 + (-z_2)$, we may apply the parallelogram rule to find the sum $z_1 + (-z_2)$. This difference $z_1 - z_2$ is indicated in Fig. 4.3. Observe that $\mathcal{O}$, z_1, $z_1 - z_2$, and $-z_2$ are vertices of a parallelogram and that $\mathcal{O}$, z_2, z_1, and $z_1 - z_2$ are vertices of a parallelogram.

A geometric representation of the product or quotient of two complex numbers will be discussed in a later chapter, after we have introduced the so-called *polar form* of a complex number.

In general, in computing with complex numbers we may apply (4.1) through (4.4), or we may simply apply the usual rules of high school algebra and then write -1 every time we have i^2.

Example 4.1

Let $z_1 = 1 + 2i$ and $z_2 = -1 + 4i$. Express the following complex numbers in the form $a + bi$, where a and b are real: $z_1 + z_2$, $z_2 + z_1$, $z_1 - z_2$, $z_2 - z_1$, z_1z_2, z_2z_1, z_1/z_2, z_2/z_1, $z_1{}^2$, and $z_2{}^2$.

Solution

Applying the definitions given in Eqs. (4.1) to (4.4), we obtain

$$z_1 + z_2 = (1 + 2i) + (-1 + 4i) = [1 + (-1)] + (2 + 4)i = 0 + 6i$$

$$z_2 + z_1 = (-1 + 4i) + (1 + 2i) = (-1 + 1) + (4 + 2)i = 0 + 6i$$

$$z_1 - z_2 = (1 + 2i) - (-1 + 4i) = [1 - (-1)] + (2 - 4)i = 2 - 2i$$

$$z_2 - z_1 = (-1 + 4i) - (1 + 2i) = (-1 - 1) + (4 - 2)i = -2 + 2i$$

$$z_1z_2 = (1 + 2i)(-1 + 4i) = [1(-1) - 2(4)] + [1(4) + 2(-1)]i = -9 + 2i$$

$$z_2z_1 = (-1 + 4i)(1 + 2i) = [(-1)1 - 4(2)] + [(-1)2 + 4(1)]i = -9 + 2i$$

$$\frac{z_1}{z_2} = \frac{1 + 2i}{-1 + 4i} = \frac{1 + 2i}{-1 + 4i} \cdot \frac{-1 - 4i}{-1 - 4i} = \frac{[1(-1) - 2(-4)] + [1(-4) + 2(-1)]i}{(-1)^2 + 4^2}$$

$$= \frac{7 - 6i}{17} = \frac{7}{17} - \frac{6}{17}i$$

$$\frac{z_2}{z_1} = \frac{-1 + 4i}{1 + 2i} = \frac{(-1 + 4i)(1 - 2i)}{(1 + 2i)(1 - 2i)}$$

$$= \frac{[(-1)(1) - 4(-2)] + [(-1)(-2) + 4(1)]i}{1^2 + 2^2} = \frac{7 + 6i}{5} = \frac{7}{5} + \frac{6}{5}i$$

$$z_1{}^2 = (1 + 2i)^2 = 1 + 4i + 4i^2 = 1 + 4i + 4(-1) = -3 + 4i$$

$$z_2{}^2 = (-1 + 4i)^2 = 1 - 8i + 16i^2 = 1 - 8i - 16 = -15 - 8i$$

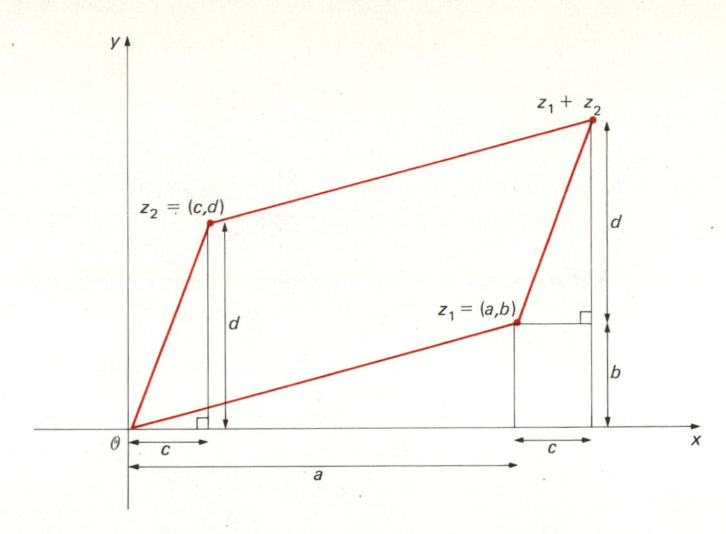

Fig. 4.2 Geometric representation of sum of complex numbers

Example 4.2 Represent the complex numbers z_1, z_2, $z_1 + z_2$, $z_1 - z_2$, and $z_2 - z_1$ as points in the xy plane, where z_1 and z_2 are as in Example 4.1. Join the origin $\mathcal{O}$ (or complex number $0 + 0i$) to each of z_1, z_2, $z_1 + z_2$, $z_1 - z_2$, and $z_2 - z_1$, and verify the parallelogram rule.

Solution Following our agreement of identifying the complex number $a + bi$ with the point (a,b), we have in view of the results of Example 4.1

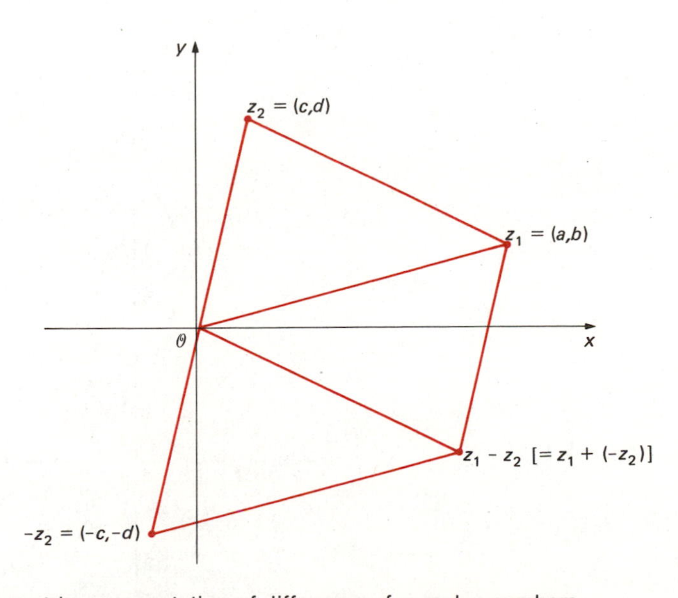

Fig. 4.3 Geometric representation of difference of complex numbers

$$z_1 = 1 + 2i = (1,2)$$
$$z_2 = -1 + 4i = (-1,4)$$
$$z_1 + z_2 = 0 + 6i = (0,6)$$
$$z_1 - z_2 = 2 - 2i = (2,-2)$$
$$z_2 - z_1 = -2 + 2i = (-2,2)$$

These points (corresponding to complex numbers) are sketched in Fig. 4.4. Observe that each of the following sets of four points represents the vertices of a parallelogram:

$\mathcal{O} \quad z_1 \quad z_2 \quad z_1 + z_2$

$\mathcal{O} \quad z_1 - z_2 \quad z_2 \quad z_1$

$\mathcal{O} \quad z_2 - z_1 \quad z_1 \quad z_2$

Since in each case above the last complex number is the sum of the remaining two nonzero complex numbers in that set, we see that the parallelogram rule is satisfied in each of the above four cases.

Example 4.3 Express each of the following in the form $a + bi$, where a and b are real (in your answers, identify $a + 0i$ by a and $0 + bi$ by bi): i^2, i^3, i^4, i^5, i^6, i^{301}, i^{199}, and i^{1000}.

Solution By definition, $i^2 = -1$; hence (by multiplying both sides by i), we get $i^3 = -i$. Moreover, $i^4 = (i^2)^2 = (-1)^2 = 1$; that is, $i^4 = 1$. Hence, $i^5 = i^4 \cdot i = 1 \cdot i = i$; that is, $i^5 = i$. Therefore (by multiplying both sides by i), we get $i^6 = i^2 = -1$; that is, $i^6 = -1$. Thus, $i^1 = i$, $i^2 = -1$, $i^3 = -i$, $i^4 = 1$, $i^5 = i$, $i^6 = -1$, and so on.

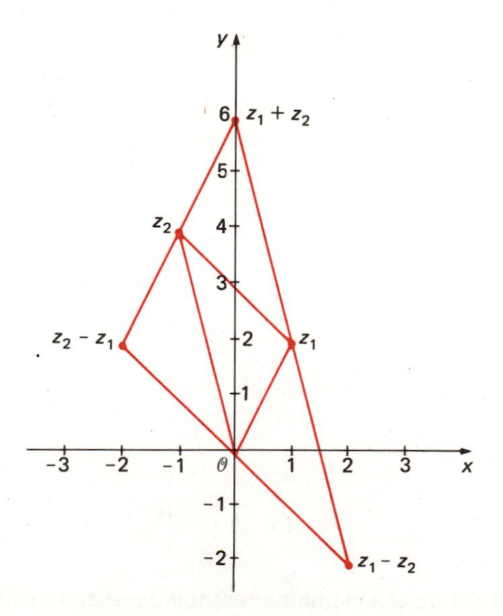

Fig. 4.4 Geometric representation of certain complex numbers

Observe that since $i^4 = 1$, it follows that $(i^4)^q = 1$ for *all* integers q; thus

$$i^{4q} = 1 \qquad \text{for all integers } q \tag{4.7}$$

This fact is very useful and suggests the following rule: Let n be any integer. To find i^n, first divide n by 4 to get $n = 4q + r$, where r is the remainder (thus $r = 0, 1, 2,$ or 3). Then $i^n = i^r$. The reason for this is the following:

$$i^n = i^{4q+r} = i^{4q}i^r = 1i^r = i^r$$

since by (4.7) $i^{4q} = 1$. For example, $i^{301} = i^1$ (since the remainder when 301 is divided by 4 is 1) $= i$; that is, $i^{301} = i$. Similarly, since the remainder when 199 is divided by 4 is 3, we have (by the above rule), $i^{199} = i^3$. But $i^3 = (i^2)(i) = (-1)(i) = -i$, and hence $i^{199} = -i$. Finally, since $1000 = 4(250)$, we obtain [by (4.7)] $i^{1000} = i^{4(250)} = 1$.

Example 4.4 Find all possible complex numbers $x + yi$ such that $(x + yi)^2 = 2i$.

Solution Expanding $(x + yi)^2$ and recalling that $i^2 = -1$, we obtain

$$(x + yi)^2 = x^2 + y^2i^2 + 2xyi = (x^2 - y^2) + (2xy)i$$

But by hypothesis $(x + yi)^2 = 2i = 0 + 2i$, and therefore

$$(x^2 - y^2) + (2xy)i = 0 + 2i \tag{4.8}$$

x and y are both *real*; it follows, by equating the corresponding real parts and the corresponding imaginary parts of both sides of (4.8), that

$$x^2 - y^2 = 0 \qquad \text{and} \qquad 2xy = 2$$

Hence $x = \pm y$ and $xy = 1$. Since xy is positive, both x and y have the *same* sign, and hence we cannot possibly have $x = -y$. Therefore $x = y$, and, moreover, $xy = 1$. This leads to $x^2 = 1$, and thus $x = \pm 1$. Therefore ($x = 1$ and $y = 1$) or ($x = -1$ and $y = -1$). Thus, all *possible* solutions of $(x + yi)^2 = 2i$ (where both x and y are real numbers) are

$$x + yi = 1 + 1i = 1 + i$$

and

$$x + yi = (-1) + (-1)i = -1 - i$$

An easy calculation shows that both of these complex numbers *are* indeed solutions.

Remark Example 4.4 shows that there are exactly two square roots of $2i$, namely,

$$\sqrt{2i} = 1 + i \qquad \text{or} \qquad \sqrt{2i} = -1 - i$$

In Chap. 10 we shall see how to extend this result, using the so-called *polar form* of a complex number.

Problem Set 4.1

4.1.1 Let z_1, z_2, and z_3 be the complex numbers given by $z_1 = -1 + i$, $z_2 = 2 - i$, and $z_3 = 3 + 2i$. Express the following in the form $a + bi$ (a and b are real): (*a*) $z_1 + z_2$, (*b*) $z_1 - z_2$, (*c*) $z_2 - z_1$, (*d*) $1/z_1$, (*e*) z_2/z_1, (*f*) z_1/z_2, (*g*) $(z_1z_2)z_3$, (*h*) $z_1(z_2z_3)$, (*i*) $z_1z_2 + z_1z_3$, and (*j*) $2z_1 - 3z_2 + 4z_3$.

4.1.2 Represent geometrically the complex numbers in Prob. 4.1.1*a, b,* and *c.* Verify that the parallelogram rule holds in each case.

4.1.3 Express the following complex numbers in the form $a + bi$ (a and b are real): (*a*) $(1 + i)^2$, (*b*) $(1 + i)^4$, (*c*) $(1 + i)^6$, and (*d*) $(1 + i)^8$.

4.1.4 Represent geometrically the complex numbers in Prob. 4.1.3.

4.1.5 Express the following complex numbers in the form $a + bi$ (a and b are real): (*a*) $(1 - i)^2$, (*b*) $(1 - i)^4$, (*c*) $(1 - i)^6$, and (*d*) $(1 - i)^8$.

4.1.6 Represent geometrically the complex numbers in Prob. 4.1.5.

4.1.7 Suppose k is any *odd* positive integer. Prove that $(1 + i)^{2k} + (1 - i)^{2k} = 0$. [*Hint*: First evaluate $(1 + i)^2$ and $(1 - i)^2$.]

4.1.8 Suppose k is any *even* positive integer. Prove that $(1 + i)^{2k} - (1 - i)^{2k} = 0$. [*Hint*: First evaluate $(1 + i)^4$ and $(1 - i)^4$.]

4.1.9 Suppose k is any *odd* positive integer. Prove that $(1 + i)^{2k} - (1 - i)^{2k}$ is pure-imaginary. [*Hint*: First evaluate $(1 + i)^2$ and $(1 - i)^2$.]

4.1.10 Express each of the following in the form $a + bi$ (a and b are real): (*a*) $(-i)^{93}$, (*b*) $(-i)^{79}$, (*c*) i^{102}, and (*d*) i^{500}.

4.1.11 Express the following complex numbers in the form $a + bi$ (a and b are real): (*a*) $1/(1 + i)$, (*b*) $1/(1 - i)$, (*c*) $(2 + 3i)/(1 - 5i)$, (*d*) $1/i$, (*e*) $1/(-i)$, (*f*) $(1 + i)^{10}$, (*g*) $(1 - i)^{10}$, (*h*) $(1 + i)^{24}$, and (*i*) $(1 - i)^{24}$. [*Hint*: For parts (*f*) to (*i*), first evaluate $(1 + i)^2$ and $(1 - i)^2$.]

4.1.12 Express each of the following in the form $a + bi$ (a, b real), and represent each answer geometrically: (*a*) $\sqrt{i}$ and (*b*) $\sqrt{-i}$. (*Hint*: Use the method described in Example 4.4. Observe that there are two answers for each one of the two parts.)

4.2 Properties of Complex Numbers

In the preceding section, we discussed the fundamental operations of addition, subtraction, multiplication, and division for complex numbers. Generally speaking, *complex numbers obey the same rules* of high school algebra which real numbers do with one important exception; namely, the relations of $<$ and $>$ *are not defined for complex numbers.* Indeed, had these relations been defined for complex numbers, then we would have, in particular,

$$i > 0 \qquad i < 0 \qquad \text{or} \qquad i = 0 \tag{4.9}$$

None of the alternatives in (4.9) is feasible. For if $i > 0$, then $i \cdot i > i \cdot 0$, and thus $i^2 > 0$. Hence $-1 > 0$ (since $i^2 = -1$), which, of course, is not true. Thus, we *cannot* possibly have $i > 0$. Similarly, if $i < 0$, then $i \cdot i > i \cdot 0$ (since i is now assumed to be *negative*), and thus $i^2 > 0$. Hence, $-1 > 0$, which is, of course, false. Thus, we cannot have $i < 0$ either. Finally, if $i = 0$, then $i^2 = 0^2 = 0$, and hence $-1 = 0$, which is again false. Thus, i *cannot* equal zero, and none of the alternatives in (4.9) holds. In view of this, we cannot expect to have the relations $<$ and $>$ be defined for complex numbers.

What about the concept of absolute value? It turns out that the absolute value of a complex number is always defined and is, in fact, always a *nonnegative* real number. Indeed, suppose that z is any complex number, say,

$$z = x + yi \qquad x, y \text{ are real numbers} \tag{4.10}$$

As we stated in the preceding section, x is called the real part of z, and y is called the imaginary part of z. We abbreviate these facts by writing

$$\text{Re } z = x \qquad \text{and} \qquad \text{Im } z = y \tag{4.11}$$

We now define the *absolute value* of z, denoted by $|z|$, as follows:

$$|z| = |x + yi| = \sqrt{x^2 + y^2} \tag{4.12}$$

($\sqrt{\ }$ denotes the *nonnegative* root.)

It should be pointed out that if the complex number $z = x + yi$ happens to be also a *real* number (that is, $y = 0$), then $z = x + 0i = x$, and (4.12) now becomes

$$|x| = |x + 0i| = \sqrt{x^2} \tag{4.13}$$

and this, in fact, is equivalent to the earlier definition of absolute value of a *real* number (see Prob. 3.1.7). Thus (4.12) may be viewed as an extension (or a generalization) of the earlier definition of absolute value (of a real number). It should be further pointed out that (4.12) could be interpreted geometrically as follows:

> $|z|$ is the nonnegative length of the line segment joining the origin and z (see Fig. 4.5). (4.14)

Thus, according to (4.12) we have

$$|2 + 3i| = \sqrt{2^2 + 3^2} = \sqrt{13}$$
$$|-1 + 5i| = \sqrt{(-1)^2 + 5^2} = \sqrt{26}$$
$$|3 - 4i| = \sqrt{3^2 + (-4)^2} = 5$$
$$|0 + 0i| = \sqrt{0^2 + 0^2} = 0$$

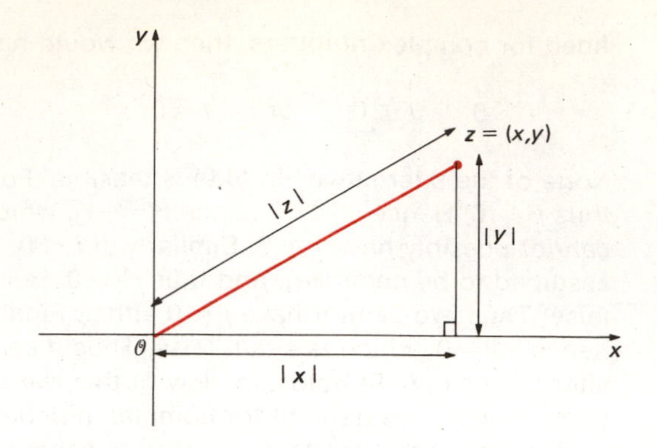

Fig. 4.5 $|z|$ interpreted geometrically

Another concept which plays an important role is that of the *conjugate*, or *complex conjugate*, of a complex number z, denoted by $\bar{z}$. This conjugate $\bar{z}$ is defined as follows:

$$\text{If } z = x + yi \qquad \text{then } \bar{z} = x - yi \qquad (x, y \text{ both real}) \tag{4.15}$$

Thus, the conjugate of z is simply the reflection (or mirror image) of z through the x axis (see Fig. 4.6). For example, we have

$$\overline{2+3i} = 2 - 3i$$
$$\overline{-1+5i} = -1 - 5i$$
$$\overline{3+4i} = 3 - 4i$$
$$\overline{-2-5i} = -2 + 5i$$
$$\overline{0+0i} = 0 - 0i = 0$$

Now, since $|z|$ is always a *real* number (in fact, a *nonnegative* real number), it is certainly meaningful to inquire, given two complex numbers z_1 and z_2,

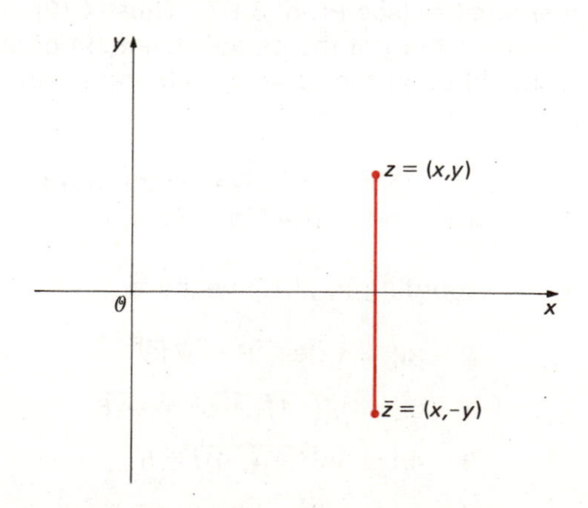

Fig. 4.6 Conjugate of a complex number

whether $|z_1|$ is less than, greater than, or equal to $|z_2|$. We shall soon direct our attention to such considerations. But first we take a closer look at the properties of the conjugate. Thus, suppose that z, z_1, and z_2 are any complex numbers. We claim that

$$\overline{(\bar{z})} = z \qquad \overline{z_1 + z_2} = \bar{z}_1 + \bar{z}_2 \qquad \overline{z_1 - z_2} = \bar{z}_1 - \bar{z}_2 \tag{4.16}$$

The proof of (4.16) is easy. Suppose that

$$z = x + yi \qquad z_1 = x_1 + y_1 i \qquad z_2 = x_2 + y_2 i \tag{4.17}$$

Then, by definition of conjugate $\bar{z} = x - yi$, and hence $\overline{(\bar{z})} = \overline{x - yi} = x + yi = z$; that is, $\overline{(\bar{z})} = z$, which proves the first equality in (4.16).

Next, observe that by (4.17) $z_1 + z_2 = (x_1 + x_2) + (y_1 + y_2)i$, and hence $\overline{z_1 + z_2} = (x_1 + x_2) - (y_1 + y_2)i = (x_1 - y_1 i) + (x_2 - y_2 i) = \bar{z}_1 + \bar{z}_2$, which proves the second equality in (4.16).

Finally, observe that by (4.17) $z_1 - z_2 = (x_1 + y_1 i) - (x_2 + y_2 i) = (x_1 - x_2) + (y_1 - y_2)i$; therefore, $\overline{z_1 - z_2} = (x_1 - x_2) - (y_1 - y_2)i = (x_1 - y_1 i) - (x_2 - y_2 i) = \bar{z}_1 - \bar{z}_2$, which proves the last equality in (4.16). In Fig. 4.7 we give a geometric representation of the second equality in (4.16). Observe that the points $\mathcal{O}$, z_1, z_2, and $z_1 + z_2$ are vertices of a parallelogram, and the points $\mathcal{O}$, $\bar{z}_1$, $\bar{z}_2$, and $\overline{z_1 + z_2}$ are also the vertices of another parallelogram.

What about the conjugate of a product or a quotient? The answer is as follows:

$$\overline{z_1 z_2} = \bar{z}_1 \bar{z}_2 \tag{4.18}$$

$$\overline{\left(\frac{z_1}{z_2}\right)} = \frac{\bar{z}_1}{\bar{z}_2} \qquad \text{if } z_2 \neq 0 + 0i \tag{4.19}$$

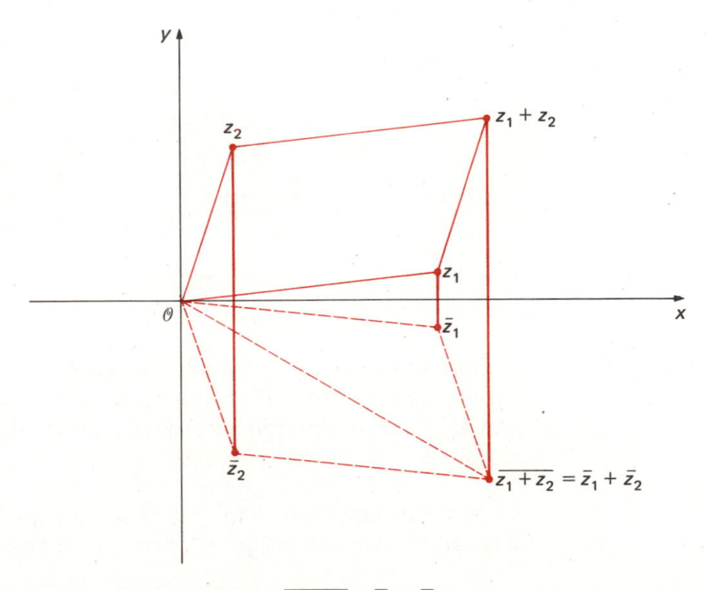

Fig. 4.7 Geometric representation of $\overline{z_1 + z_2} = \bar{z}_1 + \bar{z}_2$

To prove (4.18), let z_1 and z_2 be as in (4.17). Then, $z_1z_2 = (x_1 + y_1i)(x_2 + y_2i) = (x_1x_2 - y_1y_2) + (x_1y_2 + y_1x_2)i$, and, hence,

$$\overline{z_1z_2} = (x_1x_2 - y_1y_2) - (x_1y_2 + y_1x_2)i \tag{4.20}$$

On the other hand, we have $\bar{z}_1 = x_1 - y_1i$ and $\bar{z}_2 = x_2 - y_2i$, and therefore

$$\bar{z}_1\bar{z}_2 = (x_1 - y_1i)(x_2 - y_2i) = (x_1x_2 - y_1y_2) - (x_1y_2 + y_1x_2)i \tag{4.21}$$

Comparing (4.20) and (4.21), we obtain $\overline{z_1z_2} = \bar{z}_1\bar{z}_2$, and (4.18) is proved.

To prove (4.19), we use the same method we used in deriving Eq. (4.4). Thus, suppose z_1 and z_2 are as in (4.17), and suppose $z_2 \neq 0 + 0i$. Then,

$$\begin{aligned}\frac{z_1}{z_2} &= \frac{x_1 + y_1i}{x_2 + y_2i} = \frac{(x_1 + y_1i)(x_2 - y_2i)}{(x_2 + y_2i)(x_2 - y_2i)} \\ &= \frac{(x_1x_2 + y_1y_2) + (-x_1y_2 + y_1x_2)i}{x_2^2 + y_2^2} \\ &= \left(\frac{x_1x_2 + y_1y_2}{x_2^2 + y_2^2}\right) + \left(\frac{-x_1y_2 + y_1x_2}{x_2^2 + y_2^2}\right)i\end{aligned} \tag{4.22}$$

[Observe that since $z_2 \neq 0 + 0i$, $x_2^2 + y_2^2 \neq 0$, and hence we are not dividing by zero in (4.22).] In view of (4.22), we now have

$$\overline{\left(\frac{z_1}{z_2}\right)} = \left(\frac{x_1x_2 + y_1y_2}{x_2^2 + y_2^2}\right) - \left(\frac{-x_1y_2 + y_1x_2}{x_2^2 + y_2^2}\right)i$$

Using high school algebra, we get

$$\overline{\left(\frac{z_1}{z_2}\right)} = \left(\frac{x_1x_2 + y_1y_2}{x_2^2 + y_2^2}\right) + \left(\frac{x_1y_2 - y_1x_2}{x_2^2 + y_2^2}\right)i \tag{4.23}$$

On the other hand, recalling (4.17) and using the above method again, we obtain

$$\begin{aligned}\frac{\bar{z}_1}{\bar{z}_2} &= \frac{x_1 - y_1i}{x_2 - y_2i} = \frac{(x_1 - y_1i)(x_2 + y_2i)}{(x_2 - y_2i)(x_2 + y_2i)} \\ &= \left(\frac{x_1x_2 + y_1y_2}{x_2^2 + y_2^2}\right) + \left(\frac{x_1y_2 - y_1x_2}{x_2^2 + y_2^2}\right)i\end{aligned} \tag{4.24}$$

Comparing (4.23) and (4.24), we get $\overline{(z_1/z_2)} = \bar{z}_1/\bar{z}_2$, and (4.19) is proved. To sum up, we have

> The conjugate of a sum, difference, product, or quotient (with nonzero denominator) of any two complex numbers z_1 and z_2 is equal to the sum, difference, product, or quotient of the two conjugates of z_1 and z_2, respectively. (4.25)

We now turn our attention to the study of absolute values of complex numbers. We recall that, by definition, if $z = x + yi$ is any complex number, then

$$|z| = |x + yi| = \sqrt{x^2 + y^2} \tag{4.26}$$

Thus, the *absolute value* of *any complex number* is a *real number.* Moreover, by (4.26)

$$|z| \geq 0 \qquad \text{for all complex numbers } z \tag{4.27}$$

and

$$|z| = 0 \quad \text{if and only if} \quad z = 0 + 0i \tag{4.28}$$

In Chap. 3, we derived several equalities and inequalities concerning absolute values of real numbers. Are these equalities and inequalities still valid for absolute values of complex numbers? The answer is indeed in the affirmative, as we now proceed to show.

To begin with, suppose z is any complex number, say,

$$z = x + yi \qquad (x, y \text{ both real}) \tag{4.29}$$

We claim that

$$\text{Re } z \leq |z| \qquad \text{Im } z \leq |z| \qquad |\bar{z}| = |z| \qquad z + \bar{z} = 2 \text{ Re } z \tag{4.30}$$

The proof of (4.30) is easy. Indeed, by (4.11) and (4.12) we have

$$\text{Re } z = x \qquad \text{and} \qquad |z| = \sqrt{x^2 + y^2}$$

Because x and y are both real, $x \leq \sqrt{x^2 + y^2}$ (since $y^2 \geq 0$), and hence Re $z \leq |z|$, which proves the first part of (4.30). Similarly, by (4.11) Im $z = y$. But $y \leq \sqrt{x^2 + y^2}$ (since $x^2 \geq 0$), and hence Im $z = y \leq \sqrt{x^2 + y^2} = |z|$; that is, Im $z \leq |z|$, which proves the second part of (4.30). Next, observe that $z = x + yi$ and $\bar{z} = x - yi$, from which

$$|\bar{z}| = |x - yi| = \sqrt{x^2 + (-y)^2} = \sqrt{x^2 + y^2} = |z|$$

that is, $|\bar{z}| = |z|$, which proves the third part of (4.30). Finally, since $z = x + yi$ and $\bar{z} = x - yi$; thus,

$$z + \bar{z} = (x + yi) + (x - yi) = 2x = 2 \text{ Re } z$$

that is, $z + \bar{z} = 2$ Re z, and (4.30) is proved.

How do z, $\bar{z}$, and $|z|$ relate to each other? The next equality, which turns out to be an extremely useful one, gives the answer.

$$z\bar{z} = |z|^2 \tag{4.31}$$

The proof of (4.31) is straightforward. Thus, suppose $z = x + yi$ and hence $\bar{z} = x - yi$. Then,

$$z\bar{z} = (x + yi)(x - yi) = x^2 + y^2 = |z|^2$$

by (4.12), and (4.31) is proved.

We are now in a position to state and prove the following fundamental results, all of which were demonstrated for real numbers in Sec. 3.1 (see also Prob.

3.3.11). Thus, suppose that z_1 and z_2 are any complex numbers. Then,

$$|z_1z_2| = |z_1||z_2| \tag{4.32}$$

$$\left|\frac{z_1}{z_2}\right| = \frac{|z_1|}{|z_2|} \quad \text{if } z_2 \neq 0 + 0i \tag{4.33}$$

$$|z_1 + z_2| \leq |z_1| + |z_2| \qquad \textit{(triangle inequality)} \tag{4.34}$$

$$|z_1 - z_2| \geq ||z_1| - |z_2|| \tag{4.35}$$

Let us first prove (4.32). By (4.31), (4.18), and high school algebra, we have

$$|z_1z_2|^2 = (z_1z_2)\overline{(z_1z_2)} = z_1z_2\bar{z}_1\bar{z}_2 = (z_1\bar{z}_1)(z_2\bar{z}_2)$$
$$= |z_1|^2|z_2|^2$$

that is, $|z_1z_2|^2 = |z_1|^2|z_2|^2$. Taking the nonnegative square roots of both sides of this last equality gives (4.32).

To prove (4.33), we follow essentially the same procedure. Thus, using (4.31) and (4.19) we get

$$\left|\frac{z_1}{z_2}\right|^2 = \frac{z_1}{z_2}\overline{\left(\frac{z_1}{z_2}\right)} = \frac{z_1}{z_2}\frac{\bar{z}_1}{\bar{z}_2} = \frac{z_1\bar{z}_1}{z_2\bar{z}_2} = \frac{|z_1|^2}{|z_2|^2}$$

that is, $|z_1/z_2|^2 = |z_1|^2/|z_2|^2$. Now, taking the nonnegative square roots of both sides of this last equality, we obtain (4.33). (Observe that $z_2 \neq 0 + 0i$ in the above equalities.)

The proof of (4.34) is somewhat involved; it starts off by using (4.31) also. Thus,

$\lvert z_1 + z_2\rvert^2 = (z_1 + z_2)\overline{(z_1 + z_2)}$	By (4.31)
$= (z_1 + z_2)(\bar{z}_1 + \bar{z}_2)$	By (4.16)
$= z_1\bar{z}_1 + z_2\bar{z}_2 + (z_1\bar{z}_2 + z_2\bar{z}_1)$	Algebra
$= \lvert z_1\rvert^2 + \lvert z_2\rvert^2 + (z_1\bar{z}_2 + \bar{z}_1z_2)$	By (4.31) and algebra

We have thus shown that

$$|z_1 + z_2|^2 = |z_1|^2 + |z_2|^2 + (z_1\bar{z}_2 + \bar{z}_1z_2) \tag{4.36}$$

By (4.18), $\overline{(z_1\bar{z}_2)} = \bar{z}_1\overline{(\bar{z}_2)} = \bar{z}_1z_2$, by (4.16). Hence, $\bar{z}_1z_2 = \overline{(z_1\bar{z}_2)}$. Using this fact, together with the facts that $z + \bar{z} = 2 \text{ Re } z$ and $\text{Re } z \leq |z|$, we get

$$z_1\bar{z}_2 + \bar{z}_1z_2 = (z_1\bar{z}_2) + \overline{(z_1\bar{z}_2)} = 2 \text{ Re } z_1\bar{z}_2 \leq 2|z_1\bar{z}_2| \tag{4.37}$$

Moreover, by (4.32) and the fact that $|\bar{z}| = |z|$, we get

$$2|z_1\bar{z}_2| = 2|z_1||\bar{z}_2| = 2|z_1||z_2| \tag{4.38}$$

Hence, combining (4.37) and (4.38), we obtain

$$z_1\bar{z}_2 + \bar{z}_1z_2 \leq 2|z_1||z_2| \tag{4.39}$$

Now, combining (4.36) and (4.39), we get

$$|z_1 + z_2|^2 \leq |z_1|^2 + |z_2|^2 + 2|z_1||z_2| = (|z_1| + |z_2|)^2$$

that is, $|z_1 + z_2|^2 \leq (|z_1| + |z_2|)^2$. Hence, by taking the nonnegative square roots of both sides of this last inequality we obtain (4.34).

Finally, to prove (4.35) observe that since $z_1 = (z_1 - z_2) + z_2$, we have, using (4.34),

$$|z_1| = |(z_1 - z_2) + z_2| \leq |z_1 - z_2| + |z_2|$$

Hence,

$$|z_1 - z_2| \geq |z_1| - |z_2| \tag{4.40}$$

If we interchange z_1 and z_2 in the above argument, (4.40) becomes

$$|z_2 - z_1| \geq |z_2| - |z_1| \tag{4.41}$$

But since $z_2 - z_1 = (-1)(z_1 - z_2)$, $|z_2 - z_1| = |-1||z_1 - z_2| = (+1)|z_1 - z_2| = |z_1 - z_2|$. In other words, $|z_2 - z_1| = |z_1 - z_2|$, and hence (4.41) now becomes

$$|z_1 - z_2| \geq |z_2| - |z_1| \tag{4.42}$$

Combining (4.40) and (4.42), we obtain (4.35).

It is instructive to give a geometric interpretation of some of the equalities and inequalities we encountered in this section. To begin with, we observed in (4.14) that if z is any complex number, then $|z|$ is always a *nonnegative* real number and is, in fact, the length of the line segment joining the origin and z. This fact can actually be extended to the following:

> Let z_1 and z_2 be any complex numbers. Then $|z_2 - z_1|$ is the length of the line segment joining z_1 and z_2 (see Fig. 4.8). (4.43)

To see this, draw a parallelogram having three vertices at $\mathcal{O}$, z_1, and z_2, as indicated in Fig. 4.8. It is easy to see using the parallelogram rule that the fourth vertex P is $z_2 - z_1$. But by (4.14), $|z_2 - z_1|$ is the positive length of the line segment $\mathcal{O}P$, which in turn (by high school geometry) is equal to the positive length of the line segment joining z_1 and z_2, and (4.43) is proved. Observe that if $z_1 = 0$, Eq. (4.43) reduces to the following: $|z_2|$ is the length of the line segment joining $\mathcal{O}$ and z_2, which is the result we obtained in (4.14).

We recall that we have shown that [see (4.30)] Re $z \leq |z|$, Im $z \leq |z|$, $|\overline{z}| = |z|$, and $z + \overline{z} = 2$ Re z. In Fig. 4.9, we represent geometrically z, $\overline{z}$, Re z, Im z, $|z|$, $|\overline{z}|$, and $z + \overline{z}$. Observe that the fourth vertex of the parallelogram in the figure (for which $\mathcal{O}$, z, and $\overline{z}$ are vertices) lies on the x axis since $z + \overline{z} = 2$ Re z, which is real. Observe also that Re z and Im z may be negative (take z in the third quadrant, for example). However, $|z| \geq 0$ and $|\overline{z}| \geq 0$.

Now, consider the triangle inequality: $|z_1 + z_2| \leq |z_1| + |z_2|$. Applying the parallelogram rule, it is apparent (see Fig. 4.10) that the triangle inequality $|z_1 + z_2| \leq |z_1| + |z_2|$ can now be interpreted geometrically as follows: *The length*

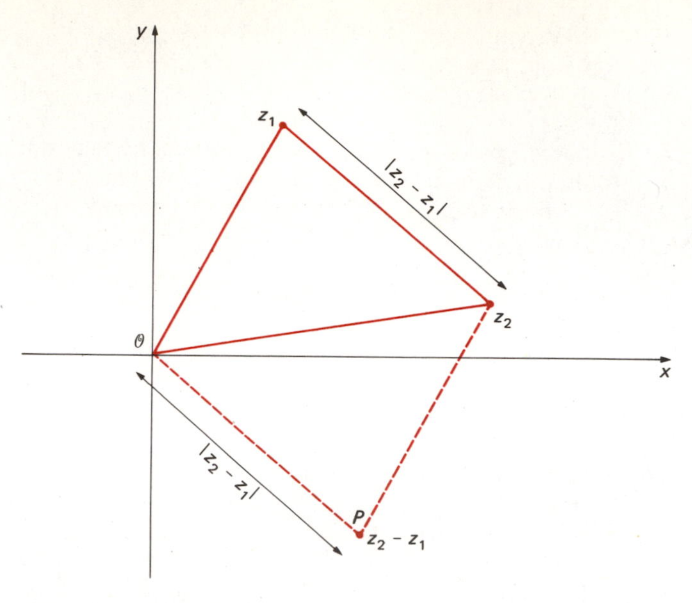

Fig. 4.8 Geometric representation of $|z_2 - z_1|$

of any side in a triangle does not exceed the sum of the lengths of the two remaining sides.

Finally, if we apply the parallelogram rule again, it is easy to see that the inequality $|z_1 - z_2| \geq ||z_1| - |z_2||$ can now be interpreted geometrically as follows: *The length of any side in a triangle is at least as big as the difference of the lengths of the two remaining sides* (see Fig. 4.11). Incidentally, in verifying that $z_1 - z_2$ is the indicated vertex of the parallelogram shown in Fig. 4.11, one may refer to Fig. 4.3 and to the discussion which led to that figure.

Example 4.5 Let $z_1 = 2 + 3i$ and $z_2 = -1 + 2i$. Then, Re $z_1 = 2$, Im $z_1 = 3$, $\bar{z}_1 = 2 - 3i$, and $|z_1| = \sqrt{2^2 + 3^2} = \sqrt{13}$. Furthermore,

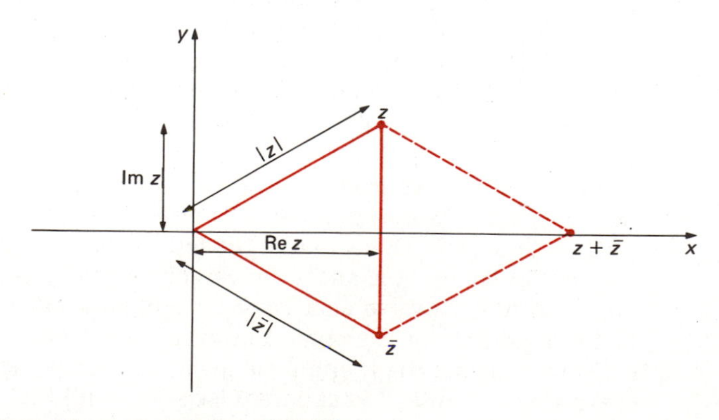

Fig. 4.9 Geometric representation of (4.30)

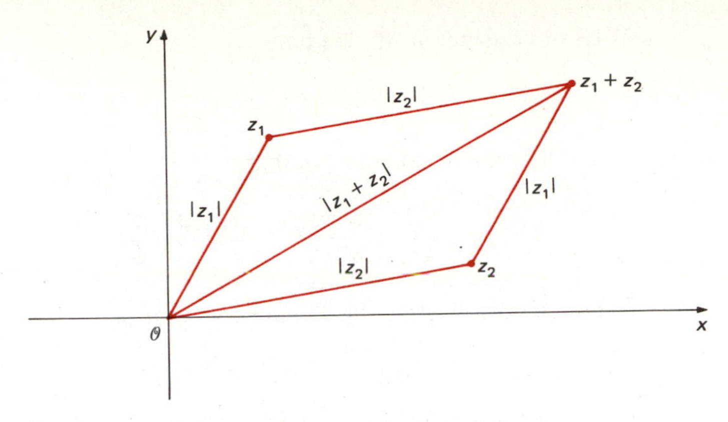

Fig. 4.10 Geometric representation of $|z_1 + z_2| \leq |z_1| + |z_2|$

$$z_1 + z_2 = [2 + (-1)] + (3 + 2)i = 1 + 5i \qquad \overline{z_1 + z_2} = 1 - 5i$$

$$z_1 - z_2 = [2 - (-1)] + (3 - 2)i = 3 + i \qquad \overline{z_1 - z_2} = 3 - i$$

$$z_1 z_2 = (2 + 3i)(-1 + 2i) = [2(-1) - 3(2)] + [2(2) + 3(-1)]i = -8 + i$$

$$\frac{z_1}{z_2} = \frac{2 + 3i}{-1 + 2i} \cdot \frac{-1 - 2i}{-1 - 2i} = \left[\frac{2(-1) + 3(2)}{(-1)^2 + 2^2}\right] + \left[\frac{2(-2) + 3(-1)}{(-1)^2 + 2^2}\right]i = \frac{4}{5} - \frac{7}{5}i$$

$$\frac{\bar{z}_1}{\bar{z}_2} = \frac{2 - 3i}{-1 - 2i} = \frac{(2 - 3i)(-1 + 2i)}{(-1 - 2i)(-1 + 2i)} = \frac{4 + 7i}{5} = \frac{4}{5} + \frac{7}{5}i$$

We now proceed to illustrate the equalities and inequalities we learned in this section. Thus, consider (4.16). Clearly,

$$\overline{(\bar{z}_1)} = \overline{2 - 3i} = 2 + 3i = z_1 \quad \text{and} \quad \overline{(\bar{z}_2)} = \overline{-1 - 2i} = -1 + 2i = z_2$$

$$\overline{z_1 + z_2} = \overline{1 + 5i} = 1 - 5i \qquad \bar{z}_1 + \bar{z}_2 = (2 - 3i) + (-1 - 2i) = 1 - 5i$$

$$\overline{z_1 - z_2} = \overline{3 + i} = 3 - i \qquad \bar{z}_1 - \bar{z}_2 = (2 - 3i) - (-1 - 2i) = 3 - i$$

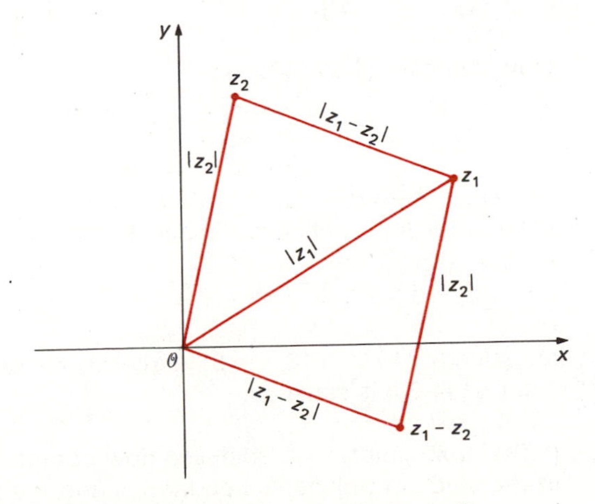

Fig. 4.11 Geometric representation of $|z_1 - z_2| \geq ||z_1| - |z_2||$

Next, consider (4.18). We have

$$\overline{z_1 z_2} = \overline{-8 + i} = -8 - i \qquad \bar{z}_1 \bar{z}_2 = (2 - 3i)(-1 - 2i) = -8 - i$$

To illustrate (4.19), observe that

$$\overline{\left(\frac{z_1}{z_2}\right)} = \overline{\frac{4}{5} - \frac{7}{5}i} = \frac{4}{5} + \frac{7}{5}i, \qquad \frac{\bar{z}_1}{\bar{z}_2} = \frac{4}{5} + \frac{7}{5}i$$

Let us now illustrate (4.30). Observe that

$$\text{Re } z_1 = 2 < |z_1| \qquad \text{Im } z_1 = 3 < |z_1| \qquad \text{since } |z_1| = \sqrt{13}$$

$$\text{Re } z_2 = -1 < |z_2| \qquad \text{Im } z_2 = 2 < |z_2| \qquad \text{since } |z_2| = \sqrt{5}$$

$$|\bar{z}_1| = |2 - 3i| = \sqrt{2^2 + (-3)^2} = \sqrt{13} = |z_1|$$

$$|\bar{z}_2| = |-1 - 2i| = \sqrt{(-1)^2 + (-2)^2} = \sqrt{5} = |z_2|$$

$$z_1 + \bar{z}_1 = (2 + 3i) + (2 - 3i) = 4 = 2 \text{ Re } z_1$$

$$z_2 + \bar{z}_2 = (-1 + 2i) + (-1 - 2i) = -2 = 2 \text{ Re } z_2$$

Next, we illustrate (4.31):

$$z_1 \bar{z}_1 = (2 + 3i)(2 - 3i) = 13 = |z_1|^2$$

$$z_2 \bar{z}_2 = (-1 + 2i)(-1 - 2i) = 5 = |z_2|^2$$

To illustrate (4.32) note that

$$|z_1 z_2| = |-8 + i| = \sqrt{(-8)^2 + 1^2} = \sqrt{65} \qquad |z_1||z_2| = \sqrt{13}\,\sqrt{5} = \sqrt{65}$$

Similarly, we illustrate (4.33). Indeed

$$\left|\frac{z_1}{z_2}\right| = \left|\frac{4}{5} - \frac{7}{5}i\right| = \sqrt{\left(\frac{4}{5}\right)^2 + \left(-\frac{7}{5}\right)^2} = \sqrt{\frac{65}{25}} = \sqrt{\frac{13}{5}} = \frac{\sqrt{13}}{\sqrt{5}} = \frac{|z_1|}{|z_2|}$$

Now consider (4.34). We have

$$|z_1 + z_2| = |1 + 5i| = \sqrt{1^2 + 5^2} = \sqrt{26} \qquad |z_1| + |z_2| = \sqrt{13} + \sqrt{5}$$

and since $\sqrt{26} < \sqrt{13} + \sqrt{5}$ (verify this by using a table of square roots), we see that (4.34) is satisfied. Finally, to illustrate (4.35), observe that

$$|z_1 - z_2| = \sqrt{3^2 + 1^2} = \sqrt{10} \qquad |z_1| - |z_2| = \sqrt{13} - \sqrt{5}$$

and since $\sqrt{10} > \sqrt{13} - \sqrt{5}$ (verify this by using a table of square roots), we see that (4.35) is satisfied.

In the next chapter, we shall see how complex numbers play an important role in the study of polynomial equations and their solution sets, whether these polynomials involve real or complex coefficients.

Problem Set 4.2

4.2.1 Let $z = -2 - 3i$. (*a*) Express the following numbers in the form $a + bi$ (a and b real): Re z, Im z, $\bar{z}$, $|z|$, $|\bar{z}|$, Re $\bar{z}$, and Im $\bar{z}$. (*b*) Represent geometrically z and all the numbers in (*a*).

4.2.2 Let z_1 and z_2 be the complex numbers given by $z_1 = -1 - 2i$ and $z_2 = 1 - 4i$. (*a*) Express the following numbers in the form $a + bi$ (a and b real): $z_1 + z_2$, $z_1 - z_2$, $\overline{z_1 + z_2}$, $\overline{z_1 - z_2}$, $|z_1 + z_2|$, and $|z_1 - z_2|$. (*b*) Represent geometrically z_1, z_2, and all the numbers in (*a*).

4.2.3 Illustrate the last two equalities in (4.16) for the complex numbers z_1 and z_2 in Prob. 4.2.2.

4.2.4 Let z_1 and z_2 be as in Prob. 4.2.2. (*a*) Express the following numbers in the form $a + bi$ (a and b real): z_1z_2 and z_1/z_2. (*b*) Illustrate (4.18) and (4.19) for the complex numbers z_1 and z_2.

4.2.5 Illustrate (4.30) and (4.31) for the complex number z given in Prob. 4.2.1.

4.2.6 Illustrate (4.32) and (4.33) for the complex numbers z_1 and z_2 given in Prob. 4.2.2. (Recall that you already evaluated z_1z_2 and z_1/z_2 in Prob. 4.2.4*a*.)

4.2.7 Illustrate (4.34) and (4.35) for the complex numbers z_1 and z_2 given in Prob. 4.2.2. (Recall that you already evaluated $|z_1 + z_2|$ and $|z_1 - z_2|$ in Prob. 4.2.2*a*.) Represent geometrically all the complex numbers involved and their absolute values.

4.2.8 Describe geometrically the following sets of complex numbers, and sketch the set in each case:

(*a*) $|z| = 1$ (*d*) $|z + 2| \leq 4$ (*g*) $1 \leq |z + 1| \leq 2$

(*b*) $|z - 1| < 1$ (*e*) $|z - 1| \geq 3$ (*h*) $1 \leq |z - 2| < 3$

(*c*) $|z + 2| > 3$ (*f*) $1 < |z - 1| < 2$ (*i*) $0 < |z + 1| \leq 1$

[*Hint*: Use (4.14) and (4.43).]

4.2.9 Express the solutions of the following equations in the form $a + bi$ (a and b real):

(*a*) $x^2 + 2 = 0$ (*c*) $2x^2 - x + 1 = 0$

(*b*) $x^2 + 2x + 2 = 0$

4.2.10 Suppose that z is a complex number such that $z^3 + 2z - 7 = 0$. Prove that $\bar{z}^3 + 2\bar{z} - 7 = 0$. [*Hint*: Use Eqs. (4.16) and (4.18).]

4.2.11 Prove that $z + \bar{z} = 2|z|$ if and only if z is a positive real number or zero.

4.2.12 Find all complex numbers z such that $1/z = \bar{z}$. Describe geometrically the set of all such solutions. [*Hint*: Use Eq. (4.31).]

4.2.13 Find all complex numbers z such that:

(*a*) $z = \bar{z}$ (*d*) $z = -|z|$

(*b*) $z = -\bar{z}$ (*e*) $z = iz$

(*c*) $z = |z|$ (*f*) $z = -iz$

4.2.14 Is it true that if z_1 and z_2 are any complex numbers such that $z_1 + iz_2 = 0$, then $z_1 = 0$ or $z_2 = 0$? Explain.

4.3 Summary

In this chapter, we have seen that the set of real numbers does not suffice for obtaining solutions of certain equations (such as $x^2 + 1 = 0$). This led naturally to the introduction of a number i [$= \sqrt{-1}$] which is not a real number, and to the extension of the set of real numbers to the set of complex numbers, that is, numbers of the form $a + bi$ (where a and b are real). We then introduced the fundamental operations of addition, subtraction, multiplication, and division for complex numbers, along with geometric interpretations of some of these operations. We also saw that every real number is at the same time a complex number and, moreover, that the complex numbers obey, in general, the same rules of high school algebra which the real numbers obey, with one notable exception: the relations of $<$ and $>$ *are not defined* for complex numbers. We then introduced the important concepts of conjugate and absolute value of a complex number and studied the properties and relationships of these concepts. We learned, for example, that $|z| \geq 0$, $z\bar{z} = |z|^2$, $\overline{(z_1 \pm z_2)} = \bar{z}_1 \pm \bar{z}_2$, $\overline{z_1 z_2} = \bar{z}_1 \bar{z}_2$, and $\overline{(z_1/z_2)} = \bar{z}_1/\bar{z}_2$. We also learned that the equalities and inequalities which we developed in Chap. 3 for the absolute values of real numbers still hold for the absolute values of complex numbers. We saw, for example, that $|z_1 + z_2| \leq |z_1| + |z_2|$, $|z_1 - z_2| \geq \big| |z_1| - |z_2| \big|$, $|z_1 z_2| = |z_1||z_2|$, and $|z_1/z_2| = |z_1|/|z_2|$. Our starting point in this direction was to first show that if z is *any* complex number, then $|z|$ is a *nonnegative real* number. In fact, $|z|$ is simply the length of the line segment joining the origin and z. More generally, $|z_1 - z_2|$ is the length of the line segment joining z_1 and z_2.

5 Polynomials

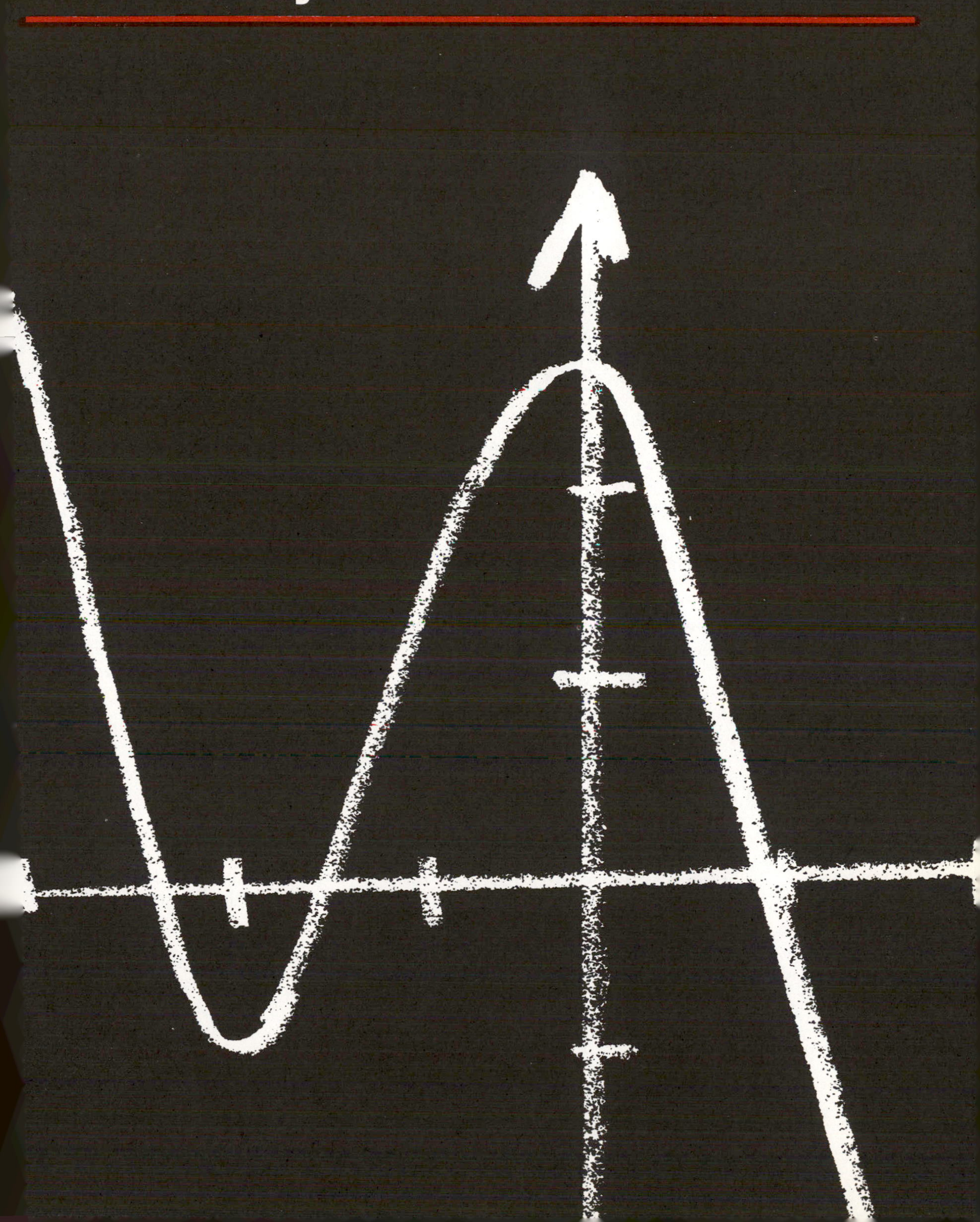

5.1 Polynomial Functions

In Chap. 2 we considered constant functions $f(x) = c$; linear functions $f(x) = ax + b$, for $a \neq 0$; and quadratic functions $f(x) = ax^2 + bx + c$, where $a \neq 0$. We now generalize this situation by introducing a concept which includes all these functions. This concept is that of a *polynomial function,* which we now proceed to define. Thus, suppose n is a nonnegative integer, and suppose that $a_0, a_1, a_2, \ldots, a_n$ are fixed real numbers (the dots stand for the omitted terms a_3, a_4, and so on). Suppose, further, $a_n \neq 0$. Let R be the set of real numbers. Now, *define* a function $f: R \to R$ as follows:

$$f(x) = a_n x^n + a_{n-1}x^{n-1} + \cdots + a_1 x + a_0 \qquad \text{for all } x \text{ in } R,\ a_n \neq 0,\ n \text{ a nonnegative integer} \qquad (5.1)$$

The function f defined in (5.1) is called a *polynomial function* of *degree n.* The real numbers $a_0, a_1, a_2, \ldots, a_n$ are called the *coefficients* of the polynomial function f, and a_n is called the *leading coefficient* of f. Each of $a_n x^n$, $a_{n-1}x^{n-1}$, $\ldots$, $a_1 x$, a_0 is called a *term*; in particular, a_0 is called the *constant term.* Finally, $f(x)$ is called a *polynomial* (expression).

So far, we have defined the degree of every polynomial function *except* the *zero* polynomial function. The zero polynomial function is defined to be the function $g: R \to R$ given by

$$g(x) = 0 \qquad \text{for } all\ x \text{ in } R$$

The degree of the zero polynomial function is not defined. However, the degree of every nonzero polynomial function f, as given in (5.1), is the nonnegative integer n in (5.1).

Now, if in (5.1) we take $n = 0$, 1, and 2, we obtain the constant function, the linear function, and the quadratic function, respectively, which we encountered earlier. The case $n = 3$ gives rise to the so-called *cubic function,* while the case $n = 4$ gives rise to the *quartic function.* We now direct our attention to recalling the four fundamental operations—addition, subtraction, multiplication, and division—on these polynomial functions.

Example 5.1 Let R be the set of real numbers, and let $f: R \to R$ and $g: R \to R$ be defined by $f(x) = 2x^3 - 7x + 2$ and $g(x) = x^2 + x - 1$. Find $f + g$, $f - g$, fg, and f/g.

Solution In Chap. 1, we defined $f + g$, $f - g$, fg, and f/g for *any* functions f and g which have the same domains and same codomains. Applying these general definitions, we get (keep in mind that the domain and codomain of every polynomial function are R)

$$\begin{aligned}(f + g)(x) = f(x) + g(x) &= (2x^3 - 7x + 2) + (x^2 + x - 1)\\ &= 2x^3 + x^2 - 6x + 1\end{aligned}$$

Therefore, $(f + g)(x) = 2x^3 + x^2 - 6x + 1$. This could be done more systematically as follows:

$$
\begin{array}{rl}
f(x) &= 2x^3 + 0x^2 - 7x + 2 \\
+ \qquad g(x) &= 0x^3 + 1x^2 + 1x - 1 \\
\hline
f(x) + g(x) &= 2x^3 + x^2 - 6x + 1
\end{array}
\qquad \text{(Add corresponding coefficients)}
$$

Thus, to add any two polynomial functions, first write the polynomials as indicated above; that is, if a term is missing, write it down with a *zero coefficient*, and then *add* the *coefficients* of *corresponding terms*. This device makes adding any two polynomial functions (regardless of what the degrees may be) very easy. We use the same device in subtraction. Thus,

$$
\begin{array}{rl}
f(x) &= 2x^3 + 0x^2 - 7x + 2 \\
- \qquad g(x) &= 0x^3 + 1x^2 + 1x - 1 \\
\hline
f(x) - g(x) &= 2x^3 - x^2 - 8x + 3
\end{array}
\qquad \text{(Subtract corresponding coefficients)}
$$

Therefore, $(f - g)(x) = 2x^3 - x^2 - 8x + 3$.

Thus, to subtract any two polynomial functions, first write the polynomials as indicated above; that is, if a term is missing, write it down with a zero coefficient, and then *subtract* the coefficients of corresponding terms.

Next, to multiply any two polynomial functions, simply *multiply each term* of *one polynomial* by *each term* of the *other* polynomial, and *add* up all these products. We can do this systematically as indicated below:

$$
\begin{array}{ll}
f(x): & 2x^3 + 0x^2 - 7x + 2 \\
g(x): & x^2 + x - 1 \\
\hline
& 2x^5 + 0x^4 - 7x^3 + 2x^2 \qquad \text{(Multiply } x^2 \text{ by top polynomial)} \\
& \qquad + 2x^4 + 0x^3 - 7x^2 + 2x \qquad \text{(Multiply } x \text{ by top polynomial)} \\
& \qquad\qquad - 2x^3 - 0x^2 + 7x - 2 \qquad \text{(Multiply } -1 \text{ by top polynomial)} \\
\hline
& 2x^5 + 2x^4 - 9x^3 - 5x^2 + 9x - 2 \qquad \text{(Add)}
\end{array}
$$

Therefore, $(fg)(x) = 2x^5 + 2x^4 - 9x^3 - 5x^2 + 9x - 2$.

Finally, to divide one polynomial function by another we apply the so-called *long-division method*, which is similar to the technique used in arithmetic. We illustrate this method below for our present polynomials $f(x)$ and $g(x)$.

$$
\begin{array}{rll}
& 2x \;\; - 2 & \\
x^2 + x - 1 \,\big) & 2x^3 + 0x^2 - 7x + 2 & \text{(Divide } 2x^3 \text{ by } x^2\text{)} \\
& 2x^3 + 2x^2 - 2x & \text{(Multiply divisor by } 2x\text{)} \\
\hline
& -2x^2 - 5x + 2 & \text{(Subtract)} \\
& & \text{(Divide } -2x^2 \text{ by } x^2\text{)} \\
& -2x^2 - 2x + 2 & \text{(Multiply divisor by } -2\text{)} \\
\hline
& -3x & \text{(Subtract)}
\end{array}
$$

Thus,

$$
\left(\frac{f}{g}\right)(x) = \frac{2x^3 - 7x + 2}{x^2 + x - 1} = 2x - 2 + \frac{-3x}{x^2 + x - 1}
$$

Observe that the sum, difference, and product of any two polynomial functions is again a polynomial function. However, the quotient f/g of two polynomial functions is *not,* in general, a polynomial function (see above example). The quotient f/g of two polynomial functions is called a *rational function* (here we assume $g \neq 0$). Rational functions will be treated in Chap. 7.

Example 5.2 Suppose that the polynomials $f(x)$ and $g(x)$ are given by $f(x) = 12x^3 + 3x^2 - 4x + 1$ and $g(x) = -4x^2 + 2x - 5$. Then $f(x)$ is a polynomial of degree 3. Its coefficients are 12, 3, −4, and 1, and its terms are $12x^3$, $3x^2$, $-4x$, and 1. The leading coefficient of the polynomial $f(x)$ is 12, and the constant term of $f(x)$ is 1. Similarly, $g(x)$ is a polynomial of degree 2. Its coefficients are −4, 2, and −5, and its terms are $-4x^2$, $2x$, and −5. The leading coefficient of the polynomial $g(x)$ is −4, and its constant term is −5.

Problem Set 5.1

5.1.1 Let R be the set of real numbers. Find the degree, the leading coefficient, and the constant term of each of the following polynomial functions:

(a) $f:R \to R \quad f(x) = 2x^5 - 3x + 4$

(b) $g:R \to R \quad g(x) = x^4 + x^2 - 2$

(c) $h:R \to R \quad h(x) = x^3 - 2x^2 + 5x$

(d) $k:R \to R \quad k(x) = 1$

5.1.2 Referring to the functions in Prob. 5.1.1, find each of the following functions: $g + f$, $f + g$, $g - h$, $h - g$, fh, hf, fg, gh, g/h, f/k, f/h. Also, find the degrees of $f + g$, $g - h$, and fh.

5.1.3 Let f and g be any two nonzero polynomial functions. Find examples to show that (a) the degree of $f \pm g \leq$ the maximum of degree of f and degree of g, assuming that $f \pm g$ is not the zero polynomial, and (b) the degree of $fg =$ the degree of $f +$ the degree of g.

5.1.4 Show that the product of two nonzero polynomial functions is a nonzero polynomial function.

5.1.5 Express the following as polynomials (if possible):

(a) $\dfrac{x^5 - 32}{x - 2}$ (assume $x \neq 2$)

(b) $\dfrac{x^5 + 32}{x + 2}$ (assume $x \neq -2$)

(c) $(x - 1)(x^{n-1} + x^{n-2} + x^{n-3} + \cdots + x + 1)$

5.2 Polynomials and Their Properties

In this section, we study some of the basic properties of polynomials such as the division algorithm and synthetic division. We also establish the Remainder and Factor theorems and discuss the applications of these theorems to finding the solutions of general polynomial equations.

In accord with Sec. 5.1, we now define a *polynomial* as an expression of the form

$$a_n x^n + a_{n-1} x^{n-1} + \cdots + a_1 x + a_0 \tag{5.2}$$

where $a_0, a_1, \ldots, a_n$ are fixed real numbers and n is a *nonnegative* integer. If in (5.2) $a_n \neq 0$, we say that the polynomial is of *degree n.* On the other hand, if all the a_i above are zero, the polynomial in (5.2) is called the *zero polynomial.* We often denote the zero polynomial by 0. We also agree that the degree of the zero polynomial is *not defined.* Observe that, by definition, every polynomial, *except the zero polynomial,* has a degree and that this degree is always a nonnegative integer. Just as in the case of a polynomial function, the real numbers $a_0, a_1, \ldots, a_n$ in (5.2) are called the *coefficients* of the polynomial. In particular, a_n is called the *leading coefficient* of the polynomial in (5.2). Moreover, each of $a_nx^n, a_{n-1}x^{n-1}, \ldots, a_1x, a_0$ is called a *term* of the polynomial in (5.2), and in particular a_0 is called the *constant term* of the polynomial.

We also saw in Sec. 5.1 that the sum, difference, and product of any two polynomials is a polynomial. However, the quotient of two polynomials is, in general, not a polynomial.

The following result, known as the *division algorithm,* is concerned with the division of two polynomials.

Division Algorithm

Suppose that $f(x)$ and $g(x)$ are any two polynomials such that $g(x) \neq 0$ [that is, $g(x)$ is not the zero polynomial]. Then there exist unique polynomials $q(x)$ and $r(x)$ such that

$$f(x) = q(x)g(x) + r(x) \qquad \text{where } r(x) = 0 \text{, or degree of } r(x) < \text{degree of } g(x) \tag{5.3}$$

The polynomial $q(x)$ is called the *quotient*, and the polynomial $r(x)$ is called the *remainder*.

Example 5.3

Find the quotient and the remainder when the polynomial $f(x)$ is divided by the polynomial $g(x)$, where $f(x) = 12x^3 + 3x^2 - 4x + 1$ and $g(x) = x^2 - 2x + 5$.

Solution

Let us use the long-division process to divide $f(x)$ by $g(x)$, as indicated below.

$$\begin{array}{r|l}
 & 12x + 27 \\ \hline
x^2 - 2x + 5 & 12x^3 + 3x^2 - 4x + 1 \\
 & 12x^3 - 24x^2 + 60x \\ \hline
 & 27x^2 - 64x + 1 \\
 & 27x^2 - 54x + 135 \\ \hline
 & -10x - 134
\end{array}$$

Hence, the quotient $q(x)$ and the remainder $r(x)$ are given by

$$q(x) = 12x + 27 \qquad r(x) = -10x - 134$$

The student should verify that Eq. (5.3) holds in this example.

Suppose once again that $f(x)$ is the polynomial given by

$$f(x) = a_nx^n + a_{n-1}x^{n-1} + \cdots + a_1x + a_0 \qquad (a_0, a_1, \ldots, a_n \text{ real}) \tag{5.4}$$

and suppose t is any real number. We define the *functional value* $f(t)$ as follows:

$$f(t) = a_n t^n + a_{n-1} t^{n-1} + \cdots + a_1 t + a_0$$

Observe that $f(t)$ is a real number. For example, if $f(x)$ is given by $f(x) = 12x^3 + 3x^2 - 4x + 1$, then

$$f(0) = 12(0)^3 + 3(0)^2 - 4(0) + 1 = 1$$

$$f(1) = 12(1)^3 + 3(1)^2 - 4(1) + 1 = 12$$

$$f(-1) = 12(-1)^3 + 3(-1)^2 - 4(-1) + 1 = -4$$

and so on. Finding functional values of a given polynomial is of considerable interest for several reasons. For example, sketching the graph of a polynomial function f involves, among other things, finding the various functional values $f(t)$, where t is a real number. Moreover, finding the solutions of an equation such as $f(x) = 0$ also involves evaluating numerous functional values $f(t)$. Furthermore, the important forthcoming Remainder and Factor theorems, too, involve the notion of functional value, and these theorems play a vital role in determining the solutions of equations involving polynomials. In view of this, we conveniently introduce a method, known as *synthetic division,* for finding functional values. Generally speaking, this method is more efficient than finding functional values by direct substitution. Before embarking on such a technique, however, we establish the Remainder theorem and its important corollary, the Factor theorem. Thus, suppose that $f(x)$ is a polynomial and c is any real number. What is the remainder when the polynomial $f(x)$ is divided by the polynomial $x - c$? The answer is given in the following theorem.

Remainder Theorem

The remainder when a polynomial $f(x)$ is divided by the polynomial $x - c$ is $f(c)$.

The proof of the Remainder theorem is quite straightforward. Thus, by the division algorithm there exist polynomials $q(x)$ and $r(x)$ such that

$$f(x) = q(x) \cdot (x - c) + r(x) \qquad (5.5)$$

Moreover, by (5.3) we know that $r(x) = 0$, or the degree of $r(x) <$ the degree of $x - c = 1$. The net result is that $r(x) = 0$ or the degree of $r(x) = 0$. In other words, $r(x)$ *is a constant polynomial,* say,

$$r(x) = k \qquad \text{(a constant)} \qquad (5.6)$$

Combining (5.5) and (5.6), we thus get

$$f(x) = q(x) \cdot (x - c) + k \qquad (5.7)$$

Hence, by substituting $x = c$ in (5.7) we get

$$f(c) = q(c) \cdot (c - c) + k = 0 + k = k$$

that is, $f(c) = k$. Therefore, (5.6) now becomes

$$\text{Remainder} = r(x) = f(c) \qquad (5.8)$$

This proves the theorem.

In view of (5.8) and (5.5), we have

$$f(x) = q(x) \cdot (x - c) + f(c) \tag{5.9}$$

which, in turn, says that when the polynomial $f(x)$ is divided by the polynomial $x - c$, then the quotient is $q(x)$, and the remainder is $f(c)$. (Here c is any real number.) Observe that the degree of $q(x)$ is one less than the degree of $f(x)$ [assuming of course that $q(x) \neq 0$].

An important case in the Remainder theorem arises when the number c happens to be such that $f(c) = 0$. This case yields the following result, known as the Factor theorem.

Factor Theorem Suppose $f(x)$ is a polynomial, and suppose c is a real number. If $f(c) = 0$, then $x - c$ is a factor of $f(x)$. Conversely, if $x - c$ is a factor of $f(x)$, then $f(c) = 0$.

This follows at once from the Remainder theorem. Indeed, by the Remainder theorem the remainder when $f(x)$ is divided by $x - c$ is $f(c)$. Hence if $f(c) = 0$, then by (5.9),

$$f(x) = q(x) \cdot (x - c) \tag{5.10}$$

that is, $x - c$ is a factor of $f(x)$. Conversely, if $x - c$ is a factor of $f(x)$, then (5.10) holds, and hence by (5.10) $f(c) = q(c) \cdot (c - c) = 0$. This proves the theorem.

The Factor theorem can also be stated in the following equivalent form.

Factor Theorem: An Alternative Form Suppose $f(x)$ is a polynomial and suppose c is a real number. Then $x - c$ is a factor of $f(x)$ if and only if $f(c) = 0$.

In the above reformulation of the Factor theorem, we encounter the phrase "if and only if." In general, suppose p and q are any statements. Then the statement

p if and only if q

means

If p then q *and* if q then p

Thus, in the Factor theorem above the last sentence means the following:

If $x - c$ is a factor of $f(x)$ then $f(c) = 0$

and

If $f(c) = 0$ then $x - c$ is a factor of $f(x)$

Example 5.4 Suppose that $f(x)$ is the polynomial given by $f(x) = 2x^3 + 3x + 5$. Find the remainder when $f(x)$ is divided by (*a*) $x - 1$, (*b*) $x + 1$, (*c*) $x - 4$, and (*d*) $x + 4$.

Solution By the Remainder theorem, we know the following things.

(*a*) The remainder when $f(x)$ is divided by $x - 1$ is

$$f(1) = 2(1)^3 + 3(1) + 5 = 10$$

(*b*) The remainder when $f(x)$ is divided by $x + 1$ $[= x - (-1)]$ is

$$f(-1) = 2(-1)^3 + 3(-1) + 5 = 0$$

(*c*) The remainder when $f(x)$ is divided by $x - 4$ is

$$f(4) = 2(4)^3 + 3(4) + 5 = 145$$

(*d*) The remainder when $f(x)$ is divided by $x + 4$ $[= x - (-4)]$ is

$$f(-4) = 2(-4)^3 + 3(-4) + 5 = -135$$

Example 5.5 Use the Factor theorem to find the factors of the polynomial $f(x)$ given by $f(x) = 2x^3 + 3x + 5$.

Solution Since, as we have seen in Example 5.4, $f(-1) = 0$, it follows by the Factor theorem that $x - (-1)$ is a factor of $f(x)$. Now, dividing $x + 1$ into $f(x)$ shows that the other factor of $f(x)$ is $2x^2 - 2x + 5$; this means that $f(x) = (x + 1) \times (2x^2 - 2x + 5)$. Since in $2x^2 - 2x + 5$, $b^2 - 4ac < 0$, $f(x)$ cannot be factored further using real coefficients only.

We have already seen in the above two examples that finding functional values of a polynomial is quite useful. We now proceed to describe a method which enables us to find such functional values with less effort (usually) than the direct-substitution method we have been using so far. This method also has the additional advantage of yielding *both* the quotient and the remainder obtained when a polynomial is divided by a linear polynomial such as $x - c$, where c is any real number.

To begin with, suppose that we take a polynomial $f(x)$ of degree 3 (say) given by

$$f(x) = a_3x^3 + a_2x^2 + a_1x + a_0 \qquad (5.11)$$

Now, suppose we divide $f(x)$ by $x - c$ using the long-division process, as indicated below. (Here c is any real number.)

$$\begin{array}{r|l}
 & b_3x^2 + b_2x \quad + b_1 \\ \hline
x - c & a_3x^3 + a_2x^2 \quad + a_1x + a_0 \\
 & b_3x^3 - b_3cx^2 \\ \hline
 & \qquad b_2x^2 \quad + a_1x \\
 & \qquad b_2x^2 \quad - b_2cx \\ \hline
 & \qquad\qquad b_1x + a_0 \\
 & \qquad\qquad b_1x - b_1c \\ \hline
 & \qquad\qquad\qquad b_0
\end{array}$$

A careful examination of the long-division process performed above shows that

$$\begin{aligned} b_3 &= a_3 \\ b_2 &= a_2 + b_3c \\ b_1 &= a_1 + b_2c \\ b_0 &= a_0 + b_1c \end{aligned} \tag{5.12}$$

The equations in (5.12) can be conveniently expressed in the following tabular form:

$$\begin{array}{cccc|l} a_3 & a_2 & a_1 & a_0 & \underline{c} \\ & b_3c & b_2c & b_1c & \\ \hline b_3 & b_2 & b_1 & b_0 & \end{array} \tag{5.13}$$

Observe that by (5.12) each of b_3, b_2, b_1, and b_0 is obtained by adding the numbers directly above it (i.e., in the same column above the horizontal line). Thus [see Eqs. (5.12) and (5.13)]

b_3 is simply a_3 itself (so b_3 is known)

b_2 is the sum of a_2 and b_3c (so b_2 is now known)

b_1 is the sum of a_1 and b_2c (so b_1 is now known)

b_0 is the sum of a_0 and b_1c (so b_0 is now known)

The above process is known as *synthetic division,* or *synthetic substitution.* Observe the following important fact, which follows from the long-division process performed above:

> The entries in the last row of (5.13), with the exception of the last entry, give the coefficients of the quotient polynomial, while the last entry gives the remainder. In other words, b_3, b_2, b_1 are the coefficients of the quotient polynomial, while b_0 is the remainder when the polynomial $f(x)$ given in (5.11) is divided by $x - c$. (5.14)

The above synthetic-division process can be performed regardless of what the degree of the polynomial $f(x)$ is. Indeed, we have the following extension of the result stated in (5.14) [for the special polynomial $f(x)$ in (5.11)]:

> When any nonconstant polynomial $f(x)$ is divided synthetically by $x - c$ (c is a real number), then the entries in the last row, with the exception of the last entry, give the coefficients of the quotient polynomial, while the last entry gives the remainder. (5.15)

Example 5.6 Use synthetic division to find the quotient and remainder when the polynomial $f(x)$ given by $f(x) = 2x^3 + 3x + 5$ is divided by $x - 4$.

Solution Before applying synthetic division, we observe that the polynomial $f(x)$ above does not involve a term in x^2. In this case, we first rewrite $f(x)$ as follows:

$$f(x) = 2x^3 + 0x^2 + 3x + 5 \tag{5.16}$$

We indicate below the computation involved in the synthetic-division process discussed above.

$$\begin{array}{llll|l} 2 & 0 & 3 & 5 & \underline{4} \\ & 2\cdot 4\ (=8) & 8\cdot 4\ (=32) & 35\cdot 4\ (=140) & \\ \hline 2 & 8 & 35 & 145 & \end{array} \tag{5.17}$$

Observe that the first row in (5.17) consists of the coefficients of $f(x)$ in (5.16) in a descending order of powers. Moreover, the first entry in the last row of (5.17) is the same as the first entry in the first row (namely, 2). Also, the first entry in the middle row is the product of the first entry in the last row (namely, 2) and 4 $[= c]$, as indicated by the arrow. Adding the two entries in the *second column above* the horizontal line, we get $0 + 8 = 8$, and this is the entry we put in the *second column below* the horizontal line. Next, we multiply the second entry in the last row (namely, 8) by 4 $[= c]$ and put this product underneath the 3, as indicated by the arrow. Then, we add the two entries in the *third column above* the horizontal line to get $3 + 32 = 35$, and this is the entry we put in the *third column below* the horizontal line. Next, we multiply this entry (namely, 35) by 4 $[= c]$ and put the product underneath the 5, as indicated by the arrow. Finally, we add the two entries in the *last column above* the horizontal line to get $5 + 140 = 145$, and this is the entry we put in the *last column below* the horizontal line.

A glance at the entries in the last row of (5.17) shows the following [see (5.15)]: The coefficients of the desired quotient are 2, 8, and 35, and hence this quotient is $2x^2 + 8x + 35$. Moreover, the remainder is 145. We have thus shown that

$$f(x) = 2x^3 + 3x + 5 = (x - 4)(2x^2 + 8x + 35) + 145$$

Observe that, as we have seen in Example 5.4*c*, the remainder $= f(4) = 145$.

Example 5.7 Use synthetic division to find the quotient and remainder when the polynomial $f(x) = 5x^3 + 2x - 7$ is divided by (*a*) $x - 2$, (*b*) $x + 3$, and (*c*) $x - 1$.

Solution By (5.15) the quotient and remainder in each case may be obtained by dividing synthetically, as indicated below (see Example 5.6).

$$\begin{array}{rrrr|l} 5 & 0 & 2 & -7 & \underline{2} \\ & 10 & 20 & 44 & \\ \hline 5 & 10 & 22 & 37 & \end{array}$$

Hence, when this $f(x)$ is divided by $x - 2$, the quotient is $5x^2 + 10x + 22$, and the remainder is 37. Thus,

$$5x^3 + 2x - 7 = (5x^2 + 10x + 22)(x - 2) + 37$$

Similarly, in dividing $f(x)$ by $x + 3$ $[= x - (-3)]$ synthetically we obtain

$$\begin{array}{rrrr|l} 5 & 0 & 2 & -7 & \underline{-3} \\ & -15 & 45 & -141 & \\ \hline 5 & -15 & 47 & -148 & \end{array}$$

Thus, when this $f(x)$ is divided by $x + 3$, the quotient is $5x^2 - 15x + 47$, and the remainder is -148. Hence,

$$5x^3 + 2x - 7 = (5x^2 - 15x + 47)(x + 3) - 148$$

Finally, if we divide $f(x)$ by $x - 1$ synthetically, we obtain

$$\begin{array}{rrrr|l} 5 & 0 & 2 & -7 & \underline{1} \\ & 5 & 5 & 7 & \\ \hline 5 & 5 & 7 & 0 & \end{array}$$

Thus, when this $f(x)$ is divided by $x - 1$, the quotient is $5x^2 + 5x + 7$, and the remainder is zero. Therefore,

$$5x^3 + 2x - 7 = (5x^2 + 5x + 7)(x - 1)$$

Now, suppose that $f(x)$ is a polynomial, and suppose that a is a complex number (real or imaginary) such that $f(a) = 0$. In this case, we say that a is a *zero* of the *polynomial function* f or that a is a *zero* of the *polynomial* $f(x)$. We also say that a is a *root* (or *solution*) of the polynomial equation $f(x) = 0$. (Strictly speaking, if a is imaginary, then we should first extend the domain of f to the set C of all *complex* numbers. For a detailed discussion of such an extension, see Sec. 6.3.)

Let us find all zeros of polynomials of degrees 0, 1, or 2. To begin with, if f is a polynomial function of degree 0, then $f(x) = c$, where c is a constant, and, moreover, $c \neq 0$ since the zero polynomial has no degree. Thus, f has no zeros in this case (since $c \neq 0$); that is, a polynomial function of *degree* 0 has *no zeros* whatsoever.

Next, consider a polynomial function f of degree 1 (i.e., linear). Then,

$$f(x) = a_1x + a_0 \qquad a_1, a_0 \text{ real; } a_1 \neq 0$$

In this case, it is easy to see that the equation $f(x) = 0$ has exactly one solution, namely, $x = -a_0/a_1$. Thus, a polynomial function f of *degree 1* has exactly *one zero*.

Next, suppose that f is a polynomial function of degree 2 (i.e., quadratic). Then,

$$f(x) = a_2x^2 + a_1x + a_0 \qquad a_2, a_1, a_0 \text{ real; } a_2 \neq 0$$

By the quadratic formula, the equation $f(x) = 0$ has exactly two solutions, given by

$$x = \frac{-a_1 \pm \sqrt{a_1^2 - 4a_2a_0}}{2a_2}$$

and hence a polynomial function f of degree 2 has exactly *two zeros* (which may happen to be equal). These examples are merely special cases of a very general result which we shall deal with in the next chapter.

In general, to find the zeros of a polynomial $f(x)$, try to *guess* one of the zeros. You can, of course, tell whether your guess, say $x = a$, is correct by evaluating $f(a)$ using synthetic division. If $f(a) = 0$, then a is indeed a zero of $f(x)$, which means by the Factor theorem that $x - a$ is a factor of $f(x)$; that is,

$f(x) = (x - a)g(x)$ for some polynomial $g(x)$ of degree one less than the degree of $f(x)$. You can now try to *guess* a zero of $g(x)$ [and, hence, also a zero of $f(x)$]. If you are successful again in guessing such a zero, you can repeat the above process to $g(x)$; etc. We illustrate this technique in the following example.

Example 5.8 Find all the zeros of the polynomial function f given by

$$f(x) = x^4 + x^3 - 13x^2 - x + 12 \qquad (5.18)$$

Solution Let us try 1 as our first guess for a possible zero of $f(x)$ above. We indicate below the computation involved in finding the quotient and remainder when $f(x)$ above is divided synthetically by $x - 1$.

$$\begin{array}{rrrrrl} 1 & 1 & -13 & -1 & 12 & \underline{|1} \\ & 1 & 2 & -11 & -12 & \\ \hline 1 & 2 & -11 & -12 & 0\ [= f(1)] & \end{array} \qquad (5.19)$$

Thus, $f(1) = 0$, and hence 1 is indeed a zero of $f(x)$. Therefore, by the Factor theorem $x - 1$ is a factor of $f(x)$, and, moreover, the first four numbers in the last row in (5.19) give us the coefficients of the quotient when $f(x)$ is divided by $x - 1$ [see (5.15)]. That is,

$$f(x) = (x - 1)(x^3 + 2x^2 - 11x - 12) \qquad (5.20)$$

Next we try 1, 2, and 3 as possible zeros of the cubic polynomial on the right side of (5.20). Thus, suppose

$$g(x) = x^3 + 2x^2 - 11x - 12 \qquad (5.21)$$

and divide synthetically $g(x)$ by $x - 1$, $x - 2$, and $x - 3$, respectively, as indicated below.

$$\begin{array}{rrrrl} 1 & 2 & -11 & -12 & \underline{|1} \\ & 1 & 3 & -8 & \\ \hline 1 & 3 & -8 & -20\ [= g(1)] & \\ \\ 1 & 2 & -11 & -12 & \underline{|2} \\ & 2 & 8 & -6 & \\ \hline 1 & 4 & -3 & -18\ [= g(2)] & \\ \\ 1 & 2 & -11 & -12 & \underline{|3} \\ & 3 & 15 & 12 & \\ \hline 1 & 5 & 4 & 0\ [= g(3)] & \end{array} \qquad (5.22)$$

Thus, $g(3) = 0$. Hence, 3 is a zero of $g(x)$, and therefore 3 is also a zero of $f(x)$ [see (5.20) and (5.21)]. Hence, by the Factor theorem $x - 3$ is a factor of $g(x)$, and moreover by (5.15) the quotient when $g(x)$ is divided by $x - 3$, is $x^2 + 5x + 4$ [see (5.22)]. Thus,

$$g(x) = (x - 3)(x^2 + 5x + 4) \qquad (5.23)$$

Now, combining (5.20), (5.21), and (5.23) yields

$$f(x) = (x - 1)(x - 3)(x^2 + 5x + 4) \qquad (5.24)$$

Moreover, since $x^2 + 5x + 4 = (x + 1)(x + 4)$, (5.24) now becomes

$$f(x) = (x - 1)(x - 3)(x + 1)(x + 4) \tag{5.25}$$

In view of (5.25) the zeros of $f(x)$ [that is, the solutions of $f(x) = 0$] are clearly given by $x = 1, 3, -1$, and -4.

In the next chapter we shall learn of methods which enable us to make more intelligent guesses for the zeros of a polynomial function. Indeed, in the above example, such intelligent guesses are all factors of 12.

Problem Set 5.2

5.2.1 Suppose that the polynomials $f(x)$ and $g(x)$ are given by $f(x) = 2x^4 - x^3 + 3x^2 + 5x - 4$ and $g(x) = x^3 - 2x^2 + x + 1$. Find $f(x) + g(x)$, $f(x) - g(x)$, $g(x) - f(x)$, $f(x)g(x)$, and $f(x)/g(x)$. Which of these are polynomials?

5.2.2 Find the degree and the constant term of each of the polynomials obtained in Prob. 5.2.1 [including $f(x)$ and $g(x)$].

5.2.3 Suppose that $f(x)$ and $g(x)$ are the polynomials given in Prob. 5.2.1. Evaluate $f(0)$, $g(0)$, $f(1)$, $g(-1)$, $f(g(2))$, and $g(f(2))$.

5.2.4 Suppose that the polynomial $f(x)$ is given by $f(x) = x^4 + 2x^3 - 13x^2 - 14x + 24$. Use the Remainder theorem to find the remainder when $f(x)$ is divided by $x - 1$, $x - 2$, $x - 3$, $x - 4$, $x + 1$, $x + 2$, $x + 3$, and $x + 4$.

5.2.5 Use the Factor theorem, together with the results you obtained in Prob. 5.2.4, to factor the polynomial $f(x)$ in Prob. 5.2.4 as a product of linear polynomials.

5.2.6 Use synthetic division to find the quotient and remainder when the polynomial $f(x) = x^4 + 2x^3 - 13x^2 - 14x + 24$ is divided by $x - 1$, $x - 2$, $x - 3$, $x - 4$, $x + 1$, $x + 2$, $x + 3$, and $x + 4$. Compare the remainders you get with those you obtained in Prob. 5.2.4.

5.2.7 Suppose that the polynomial $f(x)$ is given by $f(x) = x^4 - 2x^3 - 16x^2 + 2x + 15$. Use synthetic division to find $f(1)$, $f(2)$, $f(3)$, $f(4)$, $f(5)$, $f(-1)$, $f(-2)$, and $f(-3)$.

5.2.8 Use the results of the synthetic divisions in Prob. 5.2.7 to find the quotients when the polynomial $f(x)$ in Prob. 5.2.7 is divided by $x - 1$, $x - 2$, $x - 3$, $x - 4$, $x - 5$, $x + 1$, $x + 2$, and $x + 3$.

5.2.9 Use the results of Prob. 5.2.7 and the Factor theorem to find all the zeros of the polynomial $f(x)$ in Prob. 5.2.7. Are all these zeros real? How many zeros does the polynomial $f(x)$ have?

5.2.10 Show that the product of all the zeros of the polynomial $f(x)$ in Prob. 5.2.7 is equal to the constant term of $f(x)$.

5.2.11 Use synthetic division and the Factor theorem to find all zeros of the polynomial $f(x)$ given by $f(x) = x^3 + 5x^2 - 2x - 24$.

5.2.12 Show that the product of all the zeros of the polynomial $f(x)$ in Prob. 5.2.11 is equal to the negative of the constant term of $f(x)$.

5.2.13 Suppose a is a given constant and n is a positive integer. Use synthetic division to show that

$$x^n - a^n = (x - a)(x^{n-1} + ax^{n-2} + a^2x^{n-3} + \cdots + a^{n-2}x + a^{n-1})$$

5.2.14 Suppose a is a given constant and n is a positive odd integer. Use synthetic division to show that $x^n + a^n = (x + a)(x^{n-1} - ax^{n-2} + a^2x^{n-3} - a^3x^{n-4} + \cdots + a^{n-1})$.

5.2.15 Suppose that $f(x)$ is the polynomial given by $f(x) = a_nx^n + a_{n-1}x^{n-1} + \cdots + a_1x + a_0$ (where $a_n \neq 0$). Prove that (*a*) $f(0)$ is equal to the constant term of $f(x)$ and (*b*) $f(1)$ is equal to the sum of all the coefficients of $f(x)$.

5.2.16 Use the Factor theorem to find a polynomial of degree 3 with zeros 1, 2, and 3.

5.2.17 Suppose that the polynomial $f(x)$ is given by $f(x) = ax + b$. Suppose, further, that $f(1) = 0$ and $f(0) = 1$. Find a and b.

5.2.18 Suppose that the polynomial $f(x)$ is given by $f(x) = ax^2 + bx + c$. Suppose, further, that $f(0) = 0$, $f(1) = 2$, and $f(4) = 12$. Find a, b, and c.

5.2.19 Suppose that the polynomial $f(x)$ is given by $f(x) = 2x^3 - kx + 1$. Find k so that $f(1) = 0$.

5.2.20 Suppose that $f(x) = -2x^3 + kx + 1$, and suppose $f(-1) = 0$. Find k and, then find all the zeros of the polynomial $f(x)$. (*Hint*: Use the Factor theorem.)

5.3 Summary

In this chapter we studied properties of polynomial functions. A real polynomial function of degree n is a function $f: R \to R$ given by

$$f(x) = a_nx^n + a_{n-1}x^{n-1} + \cdots + a_1x + a_0$$

where $a_0, a_1, \ldots, a_n$ are fixed real numbers with $a_n \neq 0$. Here R is the set of real numbers, and x is in R. Real polynomial functions have the following properties:

1. *Division algorithm*. Suppose $f(x)$ and $g(x)$ are polynomials with $g(x) \neq 0$. Then there exist unique polynomials $q(x)$ and $r(x)$, called the quotient and remainder, respectively, such that

$$f(x) = q(x)g(x) + r(x)$$

where $r(x) = 0$, or the degree of $r(x) <$ the degree of $g(x)$.

2. *Remainder theorem*. The remainder when a polynomial $f(x)$ is divided by $x - a$ is $f(a)$, where a is any real number.

3. *Factor theorem*. If a is any real number, then $x - a$ divides a polynomial $f(x)$ if and only if $f(a) = 0$.

We also saw that synthetic division is very useful in computing functional values.

6 Zeros of Polynomial Functions

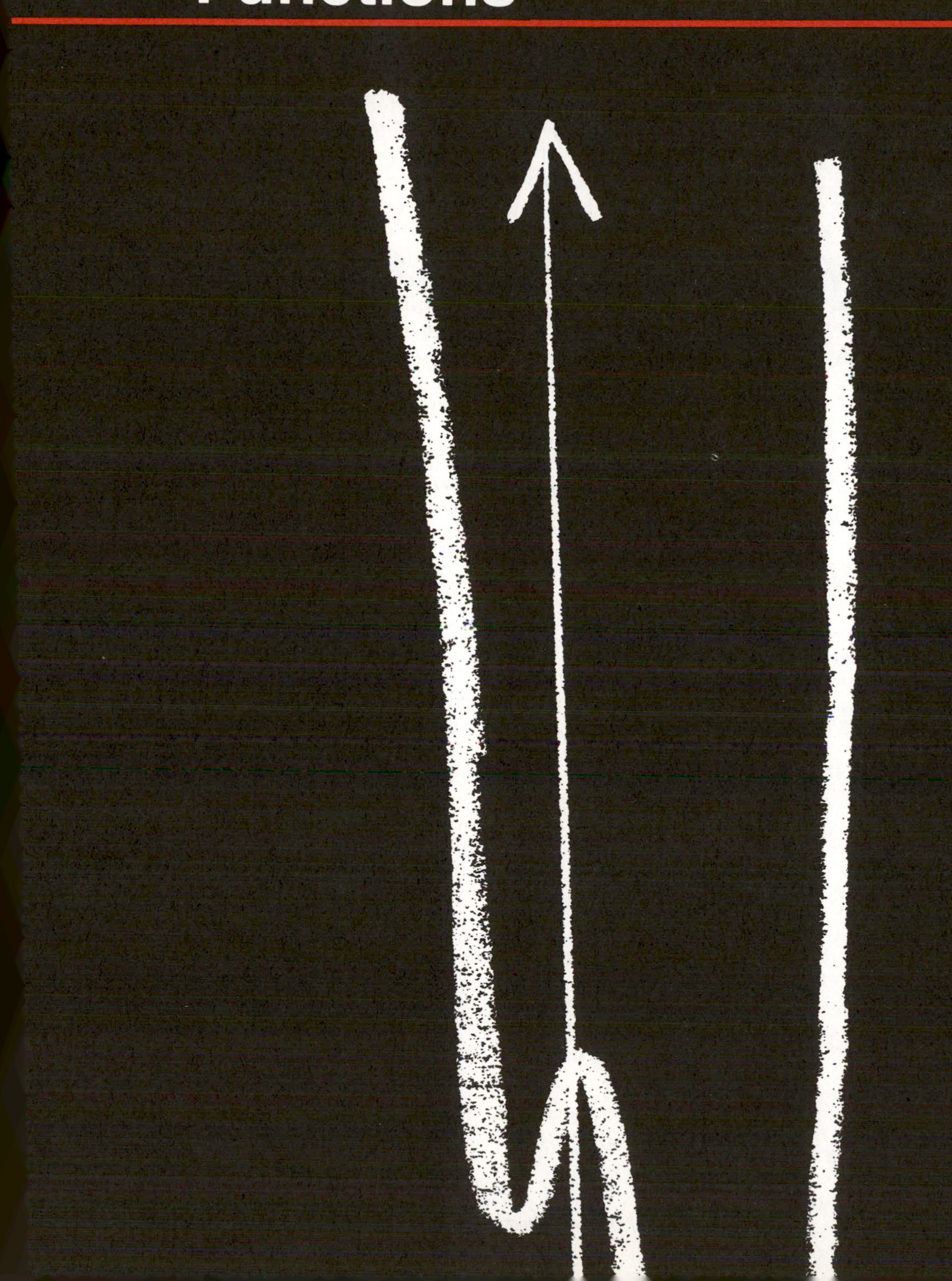

6.1 Techniques for Finding Zeros of Polynomials

In this section, we discuss some of the basic techniques for locating the zeros of a given polynomial with real coefficients. (For a definition of this concept, see Sec. 5.2.) In view of the Factor theorem, we know that if a is a zero of a polynomial $f(x)$, then $x - a$ is a factor of the polynomial $f(x)$; that is, $f(x) = (x - a)g(x)$ for some polynomial $g(x)$ of degree one less than the degree of $f(x)$. We can, of course, attempt to continue in this fashion. Thus, if we are successful in finding a zero of $g(x)$, say b, then by the Factor theorem again $x - b$ is a factor of $g(x)$; that is, $g(x) = (x - b)h(x)$ for some polynomial $h(x)$ of degree one less than the degree of $g(x)$. Hence $f(x) = (x - a)(x - b)h(x)$. We can now attempt to find a zero of $h(x)$, and if we are successful in finding such a zero, we can again apply the Factor theorem and continue as indicated above. Now, the following question naturally suggests itself: How do we go about guessing a zero of a polynomial in an intelligent way (rather than guessing at random)? In this section, we give some general techniques to help us guess zeros of polynomials intelligently. Our first result in this direction, known as the Location theorem, is quite intuitive and plausible.

Location Theorem

Suppose that $f(x)$ is a polynomial with real coefficients, and suppose that a and b are real numbers. If $f(a)$ and $f(b)$ have opposite signs, then $f(x)$ has at least one real zero between a and b.

We shall not attempt to prove the Location theorem. Instead, we give the following plausibility argument. Suppose that x_1 and x_2 are real numbers which are nearly equal. Then, intuitively, we see that $f(x_1)$ and $f(x_2)$ are very close in value. In other words, if x_1 and x_2 are sufficiently close to each other, so are $f(x_1)$ and $f(x_2)$. Geometrically speaking, this amounts to saying that the graph of the function f [or equation $y = f(x)$] *has no jumps*. (Technically, we describe this by saying that a polynomial function is *continuous*.) In view of this fact, if $f(a)$ and $f(b)$ have opposite signs, then somewhere between a and b, $f(x)$ must assume a zero value. For, otherwise, the graph of $y = f(x)$ would have to have a jump since both of the points $(a,f(a))$ and $(b,f(b))$ lie on this graph, and $f(a)$ and $f(b)$ have opposite signs.

As an illustration of the Location theorem, suppose that $f(x) = x^2 - 5x + 6$. The graph of this function is sketched in Fig. 6.1. Observe that $f(1) = 2 > 0$, $f(5/2) = -1/4 < 0$, and $f(4) = 2 > 0$. Moreover, $f(x)$ has a zero between 1 and $5/2$ (namely, 2) and another zero between $5/2$ and 4 (namely, 3).

Example 6.1

Use the Location theorem to locate the zeros of the polynomial

$$f(x) = x^3 - 12x + 1 \qquad (6.$$

between consecutive integers.

Solution

In the following list are some of the functional values of the function f: $f(4) = 17$, $f(3) = -8$, $f(2) = -15$, $f(1) = -10$, $f(0) = 1$, $f(-1) = 12$, $f(-2) = 17$, $f(-3)$

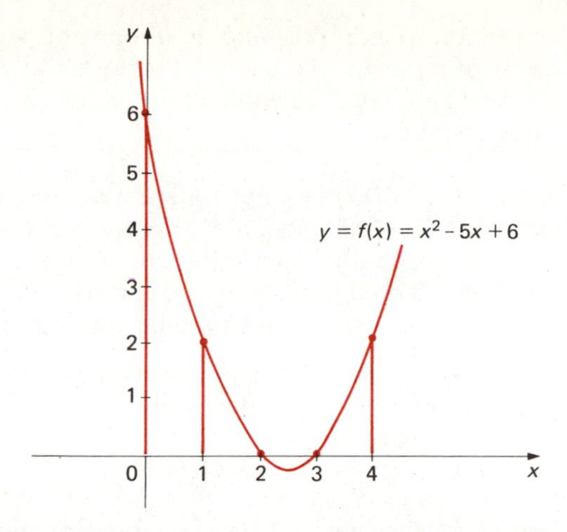

Fig. 6.1 Illustration of what can happen when $f(a)$ and $f(b)$ have the same sign. (See Remark below.)

$= 10$, and $f(-4) = -15$. Now, since $f(4) > 0$ and $f(3) < 0$, it follows by the Location theorem that the polynomial $f(x)$ has a zero between 3 and 4. Similarly, since $f(1) < 0$ and $f(0) > 0$, the Location theorem tells us that $f(x)$ has a zero between 0 and 1. Finally, since $f(-3) > 0$ and $f(-4) < 0$, it follows, using the Location theorem again, that $f(x)$ has a zero between -4 and -3. Later in this chapter, we will show that a polynomial such as $f(x)$ above has only three zeros, and hence the three zeros located above are *all* the zeros of $f(x)$. Also, we shall show later that the three zeros of $f(x)$ described above can be approximated to any desired degree of accuracy.

emark

It is noteworthy to observe that if in the Location theorem $f(a)$ and $f(b)$ have the *same* sign, then $f(x)$ may or may not have a zero between a and b, and thus *no* conclusion can be drawn one way or another regarding the existence of a zero of $f(x)$ between a and b. For example, consider the polynomial $f(x)$ given by $f(x) = x^2 - 5x + 6$ (see Fig. 6.1). We readily verify that

$$f(0) = 6 \qquad f(1) = 2 \qquad f(2) = 0 \qquad f(3) = 0 \qquad f(4) = 2 \tag{6.2}$$

Moreover, the *polynomial* $f(x)$ has *only* two zeros (being of degree 2), and these two zeros are, in fact, 2 and 3. Now, observe that $f(0)$ and $f(1)$ have the same sign but $f(x)$ has no zeros between 0 and 1. On the other hand, $f(1)$ and $f(4)$ have the same sign also. This time, however, $f(x)$ has two zeros between 1 and 4. Accordingly, when $f(a)$ and $f(b)$ have the *same* sign, we cannot be certain that $f(x)$ has a zero between a and b (see Fig. 6.1).

It is, in general, very useful if we can find an upper bound and a lower bound for the real zeros of a polynomial function with real coefficients, and our next theorem is a result in this direction. To begin with, we say that a real number c is an *upper bound* for the real zeros of a polynomial $f(x)$ if *all* the real zeros of $f(x)$ are equal to or less than c. Similarly, we say that a real number d is a *lower bound* for the real zeros of $f(x)$ if *all* the real zeros of $f(x)$ are equal to or greater than d. Knowing an upper bound and a lower bound for the real zeros of a polynomial usually shortens the computation involved in determining the real zeros of a polynomial function. The next theorem is concerned with upper bounds.

Upper Bound Theorem

Suppose that $f(x)$ is a nonconstant polynomial with real coefficients, and suppose c is a nonnegative real number. Suppose that when $x - c$ is divided into $f(x)$, all the coefficients of the quotient $q(x)$ are nonnegative and $f(c)$ is nonnegative. Then c is an upper bound for the real zeros of $f(x)$. That is, all the real zeros of $f(x)$ are less than or equal to c.

The proof of the above theorem is as follows. First, as we have seen in the proof of the Remainder theorem,

$$f(x) = (x - c)q(x) + f(c) \tag{6.3}$$

Now, suppose that x_0 is any real number such that $x_0 > c$. Then $x_0 - c > 0$. Moreover, since $f(x)$ is *not* a constant polynomial, Eq. (6.3) implies that $q(x)$ is *not* the zero polynomial. Hence at least one of the coefficients of $q(x)$ is *positive*, while the remaining coefficients of $q(x)$ are all nonnegative. Moreover, $x_0 > 0$ (since $x_0 > c \geq 0$), and hence $q(x_0) > 0$. We thus have

$$x_0 - c > 0 \qquad q(x_0) > 0 \qquad \text{and} \qquad f(c) \geq 0 \tag{6.4}$$

Combining (6.3) and (6.4), we obtain $f(x_0) = (x_0 - c)q(x_0) + f(c) > 0$; that is, $f(x_0) > 0$; hence x_0 is *not* a zero of $f(x)$ for *any* number $x_0 > c$. Thus, *all* the real zeros of $f(x)$ are less than or equal to c, and the theorem is proved.

How do we go about finding a lower bound for the real zeros of the polynomial $f(x)$? For this purpose, let us consider the polynomial $f(-x)$ instead of $f(x)$. Let $t_1, t_2, \ldots, t_k$ be all the real zeros of the polynomial $f(-x)$, and suppose c is an upper bound for these real zeros of $f(-x)$; that is,

$$t_i \leq c \qquad \text{for all } i = 1, 2, \ldots, k \tag{6.5}$$

Then, by multiplying (6.5) by the *negative* number -1 we get

$$-t_i \geq -c \qquad \text{for all } i = 1, 2, \ldots, k \tag{6.6}$$

But the real zeros of the polynomial $f(x)$ are simply the negatives of the real zeros of $f(-x)$, and hence the real zeros of $f(x)$ are precisely $-t_1, -t_2, \ldots, -t_k$. Thus, (6.6) now tells us that $-c$ is a *lower bound* for the real zeros of the polynomial $f(x)$. We have thus shown the following.

> Suppose $f(x)$ is a nonconstant polynomial with real coefficients. If c is an upper bound for the real zeros of the polynomial $f(-x)$, then $-c$ is a lower bound for the real zeros of the polynomial $f(x)$. (6.7)

In view of (6.7), then, the problem of finding a lower bound for the real zeros of $f(x)$ boils down, essentially, to finding an upper bound for the real zeros of $f(-x)$. We illustrate this in the following example.

Example 6.2 Find an upper bound and a lower bound for the zeros of the polynomial

$$f(x) = 2x^3 - 5x^2 - x + 7 \tag{6.8}$$

Solution For convenience, we use synthetic division in order to find the quotients and remainders when $f(x)$ is divided by $x - c$, where c has the various values indicated below.

2	−5	−1	7	⌊1
	2	−3	−4	
2	−3	−4	3	

2	−5	−1	7	⌊3
	6	3	6	
2	1	2	13	

2	−5	−1	7	⌊2
	4	−2	−6	
2	−1	−3	1	

We observe that when $x - 3$ is divided into the polynomial $f(x)$ in (6.8), all coefficients of the quotient, as well as the remainder, are positive. Hence, by the Upper Bound theorem 3 is an upper bound for the real zeros of the polynomial $f(x)$. In order to find a lower bound for the real zeros of the polynomial $f(x)$, we first try to find an upper bound for the real zeros of the polynomial $f(-x)$. In view of (6.8), we know that $f(-x) = 2(-x)^3 - 5(-x)^2 - (-x) + 7$, which means

$$f(-x) = -2x^3 - 5x^2 + x + 7 \tag{6.9}$$

We now try to find an upper bound for the zeros of the polynomial $f(-x)$. Since the zeros of $f(-x)$ are simply the roots (or solutions) of the equation $f(-x) = 0$ and since this equation is equivalent to the equation $-f(-x) = 0$, we conclude that the zeros of $f(-x)$ are *precisely the same* as the zeros of $-f(-x)$. By (6.9)

$$-f(-x) = 2x^3 + 5x^2 - x - 7 \tag{6.10}$$

Now, to find a nonnegative upper bound for the real zeros of the polynomial in (6.10), we use synthetic division again as indicated below.

2	5	−1	−7	⌊1
	2	7	6	
2	7	6	−1	

2	5	−1	−7	⌊2
	4	18	34	
2	9	17	27	

We see that when $x - 2$ is divided into $-f(-x)$ in (6.10), all the coefficients of the quotient, as well as the remainder, are positive. Hence 2 is an upper bound for the real zeros of $-f(-x)$, and thus 2 is also an upper bound for the real zeros of $f(-x)$. Therefore, by (6.7) -2 is a lower bound for the real zeros of the polynomial $f(x)$ in (6.8). Thus, all the real zeros of the polynomial $f(x)$ in (6.8) lie between -2 and 3.

We now direct our attention to the study of rational zeros of a given polynomial function. First, we recall that a *rational number* is simply a number of the form r/s, where r and s are integers and $s \neq 0$. In other words, a rational number is simply a fraction. The following result is extremely useful and is concerned with finding all rational zeros of a polynomial function.

Rational Zeros of a Polynomial Function Theorem

Suppose that $f(x)$ is the polynomial given by

$$f(x) = a_n x^n + a_{n-1} x^{n-1} + \cdots + a_1 x + a_0 \quad (6.11)$$

where all the a_i's are integers, $a_n \neq 0$, and $a_0 \neq 0$. Suppose that the rational number r/s, where r and s have no proper factors in common, is a zero of $f(x)$. Then r divides a_0, and s divides a_n.

The proof of the above theorem is as follows. Since by hypothesis r/s is a root (or solution) of the equation $f(x) = 0$, we have $f(r/s) = 0$, and, hence, by (6.11)

$$0 = f\left(\frac{r}{s}\right) = a_n \left(\frac{r}{s}\right)^n + a_{n-1}\left(\frac{r}{s}\right)^{n-1} + \cdots + a_1\left(\frac{r}{s}\right) + a_0$$

Multiplying this equation by s^n, we obtain

$$0 = a_n r^n + a_{n-1} r^{n-1} s + \cdots + a_1 r s^{n-1} + a_0 s^n \quad (6.12)$$

and, hence, by high school algebra

$$a_0 s^n = -a_n r^n - a_{n-1} r^{n-1} s - \cdots - a_1 r s^{n-1} \quad (6.13)$$

Now, r clearly divides the right side of (6.13); therefore r divides $a_0 s^n$; but by hypothesis r and s have no proper factors in common; therefore r must divide a_0 (since *none* of the prime factors of r is a factor of s, and thus all the prime factors of r are also factors of a_0).

To prove that s divides a_n, we solve for the term $a_n r^n$ in (6.12). Indeed, by high school algebra we have

$$a_n r^n = -a_{n-1} r^{n-1} s - \cdots - a_1 r s^{n-1} - a_0 s^n \quad (6.14)$$

Now, s certainly divides the right side of (6.14), and hence s divides $a_n r^n$. By hypothesis, s and r have no proper factors in common, which means that s divides a_n. This proves the theorem.

Remark

If in (6.11) $a_0 = 0$, then $x = 0$ is already a rational (in fact, integral) zero of $f(x)$. In this situation, we first factor x out; that is, write $f(x) = xg(x)$, and apply the Rational Zeros theorem to the polynomial $g(x)$.

Example 6.3

Find all zeros of the polynomial

$$f(x) = 6x^4 - 25x^3 - 21x^2 + 50x + 18 \quad (6.15)$$

Solution

By the Rational Zeros theorem we know that all possible rational zeros of the polynomial $f(x)$ in (6.15) are of the form r/s, where r divides 18 and s divides 6. Thus

$$\begin{array}{cccc}
\pm 1/1 & \pm 1/2 & \pm 1/3 & \pm 1/6 \\
\pm 2/1 & \cancel{\pm 2/2} & \pm 2/3 & \cancel{\pm 2/6} \\
\pm 3/1 & \pm 3/2 & \cancel{\pm 3/3} & \cancel{\pm 3/6} \\
\pm 6/1 & \cancel{\pm 6/2} & \cancel{\pm 6/3} & \cancel{\pm 6/6} \\
\pm 9/1 & \pm 9/2 & \cancel{\pm 9/3} & \cancel{\pm 9/6} \\
\pm 18/1 & \cancel{\pm 18/2} & \cancel{\pm 18/3} & \cancel{\pm 18/6}
\end{array} \tag{6.16}$$

is the totality of all possible rational zeros of $f(x)$. Observe that the rational numbers in (6.16) which we crossed out have already appeared earlier in this list. Now, it is easily seen that $f(0) = 18 > 0$ and $f(-1) = -22 < 0$, and hence by the Location theorem $f(x)$ has a zero between -1 and 0. Let us, then, test one of the rational numbers in (6.16) which lies between -1 and 0, say $-1/2$, as a possible zero of $f(x)$. Using synthetic division, we easily verify the following computation:

$$\begin{array}{rrrrr|l}
6 & -25 & -21 & 50 & 18 & \underline{-1/2} \\
 & -3 & 14 & 7/2 & -107/4 & \\
\hline
6 & -28 & -7 & 107/2 & -35/4 &
\end{array}$$

Hence $f(-1/2) = -35/4 < 0$, and with $f(0) > 0$ we know by the Location theorem that $f(x)$ has a zero between $-1/2$ and 0. This suggests that we test $-1/3$, which also appears in (6.16), as a possible zero of $f(x)$. Again, using synthetic division we obtain

$$\begin{array}{rrrrr|l}
6 & -25 & -21 & 50 & 18 & \underline{-1/3} \\
 & -2 & 9 & 4 & -18 & \\
\hline
6 & -27 & -12 & 54 & 0 &
\end{array} \tag{6.17}$$

Hence $f(-1/3) = 0$, and, thus, by the Factor theorem $x - (-1/3)$ $[= x + 1/3]$ is a factor of $f(x)$. Moreover, by (6.17) we know that the quotient when $x + 1/3$ is divided into $f(x)$ is $6x^3 - 27x^2 - 12x + 54 = 3(2x^3 - 9x^2 - 4x + 18)$; hence

$$f(x) = 3(x + 1/3)(2x^3 - 9x^2 - 4x + 18) \tag{6.18}$$

In view of (6.18) the zeros of $f(x)$ are just $-1/3$ together with all the zeros of

$$g(x) = 2x^3 - 9x^2 - 4x + 18 \tag{6.19}$$

Moreover, all the possible rational zeros of $g(x)$ are by the Rational Zeros theorem

$$\begin{array}{cccccc}
\pm 1/1 & \pm 2/1 & \pm 3/1 & \pm 6/1 & \pm 9/1 & \pm 18/1 \\
\pm 1/2 & \cancel{\pm 2/2} & \pm 3/2 & \cancel{\pm 6/2} & \pm 9/2 & \cancel{\pm 18/2}
\end{array} \tag{6.20}$$

[We crossed out repetitions in (6.20).] Also, observe that $g(4) = -14 < 0$ and $g(5) = 23 > 0$, and, hence, by the Location theorem $g(x)$ [and thus $f(x)$] has a zero between 4 and 5. Now, the *only* number in (6.20) which lies between 4 and 5 is $9/2$; so let us test $9/2$ as a possible zero of $g(x)$ using synthetic division, as indicated below.

$$\begin{array}{rrrr|l} 2 & -9 & -4 & 18 & \underline{9/2} \\ & 9 & 0 & -18 & \\ \hline 2 & 0 & -4 & 0 & \end{array} \tag{6.21}$$

Hence $g(9/2) = 0$, and thus $9/2$ is indeed a zero of $g(x)$ [and therefore a zero of $f(x)$ also, see (6.18) and (6.19)]. By the Factor theorem, $(x - 9/2)$ is thus a factor of $g(x)$, and, moreover, by (6.21) we have $g(x) = (x - 9/2)(2x^2 + 0x - 4) = 2(x - 9/2) \times (x^2 - 2)$, that is,

$$g(x) = 2(x - 9/2)(x^2 - 2) \tag{6.22}$$

Combining (6.18), (6.19), and (6.22), we obtain

$$f(x) = 6(x + 1/3)(x - 9/2)(x^2 - 2) \tag{6.23}$$

Hence the zeros of $f(x)$ are $-1/3$, $9/2$, $\sqrt{2}$, and $-\sqrt{2}$.

Remark

In the last example, we found that one of the zeros of the polynomial $f(x)$ is $\sqrt{2}$. We claim that this zero of $f(x)$ is *not* a rational number. To prove this, consider the zeros of the polynomial

$$h(x) = x^2 - 2 \tag{6.24}$$

By the Rational Zeros theorem, the only *possible* rational zeros of $h(x)$ are $\pm 1/1$ and $\pm 2/1$. It is easy to check that none of these numbers is a zero of $h(x)$, which means that $h(x)$ has no rational zeros. Since $\sqrt{2}$ is a zero of $h(x)$, it follows that $\sqrt{2}$ is not rational.

The student is probably familiar with the fact that every *irrational* number can be approximated to any desired degree of accuracy by a rational number. Later in this chapter we shall see how the Location theorem can be used to obtain rational approximations of irrational zeros of a polynomial function.

Returning to the Rational Zeros theorem, we now single out an extremely interesting special case, namely, the case in which the leading coefficient a_n of the polynomial $f(x)$ in (6.11) is 1. In this case, we have the following important corollary.

Corollary

Suppose that

$$f(x) = x^n + a_{n-1}x^{n-1} + \cdots + a_1x + a_0 \tag{6.25}$$

is a polynomial with integer coefficients, and suppose $a_0 \neq 0$. Then all the rational zeros of $f(x)$ (if any) are integers which divide a_0.

This follows at once from the Rational Zeros theorem. Indeed, if r/s is *any* rational zero of $f(x)$, then r divides a_0, and s divides 1 (since the leading coefficient a_n is now equal to 1). But the only integers s which divide 1 are ± 1, and hence $s = \pm 1$. Thus *any* rational zero of $f(x)$ is of the form $r/(\pm 1) = \pm r$, which, of course, is an integer. Moreover, since r divides a_0, it follows that *any* rational zero of $f(x)$ divides a_0.

Remark 1 If $a_0 = 0$ in the polynomial $f(x)$ in (6.25), then zero is a zero of $f(x)$, and hence we may write $f(x) = xg(x)$ and apply the above Corollary to $g(x)$.

Remark 2 The above Corollary is also true if the leading coefficient is -1 (instead of 1). In fact, the above proof is still valid in this case.

Example 6.4 Find the zeros of the polynomial

$$f(x) = x^4 - 2x^3 - 8x^2 + 10x + 15 \tag{6.26}$$

Solution Observe that by the above Corollary all the rational zeros of the polynomial $f(x)$ in Eq. (6.26) must be integers which divide 15. Thus, the only possible rational zeros of $f(x)$ are

$$\pm 1 \qquad \pm 3 \qquad \pm 5 \qquad \pm 15 \tag{6.27}$$

Now, since $f(1) = 16 \neq 0$, it follows that 1 is *not* a zero of $f(x)$. However, we observe that $f(-1) = 0$, and thus -1 is a zero of $f(x)$. Hence, by the Factor theorem $x - (-1)$ is a factor of $f(x)$. In fact, dividing $x + 1$ synthetically into $f(x)$, we get

$$\begin{array}{rrrrr|l} 1 & -2 & -8 & 10 & 15 & \underline{-1} \\ & -1 & 3 & 5 & -15 & \\ \hline 1 & -3 & -5 & 15 & 0 & \end{array}$$

Hence,

$$f(x) = (x + 1)(x^3 - 3x^2 - 5x + 15) \tag{6.28}$$

Let

$$g(x) = x^3 - 3x^2 - 5x + 15 \tag{6.29}$$

and let us test one of the integers in (6.27), say 3, as a possible zero of $g(x)$. If we divide $g(x)$ synthetically by $x - 3$, we obtain

$$\begin{array}{rrrr|l} 1 & -3 & -5 & 15 & \underline{3} \\ & 3 & 0 & -15 & \\ \hline 1 & 0 & -5 & 0 & \end{array}$$

Thus, $x - 3$ is a factor of $g(x)$, and

$$g(x) = (x - 3)(x^2 + 0x - 5) = (x - 3)(x^2 - 5) \tag{6.30}$$

Combining (6.28) to (6.30), we obtain $f(x) = (x + 1)(x - 3)(x^2 - 5)$. Hence the zeros of $f(x)$ are -1, 3, $\sqrt{5}$, and $-\sqrt{5}$. Observe that $\sqrt{5}$ is irrational since $x^2 - 5$ has *no rational* zeros. (Why?)

Example 6.5 Show that the polynomial

$$f(x) = x^6 + x^5 - 3x^4 - 5x^3 - 4x^2 + 6x + 12 \tag{6.31}$$

has no rational zeros.

Solution We know that all rational zeros (if any) of $f(x)$ must be integers which divide 12. Thus, the *only possible* rational zeros of $f(x)$ are

$$\pm 1 \quad \pm 2 \quad \pm 3 \quad \pm 4 \quad \pm 6 \quad \pm 12 \tag{6.32}$$

Let us use synthetic division to evaluate a few functional values of $f(x)$.

$$\begin{array}{rrrrrrrl} 1 & 1 & -3 & -5 & -4 & 6 & 12 & \underline{|1} \\ & 1 & 2 & -1 & -6 & -10 & -4 & \\ \hline 1 & 2 & -1 & -6 & -10 & -4 & 8 & [=f(1)] \end{array}$$

$$\begin{array}{rrrrrrrl} 1 & 1 & -3 & -5 & -4 & 6 & 12 & \underline{|2} \\ & 2 & 6 & 6 & 2 & -4 & 4 & \\ \hline 1 & 3 & 3 & 1 & -2 & 2 & 16 & [=f(2)] \end{array}$$

$$\begin{array}{rrrrrrrl} 1 & 1 & -3 & -5 & -4 & 6 & 12 & \underline{|3} \\ & 3 & 12 & 27 & 66 & 186 & 576 & \\ \hline 1 & 4 & 9 & 22 & 62 & 192 & 588 & [=f(3)] \end{array}$$

Now, since all the entries in the last row above are *positive,* we know by the Upper Bound theorem that 3 is an upper bound for the real zeros of $f(x)$. That is, all real zeros of $f(x)$ are less than or equal to 3. Next, let us try to find a lower bound for the zeros of $f(x)$. Thus we consider the polynomial $f(-x)$. In view of (6.31), we know that $f(-x) = (-x)^6 + (-x)^5 - 3(-x)^4 - 5(-x)^3 - 4(-x)^2 + 6(-x) + 12$, and, hence,

$$f(-x) = x^6 - x^5 - 3x^4 + 5x^3 - 4x^2 - 6x + 12$$

We now seek to find an upper bound for the real zeros of $f(-x)$, and, as a reasonable guess, we test 3 as a possibility for such an upper bound. For this purpose, we perform the synthetic division indicated below.

$$\begin{array}{rrrrrrrl} 1 & -1 & -3 & 5 & -4 & -6 & 12 & \underline{|3} \\ & 3 & 6 & 9 & 42 & 114 & 324 & \\ \hline 1 & 2 & 3 & 14 & 38 & 108 & 336 & \end{array}$$

Since all the entries in the last row above are positive, we know by the Upper Bound theorem that 3 is an upper bound for the real zeros of $f(-x)$. Hence, by (6.7), -3 is a lower bound for the real zeros of the polynomial $f(x)$. The net result, then, is the following:

> 3 is an upper bound for the real zeros of $f(x)$, and -3 is a lower bound for the real zeros of $f(x)$. (6.33)

In other words, all the real zeros of $f(x)$ lie between -3 and 3. In view of this, the only possible rational zeros of $f(x)$ are ±1, ±2, and ±3. We have already seen in the above synthetic divisions that $f(1) = 8$, $f(2) = 16$, and $f(3) = 588$, and hence *none* of 1, 2, and 3 is a zero of $f(x)$. Moreover, we easily verify that $f(-1) = 4$, $f(-2) = 8$, and $f(-3) = 336$, and hence *none* of -1, -2, and -3 is a zero of $f(x)$. In view of (6.33), we conclude that the polynomial $f(x)$ has *no* rational zeros.

Remark The above example shows that knowing an upper bound and a lower bound for the real zeros of a polynomial can shorten the computation involved in seeking its zeros. Thus, in the above example we did *not* have to test ±4, ±6, and ±12 as possible zeros of $f(x)$ because we found that 3 is an upper bound for the real zeros of $f(x)$ and -3 is a lower bound for the real zeros of $f(x)$.

Problem Set 6.1

6.1.1 Suppose that $f(x) = 24x^3 - 26x^2 - 13x + 10$. Use synthetic division to find $f(-2)$, $f(-1)$, $f(0)$, $f(1)$, and $f(2)$.

6.1.2 Suppose that $f(x)$ is the polynomial given in Prob. 6.1.1. Use the synthetic-division results you obtained in Prob. 6.1.1 to find the quotient and remainder when $f(x)$ is divided by each of $x + 2$, $x + 1$, x, $x - 1$, and $x - 2$.

6.1.3 Use the Location theorem to locate the real zeros of the polynomial $f(x)$ in Prob. 6.1.1 between consecutive integers.

6.1.4 Use the Rational Zeros theorem to find *all* rational zeros of the polynomial $f(x)$ in Prob. 6.1.1. Are any of these zeros integers?

6.1.5 Suppose $f(x)$ is a polynomial of any degree, and suppose $g(x) = ax + b$ is a linear polynomial. To divide $f(x)$ by $g(x)$, first write $g(x) = a[x - (-b/a)]$, and then divide $f(x)$ by $x - (-b/a)$ to obtain

$$f(x) = \left(x - \frac{-b}{a}\right)q(x) + r \qquad (r = \text{constant})$$

Hence,

$$f(x) = (ax + b)\,\frac{q(x)}{a} + r$$

Thus, when $ax + b$ is divided into $f(x)$, the quotient is $q(x)/a$, and the remainder is r. Use this technique to find the quotient and remainder when (*a*) $f(x) = 2x^3 - 3x^2 + 7x + 10$ is divided by $2x - 1$ and (*b*) $f(x) = 2x^3 - 3x^2 + 7x + 10$ is divided by $3x + 1$.

6.1.6 Suppose that $f(x)$ is a polynomial and a and b are real numbers, $a \neq 0$. Show that the remainder when $f(x)$ is divided by $x + b/a$ is the same as the remainder when $f(x)$ is divided by $ax + b$. How do the two quotients compare?

6.1.7 Suppose that $f(x) = x^4 + ax^3 + 2x^2 - 3x - 5b$, and suppose $f(1) = -14$ and $f(-1) = 0$. Find a and b.

6.1.8 Suppose that $f(x) = x^3 - ax + b$ is divisible by both $x + 1$ and $x - 2$. Find *all* the zeros of $f(x)$. Are these zeros distinct?

6.1.9 Find all rational zeros of $f(x) = x^n - 1$ if (*a*) n is even and (*b*) n is odd.

6.1.10 Find a polynomial of degree 4 with zeros $1/2$, $-2/3$, 0, and 1.

6.1.11 Use the Rational Zeros theorem to find all *rational* zeros of the following polynomials. Use this information together with the Factor theorem to find *all* zeros of each of these polynomials.

(*a*) $f(x) = 30x^3 + 31x^2 - 24x - 16$

(*b*) $f(x) = 9x^5 - 18x^4 + 8x^3 - 16x^2 - x + 2$

(*c*) $f(x) = x^3 + 2x^2 - 5x - 6$

(*d*) $f(x) = x^4 - x^3 - 17x^2 - 3x - 60$

(*e*) $f(x) = 4x^4 + 8x^3 - 21x^2 - 18x + 27$

(*f*) $f(x) = x^4 - 3x^3 - 6x^2 + 6x + 8$

(*g*) $f(x) = x^4 + 2x^3 - 10x^2 - 11x - 12$

(*h*) $f(x) = x^4 + 3x^3 - x^2 - 8x - 4$

(*i*) $f(x) = x^3 + 1$

(*j*) $f(x) = x^3 - 8$

6.1.12 Find an upper bound and a lower bound for the real zeros of each of the following polynomials:

(*a*) $f(x) = x^3 + x - 3$

(*b*) $f(x) = x^4 + 2x^3 - x^2 - 5$

(*c*) $f(x) = x^3 + x^2 + x - 9$

6.1.13 Find all integral zeros of the polynomials below. Use this information together with the Factor theorem to find *all* the zeros of each of these polynomials.

(*a*) $f(x) = x^4 - x^3 - 3x^2 - 7x - 6$

(*b*) $f(x) = x^4 + 3x^3 - 8x^2 + 9x - 5$

(*c*) $f(x) = x^3 + 2x^2 - 4x - 3$

(*d*) $f(x) = x^3 - 4x^2 - 4x - 5$

6.1.14 Prove that the following polynomials have no rational zeros:

(*a*) $f(x) = x^3 - x^2 - 3x + 1$

(*b*) $f(x) = x^3 + 2x^2 - 4x - 6$

(*c*) $f(x) = -x^3 + x^2 - 2x + 5$

(*d*) $f(x) = -x^4 + 2x^3 - x^2 - 4$

6.1.15 Find a polynomial $f(x)$ of degree 4 such that $f(1) = 0$, $f(2) = 0$, $f(-1) = 0$, $f(0) = 2$, and $f(-2) = 108$. (*Hint*: Use the Factor theorem.)

6.1.16 Suppose that the zeros of $f(x) = x^3 + 2x + 5$ are a, b, and c. Find a polynomial whose zeros are $a + 1$, $b + 1$, and $c + 1$.

6.1.17 Suppose that the zeros of $f(x) = x^3 + 2x + 5$ are a, b, and c. Find a polynomial whose zeros are $3a - 1$, $3b - 1$, and $3c - 1$.

6.1.18 We recall that a positive integer $p > 1$ is *prime* if its only divisors are ± 1 and $\pm p$. Suppose that p is any positive prime and n is an integer > 1. Prove that $\sqrt[n]{p}$ is irrational. (*Hint*: Consider the rational zeros of $x^n - p$.)

6.1.19 Suppose $p_1, p_2, \ldots, p_t$ are *distinct* positive primes, and suppose that n is a positive integer > 1. Prove that $\sqrt[n]{p_1 p_2 \cdots p_t}$ is irrational. (*Hint*: Consider the rational zeros of $x^n - p_1 p_2 \cdots p_t$.)

6.1.20 Prove that $\sqrt{3}-\sqrt{2}$ is irrational. [*Hint*: Let $x=\sqrt{3}-\sqrt{2}$. Then $(x+\sqrt{2})^2=3$. Now, clear radicals in order to obtain $f(x)=0$, where *all the coefficients of $f(x)$ are integers*. Then show that $f(x)$ has *no* rational zeros.]

6.2 Irrational and Imaginary Zeros of Polynomials

In this section, we give a brief discussion of irrational as well as imaginary zeros of polynomials with real coefficients. We also discuss graphs of certain polynomial functions.

We have already seen that some (or even all) of the zeros of a polynomial may be irrational. For example, the zeros of the polynomial x^2-2 are $\pm\sqrt{2}$, and both of these zeros are irrational. We now direct our attention to a brief study of irrational zeros of polynomials. Thus, suppose that $f(x)$ is a polynomial with real coefficients, and suppose that for some real numbers a and b, $f(a)$ and $f(b)$ are of *opposite* signs. Then, by the Location theorem $f(x)$ has at least one zero between a and b. If we wish to get a closer approximation for such a zero of $f(x)$, we might try to take two real numbers a_1 and b_1, lying between a and b. Now, if $f(a_1)$ and $f(b_1)$ still have *opposite* signs, then, by the Location theorem again, $f(x)$ has a zero between a_1 and b_1. Clearly, we can continue this process as long as we keep getting functional values which are of opposite signs. We illustrate this in the following example.

Example 6.6 Find, correct to one decimal place, one of the zeros of the polynomial

$$f(x)=x^3-5x+1 \tag{6.34}$$

Solution First observe that the only *possible* rational zeros of $f(x)$ are integers which divide 1, that is, ± 1, and neither of these is a zero of $f(x)$. Thus $f(x)$ has *no rational* zeros, and we now search for irrational zeros. Observe that

$$\begin{array}{llll} f(0)=1>0 & f(1)=-3<0 & f(2)=-1<0 & f(3)=13>0 \\ f(-1)=5>0 & f(-2)=3>0 & f(-3)=-11<0 & \end{array} \tag{6.35}$$

With $f(0)>0$ and $f(1)<0$, the Location theorem tells us that $f(x)$ has a zero a, where $0<a<1$. Similarly, since $f(2)<0$ and $f(3)>0$, $f(x)$ has a zero b, where $2<b<3$. Finally, since $f(-2)>0$ and $f(-3)<0$, $f(x)$ has a zero c, where $-3<c<-2$. We have thus established that

$$f(x) \text{ has three real zeros } a, b, \text{ and } c, \text{ where } 0<a<1,\ 2<b<3, \text{ and } -3<c<-2. \tag{6.36}$$

Now, if we wish to get a better approximation for one of these zeros, say a, we proceed as follows: Since, by (6.35), $f(0)=1$ and $f(1)=-3$, it is intuitively plausible that a is closer to 0 than it is to 1. In fact, by interpolating we expect that a is probably close to 0.25; so we use synthetic division to evaluate $f(0.2)$ and $f(0.3)$, as indicated below. [Observe that the coefficient of x^2 in $f(x)$ is 0.]

1	0	−5	1	$\lfloor 0.2$
	0.2	0.04	−0.992	
1	0.2	−4.96	0.008 [= $f(0.2)$]	

1	0	−5	1	$\lfloor 0.3$
	0.3	0.09	−1.473	
1	0.3	−4.91	−0.473 [= $f(0.3)$]	

Since $f(0.2) = 0.008$ and $f(0.3) = -0.473$, it is intuitively plausible that the zero a is much closer to 0.2 than it is to 0.3; so we try to find $f(0.21)$, say, as indicated below.

1	0	−5	1	$\lfloor 0.21$
	0.21	0.0441	−1.040739	
1	0.21	−4.9559	−0.040739 [= $f(0.21)$]	

Thus $f(0.21) = -0.040739$, and it is now apparent [since $f(0.2) = 0.008$] that $0.2 < a < 0.21$. Indeed, we expect a to be closer to 0.2 than it is to 0.21, but, at any rate, we know that the zero a, correct to one decimal place, is 0.2. Similarly, we could find approximate values for the zeros b and c.

Finding approximate values for irrational zeros of a polynomial is usually quite laborious if done without the aid of a calculator. Moreover, there are more efficient methods, such as *Newton's method* and *Horner's method,* for finding approximate values for irrational zeros of a polynomial, and the interested reader can consult more advanced texts for such methods. (It is recommended that the present method be accompanied by a calculator.)

We have already seen that a polynomial $f(x)$ with real coefficients may have some (or even all) of its zeros *imaginary.* For example, the zeros of the polynomial $x^2 + 1$ are i and $-i$, and both of these zeros are imaginary. Similarly, the zeros of the polynomial $x^2 + 2x + 2$ are, by the quadratic formula, $-1 + i$ and $-1 - i$, and both of these zeros are also imaginary. In both of these examples, we observe that *imaginary zeros always occur in conjugate pairs*; that is, if $a + bi$ (a and b real) is a zero of a polynomial $f(x)$ *with real coefficients*, so is $a - bi$. Is this a coincidence? The answer is "no." Indeed, this is always true, as the following theorem asserts.

Complex Conjugates Theorem

Suppose that $f(x)$ is a polynomial with real coefficients, and suppose z is any complex zero of $f(x)$. Then $\bar{z}$, the complex conjugate of z, is also a zero of $f(x)$.

The proof of this theorem leans heavily on the important properties of conjugates which we established in Chap. 4. Thus, suppose that

$$f(x) = a_nx^n + a_{n-1}x^{n-1} + \cdots + a_1x + a_0 \qquad \text{all } a_i \text{ real} \tag{6.37}$$

and suppose z is a complex zero of $f(x)$. Then $f(z) = 0$. Hence, using (6.37) we get

$$0 = f(z) = a_nz^n + a_{n-1}z^{n-1} + \cdots + a_1z + a_0 \tag{6.38}$$

Recall that in Chap. 4 we proved that the conjugate of a sum is equal to the sum of the conjugates and the conjugate of a product is equal to the product of the conjugates. Moreover, since all the coefficients a_i in (6.38) are *real*, we have

$$\overline{a}_i = a_i \qquad \text{for each } i = 0, 1, 2, \ldots, n \tag{6.39}$$

Hence, by taking the conjugates of both sides of (6.38), keeping in mind (6.39) and the above facts about conjugates, we obtain

$$\overline{0} = \overline{a_n z^n + a_{n-1} z^{n-1} + \cdots + a_1 z + a_0} \qquad \text{by (6.38)}$$

$$0 = \overline{a_n z^n} + \overline{a_{n-1} z^{n-1}} + \cdots + \overline{a_1 z} + \overline{a}_0 \qquad \text{by conjugate-of-a-sum rule}$$

$$0 = \overline{a}_n \overline{z^n} + \overline{a}_{n-1} \overline{z^{n-1}} + \cdots + \overline{a}_1 \overline{z} + \overline{a}_0 \qquad \text{by conjugate-of-a-product rule}$$

$$0 = \overline{a}_n (\overline{z})^n + \overline{a}_{n-1} (\overline{z})^{n-1} + \cdots + \overline{a}_1 \overline{z} + \overline{a}_0 \qquad \text{by conjugate-of-a-product rule}$$

$$0 = a_n (\overline{z})^n + a_{n-1} (\overline{z})^{n-1} + \cdots + a_1 \overline{z} + a_0 \qquad \text{by (6.39)}$$

Hence,

$$a_n (\overline{z})^n + a_{n-1} (\overline{z})^{n-1} + \cdots + a_1 (\overline{z}) + a_0 = 0 \tag{6.40}$$

But in view of (6.37) the left side of (6.40) is precisely $f(\overline{z})$, and hence (6.40) now becomes $f(\overline{z}) = 0$; that is, $\overline{z}$ is a zero of $f(x)$, and the theorem is proved.

Example 6.7 Find a polynomial $f(x)$ with real coefficients and of degree 4 which has $1 + i$ and $2 - 3i$ as zeros.

Solution Since the polynomial $f(x)$ has *real* coefficients and since $1 + i$ is a zero of $f(x)$, it follows by the Complex Conjugates theorem that $1 - i$ is also a zero of $f(x)$. Similarly, since $2 - 3i$ is a zero of $f(x)$, this theorem also tells us that $2 + 3i$ is a zero of $f(x)$. Thus, $1 + i$, $1 - i$, $2 - 3i$, and $2 + 3i$ are all zeros of $f(x)$. Hence, by the Factor theorem $x - (1 + i)$, $x - (1 - i)$, $x - (2 - 3i)$, and $x - (2 + 3i)$ are all factors of $f(x)$. Moreover, since $f(x)$ is, by hypothesis, of degree 4, we may try to take

$$f(x) = [x - (1 + i)]\,[x - (1 - i)]\,[x - (2 - 3i)]\,[x - (2 + 3i)] \tag{6.41}$$

Clearly, $f(x)$ is of degree 4 and has the desired zeros, and there remains only to verify that the coefficients of $f(x)$ are all real. By multiplying the first two factors of the product in (6.41) together and the last two factors together, we obtain $f(x) = (x^2 - 2x + 2)\,(x^2 - 4x + 13)$, and hence

$$f(x) = x^4 - 6x^3 + 23x^2 - 34x + 26$$

Thus, all the coefficients of $f(x)$ are real, and $f(x)$ is our desired polynomial.

Example 6.8 The polynomial in (6.42)

$$f(x) = x^4 - 6x^3 + 20x^2 + 30x - 125 \tag{6.42}$$

has $3 - 4i$ as a zero. Find all the zeros of $f(x)$.

Solution Since $3 - 4i$ is a zero of $f(x)$, we know by the Complex Conjugates theorem that $3 + 4i$ is also a zero of $f(x)$. Hence, by the Factor theorem both $x - (3 - 4i)$ and $x - (3 + 4i)$ are factors of $f(x)$. Thus the product of these factors, namely, $[x - (3 - 4i)][x - (3 + 4i)]$, divides $f(x)$. Now, using the usual rules of high school algebra we obtain

$$[x - (3 - 4i)][x - (3 + 4i)] = (x - 3)^2 + 4^2 = x^2 - 6x + 25 \tag{6.43}$$

We have thus shown that $x^2 - 6x + 25$ divides $f(x)$, and to find the other factor of $f(x)$ we use the long-division method, as indicated below.

$$\begin{array}{r} x^2 - 5 \\ x^2 - 6x + 25 \overline{\smash{)}\, x^4 - 6x^3 + 20x^2 + 30x - 125} \\ \underline{x^4 - 6x^3 + 25x^2 } \\ -5x^2 + 30x - 125 \\ \underline{-5x^2 + 30x - 125} \\ 0 \end{array}$$

Hence,

$$f(x) = (x^2 - 6x + 25)(x^2 - 5) \tag{6.44}$$

Now, combining (6.43) and (6.44) we get

$$f(x) = [x - (3 - 4i)][x - (3 + 4i)](x^2 - 5)$$

and hence the zeros of $f(x)$ are $3 - 4i$, $3 + 4i$, $\sqrt{5}$, and $-\sqrt{5}$.

Warning In the Complex Conjugates theorem, it is essential that we assume that all the *coefficients of $f(x)$ are real.* Indeed, this theorem need not be true if the coefficients of $f(x)$ are not all real. For example, the zeros of $f(x) = x^2 - 2i$ are $1 + i$ and $-1 - i$ (this can be shown by direct substitution), but these zeros are *not* complex conjugates. Similarly, it is easily verified that the zeros of $f(x) = (1 + i)x^2 + (1 - i)x - 2i$ are -1 and $1 + i$, and, again, these zeros are *not* complex conjugates.

We conclude this section with a brief discussion of graphs of polynomial functions. In general, to sketch the graph of a given polynomial function, we make a table of values, sketch the points represented in this table, and then join these points by means of a smooth curve. We illustrate this in the following examples. (It should be pointed out, however, that a thorough treatment of curve sketching requires the use of calculus.)

Example 6.9 Sketch the graphs of the polynomial functions given by (*a*) $f(x) = x^4$ and (*b*) $f(x) = x^3$.

Solution The following tables of values correspond to the functions given in (*a*) and (*b*), respectively, and the graphs appear in Figs. 6.2 and 6.3.

(*a*)

x	0	1	2	−1	−2
y	0	1	16	1	16

(*b*)

x	0	1	2	−1	−2
y	0	1	8	−1	−8

Example 6.10 Sketch the graph of the polynomial function f given by $f(x) = x^3 - 3x - 4$.

Solution We proceed by making a table of values, as indicated below.

x	0	1	2	3	−1	−2
y	−4	−6	−2	14	−2	−6

Next, we plot these points and join them by means of a smooth curve. The graph is sketched in Fig. 6.4. Observe that since $f(2) < 0$ and $f(3) > 0$, $f(x)$ has a real zero somewhere between 2 and 3. Also, since the graph of f intersects the x axis only at one point, $f(x)$ has only one real zero. This real zero of $f(x)$ is, in fact, *irrational.* (Why?) Moreover, the other two zeros of $f(x)$ are complex conjugates. [Note that the complex-conjugate zeros of $f(x)$ cannot be located on the given graph because both axes represent *real* numbers.]

Example 6.11 Sketch the graph of the polynomial function f given by $f(x) = \frac{3}{4}x^4 - 5x^3 + \frac{21}{2}x^2 - 9x + \frac{5}{2}$.

Solution We proceed by making a table of values, as indicated below.

x	0	1	2	3	4
y	$\frac{5}{2}$	$-\frac{1}{4}$	$-\frac{3}{2}$	$-\frac{17}{4}$	$\frac{13}{2}$

Next, we plot these points and join them by means of a smooth curve. The graph is sketched in Fig. 6.5. Observe that since $f(0) > 0$ and $f(1) < 0$, $f(x)$ has a real zero somewhere between 0 and 1. Moreover, since $f(3) < 0$ and

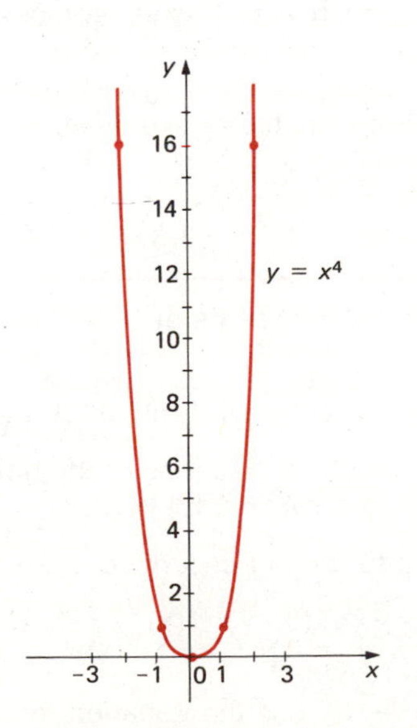

Fig. 6.2 Graph of $y = f(x) = x^4$

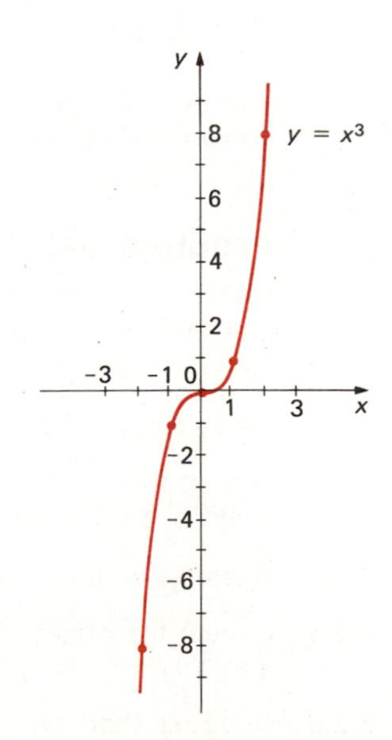

Fig. 6.3 Graph of $y = f(x) = x^3$

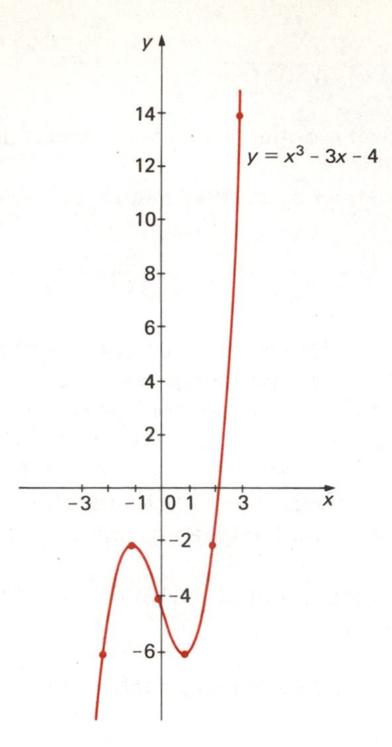

Fig. 6.4 Graph of $y = f(x) = x^3 - 3x - 4$

$f(4) > 0$, $f(x)$ has another real zero somewhere between 3 and 4. Also, since the graph of f intersects the x axis only at two points, $f(x)$ has exactly two real zeros. The other two zeros of $f(x)$ are complex conjugates, which, of course, cannot be located on the given graph since both axes represent *real* numbers.

You should check a value between each two values given in the chart above to convince yourself that the graph is reasonable as sketched.

Problem Set 6.2

6.2.1 Use the Location theorem to locate the real zeros of $f(x) = -x^3 - 3x^2 + 3$ between consecutive integers.

6.2.2 Prove that all the zeros of the polynomial $f(x)$ in Prob. 6.2.1 are irrational.

6.2.3 Find, correct to one decimal place, each of the zeros of the polynomial $f(x)$ in Prob. 6.2.1. (*Hint*: Use the results of Prob. 6.2.1.)

6.2.4 Sketch the graph of the polynomial function f described in Prob. 6.2.1.

6.2.5 Sketch the graph of (a) $y = f(x) = -x^3$, (b) $y = f(x) = -x^5$, (c) $y = f(x) = -x^4$ and (d) $y = f(x) = -x^6$.

6.2.6 Suppose that $f(x) = x^4 - \frac{4}{3}x^3 - 4x^2 + 10$. Use the Location theorem to show that $f(x)$ has a real zero between 1 and 2 and a real zero between 2 and 3.

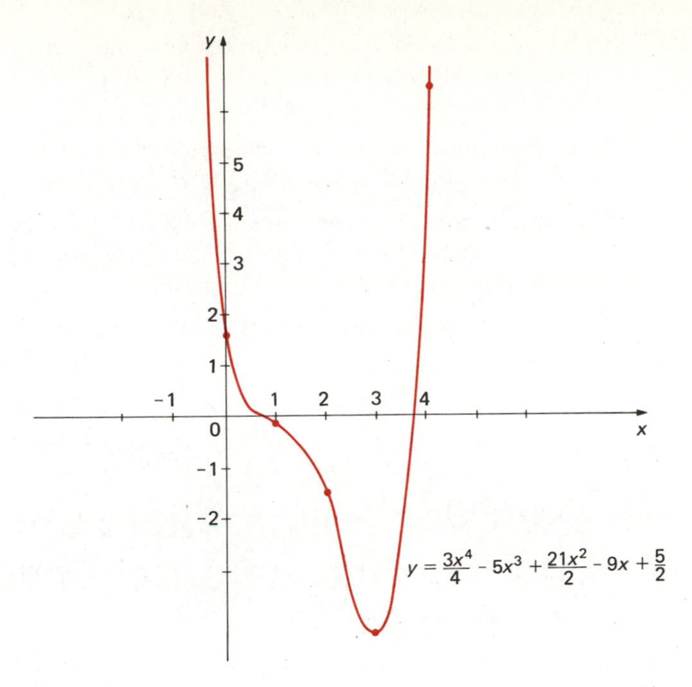

Fig. 6.5 Graph of $y = f(x) = 3x^4/4 - 5x^3 + 21x^2/2 - 9x + 5/2$. Observe that $f(x) > 0$ when $x < 0$.

6.2.7 Use the Upper Bound and Lower Bound theorems for real zeros of a polynomial function to show that *all* the real zeros of the polynomial $f(x)$ in Prob. 6.2.6 are between -2 and 3. Use this information together with the Rational Zeros theorem to show that $f(x)$ has *no* rational zeros. [*Hint*: Consider $3f(x)$ to get integer coefficients.]

6.2.8 Use the information in Probs. 6.2.6 and 6.2.7, together with a table of values, to sketch the graph of $y = f(x) = x^4 - \frac{4}{3}x^3 - 4x^2 + 10$.

6.2.9 Find, correct to one decimal place, each of the two zeros of the polynomial $f(x)$ you located in Prob. 6.2.6. Compare your results with the zeros of $f(x)$ suggested by the graph in Prob. 6.2.8.

6.2.10 Find a polynomial $f(x)$ of degree 4 with real coefficients given that $2 + 3i$ and $-1 + 7i$ are zeros of $f(x)$.

6.2.11 Suppose that a and b are rational numbers, and define the *conjugate* of $a + b\sqrt{2}$, denoted by $(a + b\sqrt{2})'$, by $(a + b\sqrt{2})' = a - b\sqrt{2}$. Prove that if a, b, c, and d are all rational, then (*a*) $[(a + b\sqrt{2}) + (c + d\sqrt{2})]' = (a + b\sqrt{2})' + (c + d\sqrt{2})'$ and (*b*) $[(a + b\sqrt{2})(c + d\sqrt{2})]' = (a + b\sqrt{2})' \times (c + d\sqrt{2})'$.

6.2.12 Suppose that $f(x)$ is a polynomial with rational coefficients. Use the results of Prob. 6.2.11 to show that if $a + b\sqrt{2}$ [a and b rational] is a zero of $f(x)$, then $a - b\sqrt{2}$ is also a zero of $f(x)$. (*Hint*: Imitate the proof of the Complex Conjugates theorem.)

6.2.13 Find a polynomial $f(x)$ of degree 4 with rational coefficients given that $1 + \sqrt{2}$ and $-1 + 3\sqrt{2}$ are zeros of $f(x)$. (*Hint*: Use the result of Prob. 6.2.12 and the Factor theorem.)

6.2.14 The polynomial $f(x) = x^4 - 6x^3 + 8x^2 - 6x + 7$ has $3 + \sqrt{2}$ as a zero. Use the result of Prob. 6.2.12 to find all the zeros of $f(x)$.

6.2.15 State and prove a result similar to the one given in Prob. 6.2.12 except that 2 is now replaced by p (and thus $a + b\sqrt{2}$ is replaced by $a + b\sqrt{p}$), where p is any positive prime.

6.2.16 Show that both of the real zeros of the function f in Example 6.11 are irrational.

6.3 Complex Polynomial Functions and the Fundamental Theorem of Algebra

We recall that in Sec. 5.1 we defined a polynomial function of degree n to be a function $f:R \to R$, where R is the set of real numbers and where

$$f(x) = a_n x^n + a_{n-1}x^{n-1} + \cdots + a_1 x + a_0 \qquad \text{for all } x \text{ in } R \tag{6.45}$$

Here $a_0, a_1, \ldots, a_n$ are fixed *real* numbers and $a_n \neq 0$. We now extend this definition in two ways. First, we agree to allow the *coefficients* $a_0, a_1, \ldots, a_n$ to be any *fixed complex numbers.* Secondly, we now view x to be *any complex number.* In other words, starting with any fixed complex numbers $a_0, a_1, \ldots, a_n$ where $a_n \neq 0$, we define a function $f:C \to C$, where C is the set of complex numbers, as follows:

$$f(x) = a_n x^n + a_{n-1}x^{n-1} + \cdots + a_1 x + a_0 \qquad \text{for all } x \text{ in } C \tag{6.46}$$

This function $f:C \to C$ is now called a *complex polynomial function* of degree n, while the function $f:R \to R$ in (6.45) is called a *real polynomial function* of degree n.

The importance of this extension of real polynomial functions to complex polynomial functions is due to the so-called *Fundamental Theorem of Algebra.* This theorem essentially states that *every nonconstant complex polynomial function must have at least one complex zero.* This could also be stated as follows.

Fundamental Theorem of Algebra

Suppose that $f(x)$ is a nonconstant polynomial with complex (i.e., real or imaginary) coefficients. Then there exists at least one complex number x_0 such that $f(x_0) = 0$. That is, the complex polynomial function $f:C \to C$ has at least one complex zero.

We shall not attempt to prove the Fundamental Theorem of Algebra. For a proof of this theorem, the reader should consult more advanced texts. Instead, we illustrate the Fundamental Theorem of Algebra by some examples.

Example 6.12 Suppose that $f: C \to C$ is the complex polynomial function given by $f(x) = (1+i)x - (1-i)$. Then $-i$ is a zero of f because

$$f(-i) = (1+i)(-i) - (1-i) = -i - i^2 - 1 + i = 0$$

Example 6.13 Suppose that $f: C \to C$ is the complex polynomial function given by $f(x) = x^2 - 2i$. Then $1+i$ and $-1-i$ are both zeros of f:

$$f(1+i) = (1+i)^2 - 2i = (1 + i^2 + 2i) - 2i = 0$$

$$f(-1-i) = (-1-i)^2 - 2i = (1 + i^2 + 2i) - 2i = 0$$

Example 6.14 Suppose that $f: C \to C$ is the complex polynomial function given by $f(x) = (1+i)x^2 + (1-i)x - 2i$. Then -1 and $1+i$ are both zeros of f since

$$f(-1) = (1+i)(-1)^2 + (1-i)(-1) - 2i = 1 + i - 1 + i - 2i = 0$$

$$\begin{aligned} f(1+i) &= (1+i)(1+i)^2 + (1-i)(1+i) - 2i \\ &= (1+i)(1+i^2+2i) + (1-i^2) - 2i \\ &= (1+i)2i + 2 - 2i \\ &= 2i + 2i^2 + 2 - 2i \\ &= 0 \end{aligned}$$

These examples suggest that a complex polynomial function f of degree n has exactly n complex zeros. This is indeed the case, provided we agree to count a zero of f which occurs k times as k zeros of f. Let us make this notion more precise. Suppose that a is a complex number, and suppose a polynomial $f(x)$ satisfies the following two conditions: (1) $f(x) = (x-a)^k g(x)$ and (2) $g(a) \neq 0$, where k is a positive integer. In this case, we say that a is a *zero* of $f(x)$ of *multiplicity* k. Observe that since $g(a) \neq 0$, $x - a$ does *not* divide $g(x)$. Moreover, if $k = 1$, we say that a is a *simple zero* of $f(x)$, while if $k > 1$, we say that a is a *multiple zero* of $f(x)$. A *double zero* of $f(x)$ is simply a zero of multiplicity 2, while a *triple zero* of $f(x)$ is just a zero of multiplicity 3. We now have the following important result.

Fundamental Theorem of Algebra: General Form Suppose that $f(x)$ is a polynomial with complex (i.e., real or imaginary coefficients and of positive degree n. Then $f(x)$ has exactly n complex zeros, provided a zero of $f(x)$ of multiplicity k is counted as k zeros.

A formal proof of this theorem requires the use of *mathematical induction* (see Chap. 12). An intuitive proof can be given as follows: By the Fundamental Theorem of Algebra, the polynomial $f(x)$ has a complex zero a_1. Hence, by the Factor theorem $x - a_1$ is a factor of $f(x)$, and thus

$$f(x) = (x - a_1)g(x) \tag{6.47}$$

where $g(x)$ is a polynomial of degree $n-1$. Now, applying the Fundamental Theorem of Algebra to the polynomial $g(x)$, we know that $g(x)$ has a complex zero a_2 (a_2 may well be equal to a_1). Then, by the Factor theorem $x - a_2$ is a factor of $g(x)$, and hence [assuming $g(x)$ is of positive degree]

$$g(x) = (x - a_2)h(x) \tag{6.48}$$

where $h(x)$ is a polynomial of degree $n-2$. Combining (6.47) and (6.48), we obtain

$$f(x) = (x - a_1)(x - a_2)h(x) \tag{6.49}$$

Clearly, we can continue this process. Thus, if $h(x)$ is of positive degree (i.e., nonconstant), then by the Fundamental Theorem of Algebra the polynomial $h(x)$ has a complex zero a_3 (a_3 may equal a_1 or a_2), which means that by the Factor theorem $x - a_3$ divides $h(x)$; that is,

$$h(x) = (x - a_3)m(x) \tag{6.50}$$

where $m(x)$ is a polynomial of degree one less than the degree of $h(x)$. Equations (6.49) and (6.50) give $f(x) = (x - a_1)(x - a_2)(x - a_3)m(x)$. Clearly, we can continue this process until we eventually obtain

$$f(x) = (x - a_1)(x - a_2)(x - a_3) \cdots (x - a_n)c \qquad (c \neq 0) \tag{6.51}$$

where c is a constant polynomial, since by hypothesis the degree of $f(x)$ is n. Observe that $c \neq 0$ because $f(x)$ is *not* the zero polynomial. Now, in view of (6.51) we easily see that $a_1, a_2, a_3, \ldots, a_n$ are all zeros of $f(x)$. Of course, some of (or all) these zeros of $f(x)$ may be equal. If this happens to be the case, we count every zero according to its multiplicity. When we do this, we see that $f(x)$ has at least n zeros (not necessarily distinct). Could $f(x)$ have more than n zeros? The answer is "no" for the following reason. Suppose that t is *any* zero of $f(x)$. Then $f(t) = 0$. Hence, using (6.51) we get

$$0 = f(t) = (t - a_1)(t - a_2)(t - a_3) \cdots (t - a_n)c \tag{6.52}$$

Now, since a product of (real or imaginary) numbers is zero if and only if one of the factors is zero, (6.52) implies that $t - a_1 = 0$, $t - a_2 = 0$, $t - a_3 = 0, \ldots,$ or $t - a_n = 0$ [recall that $c \neq 0$, see (6.51)]. Hence

$$t = a_1 \qquad t = a_2 \qquad t = a_3, \ldots, \qquad \text{or} \qquad t = a_n$$

In other words, *any* zero of $f(x)$ must be equal to $a_1, a_2, a_3, \ldots,$ or a_n, and thus $f(x)$ cannot possibly have more than n zeros. This completes our intuitive proof.

An extremely interesting corollary of the general form of the Fundamental Theorem of Algebra is the following.

Identity of Polynomial Functions Theorem

If two polynomials $f(x)$ and $g(x)$ (with real or imaginary coefficients) have the same positive degree n and if there exist $n+1$ complex numbers $a_1, a_2, \ldots, a_{n+1}$ such that $f(a_i) = g(a_i)$ for each $i = 1, 2, \ldots, n+1$, then $f(x)$ and $g(x)$ are identical polynomials (and thus $f = g$).

The proof of this corollary is as follows: Suppose $h(x) = f(x) - g(x)$. If $f(x) \neq g(x)$, then $h(x)$ is *not* the zero polynomial, and hence $h(x)$ has a degree m, say. Clearly, $m \leq n$ since each of $f(x)$ and $g(x)$ is of degree n. By the general form of the Fundamental Theorem of Algebra, $h(x)$ has exactly m complex zeros (not necessarily distinct); and since $m \leq n$, we conclude that

$$h(x) \text{ has at most } n \text{ zeros} \tag{6.53}$$

On the other hand, $h(x)$ must have at least $n+1$ complex zeros, namely, $a_1, a_2, \ldots, a_{n+1}$, because

$$h(a_i) = f(a_i) - g(a_i) = 0 \qquad \text{for } i = 1, 2, \ldots, n+1$$

This, however, contradicts (6.53). This contradiction shows that $h(x)$ must be the *zero polynomial*; hence $f(x) = g(x)$. This proves the corollary.

The above corollary essentially says that if two polynomials agree in value "often enough," then they are identical polynomials.

Example 6.15 Find all zeros of the polynomial

$$f(x) = x^6 + x^5 - x^4 - 2x^3 - x^2 + x + 1 \tag{6.54}$$

Determine whether there are multiple roots.

Solution First, let us test for rational zeros. By the Rational Zeros theorem we know that the only possible rational zeros of $f(x)$ are 1 and -1. Using synthetic division, we obtain

1	1	−1	−2	−1	1	1	⌊1
	1	2	1	−1	−2	−1	
1	2	1	−1	−2	−1	0 [$= f(1)$]	

Thus 1 is indeed a zero of $f(x)$, and, moreover,

$$f(x) = (x-1)(x^5 + 2x^4 + x^3 - x^2 - 2x - 1) \tag{6.55}$$

Is 1 a multiple zero of $f(x)$? To answer this question, we may divide synthetically $x - 1$ into the fifth-degree polynomial on the right side of (6.55), as indicated below.

1	2	1	−1	−2	−1	⌊1
	1	3	4	3	1	
1	3	4	3	1	0	

Yes, $x - 1$ does indeed divide the fifth-degree polynomial in (6.55), and, in fact (see above),

$$x^5 + 2x^4 + x^3 - x^2 - 2x - 1 = (x-1)(x^4 + 3x^3 + 4x^2 + 3x + 1) \tag{6.56}$$

Combining (6.55) and (6.56) yields

$$f(x) = (x-1)^2(x^4 + 3x^3 + 4x^2 + 3x + 1) \tag{6.57}$$

Since all the coefficients of the polynomial

$$x^4 + 3x^3 + 4x^2 + 3x + 1 \tag{6.58}$$

are positive, this polynomial *cannot* possibly have any nonnegative zeros. Thus, all the real zeros (if any) of the polynomial in (6.58) are negative. Let us try our other candidate for a rational zero, namely, -1. Dividing $x - (-1)$ synthetically into the polynomial in (6.58), we get

$$\begin{array}{rrrrr|r} 1 & 3 & 4 & 3 & 1 & -1 \\ & -1 & -2 & -2 & -1 & \\ \hline 1 & 2 & 2 & 1 & 0 & \end{array}$$

Thus, -1 is indeed a zero of the polynomial in (6.58) [and hence is also a zero of $f(x)$, see (6.57)]. Moreover,

$$x^4 + 3x^3 + 4x^2 + 3x + 1 = (x+1)(x^3 + 2x^2 + 2x + 1) \tag{6.59}$$

Combining (6.57) and (6.59), we obtain

$$f(x) = (x-1)^2(x+1)(x^3 + 2x^2 + 2x + 1) \tag{6.60}$$

Is -1 a multiple zero of $f(x)$? In view of (6.60), we now divide $x - (-1)$ synthetically into the polynomial $x^3 + 2x^2 + 2x + 1$, as indicated below.

$$\begin{array}{rrrr|r} 1 & 2 & 2 & 1 & -1 \\ & -1 & -1 & -1 & \\ \hline 1 & 1 & 1 & 0 & \end{array}$$

Hence,

$$x^3 + 2x^2 + 2x + 1 = (x+1)(x^2 + x + 1) \tag{6.61}$$

Combining (6.60) and (6.61), we get

$$f(x) = (x-1)^2(x+1)^2(x^2 + x + 1) \tag{6.62}$$

Now, by the quadratic formula the zeros of the polynomial $x^2 + x + 1$ are $(-1 + \sqrt{3}\,i)/2$ and $(-1 - \sqrt{3}\,i)/2$. Combining this fact with (6.62), we easily see that the zeros of $f(x)$ are

$$1 \qquad 1 \qquad -1 \qquad -1 \qquad \frac{-1+\sqrt{3}\,i}{2} \qquad \frac{-1-\sqrt{3}\,i}{2}$$

Observe that 1 is a double zero and -1 is also a double zero, while the remaining two zeros are simple zeros. Observe also that the last two zeros above are complex conjugates, and this complies with the Complex Conjugates theorem. Moreover, observe that $f(x)$ has exactly six zeros (not all distinct), and this complies with the general form of the Fundamental Theorem of Algebra.

Example 6.16 A polynomial $f(x)$ is of degree 7 and has real coefficients. Moreover, $1 - i$ is a double zero of $f(x)$, while -2 is a triple zero of $f(x)$. Find $f(x)$. Is this $f(x)$ unique?

Solution Since $1 - i$ is a zero of $f(x)$ and $f(x)$ has real coefficients, we know by the Complex Conjugates theorem that $1 + i$ is also a zero of $f(x)$. Hence, by the Factor theorem both $x - (1 - i)$ and $x - (1 + i)$ are factors of $f(x)$; thus

$$[x - (1 - i)][x - (1 + i)] = x^2 - 2x + 2 \tag{6.63}$$

divides $f(x)$; that is,

$$f(x) = (x^2 - 2x + 2)g(x) \tag{6.64}$$

for some polynomial $g(x)$. Clearly, all the coefficients of $g(x)$ are real [since all the coefficients of $f(x)$ are real], and, moreover, the degree of $g(x)$ is 5 [since $f(x)$ is of degree 7]. Now, since $1 - i$ is a *double* zero of $f(x)$, it follows [see (6.63) and (6.64)] that $1 - i$ is a simple zero of $g(x)$. Since, moreover, all the coefficients of $g(x)$ are real, we know by the Complex Conjugates theorem that $1 + i$ is also a zero of $g(x)$. Thus, as in the above argument $[x - (1 - i)] \times [x - (1 + i)]$ (which equals $x^2 - 2x + 2$) divides $g(x)$; that is,

$$g(x) = (x^2 - 2x + 2)h(x) \tag{6.65}$$

for some polynomial $h(x)$ with real coefficients and of degree 3 [since $g(x)$ is of degree 5]. Now, combining (6.64) and (6.65), we obtain

$$f(x) = (x^2 - 2x + 2)^2 h(x) \tag{6.66}$$

But, by hypothesis, -2 is a zero of $f(x)$ of multiplicity 3. Hence, in view of (6.66) we easily see that

$$h(x) = [x - (-2)]^3 q(x) \tag{6.67}$$

for some polynomial $q(x)$. However, since $h(x)$ is of degree 3 (as we remarked above), it follows that $q(x)$ is a nonzero constant polynomial (of degree 0); that is,

$$q(x) = c \qquad (c \text{ is a nonzero constant}) \tag{6.68}$$

Combining (6.66) to (6.68), we get

$$f(x) = c(x^2 - 2x + 2)^2(x + 2)^3 \qquad (c \text{ a nonzero constant})$$

Thus, taking $c = 1$ (for convenience), we obtain $f(x) = (x^2 - 2x + 2)^2(x + 2)^3$. Observe that $f(x)$ is *not* unique because c may be *any* nonzero constant. We conclude with the following comment.

[R]emark All the properties given in Sec. 5.2 for real polynomials are true for complex polynomials. Indeed, the division algorithm, the Remainder theorem, the Factor theorem, and synthetic division are all valid for complex polynomials.

Problem Set 6.3

6.3.1 Suppose C is the set of complex numbers, and suppose $f:C \to C$ is the complex polynomial function given by $f(x) = x^3 + (i+1)x^2 + (i+2)x + 2$. Use synthetic division to evaluate $f(i)$, $f(2i)$, $f(-i)$, $f(-2i)$, $f(1)$, and $f(-1)$.

6.3.2 Use the results of Prob. 6.3.1 to find *all* the zeros of the complex polynomial function $f:C \to C$ in Prob. 6.3.1.

6.3.3 Factor the complex polynomial $f(x)$ in Prob. 6.3.1 into a product of linear polynomials with complex coefficients.

6.3.4 Can the complex polynomial $f(x)$ in Prob. 6.3.1 be factored into a product of linear and/or quadratic polynomials using real coefficients only? Explain.

6.3.5 Let $f:C \to C$ be the complex polynomial function given by $f(x) = x^3 + (3-i)x^2 - (1+2i)x - 3(1-i)$. Use synthetic division to show that the zeros of $f(x)$ are 1, -3, and $-1+i$. Does the fact that the conjugate of the zero $-1+i$ is *not* a zero of $f(x)$ contradict the Complex Conjugates theorem? Explain.

6.3.6 Find all complex numbers z such that (*a*) $z^2 = -2$, (*b*) $z^2 = i$, and (*c*) $z^2 = -i$. Verify the general form of the Fundamental Theorem of Algebra in each case. [*Hint*: Let $z = x + yi$ (x and y real) be a solution of $z^2 = -2$, for example. Then $(x+yi)^2 = -2$, and hence $(x^2 - y^2) + 2xyi = -2 + 0i$. Thus, $x^2 - y^2 = -2$ and $2xy = 0$. Now, find all *real* numbers x and y which satisfy these last two equations.]

6.3.7 Find all complex numbers z such that $4z^2 - 4z + (1-i) = 0$. (*Hint*: Use the quadratic formula first, and then use the result you obtained in Prob. 6.3.6*b*.)

6.3.8 Find a polynomial $f(x)$ of least degree and with *real* coefficients given that $1+i$ and $5-2i$ are zeros of the polynomial $f(x)$.

6.3.9 Find a polynomial $f(x)$ of least degree and with *complex* coefficients given that $1+i$ and $5-2i$ are zeros of the polynomial $f(x)$.

6.3.10 Suppose that $f:C \to C$ is the complex polynomial given by $f(x) = x^n + a_{n-1}x^{n-1} + \cdots + a_1x + a_0$. By the general form of the Fundamental Theorem of Algebra, we know that $f(x)$ has exactly n zeros (not necessarily distinct). Let $t_1, t_2, \ldots, t_n$ be the zeros of $f(x)$. Show that (*a*) $t_1 + t_2 + \cdots + t_n = -a_{n-1}$ and (*b*) $t_1t_2 \cdots t_n = (-1)^n a_0$. [*Hint*: Write $f(x)$ as a product of linear factors, and then compare the coefficients of x^{n-1}, and also the constant terms.]

6.3.11 Suppose that $f(x)$ is a polynomial of odd degree and with *real* coefficients. Prove that $f(x)$ has at least one real zero. (*Hint*: Use the general form of the Fundamental Theorem of Algebra, together with the Complex Conjugates theorem.)

6.3.12 The polynomial $f(x) = x^3 - 2x + 4$ has $1+i$ as a zero. Use the Complex Conjugates theorem and the result of Prob. 6.3.10*a* to find the other zeros of $f(x)$.

6.3.13 The polynomial $f(x) = x^4 - 23x^2 + 68x - 60$ has $2-i$ as a zero. Use the Complex Conjugates theorem and the results of Prob. 6.3.10 to find the other zeros of $f(x)$.

6.3.14 The polynomial $f(x) = ax^4 + bx^3 + cx^2 + dx + e$, where a, b, c, d, and e are complex numbers, has at least five zeros. What can you say about a, b, c, d, and e? Explain.

6.3.15 The polynomial $f(x) = x^5 - x^4 - 5x^3 + x^2 + 8x + 4$ has multiple zeros. Find each zero and its multiplicity.

6.3.16 Find a polynomial $f(x)$ of least degree and with *real* coefficients given that $1 + i$ is a zero of $f(x)$ of multiplicity 2 and $3 - 2i$ is a zero of $f(x)$ of multiplicity 3.

6.3.17 Find a polynomial $f(x)$ of least degree and with *complex* coefficients given that $1 + i$ is a zero of $f(x)$ of multiplicity 2 and $3 - 2i$ is a zero of $f(x)$ of multiplicity 3.

6.3.18 Can the following polynomials be factored "over the reals," that is, using *real numbers only* for coefficients? If so, find these factors.

(*a*) $x^2 + x - 2$ (*c*) $x^2 - x + 3$

(*b*) $x^2 + x - 5$ (*d*) $-2x^2 + 3x - 4$

(*Hint*: First, use the quadratic formula to find the zeros of each polynomial, and, then, use the Factor theorem.)

6.3.19 Factor all the polynomials in Prob. 6.3.18 "over the complexes" (that is, using complex numbers for coefficients).

6.3.20 Factor the following polynomials "over the reals":

(*a*) $f(x) = x^4 - 2x^2 - 3x - 2$

(*b*) $f(x) = x^4 - 3x^3 - 2x^2 + 5x + 3$

(*c*) $f(x) = x^4 + 4x^2 - 5$

(*Hint*: Find the zeros of each polynomial, and apply the Factor theorem.)

6.3.21 Factor the polynomials in Prob. 6.3.20 "over the complexes." (*Hint*: Use the results of Prob. 6.3.20.)

6.3.22 The following polynomials have multiple zeros. Find each zero together with its multiplicity. In each case verify the general form of the Fundamental Theorem of Algebra.

(*a*) $f(x) = -x^3 + 3x - 2$

(*b*) $f(x) = -x^3 + 3x + 2$

(*c*) $f(x) = x^4 + 2x^3 - 11x^2 - 12x + 36$

6.3.23 Use the Fundamental Theorem of Algebra to show that any *complex* polynomial function $f: C \to C$ (C is the set of complex numbers) is *onto* (or *surjective*).

6.3.24 Find a polynomial of least degree and with *rational* coefficients for which $1 + \sqrt{2}$ is a zero. (*Hint*: Use the result of Prob. 6.2.12.)

6.3.25 Find a polynomial of least degree and with *rational* coefficients for which $1 + \sqrt{2}$ and $2 - 3i$ are zeros. (*Hint*: Use the Complex Conjugates theorem and the result of Prob. 6.2.12.)

6.3.26 One of the zeros of the polynomial $f(x) = x^4 - 10x^3 + 20x^2 + 30x - 69$ is $5 - \sqrt{2}$. Find the other zeros of $f(x)$.

6.3.27 There is a theorem, known as *Descartes' rule of signs,* which says that if $f(x)$ is a polynomial with real coefficients, then the number of positive zeros of $f(x)$ cannot exceed the number of variations in sign of the coefficients of $f(x)$.

A "variation in sign" means that the sign of a coefficient of $f(x)$ differs from the sign of the next nonzero coefficient. For example, the polynomial $f(x) = x^3 - x + 1$ has exactly two variations in sign. Because the zeros of $f(-x)$ are simply the negatives of the zeros of $f(x)$, it follows that the number of negative zeros of $f(x)$ cannot exceed the number of variations in sign of the coefficients of $f(-x)$. Use these facts to find the maximum number of positive and negative zeros of the polynomials below:

(*a*) $f(x) = x^3 + x - 1$

(*b*) $f(x) = x^5 + 2x^3 - 5x^2 - 3x + 1$

(*c*) $f(x) = x^7 - 1$

(*d*) $f(x) = x^7 + 1$

(*e*) $f(x) = x^6 - 10$

(*f*) $f(x) = x^6 + 10$

(*g*) $f(x) = -x^3 - 2x - 4$

6.3.28 Use the results you obtained in Prob. 6.3.27 together with the general form of the Fundamental Theorem of Algebra to show that each polynomial in Prob. 6.3.27 has at least two nonreal zeros.

6.4 Summary

In this chapter we studied zeros of polynomial functions. Thus, suppose that

$$f(x) = a_n x^n + a_{n-1} x^{n-1} + \cdots + a_1 x + a_0 \tag{6.69}$$

where $a_0, a_1, \ldots, a_n$ are fixed real or imaginary numbers with $a_n \neq 0$. A real or imaginary number a is called a *zero* of $f(x)$ if $f(a) = 0$. In order to help us find the zeros of a polynomial $f(x)$, we learned the following theorems:

1. *Location theorem*. Suppose $f(x)$ is a polynomial with real coefficients, and suppose that a and b are real numbers. If $f(a)$ and $f(b)$ have opposite signs, then $f(x)$ has at least one real zero between a and b.
2. *Rational Zeros theorem*. If all the coefficients $a_0, a_1, \ldots, a_n$ of the polynomial $f(x)$ in (6.69) are integers with $a_n \neq 0$ and $a_0 \neq 0$ and if r/s is a rational zero of $f(x)$, then r divides a_0, and s divides a_n (here we assume that r and s have no proper factors in common). In particular if $a_n = \pm 1$, then all rational zeros of $f(x)$ (if any) are integers which divide a_0.
3. *Complex Conjugates theorem*. If all the coefficients $a_0, a_1, \ldots, a_n$ of the polynomial $f(x)$ in (6.69) are real and if $a + ib$ (a and b real) is a complex zero of $f(x)$, so is $a - ib$.
4. *Fundamental Theorem of Algebra (general form)*. Every polynomial (with real or imaginary coefficients) of positive degree n has exactly n zeros (not necessarily distinct).

We also established the Upper and Lower Bound theorems for the real zeros of a polynomial with real coefficients. Moreover, we saw how the Location theorem can be used to get rational approximations of irrational zeros of polynomials with real coefficients.

7 Rational Functions

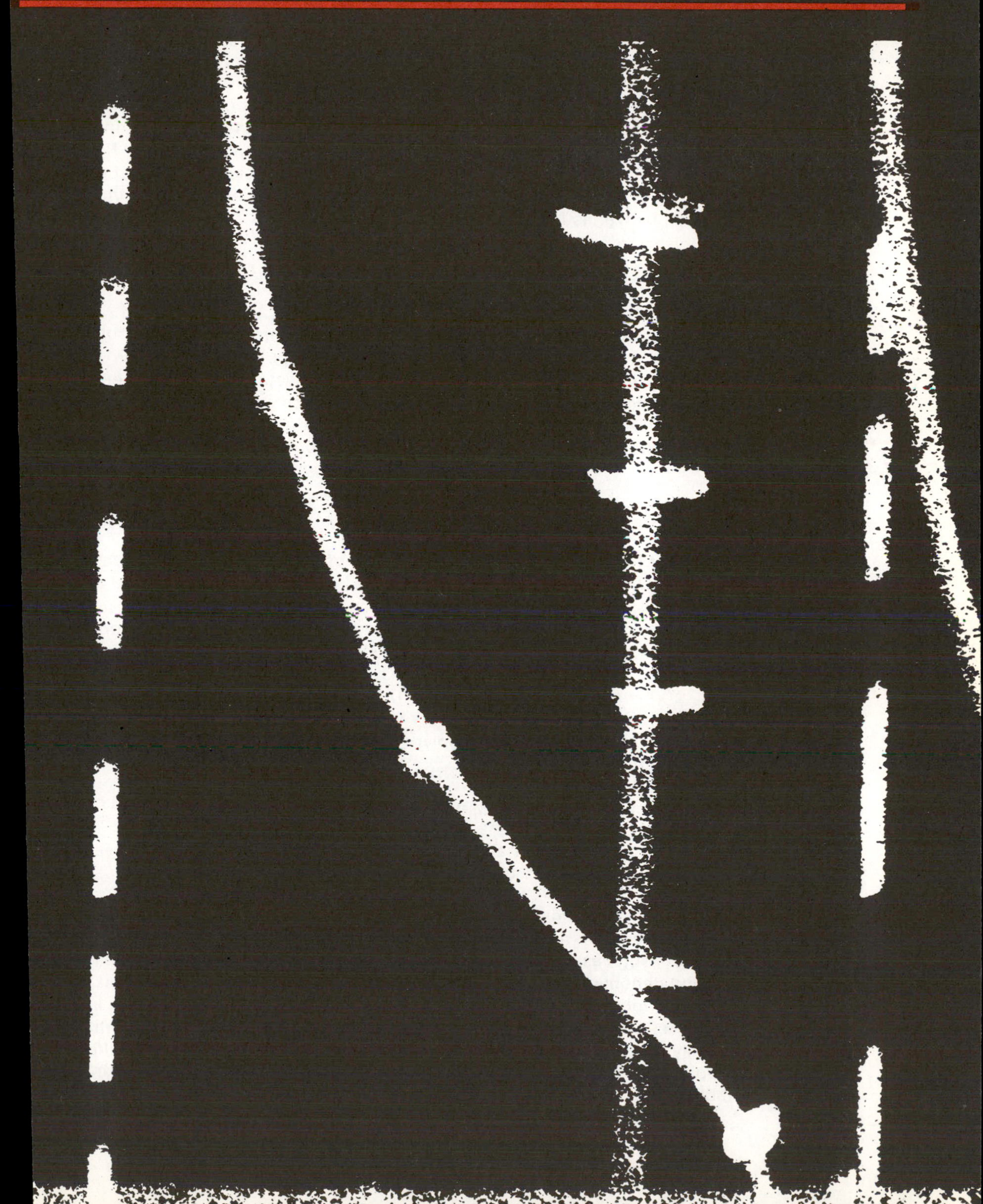

7.1 Properties of Rational Functions

In Chaps. 5 and 6, we considered polynomial functions. In this chapter, we study another important class of functions, called *rational functions.* We have already encountered many examples of rational functions in the preceding chapters. In this section, we first give the definition of a rational function and then proceed to consider some examples and properties of rational functions.

Suppose that R denotes the set of all real numbers, and suppose $f:R \to R$ and $g:R \to R$ are polynomial functions with real coefficients such that $g \neq 0$ (that is, g is not the zero polynomial function). We define a *rational function,* denoted by f/g, as follows:

$$\left(\frac{f}{g}\right)(x) = \frac{f(x)}{g(x)} \qquad g(x) \neq 0 \qquad (x \in R) \tag{7.1}$$

In other words, a *rational function* is simply the quotient of two real polynomial functions in which the denominator is not the zero polynomial function. Observe that the domain of the rational function f/g is the set of all real numbers with the exception of all those real numbers x for which $g(x) = 0$ [see (7.1)]. Thus if S denotes the set of all real zeros of the polynomial function g and if $R\backslash S$ denotes the set of all real numbers which are *not* in S, then

$$\frac{f}{g}: R\backslash S \to R \tag{7.2}$$

In view of (7.2), f/g is a real-valued function of a real variable. Incidentally, we call $f(x)/g(x)$ in (7.1) a *rational expression.*

Returning to Eq. (7.1), it may well happen that the polynomials $f(x)$ and $g(x)$ have some common factors of positive degree. If $h(x)$ denotes the product of all these common factors of $f(x)$ and $g(x)$, then we can write

$$f(x) = f_1(x)h(x) \qquad g(x) = g_1(x)h(x) \tag{7.3}$$

Here, $f_1(x)$ and $g_1(x)$ are polynomials which have no common factors of positive degree. (7.4)

Now, in view of (7.3) and (7.4) we have

$$\frac{f(x)}{g(x)} = \frac{f_1(x)}{g_1(x)} \qquad \text{for all real numbers } x \text{ such that } g(x) \neq 0,\ g_1(x) \neq 0;\ g_1(x) \text{ and } f_1(x) \text{ are as in (7.3) and (7.4)} \tag{7.5}$$

Two polynomial functions, such as f_1 and g_1 in (7.5), are said to be *relatively prime* if they have no common factors of positive degree. Moreover, a rational function, such as f_1/g_1 in (7.5), in which the polynomial functions in the numerator and denominator have no common factors of positive degree is said to be *in lowest terms*. Thus, *every rational function f/g* can be *reduced* to one in lowest terms as indicated in (7.5). For this reason, in studying rational functions we usually *assume* that the rational function f/g is in *lowest terms*. It should be emphasized that the domains of f/g and f_1/g_1 are, in general, *not* the same, even though (7.5) is valid.

Now, suppose that f/g is a rational function in lowest terms, and suppose g is of *positive* degree. In this case, we say that f/g is a *fractional function*. If, further, the degree of f is smaller than the degree of g, then f/g is called a *proper fractional function*; otherwise, f/g is called an *improper fractional function*. Of course, if the degree of f is equal to or greater than the degree of g, then we can divide f by g (using the long-division process usually) to obtain

$$f(x) = q(x)g(x) + r(x) \qquad \text{degree of } r(x) < \text{degree of } g(x) \tag{7.6}$$

Observe that $r(x) \neq 0$ in (7.6) since if $r(x) = 0$, then $f(x)/g(x) = q(x)/1$, which contradicts the hypothesis that f/g is in lowest terms. Hence, for all x in the domain of f/g

$$\frac{f(x)}{g(x)} = q(x) + \frac{r(x)}{g(x)} \tag{7.7}$$

therefore,

$$\frac{f}{g} = q + \frac{r}{g} \qquad \text{degree of } r < \text{degree of } g \tag{7.8}$$

In view of (7.8), we see that *every improper fractional function* can be written as a *sum* of a *polynomial function* and a *proper fractional function.*

It should be pointed out that *every polynomial function* f is also a *rational function* since $f = f/1$. Moreover, any rational function f/g in which g is a nonzero constant polynomial [say, $g(x) = c$] is, in fact, a *polynomial* function because

$$\left(\frac{f}{g}\right)(x) = \frac{f(x)}{g(x)} = \frac{f(x)}{c}$$

and, of course, $f(x)/c$ is a polynomial since c is a constant (i.e., a real number). In view of this, we shall emphasize in this chapter studying *fractional* functions (i.e., rational functions f/g in which g is of *positive* degree).

We operate on rational functions by applying the usual rules we learned in high school algebra. Recalling these rules, we can very easily see that the sum, difference, product, and quotient (with nonzero denominator) of any two rational functions is again a rational function.

xample 7.1 Suppose that f/g is the rational function given by Eq. (7.9):

$$\left(\frac{f}{g}\right)(x) = \frac{2x + 1}{x^2 - x - 2} \tag{7.9}$$

Find the domain of f/g. Is f/g in lowest terms? Is f/g a fractional function? Is f/g a proper fractional function?

olution The denominator in the right-hand side of 7.9 is $g(x) = x^2 - x - 2 = (x - 2) \times (x + 1)$; hence, the zeros of the polynomial function g are -1 and 2. Thus, the domain of the rational function f/g is the set of all real numbers *except* -1 and 2. Also, since the numerator of $(f/g)(x)$ (namely, $2x + 1$) and the denominator of $(f/g)(x)$ [namely, $(x - 2)(x + 1)$] have no common factors of positive degree,

f/g is indeed in lowest terms. Moreover, since g has a positive degree (namely, 2), f/g is a fractional function. Finally, since the degree of f is less than the degree of g, f/g is a *proper fractional* function.

The following factorization formulas are extremely useful in problems involving operations on rational expressions. Presumably, the student has seen these formulas before. The formulas (or identities) below are very easy to verify by multiplying the right side of each formula to obtain the left side.

$$x^2 - a^2 = (x + a)(x - a) \qquad \textit{(difference-of-two-squares formula)} \tag{7.10}$$

$$x^3 - a^3 = (x - a)(x^2 + ax + a^2) \qquad \textit{(difference-of-two-cubes formula)} \tag{7.11}$$

$$x^3 + a^3 = (x + a)(x^2 - ax + a^2) \qquad \textit{(sum-of-two-cubes formula)} \tag{7.12}$$

(It should also be pointed out that in Probs. 5.2.13 and 5.2.14 we obtained some generalizations of these formulas.)

Example 7.2 Express the following sum of rational expressions as a single rational expression:

$$\frac{5x+3}{x^2-4} + \frac{1-x}{x+2} + \frac{3}{2-x} \tag{7.13}$$

Solution By the difference-of-two-squares formula, we have $x^2 - 4 = x^2 - 2^2 = (x + 2) \times (x - 2)$. Hence the least common denominator in (7.13) is $(x + 2)(x - 2)$ [observe that $2 - x = -(x - 2)$], and we now proceed to make all the denominators in (7.13) $(x + 2)(x - 2)$, as indicated below.

$$\frac{5x+3}{x^2-4} = \frac{5x+3}{(x+2)(x-2)}$$

$$\frac{1-x}{x+2} = \frac{(1-x)(x-2)}{(x+2)(x-2)}$$

$$\frac{3}{2-x} = \frac{-3}{x-2} = \frac{-3(x+2)}{(x+2)(x-2)}$$

Hence, now that we have found the necessary common denominator, we may add to obtain

$$\begin{aligned}
\frac{5x+3}{x^2-4} + \frac{1-x}{x+2} + \frac{3}{2-x} &= \frac{5x+3}{(x+2)(x-2)} + \frac{(1-x)(x-2)}{(x+2)(x-2)} + \frac{(-3)(x+2)}{(x+2)(x-2)} \\
&= \frac{5x+3+(1-x)(x-2)-3(x+2)}{(x+2)(x-2)} \\
&= \frac{5x+3+x-2-x^2+2x-3x-6}{(x+2)(x-2)} \\
&= \frac{-x^2+5x-5}{(x+2)(x-2)} \\
&= \frac{-x^2+5x-5}{x^2-4} \qquad \text{if } x \neq 2, -2
\end{aligned}$$

Thus, the desired single rational expression in (7.13) is $(-x^2+5x-5)/(x^2-4)$.

Example 7.3 Express the following product of rational expressions as a rational expression:

$$\frac{x^2-4x-5}{x^2-25}\cdot\frac{8x^3-27}{2x^2-x-3} \tag{7.14}$$

Solution We first try to factor the polynomial expressions in the numerators and denominators above since there may be some common factors. Now, it is easily seen that $x^2-25=x^2-5^2=(x+5)(x-5)$, $8x^3-27=(2x)^3-3^3=(2x-3)(4x^2+6x+9)$, $x^2-4x-5=(x+1)(x-5)$, and $2x^2-x-3=(x+1)(2x-3)$. Substituting the above identities in (7.14), we get

$$\frac{x^2-4x-5}{x^2-25}\cdot\frac{8x^3-27}{2x^2-x-3}=\frac{(x+1)(x-5)}{(x+5)(x-5)}\cdot\frac{(2x-3)(4x^2+6x+9)}{(x+1)(2x-3)} \tag{7.15}$$

Canceling common factors in (7.15) gives

$$\frac{x^2-4x-5}{x^2-25}\cdot\frac{8x^3-27}{2x^2-x-3}=\frac{4x^2+6x+9}{x+5} \qquad \text{if } x\neq 5, -5, -1, 3/2$$

Example 7.4 Express the following quotient of rational expressions as a rational expression:

$$\frac{x^3+27}{x^2-5x+6}\div\frac{x^2-9}{x^3-8} \tag{7.16}$$

Solution By Eqs. (7.10) to (7.12), we have

$$x^2-9=x^2-3^2=(x+3)(x-3)$$
$$x^3-8=x^3-2^3=(x-2)(x^2+2x+4)$$
$$x^3+27=x^3+3^3=(x+3)(x^2-3x+9)$$

Moreover, it is easily checked that $x^2-5x+6=(x-2)(x-3)$. Hence, if we substitute the above identities in (7.16), we get

$$\frac{x^3+27}{x^2-5x+6}\div\frac{x^2-9}{x^3-8}=\frac{(x+3)(x^2-3x+9)}{(x-2)(x-3)}\div\frac{(x+3)(x-3)}{(x-2)(x^2+2x+4)} \tag{7.17}$$

We recall the familiar rule, namely, that to divide, just invert the divisor and multiply. Hence, applying this rule, all we need to do now is to invert (that is, take the reciprocal of) the last fraction on the right side of (7.17) and change $\div$ to $\times$. When we do this and cancel out common factors, we get

$$\begin{aligned}\frac{x^3+27}{x^2-5x+6}\div\frac{x^2-9}{x^3-8}&=\frac{(x+3)(x^2-3x+9)}{(x-2)(x-3)}\times\frac{(x-2)(x^2+2x+4)}{(x+3)(x-3)}\\&=\frac{(x^2-3x+9)(x^2+2x+4)}{(x-3)^2}\\&=\frac{x^4-x^3+7x^2+6x+36}{x^2-6x+9}\end{aligned}$$

Hence,

$$\frac{x^3+27}{x^2-5x+6} \div \frac{x^2-9}{x^3-8} = \frac{x^4-x^3+7x^2+6x+36}{x^2-6x+9} \qquad \text{if } x \neq 2, 3, -3$$

Next, we consider solutions of equations involving rational expressions. In general, we have the following useful fact:

> The solutions of the fractional equation $f(x)/g(x) = 0$ are simply all those, and only those, numbers x such that $f(x) = 0$ but $g(x) \neq 0$, if such x's exist. (7.18)

Example 7.5 Find all solutions of the fractional equation

$$\frac{2x^3-x^2-7x+6}{x^2-x-6} = 0 \tag{7.19}$$

Solution In view of the above fact, to obtain the solutions of Eq. (7.19) we first find the solutions of

$$2x^3-x^2-7x+6=0 \tag{7.20}$$

Now, (7.20) is simply a polynomial equation, and we can, therefore, apply the techniques we learned in Chaps. 5 and 6 to find the solutions of (7.20). Let us first search for a (possible) rational solution. By the Rational Zeros theorem, any rational solution of (7.20) is of the form r/s, where r is an integer which divides 6 and s is an integer which divides 2. Thus, one possibility for a rational solution of (7.20) is $1/1$ $[= 1]$. Substituting $x = 1$ in (7.20), we see that 1 *is* indeed a solution. Hence, by the Factor theorem $x - 1$ is a factor of the polynomial on the left side of (7.20). In fact, as can easily be checked by synthetic division, we have

$$\begin{aligned} 2x^3-x^2-7x+6 &= (x-1)(2x^2+x-6) \\ &= (x-1)(x+2)(2x-3) \end{aligned}$$

and (7.20) is thus equivalent to $(x-1)(x+2)(2x-3) = 0$. The solutions of this equation are clearly 1, -2, and $3/2$. Hence, by (7.18) the solutions of (7.19) are those numbers in $\{1, -2, 3/2\}$ which are not solutions of $x^2 - x - 6 = 0$. Thus, the solutions of (7.19) are 1 and $3/2$ (since -2 is a solution of $x^2 - x - 6 = 0$).

Remark 1 It is very important to stress here that -2 is *not* a solution of (7.19). Indeed, substituting $x = -2$ in the left side of (7.19) gives 0/0, which is not defined. Thus, in finding the solutions of $f(x)/g(x) = 0$ we must always test every candidate for a solution [i.e., a solution of $f(x) = 0$] by substituting this candidate in the denominator $g(x)$. If we get zero as a result, we must reject this candidate, and move on to the next candidate.

Remark 2 The fact that -2 turns out to be a zero of *both* polynomials in (7.19) implies that the rational expression on the left side of (7.19) is *not* in lowest terms.

Indeed, by the Factor theorem $x - (-2)$ $[= x + 2]$ is a factor of both of these polynomials, and, in fact,

$$\frac{2x^3 - x^2 - 7x + 6}{x^2 - x - 6} = \frac{(x-1)(x+2)(2x-3)}{(x-3)(x+2)}$$

$$= \frac{(x-1)(2x-3)}{x-3} \qquad \text{if } x \neq 3, -2 \tag{7.21}$$

In view of (7.21), the solutions of (7.19) are easily seen to be 1 and $3/2$.

Example 7.6 Find all solutions (if any) of the fractional equation

$$\frac{x-1}{x^2-9} - \frac{1}{3(x-3)} = \frac{1}{x+3} \tag{7.22}$$

Solution It is indeed tempting to multiply both sides of (7.22) by $3(x+3)(x-3)$ to clear denominators. If we do this, then [recalling that $x^2 - 9 = (x+3)(x-3)$] Eq. (7.22) becomes

$$3(x-1) - (x+3) = 3(x-3) \tag{7.23}$$

or $2x - 6 = 3x - 9$; that is, $x = 3$. It thus appears that 3 is a solution of (7.22). This, however, is false, because if we substitute $x = 3$ in (7.22), we get something which is utterly meaningless! This situation could be avoided by proceeding a little more cautiously, as follows: First, it is clear that we can write (7.22) in the form

$$\frac{x-1}{x^2-9} - \frac{1}{3(x-3)} - \frac{1}{x+3} = 0$$

If we express these fractions with the common denominator $3(x+3)(x-3)$, we obtain

$$\frac{3(x-1) - (x+3) - 3(x-3)}{3(x+3)(x-3)} = 0$$

or, equivalently,

$$\frac{-(x-3)}{3(x+3)(x-3)} = 0 \tag{7.24}$$

Recalling (7.18), we see that the *only possible* candidate for a solution of Eq. (7.24) [and hence (7.22)] is given by $x = 3$ [this is obtained by setting the numerator in Eq. (7.24) to zero]. However, if $x = 3$, the denominator in (7.24) becomes zero, and hence according to (7.18) 3 is *not* a solution of (7.24) [or, (7.22)]. Thus, Eq. (7.22) has no solutions whatsoever.

Remark By taking a closer look at the first method we considered in attempting to solve (7.22), we observe that all we have shown by that method is the following: If (7.22) has a solution, then this solution is 3. In view of this, we still have to check if 3 is a solution of (7.22). In fact, as we have already seen, 3 is *not* a solution of (7.22). The reason this erroneous solution arose is that Eq. (7.23) is not equivalent to Eq. (7.22). Indeed, if we start with (7.23) and wish to obtain (7.22), we find ourselves forced to divide both sides of (7.23) by $3(x+3)(x-3)$, and this expression is equal to zero when $x=3$! In order to obtain (7.22) from (7.23), then, we will be dividing by zero when $x=3$, and such division, of course, is not valid at all.

Is the composition of two rational expressions always a rational expression? The answer is yes, as illustrated in the following example.

Example 7.7 Suppose that

$$r(x)=\frac{x+3}{2x-5} \quad \text{and} \quad s(x)=\frac{x^2+5x-2}{3x+4} \tag{7.25}$$

Find $(r \circ s)(x)$ and $(s \circ r)(x)$.

Solution By definition of composition, we have the following:

$$\begin{aligned}(r \circ s)(x) &= r(s(x))\\ &= \frac{s(x)+3}{2s(x)-5}\\ &= \frac{(x^2+5x-2)/(3x+4)+3}{2(x^2+5x-2)/(3x+4)-5}\\ &= \frac{x^2+5x-2+3(3x+4)}{2(x^2+5x-2)-5(3x+4)}\\ &= \frac{x^2+14x+10}{2x^2-5x-24}\end{aligned}$$

Thus, if we assume that x is such that none of the above denominators is zero,

$$(r \circ s)(x)=\frac{x^2+14x+10}{2x^2-5x-24}$$

Moreover, by definition of composition

$$\begin{aligned}(s \circ r)(x) &= s(r(x))\\ &= \frac{[r(x)]^2+5r(x)-2}{3r(x)+4}\\ &= \frac{[(x+3)/(2x-5)]^2+5(x+3)/(2x-5)-2}{3(x+3)/(2x-5)+4}\\ &= \frac{(x+3)^2+5(x+3)(2x-5)-2(2x-5)^2}{3(x+3)(2x-5)+4(2x-5)^2}\\ &= \frac{3x^2+51x-116}{22x^2-77x+55}\end{aligned}$$

Thus, if we assume that x is such that none of the above denominators is zero,

$$(s \circ r)(x) = \frac{3x^2 + 51x - 116}{22x^2 - 77x + 55}$$

Observe that $(r \circ s)(x) \neq (s \circ r)(x)$, and thus the order of composition *cannot* be ignored, in general.

Problem Set 7.1

7.1.1 Suppose that R is the set of real numbers and $f: R \to R$ and $g: R \to R$ are the polynomial functions defined by $f(x) = x^3 - 2x + 1$ and $g(x) = x - 2$. Find $(f \circ g)(x)$ and $(g \circ f)(x)$. What are the domains of $f \circ g$ and $g \circ f$? Are $f \circ g$ and $g \circ f$ equal?

7.1.2 Suppose that f and g are the polynomial functions given in Prob. 7.1.1. Write down the rational functions f/g and g/f. Are f/g and g/f in lowest terms? Are f/g and g/f fractional functions? Are f/g and g/f proper fractional functions?

7.1.3 Suppose that f and g are the polynomial functions given in Prob. 7.1.1. Express f/g as a sum of a polynomial function and a proper fractional function.

7.1.4 Suppose that f and g are the polynomial functions given in Prob. 7.1.1. Find the domains of f/g and g/f.

7.1.5 Express each of the following sums and differences of rational expressions as a single rational expression:

(*a*) $\dfrac{2x+3}{x^2-1} + \dfrac{1}{x-1} - \dfrac{5}{x+1}$

(*b*) $\dfrac{13-4x}{x^2-1} - \dfrac{2x+5}{x^2-x-2} + \dfrac{7}{x-2}$

(*c*) $\dfrac{x^2+x-5}{x^3-8} - \dfrac{7x}{2-x} + \dfrac{3x+8}{x^2-4}$

(*d*) $\dfrac{1-x+x^2}{x^3+8} + \dfrac{3+5x}{-x-2} - \dfrac{4x+9}{(x+2)^2}$

7.1.6 Express each of the following products of rational expressions as a single rational expression:

(*a*) $\dfrac{x^2-1}{x^3-1} \cdot \dfrac{x-1}{x+1} \cdot \dfrac{x^2+x+1}{x^2+x-2}$

(*b*) $\dfrac{x^3-27}{x-5} \cdot \dfrac{x^2-9}{x^2-1} \cdot \dfrac{x^2-6x+5}{x^2-x-12}$

(*c*) $\dfrac{x^5-32}{x+2} \cdot \dfrac{x^5+32}{x-2}$

(*d*) $\dfrac{2x^2-7x+3}{1-4x^2} \cdot \dfrac{x^2+2x-3}{9-x^2}$

7.1.7 Express each of the following quotients of rational expressions as a single rational expression:

(a) $\dfrac{8x^3 - 27}{1 - 9x^2} \div \dfrac{3 - 2x}{1 + 6x + 9x^2}$ (b) $\dfrac{x^2 - x - 2}{x^2 + 3x - 4} \div \dfrac{x^3 + 1}{16 - x^2}$

7.1.8 Suppose that n is any *odd* positive integer and a is a constant (i.e., real number). Prove that $(x^n + a^n)/(x + a)$ is a polynomial, and find this polynomial. (Assume $x \neq -a$.)

7.1.9 Suppose that n is any *even* positive integer and a is a nonzero constant. Prove that $(x^n + a^n)/(x + a)$ is not a polynomial. Express $(x^n + a^n)/(x + a)$ as a sum of a polynomial and a proper fractional expression.

7.1.10 Suppose n is *any* positive integer and a is a constant. Prove that $(x^{2n} - a^{2n})/(x^2 - a^2)$ is a polynomial, and find this polynomial. (Assume $x \neq \pm a$.)

7.1.11 Suppose n is *any* positive integer and a is a nonzero constant. Prove that $(x^n + a^n)/(x - a)$ is not a polynomial. Express $(x^n + a^n)/(x - a)$ as a sum of a polynomial and a proper fractional expression.

7.1.12 Suppose n is any *odd* positive integer and a is a nonzero constant. Prove that $(x^n - a^n)/(x + a)$ is not a polynomial. Express $(x^n - a^n)/(x + a)$ as a sum of a polynomial and a proper fractional expression.

7.1.13 Do the following equations have solutions? If so, find them.

(a) $\dfrac{1}{x - 4} - \dfrac{8}{x^2 - 16} = \dfrac{8}{x + 4}$ (c) $\dfrac{2}{x - 1} - \dfrac{2}{x + 1} - \dfrac{1}{x^2 - 1} = 1$

(b) $\dfrac{2x + 3}{x^2 - 9} + \dfrac{15}{2(3 - x)} = \dfrac{1}{2(3 + x)}$

7.1.14 Do the following equations have solutions? If so, find them.

(a) $\dfrac{x^2 + x - 2}{x^2 + 1} = 0$ (c) $\dfrac{x^3 - 2x + 1}{x - 1} = 0$

(b) $\dfrac{x^2 + x - 2}{x^2 - 1} = 0$

7.1.15 Are the following fractional expressions in lowest terms? If not, reduce them to lowest terms. Also, state which of these expressions reduce to polynomial expressions (assuming that x is such that none of the denominators is zero).

(a) $\dfrac{x^2 + x - 2}{x^2 - 1}$

(b) $\dfrac{x^3 - 2x + 1}{x - 1}$

(c) $\dfrac{x^{10} - a^{10}}{x - a}$ (a is a constant)

(d) $\dfrac{x - a}{x^9 + a^9}$ (a is a constant)

(e) $\dfrac{x^3 + x^2 - 10x + 8}{x^2 + x - 6}$

7.2 Linear Rational Functions

In this section, we study an important class of rational functions—*linear rational* functions. Just as *linear polynomial* functions are essentially the simplest type of polynomial functions, it turns out that *linear rational* functions are among the simplest and most interesting types of rational functions. We also consider the graphs of rational functions in this special class, and, as we shall soon see, this leads in a natural way to the concepts of vertical and horizontal asymptotes.

Consider for a moment the rational function f defined by

$$f(x) = \frac{ax + b}{cx + d} \qquad a, b, c, d \text{ fixed real numbers; } c, d \text{ not both zero} \tag{7.26}$$

It may well happen that $f(x)$ above reduces to a constant polynomial (with x restricted). This is indeed the case if, for example, $f(x) = (5x + 1)/(15x + 3)$, $f(x) = (0x - 3)/(0x + 7)$, or $f(x) = (2x + 0)/(9x + 0)$, and so on. This is a rather trivial situation, which we would certainly like to avoid. Thus, the following question suggests itself: Under what conditions does the rational function f given in (7.26) reduce to a constant function? The answer to this question turns out to be the following:

> The rational function f given by
>
> $$f(x) = \frac{ax + b}{cx + d} \qquad \begin{array}{l} a, b, c, d \text{ fixed real numbers,} \\ c \text{ and } d \text{ not both zero, } x \neq -d/c \end{array} \tag{7.27}$$
>
> is a constant function if and only if $ad - bc = 0$.

The proof of this is as follows. First, suppose that

$$\frac{ax + b}{cx + d} = k \qquad k \text{ a constant} \tag{7.28}$$

Then, multiplying both sides of (7.28) by $cx + d$ gives

$$ax + b = (kc)x + (kd) \tag{7.29}$$

for *all* real numbers in the domain of f [i.e., all real x such that $x \neq -d/c$ (if $c \neq 0$); see (7.26)]. Thus, by the theorem on identity of polynomial functions established in Chap. 6, Eq. (7.29) readily implies

$$a = kc \qquad \text{and} \qquad b = kd \tag{7.30}$$

Multiplying the first equation in (7.30) by d and the second one by c, we get $ad = kcd$ and $bc = kcd$. Hence $ad - bc = 0$.

Conversely, if $ad - bc = 0$, then, assuming that $c \neq 0$, we obtain $b = ad/c$, and hence

$$f(x) = \frac{ax + b}{cx + d} = \frac{ax + ad/c}{cx + d} = \frac{acx + ad}{c(cx + d)} = \frac{a(cx + d)}{c(cx + d)} = \frac{a}{c} \tag{7.31}$$

Thus, $f(x) = (ax + b)/(cx + d) = a/c$ is a constant function in this case. Next, suppose that $c = 0$. Then $d \neq 0$, by hypothesis [see (7.26)]. Moreover, since, by hypothesis, $ad = bc$, $ad = 0$. But $d \neq 0$, and hence $a = 0$. The net result, then, is that $a = 0$ *and* $c = 0$. Therefore, in this case

$$f(x) = \frac{ax + b}{cx + d} = \frac{b}{d}$$

which is again a constant. Thus, in *any* case, f is a constant function (if $ad - bc = 0$). This establishes (7.27).

In view of (7.27), we shall be interested in the remainder of this section only in the case in which

$$f(x) = \frac{ax + b}{cx + d} \qquad a, b, c, d \text{ fixed real numbers, } ad - bc \neq 0,\ x \neq \frac{-d}{c} \tag{7.32}$$

The rational function f in (7.32) is called a *linear rational function.* We recall that in Chap. 2 we defined a linear polynomial function to be a function $f: R \to R$ of the set R of real numbers into R such that

$$f(x) = ax + b \qquad a, b \text{ fixed real numbers, } a \neq 0 \tag{7.33}$$

Now, because $f(x) = (ax + b)/(0x + 1)$, we see that every linear polynomial function f is also a linear rational function. Thus, the set of all linear rational functions contains the set of linear polynomial functions. It is also easy to see that every linear polynomial function $f: R \to R$ is both one-to-one and onto, and hence f has an inverse. Indeed, if one were to solve the equation $y = f(x) = ax + b$ for x, one would get $x = (1/a)y + (-b/a)$ [recall that $a \neq 0$, see (7.33)]. Hence the inverse of the linear polynomial function $f: R \to R$ given in (7.33) is the linear polynomial function $g: R \to R$ given by

$$f^{-1}(x) = g(x) = \frac{1}{a}x + \frac{-b}{a} \qquad a \neq 0 \tag{7.34}$$

In fact, the student will have no difficulty in verifying that if $f \circ g$ denotes the composition of f and g as defined in Chap. 1, then

$$(f \circ g)(x) = (g \circ f)(x) = x \qquad \text{for all real numbers } x$$

The above facts give rise to the following question: Does every nonconstant linear rational function have an inverse? Before answering this question, let us assume once and for all that f is a nonconstant linear rational function which is not a linear polynomial function (since we already answered this question in the affirmative when f is a linear polynomial function). Thus, in view of (7.27) f now satisfies the following conditions:

$$f(x) = \frac{ax + b}{cx + d} \qquad \begin{array}{l} a, b, c, d \text{ fixed real numbers;} \\ ad - bc \neq 0,\ c \neq 0,\ x \neq -d/c \end{array} \tag{7.35}$$

Now, if we solve the equation $y = f(x) = (ax + b)/(cx + d)$ for x, we easily see that $y(cx + d) = ax + b$; hence, $x = (-dy + b)/(cy - a)$. This suggests that the linear rational function f given in (7.35) has an inverse g, where g is given by

$$g(x) = \frac{-dx + b}{cx - a} \qquad a, b, c, d \text{ as in (7.35)} \tag{7.36}$$

Indeed, if we keep in mind that $ad - bc \neq 0$, it can be verified that if f and g are as in (7.35) and (7.36), then (see Example 7.8 below)

$$(f \circ g)(x) = f(g(x)) = x \qquad \begin{array}{l}\text{for all } x \text{ in the domain of } g\\ \text{(that is, } x \neq a/c)\end{array} \tag{7.37}$$

and

$$(g \circ f)(x) = g(f(x)) = x \qquad \begin{array}{l}\text{for all } x \text{ in the domain of } f\\ \text{(that is, } x \neq -d/c)\end{array} \tag{7.38}$$

Moreover, the computation which led to (7.36) shows that for any real number y, $y \neq a/c$, there exists a real number x such that $f(x) = y$. [In fact, one may take $x = (-dy + b)/(cy - a)$.] This shows that

$$\begin{aligned}\text{Range of } f &= \text{set of all real number except } \frac{a}{c}\\ &= \text{domain of } g\end{aligned} \tag{7.39}$$

[Observe that a/c is *not* in the range of f since $f(x) = a/c$ implies $(ax + b)/(cx + d) = a/c$, and hence $ad = bc$, contrary to what is given in (7.35).] Similarly, we show that

$$\begin{aligned}\text{Range of } g &= \text{set of all real numbers except } \frac{-d}{c}\\ &= \text{domain of } f\end{aligned} \tag{7.40}$$

Now, in view of Eqs. (7.37) to (7.40) we say that g is an *inverse* of f. We leave it to the student to further verify that both f and g are one-to-one functions since $ad - bc \neq 0$. We summarize our results above as follows:

(7.41)

> The linear rational function f given by
>
> $$f(x) = \frac{ax + b}{cx + d} \qquad a, b, c, d \text{ real; } ad - bc \neq 0;\ c \neq 0$$
>
> has a domain consisting of the set R of real numbers except $-d/c$ and has a range consisting of R except a/c. Moreover, f has an inverse g given by
>
> $$g(x) = \frac{-dx + b}{cx - a}$$
>
> The domain of g is equal to the range of f, and the range of g is equal to the domain of f. Finally, both f and g are one-to-one functions.

It is noteworthy to observe also that the function g given in (7.36) is of the same "type" as the function f given in (7.35) in the sense that the coefficients involved in g (namely, $-d$, b, c, and $-a$) do satisfy the conditions given in (7.35); that is, $(-d)(-a) - bc \neq 0$, and $c \neq 0$.

Remark We have encountered above the concept of inverse of a function in a slightly more general setting than we did in Chap. 1, where we introduced the concept of inverse of a function for the first time. However, our discussion of an inverse of a function (in Chap. 1) remains valid here with only minor (and obvious) modifications. Indeed, if $D(f)$ and $R(f)$ denote the domain of f and range of f, respectively, then, as we have already seen above,

$f:D(f) \to R(f)$ is both one-to-one and onto

$g:R(f) \to D(f)$ is both one-to-one and onto

and our discussion of inverse of a function as given in Chap. 1 holds here also, with the obvious change in notation because $D(f)$ and $R(f)$ are now not quite the same set (as was the case in Chap. 1).

Example 7.8 Find the domain, range, and inverse of the linear rational function f given by

$$f(x) = \frac{-3x+4}{2x-3} \tag{7.42}$$

Solution Here $a = -3$, $b = 4$, $c = 2$, and $d = -3$, and hence $ad - bc = 1 \neq 0$. Thus, by (7.27), f is a nonconstant linear rational function. Now, by (7.41) we know that

Domain of f is the set R of real numbers except $\dfrac{-d}{c} = \dfrac{3}{2}$

Range of f is the set R of real numbers except $\dfrac{a}{c} = \dfrac{-3}{2}$

Moreover, if we solve the equation $y = (-3x+4)/(2x-3)$ for x, we get $x = (3y+4)/(2y+3)$, and hence [see (7.41)] the inverse of f is the linear rational function g given by

$$g(x) = \frac{3x+4}{2x+3} \tag{7.43}$$

Observe that

$$\text{Domain of } g = \text{range of } f = \{x | x \text{ real}, x \neq -3/2\} \tag{7.44}$$

$$\text{Range of } g = \text{domain of } f = \{x | x \text{ real}, x \neq 3/2\} \tag{7.45}$$

It is instructive to verify directly our above claim that g is an inverse of f. Indeed, using (7.42) and (7.43), we have

$$(f \circ g)(x) = f(g(x)) = \frac{-3g(x) + 4}{2g(x) - 3} = \frac{-3(3x+4)/(2x+3) + 4}{2(3x+4)/(2x+3) - 3}$$

$$= \frac{-3(3x+4) + 4(2x+3)}{2(3x+4) - 3(2x+3)} = \frac{-x}{-1} = x$$

for all $x \neq -3/2$. We have thus shown that

$$(f \circ g)(x) = x \qquad \text{for all } x \text{ in the domain of } g \tag{7.46}$$

Moreover, using (7.42) and (7.43) again we get

$$(g \circ f)(x) = g(f(x)) = \frac{3f(x) + 4}{2f(x) + 3} = \frac{3(-3x+4)/(2x-3) + 4}{2(-3x+4)/(2x-3) + 3}$$

$$= \frac{3(-3x+4) + 4(2x-3)}{2(-3x+4) + 3(2x-3)} = \frac{-x}{-1} = x$$

for all $x \neq 3/2$. We have thus shown that

$$(g \circ f)(x) = x \qquad \text{for all } x \text{ in the domain of } f \tag{7.47}$$

In view of Eqs. (7.44) to (7.47), g is indeed an inverse of f.

We now direct our attention to the graphs of linear rational functions. Thus, suppose that f is the linear rational function given by

$$f(x) = \frac{ax+b}{cx+d} \qquad a, b, c, d \text{ fixed real numbers; } ad - bc \neq 0;\ c \neq 0 \tag{7.48}$$

What is the approximate value of $f(x)$ when x is very large? To answer this question, let us first divide both numerator and denominator of the fraction in (7.48) by x; we get

$$f(x) = \frac{a + b/x}{c + d/x} \qquad (x \neq 0) \tag{7.49}$$

When x is very large, both b/x and d/x become very close to zero since a fraction with a fixed numerator and very large denominator is close to zero. Hence, by Eq. (7.49) $f(x)$ becomes very close to $(a+0)/(c+0) = a/c$. We have thus shown that

$$f(x) \text{ is very close to } \frac{a}{c} \text{ when } x \text{ is very large} \tag{7.50}$$

We describe the statement in (7.50) by saying that

$$f(x)\ \textit{approaches}\ \frac{a}{c} \text{ as } x \text{ grows large } \textit{without bound} \tag{7.51}$$

Thus, if we set $y = f(x)$, then (7.51) becomes

$$y \text{ approaches } \frac{a}{c} \text{ as } x \text{ grows large without bound} \tag{7.52}$$

We describe this situation by saying that the horizontal line $y = a/c$ is a *horizontal asymptote* of the graph of $y = f(x) = (ax + b)/(cx + d)$.

Next, we ask the following question: For what values of x does $f(x)$ become very large? To answer this question, observe that when x is very close to $-d/c$, $cx + d$ is very close to $c(-d/c) + d = 0$. On the other hand, when x is very close to $-d/c$, $ax + b$ is very close to $a(-d/c) + b = (-ad + bc)/c \neq 0$ since $ad - bc \neq 0$ [see (7.48)]. The net result, then, is the following:

> When x is very close to $-d/c$, $cx + d$ is very close to zero, but $ax + b$ is not close to zero. Hence in this case $(ax + b)/(cx + d)$ becomes very large since a fraction whose numerator is not close to zero but whose denominator is close to zero is very large. (7.53)

The assertion made in (7.53) is usually expressed by saying that

$$\text{When } x \text{ approaches } \frac{-d}{c},\ f(x) = \frac{ax + b}{cx + d} \text{ grows large without bound.} \tag{7.54}$$

We describe this situation by saying that the vertical line $x = -d/c$ is a *vertical asymptote* of the graph of $y = f(x) = (ax + b)/(cx + d)$. We summarize our results as follows:

> The graph of the linear rational function f given by
>
> $$f(x) = \frac{ax + b}{cx + d} \qquad a, b, c, d \text{ fixed real numbers, } ad - bc \neq 0,\ c \neq 0 \tag{7.55}$$
>
> has the line $y = a/c$ as a horizontal asymptote and has the line $x = -d/c$ as a vertical asymptote.

The following example illustrates the above ideas.

Example 7.9 Find the domain, range, and horizontal and vertical asymptotes arising from (*a*) $f(x) = 1/x$; (*b*) $f(x) = -2/x$; (*c*) $f(x) = 1/(x + 1)$; and (*d*) $f(x) = (4x + 7)/(-2x - 6)$. Also, sketch each graph.

Solution (*a*) The domain of f is clearly the set R of all real numbers except zero. Moreover, since $y = f(x) = 1/x$ implies that $x = 1/y$, the range of f is also the set R of all real numbers except zero. Now, using (7.55), we know that $y = 0$ is a horizontal asymptote of the graph of f and $x = 0$ is a vertical asymptote of the graph of f. Let us construct a table of values as indicated below.

x	1	2	3	−1	−2	−3	1/2	1/3	−1/2	−1/
$y = 1/x$	1	1/2	1/3	−1	−1/2	−1/3	2	3	−2	−

A sketch of the graph of $y = f(x) = 1/x$ appears in Fig. 7.1. Observe that as $|x|$ becomes very large, $f(x)$ approaches zero, and this is what is meant by saying that $y = 0$ (i.e., the x axis) is a horizontal asymptote of the graph of f. Similarly, we observe that as x approaches zero (through positive or negative values), $f(x)$ becomes very large numerically, and this is what is meant by $x = 0$ (i.e., the y axis) is a vertical asymptote of the graph of f.

(*b*) Arguing exactly as in part (*a*), we easily see that with $f(x) = -2/x$ the domain of $f = \{x|x \text{ real}, x \neq 0\}$, the range of $f = \{x|x \text{ real}, x \neq 0\}$, the horizontal asymptote of the graph of f is $y = 0$, and the vertical asymptote of the graph of f is $x = 0$. Let us now construct a table of values as indicated below.

x	1	2	3	4	−1	−2	−3	−4	½	⅓	−½	−⅓
$y = -2/x$	−2	−1	−⅔	−½	2	1	⅔	½	−4	−6	4	6

A sketch of the graph of $y = f(x) = -2/x$ appears in Fig. 7.2. Observe that as x becomes very large numerically, $f(x)$ approaches zero, and hence $y = 0$ (i.e., the x axis) is a horizontal asymptote of the graph of f. Similarly, we observe that as x approaches zero (through positive or negative values), $f(x)$ becomes very large numerically, and hence $x = 0$ (i.e., the y axis) is a vertical asymptote of the graph of f.

(*c*) Since $f(x) = (0x + 1)/(1x + 1)$, we have $a = 0$, $b = 1$, $c = 1$, and $d = 1$, and hence by (7.41) we know that the domain of $f = \{x|x \text{ real}, x \neq -1\}$ $[-1 = -d/c]$ and the range of $f = \{x|x \text{ real}, x \neq 0\}$ $[0 = a/c]$. Moreover, by (7.55) we also know that $y = a/c = 0$ is a horizontal asymptote of the graph of f and $x = -d/c = -1$ is a vertical asymptote of the graph of f. Let us now construct a table of values as indicated below.

x	0	1	2	3	−2	−3	−3/2	−5/4	−½
$y = 1/(x + 1)$	1	½	⅓	¼	−1	−½	−2	−4	2

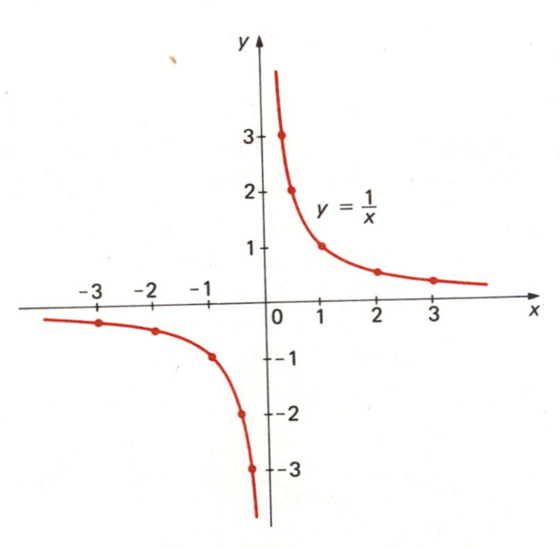

Fig. 7.1 Graph of $y = f(x) = 1/x$. The asymptotes are $y = 0$ (x axis) and $x = 0$ (y axis).

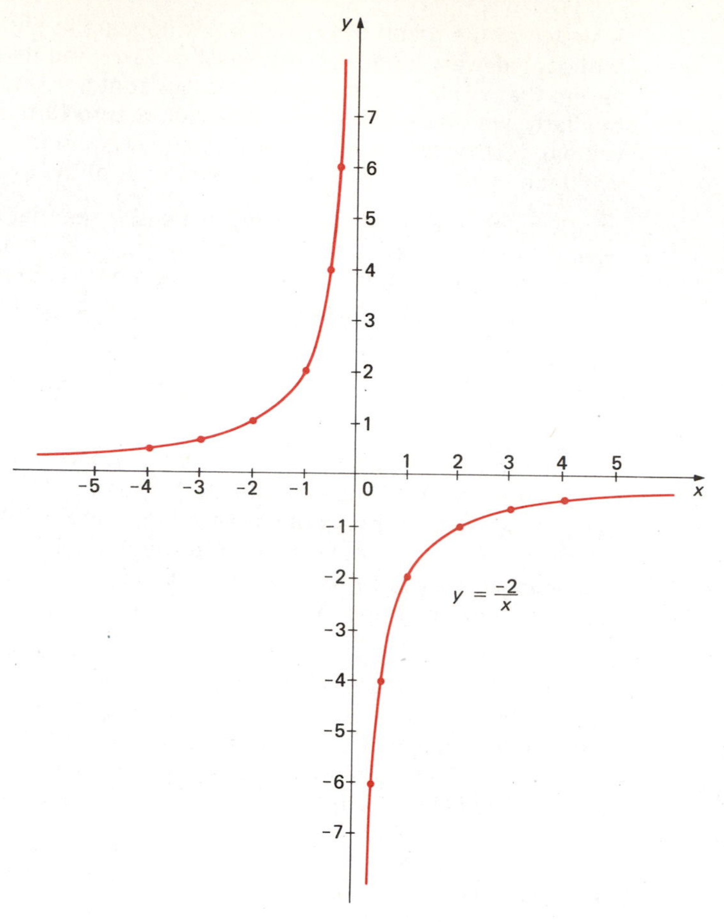

Fig. 7.2 Graph of $y = f(x) = -2/x$. The asymptotes are $y = 0$ (x axis) and $x = 0$ (y axis).

A sketch of the graph of $y = f(x) = 1/(x+1)$ appears in Fig. 7.3. Observe that as x grows large numerically, $f(x)$ approaches zero, and as x approaches -1 (from either side), $|f(x)|$ grows large without bound.

(*d*) Since $f(x) = (4x+7)/(-2x-6)$, we have $a = 4$, $b = 7$, $c = -2$, and $d = -6$, which by (7.41) tells us that the domain of $f = \{x | x \text{ real}, x \neq -3\}$ $[-3 = -d/c]$ and the range of $f = \{x | x \text{ real}, x \neq -2\}$ $[-2 = a/c]$. Moreover, by (7.55) we also know that $y = a/c = -2$ is a horizontal asymptote of the graph of f and $x = -d/c = -3$ is a vertical asymptote of the graph of f. We now construct a table of values as indicated below.

x	0	1	2	-1	-2	$-5/2$	$-7/2$	-4	-5	-6
$y = (4x+7)/(-2x-6)$	$-7/6$	$-11/8$	$-3/2$	$-3/4$	$1/2$	3	-7	$-9/2$	$-13/4$	$-17/$

A sketch of the graph of $y = f(x) = (4x+7)/(-2x-6)$ appears in Fig. 7.4. Observe that as x grows large numerically, $f(x)$ approaches $4/(-2) = -2$.

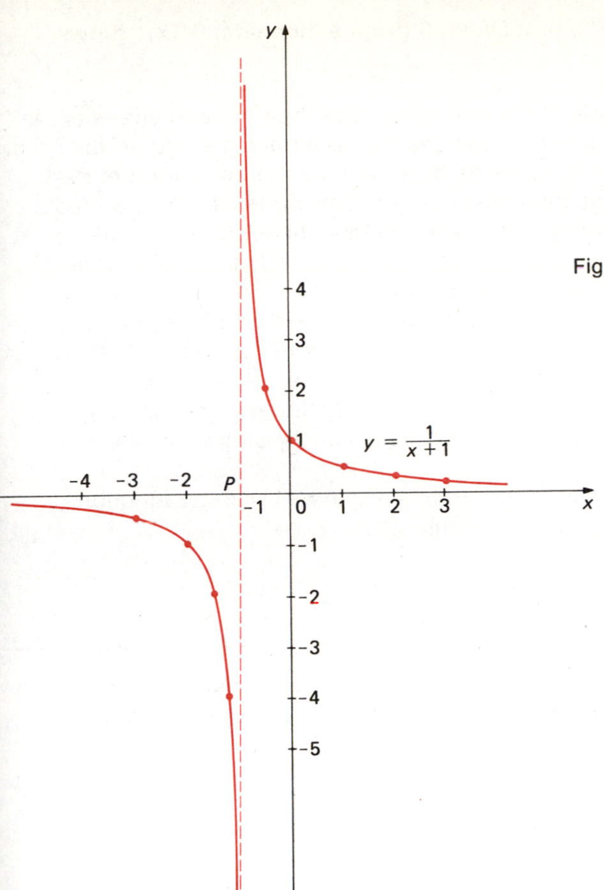

Fig. 7.3 Graph of $y = f(x) = 1/(x+1)$. The asymptotes are $y = 0$ (x axis) and $x = -1$. Observe that (0,1) is the y intercept point but there is *no* x intercept point. (Why?)

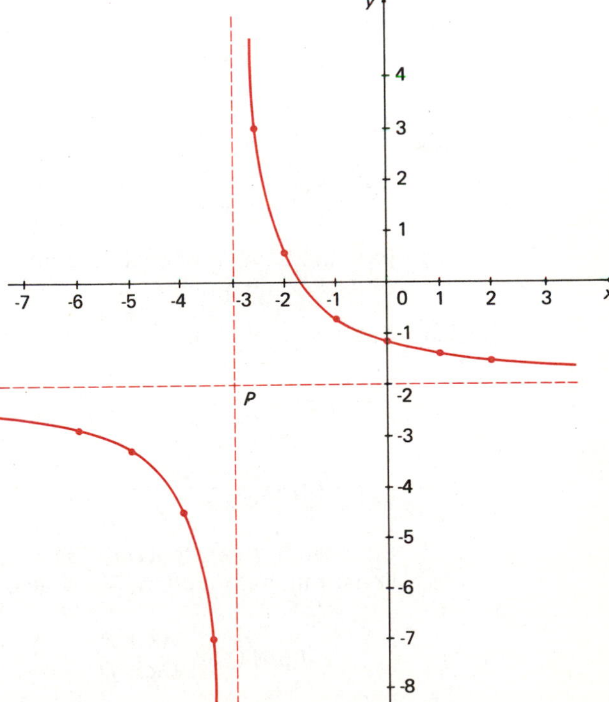

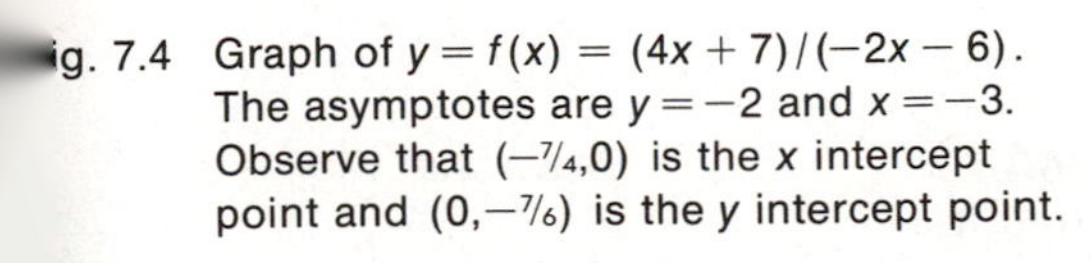
ig. 7.4 Graph of $y = f(x) = (4x+7)/(-2x-6)$. The asymptotes are $y = -2$ and $x = -3$. Observe that $(-7/4, 0)$ is the x intercept point and $(0, -7/6)$ is the y intercept point.

Moreover, observe that as x approaches -3 (from either side), $|f(x)|$ grows large without bound.

A glance at the graphs in Figs. 7.1 to 7.4 shows that their general shape is the same. We also observe that in each case the two asymptotes separate the graph into two parts and that each of these parts is the reflection of the other part through the point P of intersection of these two asymptotes. (This is, in fact, true; however, we omit the proof.) We describe this phenomenon by saying that the graph is *symmetric* with respect to this point P. The above graphs of the linear rational functions given in Example 7.9 are called *hyperbolas.* A hyperbola is a special type of a conic section. A detailed treatment of conic sections will appear in Chap. 11. In the meantime, it suffices to say that as soon as we know the horizontal and vertical asymptotes of the graph of a linear rational function [see (7.55)], then we already know its general shape (see above graphs), and a brief table of values is usually sufficient to sketch the graph.

In the next section, we consider the graphs of general rational functions, with particular emphasis on locating asymptotes.

Problem Set 7.2

7.2.1 Determine whether the linear rational functions given below are constant functions. If so, find such constants. Assume that each denominator is nonzero.

(a) $f(x) = \dfrac{2x - 3}{-2x + 3}$

(b) $f(x) = \dfrac{-x + 1/2}{1/3x - 1/6}$

(c) $f(x) = \dfrac{\sqrt{2}x - 3}{x - 3/\sqrt{2}}$

(d) $f(x) = \dfrac{x - 1}{x + 1}$

7.2.2 Find the domain of each function given in Prob. 7.2.1, and sketch the graph of each function.

7.2.3 Suppose that f and g are linear rational functions given by

$$f(x) = \frac{ax + b}{cx + d}$$

$$g(x) = \frac{a'x + b'}{c'x + d'}$$

Suppose, further, that $ad - bc = 1$ and $a'd' - b'c' = 1$. Prove that $f \circ g$ is a linear rational function. Show also that if

$$(f \circ g)(x) = \frac{Ax + B}{Cx + D}$$

then $AD - BC = 1$. What is the domain of $f \circ g$?

7.2.4 Suppose that f and g are linear rational functions given by

$$f(x) = \frac{ax + b}{cx + d}$$

$$g(x) = \frac{a'x + b'}{c'x + d'}$$

and suppose $ad - bc \neq 0$ and $a'd' - b'c' \neq 0$. Prove that $f \circ g$ is a nonconstant linear rational function. (We are *not* now assuming that $c \neq 0$ or that $c' \neq 0$.) What is the domain of $f \circ g$?

7.2.5 Explain why a constant function can never have an inverse which is a function.

7.2.6 Prove that every nonconstant linear rational function is one-to-one. Is this true for constant linear rational functions?

7.2.7 Find the domain, range, and inverse of each of the linear rational functions given by:

(*a*) $f(x) = \dfrac{5}{x}$

(*b*) $f(x) = \dfrac{-4}{x}$

(*c*) $f(x) = \dfrac{3}{x - 1}$

(*d*) $f(x) = \dfrac{2}{3x - 5}$

(*e*) $f(x) = \dfrac{2x}{3x + 7}$

(*f*) $f(x) = \dfrac{5x - 2}{3x}$

(*g*) $f(x) = \dfrac{4x - 3}{7x + 1}$

(*h*) $f(x) = \dfrac{2x - 1}{-3x + 2}$

(*i*) $f(x) = \dfrac{3x - 5}{x - 2}$

7.2.8 Sketch the graph of each linear rational function in Prob. 7.2.7. In each case, find the x and y intercept points (if they exist). Also, find the asymptotes of each graph.

7.2.9 Suppose that $f(x) = (ax + b)/(cx + d)$, where a, b, c, and d are real, $ad - bc \neq 0$, and $c \neq 0$. Prove that f^{-1} exists and, moreover, $f = f^{-1}$ if and only if $d = -a$.

7.2.10 Suppose that f is as given in Prob. 7.2.9 except that we now assume $c = 0$ and $b \neq 0$. Prove that f^{-1} exists and, moreover, $f = f^{-1}$ if and only if $d = -a$.

7.2.11 Suppose that f is as given in Prob. 7.2.9 except that we now assume $c=0$ and $b=0$. Prove that f^{-1} exists and, moreover, $f=f^{-1}$ if and only if $d=\pm a$.

7.2.12 Suppose that f is as given in Prob. 7.2.9. Prove that if $g=f^{-1}$, then the graphs of $y=f(x)$ and $y=g(x)$ have the same asymptotes if and only if $d=-a$.

7.2.13 Suppose that f is a linear rational function which is *not* a polynomial function. Show that f^{-1} exists and, moreover, $f=f^{-1}$ if and only if the graphs of $y=f(x)$ and $y=f^{-1}(x)$ have the same asymptotes.

7.2.14 Find a linear rational function f such that the graph of $y=f(x)$ passes through the points $(0,1)$, $(1,0)$, and $(2,-2)$.

7.2.15 Find the domain and range of the function f you obtained in Prob. 7.2.14. Also, find the asymptotes of the graph of $y=f(x)$, and sketch the graph.

7.2.16 Find the inverse of the function f you obtained in Prob. 7.2.14. Also, find the domain and range of f^{-1}. What are the asymptotes of the graph of $y=f^{-1}(x)$? Sketch this graph.

7.2.17 Find a linear rational function f such that the asymptotes of $y=f(x)$ are $x=1$ and $y=2$ given that the point $(3,-2)$ lies on the graph of $y=f(x)$.

7.3 Rational Functions and Their Graphs

In this section, we discuss some of the properties which relate to the graphs of arbitrary rational functions. We recall that in the previous section we studied in some detail the properties of linear rational functions. In particular, we found that the graph of every linear rational function is either a straight line or a hyperbola and that in the latter case the graph has exactly one horizontal asymptote and one vertical asymptote. For the graph of an arbitrary rational function, the situation may not be that simple. Indeed, there may be many asymptotes, not all of which are horizontal or vertical lines. Also, we consider in this section the question of whether the graph of a rational function is symmetric with respect to the y axis or the origin, and this, in turn, gives rise in a natural way to the concept of *even* and *odd* functions.

Now, suppose that

$$f(x)=\frac{a_nx^n+a_{n-1}x^{n-1}+\cdots+a_1x+a_0}{b_mx^m+b_{m-1}x^{m-1}+\cdots+b_1x+b_0}=\frac{g(x)}{h(x)} \qquad \text{all } a_i \text{ and all } b_j \text{ are real, } a_n\neq 0,\ b_m\neq 0 \qquad (7.56)$$

is any rational function *in lowest terms.* We certainly may assume this since, as we have seen in Sec. 7.1, every rational function can be expressed in lowest terms (by the simple device of canceling out all the common factors of positive degree). Our present objective is to gain as much information as possible to facilitate the sketching of the graph of the function given in (7.56). To begin with, observe that if $m = 0$, then $f(x)$ in (7.56) is simply a polynomial. Also, if $n = 1$ and $m = 1$, then $f(x)$ in (7.56) reduces to a linear rational function, which we studied in Sec. 7.2. In fact, $f(x)$ in (7.56) represents the most general form of a rational function.

What are the intercepts of the graph of the function given in (7.56)? The answer is quite simple. Indeed, we have the following facts:

> The x intercepts of the graph of the function f given in (7.56) are all those (if any) and only those real numbers x such that
>
> $$a_nx^n + a_{n-1}x^{n-1} + \cdots + a_1x + a_0 = 0 \tag{7.57}$$
>
> Moreover, the y intercept point is precisely the point $(0, a_0/b_0)$ if $b_0 \neq 0$. However, if $b_0 = 0$, then there is no y intercept point (i.e., the graph does not intersect the y axis at all).

The verification of (7.57) is straightforward if one keeps in mind that because the rational function $f(x) = g(x)/h(x)$ is *in lowest terms*, $g(x)$ and $h(x)$ cannot possibly have a zero x_0 in common (for if they did, then by the Factor theorem they would both have $x - x_0$ as a common factor).

As we have seen in Sec. 7.1, if R denotes the set of all real numbers, S denotes the set of all real zeros of $h(x)$, and $R\backslash S$ denotes the complement of S in R (i.e., the set of all x in R but not in S), then $f: R\backslash S \to R$; that is, the domain of f is $R\backslash S$, and the codomain of f is R.

Next, we consider symmetry. By definition, the graph of a function f [or equation $y = f(x)$] is said to be *symmetric* with respect to the y axis if the following is true (see Fig. 7.5):

$$\text{Whenever } (x,y) \text{ is on the graph of } y = f(x), \text{ so is } (-x,y). \tag{7.58}$$

This leads to the following test for symmetry with respect to the y axis (see Fig. 7.5):

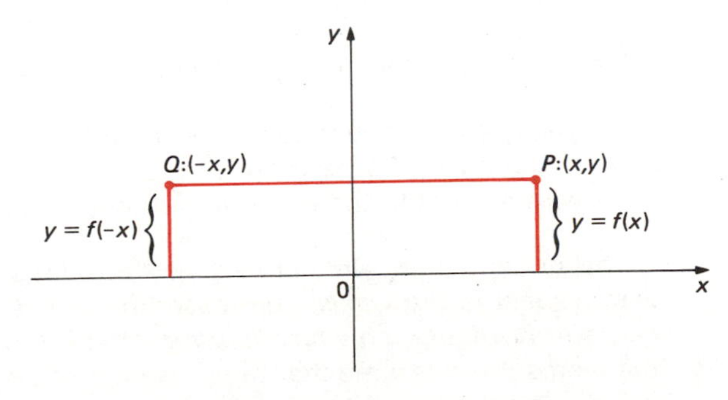

Fig. 7.5 Symmetry with respect to the y axis. Both P and Q are on the graph of $y = f(x)$. Observe that the y axis is the perpendicular bisector of the line segment PQ.

The graph of $y = f(x)$ is symmetric with respect to the y axis if $f(-x) = f(x)$ for all x in the domain of f. Thus, the part of the graph of $y = f(x)$ in the second and third quadrants is simply the reflection with respect to the y axis of the part of the graph of $y = f(x)$ in the first and fourth quadrants; and vice versa. (7.59)

In view of this, if a graph of a given function f happens to be symmetric with respect to the y axis, then as soon as we determine the part of the graph of f which lies in the first and fourth quadrants, say, the rest of the graph is simply the reflection of this part with respect to the y axis.

At this point we consider symmetry with respect to the origin. By definition, the graph of a function f [or equation $y = f(x)$] is said to be *symmetric* with respect to the *origin* if the following is true (see Fig. 7.6):

Whenever (x,y) lies on the graph of $y = f(x)$, so does $(-x,-y)$.

This leads to the following test for symmetry with respect to the origin (see Fig. 7.6):

The graph of $y = f(x)$ is symmetric with respect to the origin if $f(-x) = -f(x)$ for all x in the domain of f. Thus, the part of the graph of $y = f(x)$ in the second and third quadrants is simply the reflection with respect to the origin of the part of the graph of $y = f(x)$ in the first and fourth quadrants; and vice versa. (7.60)

In view of this, if a graph of a given function f happens to be symmetric with respect to the origin, then as soon as we determine the part of the graph of f which lies in the first and fourth quadrants, say, the rest of the graph is simply the reflection of this part with respect to the origin.

In the above discussion, we encountered what are known as even and odd functions, in the sense of the following definition:

A function f is ***even*** *if* $f(-x) = f(x)$ *for all x in the domain of f. Moreover, a function f is* ***odd*** *if* $f(-x) = -f(x)$ *for all x in the domain of f.* (7.61)

Thus, we may summarize our results regarding symmetry as follows:

The graph of a function f is symmetric with respect to the y axis if and only if f is an even function. The graph of f is symmetric with respect to the origin if and only if f is an odd function. (7.62)

The student might wonder why we have not said anything about symmetry *with respect to the x axis.* The reason for this is that *no nonzero function f can ever have a graph which is symmetric with respect to the x axis*! Indeed, the moment we assume that the presence of the point (x,y) on the graph of $y = f(x)$ always forces the point $(x,-y)$ to also lie on the graph of $y = f(x)$ (which is what symmetry with respect to the x axis means), then we must have

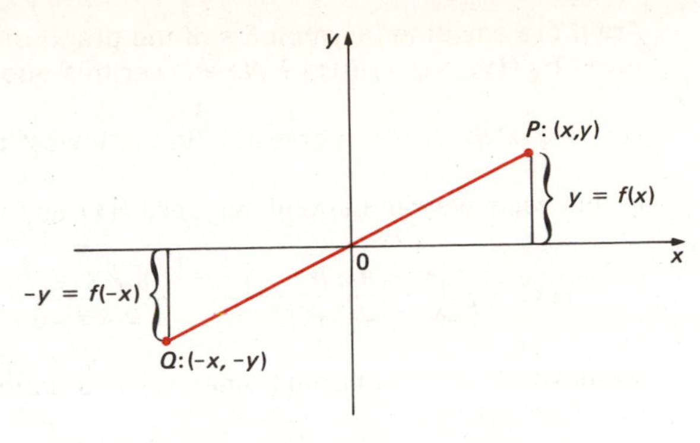

Fig. 7.6 Symmetry with respect to the origin. Both P and Q lie on the graph of $y = f(x)$. Observe that the origin is the midpoint of the line segment PQ.

$y = f(x)$ because (x,y) is on the graph of $y = f(x)$ and $-y = f(x)$ since $(x,-y)$ is on the graph of $y = f(x)$. Thus, $f(x) = -f(x)$, and hence $f(x) = 0$ for *all* x in the domain of f, contrary to our assumption that f is a *nonzero* function. Incidentally, in Chap. 11, we will discuss the graphs of certain *equations* (such as $x^2 + y^2 = 1$), and in such a situation we will be able to talk about symmetry with respect to the x axis. Observe, however, that the equation $x^2 + y^2 = 1$, as it stands, does *not* define a *function* [since, for example, *both* (0,1) and (0,−1) lie on the graph of this equation].

We now direct our attention to finding the asymptotes of the rational function given by

$$f(x) = \frac{g(x)}{h(x)} \qquad h(x) \text{ of positive degree; all coefficients of the polynomials } g(x),\ h(x) \text{ real; } f(x) \text{ in lowest terms}$$

[Observe that if $h(x)$ is of degree 0, then $f(x)$ is a polynomial; therefore, $f(x)$ has *no asymptotes* in this case. For this reason, we have assumed that $h(x)$ is of positive degree.] First, let us consider vertical asymptotes. Suppose that x_0 is a zero of $h(x)$; that is, $h(x_0) = 0$. Since, by hypothesis $g(x)/h(x)$ is in lowest terms, $g(x_0) \neq 0$ [for if $g(x_0) = 0$, then by the Factor theorem $x - x_0$ is a common factor of *both* $g(x)$ and $h(x)$ — contradiction]. In view of this, $f(x) = g(x)/h(x)$ becomes very large as x approaches x_0, and hence the line $x = x_0$ is a vertical asymptote of the graph of $y = f(x) = g(x)/h(x)$. Clearly, there are *no* other vertical asymptotes [since all other choices for x_0 make $f(x) = g(x)/h(x)$ assume a *finite* value]. We have thus shown that

> The vertical asymptotes of the graph of the rational function f given by
>
> $$f(x) = \frac{g(x)}{h(x)} \qquad f(x) \text{ is real and in lowest terms} \tag{7.63}$$
>
> are precisely the vertical lines $x = x_0$, where x_0 is any real zero of $h(x)$, if such zeros exist. If $h(x)$ has no real zeros, then the graph of f has no vertical asymptote.

Are there any other asymptotes of the graph of the above rational function f given by $f(x) = g(x)/h(x)$? We answer this question in two stages.

Case 1

$f(x) = g(x)/h(x)$ is a *proper* fraction in lowest terms

In this case, we can certainly express $f(x)$ as follows:

$$f(x) = \frac{a_n x^n + a_{n-1}x^{n-1} + \cdots + a_1 x + a_0}{b_m x^m + b_{m-1}x^{m-1} + \cdots + b_1 x + b_0} \qquad (n < m,\ b_m \neq 0)$$

Moreover, by dividing both numerator and denominator of the above fraction by x^m we get

$$f(x) = \frac{\dfrac{a_n}{x^{m-n}} + \dfrac{a_{n-1}}{x^{m-n+1}} + \cdots + \dfrac{a_1}{x^{m-1}} + \dfrac{a_0}{x^m}}{b_m + \dfrac{b_{m-1}}{x} + \cdots + \dfrac{b_1}{x^{m-1}} + \dfrac{b_0}{x^m}} \tag{7.64}$$

Now, as x becomes very large, every little fraction (with x in the denominator) on the right side of (7.64) becomes very close to zero, and hence the right side of (7.64) approaches $0/b_m = 0$ (since $b_m \neq 0$). In other words,

$f(x)$ approaches 0 as x becomes very large (7.65)

In view of (7.65), we have (see Sec. 7.2)

$y = 0$ (i.e., the x axis) is a horizontal asymptote of the graph of $y = f(x) = g(x)/h(x)$ when $f(x)$ is a *proper* fraction in lowest terms. (7.66)

Case 2

$f(x) = g(x)/h(x)$ is an *improper* fraction in lowest terms

In this case, we can, using long division, write

$$\frac{g(x)}{h(x)} = p(x) + \frac{r(x)}{h(x)} \qquad p(x) \text{ is a polynomial, } \frac{r(x)}{h(x)} \text{ is a } \textit{proper} \text{ fraction in lowest terms} \tag{7.67}$$

[Observe that $r(x) \neq 0$ since if $r(x) = 0$, then $g(x)/h(x) = p(x)/1$, which contradicts the hypothesis that $g(x)/h(x)$ is in lowest terms.] Now, what happens when x becomes very large in (7.67)? By what we have already shown in Case 1 the proper fraction $r(x)/h(x)$ approaches zero as x becomes very large [see (7.65)], and, hence, by (7.67) we have

$f(x)\ \left[= \dfrac{g(x)}{h(x)}\right]$ approaches $p(x)$ as x becomes very large (7.68)

In view of (7.68), we say that the graph of $y = p(x)$ is an *asymptotic curve* of the graph of $y = f(x)$.

We summarize our results as follows:

If $f(x) = g(x)/h(x)$ is a proper fraction in lowest terms, then $y = 0$ (i.e., the x axis) is a horizontal asymptote of the graph of $y = f(x)$; while if $f(x) = g(x)/h(x)$ is improper, then $y = p(x)$ is an asymptotic curve of the graph of $y = f(x)$, where the polynomial $p(x)$ is the quotient obtained when $g(x)$ is divided by $h(x)$ as indicated in (7.67). (7.69)

Observe that if in (7.67) $p(x)$ is a constant, then we obtain a *horizontal asymptote*. If, on the other hand, $p(x)$ is a linear polynomial, then we obtain in this case what is called an *oblique asymptote*. If, further, $p(x)$ is a quadratic polynomial, we say that we have an *asymptotic parabola* (see the examples below).

Example 7.10 Use (7.63) and (7.69) to obtain the vertical and horizontal asymptotes of the graph of the linear rational function f given by

$$f(x) = \frac{ax + b}{cx + d} \qquad a, b, c, d \text{ fixed real numbers; } ad - bc \neq 0; c \neq 0 \tag{7.70}$$

Compare your results with those obtained in Sec. 7.2.

Solution Since $ad - bc \neq 0$, $f(x)$ is not a constant function, and hence $f(x)$ is in lowest terms. Therefore, by (7.63), the vertical asymptote of the graph of $y = f(x)$ [where $f(x)$ is as given in (7.70)] is $x = x_0$, where x_0 is a real number such that $cx_0 + d = 0$. Hence $x = -d/c$ is the one and only vertical asymptote of the graph of $y = f(x)$. Now, to apply (7.69) we divide $ax + b$ by $cx + d$ to get

$$f(x) = \frac{ax + b}{cx + d} = \frac{a}{c} + \frac{b - ad/c}{cx + d} \tag{7.71}$$

[Observe that (7.71) is trivially satisfied if $a = 0$.] Thus, in view of (7.69) the equality (7.71) tells us that

$$y = \frac{a}{c} \text{ is an asymptote of the graph of } y = f(x) \tag{7.72}$$

Hence, $y = a/c$ is the one and only horizontal asymptote of the graph of $y = f(x)$. [If $a = 0$, this horizontal asymptote becomes $y = 0$ (that is, x axis).] Observe that these results agree with those obtained in (7.55).

Example 7.11 Find the x intercepts, y intercepts, and asymptotes of the graph of

$$y = f(x) = \frac{-4x^2 + 1}{x^2} \tag{7.73}$$

Is the graph symmetric with respect to the y axis? Is it symmetric with respect to the origin? Also, sketch the graph of $y = f(x)$.

Solution Since $f(0)$ is *not* defined, there is *no* y intercept point; that is, the graph of (7.73) does *not* intersect the y axis. Now, setting $y = 0$ in (7.73) and solving for x we get $x = \pm 1/2$, and hence

The x intercept points are $(1/2, 0)$ and $(-1/2, 0)$. (7.74)

By (7.63), we know that

$x = 0$ is the one and only vertical asymptote of the graph of $y = f(x)$ (7.75)

Now, since the fraction in (7.73) is improper, we divide $-4x^2 + 1$ by x^2 first to get

$$f(x) = \frac{-4x^2 + 1}{x^2} = -4 + \frac{1}{x^2} \quad (7.76)$$

Hence, by (7.69) the equality in (7.76) tells us that

$y = -4$ is a horizontal asymptote of the graph of $y = f(x)$ (7.77)

In addition, since $f(-x) = f(x)$ for all x in the domain of f, we know that

The graph of $y = f(x)$ is symmetric with respect to the y axis. (7.78)

However, since $f(-x) \neq -f(x)$ for all x in the domain of f, the graph of $y = f(x)$ is *not* symmetric with respect to the origin. Now, let us construct a brief table of values, as indicated below.

x	1	2	3	$1/2$	$1/4$
y	-3	$-15/4$	$-35/9$	0	12

Observe that in the above table we confined ourselves only to positive values of x since the graph is symmetric with respect to the y axis. Using the information we obtained in (7.74), (7.75), (7.77), and (7.78) together with the above table of values, we can now easily sketch the desired graph, as indicated in Fig. 7.7. Observe that, as this graph indicates, the domain of f, is the set of all *nonzero* real numbers, while the range of f is the set of all real numbers greater than -4 [see (7.76)].

Example 7.12 Find the x intercepts, y intercepts, and asymptotes of the graph of

$$y = f(x) = \frac{3x + 2}{(x + 1)(x - 2)} \quad (7.79)$$

Also, discuss symmetry, and sketch the graph of $y = f(x)$.

Solution Since $f(0) = -1$, the y intercept point is $(0, -1)$. Moreover, setting $y = 0$ in (7.79) and solving for x give $x = -2/3$. Thus $(-2/3, 0)$ is the x intercept point. Hence, we have shown that

The x intercept point is $(-2/3, 0)$, and the y intercept point is $(0, -1)$. (7.80)

Now, by (7.63) and (7.69) we know that

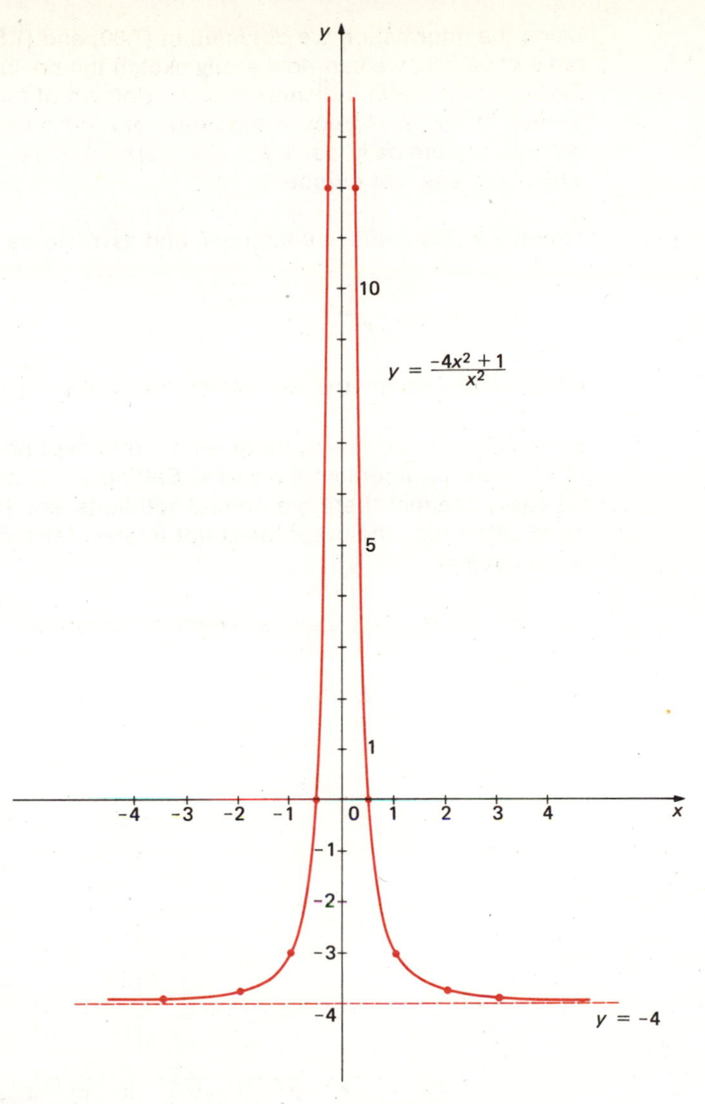

Fig. 7.7 Graph of $y = f(x) = (-4x^2 + 1)/x^2$. The lines $y = -4$ and $x = 0$ (i.e., the y axis) are asymptotes.

$x = -1$ and $x = 2$ are the vertical asymptotes, while $y = 0$ (i.e., the x axis) is the horizontal asymptote. (7.81)

Since, moreover, $f(-x) \neq f(x)$ and $f(-x) \neq -f(x)$ for all values x in the domain of f, the graph of $y = f(x)$ is *not* symmetric with respect to the y axis and is *not* symmetric with respect to the origin either. Now, let us construct a brief table of values, as indicated below.

x	0	1	$3/2$	$5/2$	3	4	5	$-8/9$	-2	-3	$-3/2$	$-7/2$
y	-1	$-5/2$	$-26/5$	$38/7$	$11/4$	$7/5$	$17/18$	$27/13$	-1	$-7/10$	$-10/7$	$-34/55$

Using the information we obtained in (7.80) and (7.81) together with the above table of values, we can now easily sketch the desired graph, as indicated in Fig. 7.8. This graph indicates that the domain of f is the set of all real numbers *except* -1 and 2. Moreover, the range of f is the set of *all* real numbers. [To see this algebraically, solve $y = (3x+2)/(x+1)(x-2)$ and $y = k$ simultaneously, where k is *any* real number.]

Example 7.13 Find the x intercepts, y intercepts, and asymptotes of the graph of

$$y = f(x) = \frac{x^2+1}{x} \tag{7.82}$$

Also, discuss symmetry, and sketch the graph of $y = f(x)$.

Solution Since $f(0)$ is *not* defined, there is *no* y intercept point; that is, the graph of (7.82) does not intersect the y axis. Setting $y = 0$ in (7.82) and solving for x, we easily see that there are *no* real solutions, and hence there is *no* x intercept point either (i.e., the graph does not intersect the x axis). Moreover, by (7.63) we know that

$$x = 0 \text{ (i.e., the } y \text{ axis) is a vertical asymptote} \tag{7.83}$$

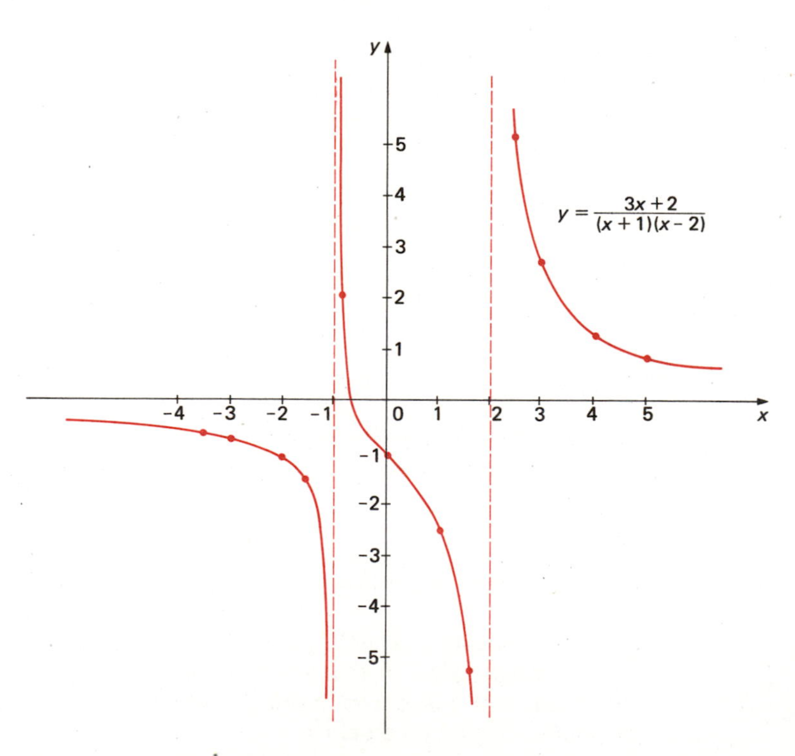

Fig. 7.8 Graph of $y = f(x) = (3x+2)/(x+1)(x-2)$. Observe that the lines $x = -1$, $x = 2$ and $y = 0$ (i.e., the x axis) are asymptotes.

Now, since the fraction in (7.83) is improper, we divide $x^2 + 1$ by x first to get

$$y = x + \frac{1}{x} \tag{7.84}$$

Hence, by (7.69) the equality in (7.84) tells us that

$$y = x \text{ is an oblique asymptote} \tag{7.85}$$

Next, since $f(-x) = -f(x)$ for all x in the domain of f, the graph of $y = f(x)$ is symmetric with respect to the origin. However, since $f(-x) \neq f(x)$ for all x in the domain of f, the graph of $y = f(x)$ is not symmetric with respect to the y axis. Let us construct a brief table of values, as indicated below.

x	1	2	3	4	$1/2$	$1/4$
y	2	$5/2$	$10/3$	$17/4$	$5/2$	$17/4$

Observe that in the above table we confined ourselves only to positive values of x, since the graph is symmetric with respect to the origin. If we use the information we obtained about the graph thus far together with the above table of values, we can now easily sketch the desired graph, as indicated in Fig. 7.9. Note that there are two points where the graph "turns" and these appear to be the points (1,2) and (−1,−2).

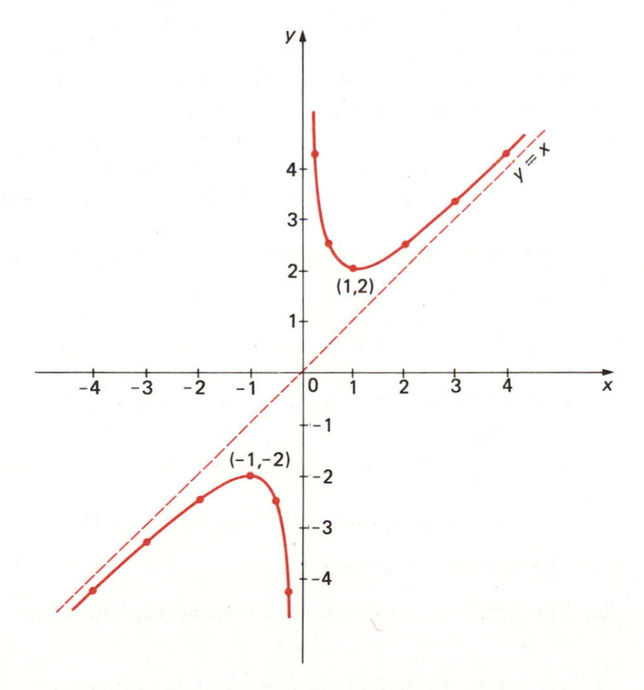

Fig. 7.9 Graph of $y = f(x) = (x^2 + 1)/x$. The lines $x = 0$ (i.e., the y axis) and $y = x$ are asymptotes. The graph has turning points at (1,2) and (−1,−2).

To verify that (1,2) and (–1,–2) are *turning points,* suppose k is a real number, and suppose we try to find the points of intersection (if any) of

$$y = \frac{x^2 + 1}{x} \quad \text{and} \quad y = k \text{ (a horizontal line)}$$

Eliminating y, we obtain $(x^2 + 1)/x = k$; therefore,

$$x^2 - kx + 1 = 0 \tag{7.86}$$

If we use the quadratic formula, we obtain

$$x = \frac{k \pm \sqrt{k^2 - 4}}{2} \tag{7.87}$$

Since the only points that appear on a graph are just those with *real* coordinates, (7.87) tells us the following things:

> For any real number k such that $k^2 - 4 > 0$, the line $y = k$ intersects the graph of $y = (x^2 + 1)/x$ at two distinct points; while if $k^2 - 4 < 0$, then the line $y = k$ does not intersect the graph of $y = (x^2 + 1)/x$. Finally, if $k^2 - 4 = 0$, then the line $y = k$ intersects the graph of $y = (x^2 + 1)/x$ at exactly one point. (7.88)

In view of this, the graph "turns" (see Fig. 7.9) when $k^2 = 4$, that is, when $k = 2$ or $k = -2$. Substituting these values of k in (7.87), we obtain $x = 1$ or $x = -1$. Hence the graph "turns" at the points (1,2) and at (–1,–2). Observe also that by (7.88) the graph of $y = (x^2 + 1)/x$ does *not* have any points in which the y coordinate is strictly between -2 and 2 (and thus $k^2 - 4 < 0$). The graph in Fig. 7.9 confirms the above facts. It is clear from this graph that the domain of f is the set of all *nonzero* real numbers while the range of f is the set of all real numbers *except* those strictly between -2 and 2.

Example 7.14 Find the x intercepts, y intercepts, and asymptotes of the graph of

$$y = f(x) = \frac{x^4 - 1}{x^2} \tag{7.89}$$

Also, discuss symmetry and sketch the graph of $y = f(x)$.

Solution Arguing exactly as in the above examples, we easily verify the following:

1. The x intercept points are (1,0) and (–1,0).
2. There is no y intercept point.
3. The graph is symmetric with respect to the y axis but not with respect to the origin.
4. $x = 0$ (i.e., the y axis) is a vertical asymptote.
5. $y = x^2$ is an asymptotic parabola [since $f(x) = x^2 - 1/x^2$, see (7.69)].

Let us now construct a brief table of values, as indicated below.

x	1	2	3	$\frac{1}{2}$
y	0	$\frac{15}{4}$	$\frac{80}{9}$	$-\frac{15}{4}$

Taking advantage of the information we obtained in statements 1 to 5 above, together with the above table of values, we can now easily sketch the graph of $y = f(x)$, as indicated in Fig. 7.10. A glance at this graph shows that the domain of f is the set of all *nonzero* real numbers while the range of f is the set of *all* real numbers. [To see this algebraically, solve $y = (x^4 - 1)/x^2$ and $y = k$ simultaneously, where k is *any* real number.]

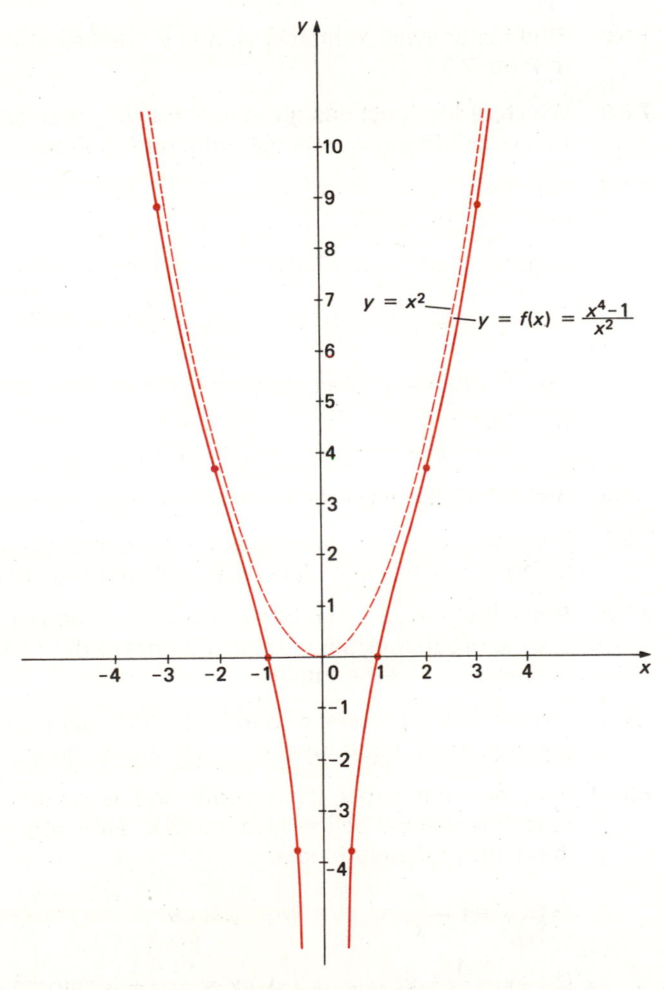

Fig. 7.10 Graph of $y = f(x) = (x^4 - 1)/x^2$. This graph has the vertical asymptote $x = 0$ (i.e., the y axis) and an asymptotic parabola, namely, $y = x^2$.

Problem Set 7.3

7.3.1 Are the functions defined below even, odd, or neither even nor odd?

(*a*) $f(x) = \frac{1}{x} - x$

(*b*) $f(x) = \frac{1}{x^2} + x^2$

(*c*) $f(x) = |x|$

(*d*) $f(x) = |x| + x$

(*e*) $f(x) = |x| - x$

(*f*) $f(x) = |x|/x$

(*g*) $f(x) = (1 - x)/(1 + x)$

(*h*) $f(x) = \frac{1 - x^2}{1 + x}$

(*i*) $f(x) = \frac{1}{x} + \frac{1}{x^3}$

(*j*) $f(x) = 1$

(*k*) $f(x) = 0$

(*l*) $f(x) = x|x|$

7.3.2 Find the domain, x intercepts, and y intercepts of each of the functions given in Prob. 7.3.1.

7.3.3 Which of the functions given in Prob. 7.3.1 are, or can be reduced to, rational functions? For each rational function f in Prob. 7.3.1 find the asymptotes.

7.3.4 Use the information in Probs. 7.3.1 to 7.3.3 to sketch the graph of each function given in Prob. 7.3.1.

7.3.5 Suppose that f is any function, and suppose that

$$g(x) = \frac{f(x) + f(-x)}{2} \qquad h(x) = \frac{f(x) - f(-x)}{2}$$

(*a*) Prove that g is an even function and h is an odd function.

(*b*) Use part (*a*) to show that any function f can always be written as a sum of an even function and an odd function.

7.3.6 Show that the only function which is both even and odd is the zero function.

7.3.7 Prove that the sum, difference, product, and quotient (with nonzero denominator) of any two even functions are even functions.

7.3.8 Prove that the sum and difference of any two odd functions are odd functions but the product or quotient (with nonzero denominator) of any two odd functions is an *even* function.

7.3.9 Prove that the product or quotient (with nonzero denominator) of an even function and an odd function is an odd function.

7.3.10 Find the x intercepts, y intercepts, and asymptotes of the graph of each of the functions defined below. Also, discuss symmetry, and sketch each graph. Give the domain of each function.

(*a*) $y = 1 - \frac{1}{x^2}$

(*b*) $y = \frac{1}{x^3} - x$

(*c*) $y = \frac{x^2 - 6x + 5}{x - 2}$

(*d*) $y = -x^3 + \frac{1}{x}$

(*e*) $y = \frac{x^2 - 1}{x}$ (compare with Example 7.13)

7.3.11 Find the x intercepts, y intercepts, asymptotes, and turning points of the graphs of the functions defined below (as shown in Example 7.13). Also, give the domains and ranges of these functions, and sketch the graphs.

(a) $y = \dfrac{x+3}{x^2+2x-2}$ (c) $y = \dfrac{x-1}{(x-2)(x-3)}$

(b) $y = \dfrac{x^2+2x-2}{x+3}$

7.3.12 Find constants A and B such that

$$\frac{x+1}{(x-1)(x-3)} = \frac{A}{x-1} + \frac{B}{x-3}$$

for all real numbers x different from 1 and 3. (*Hint*: Clear denominators.)

7.3.13 Find constants A and B such that

$$\frac{x-1}{(x+1)^2} = \frac{A}{x+1} + \frac{B}{(x+1)^2}$$

for all real numbers x different from -1. (*Hint*: Clear denominators.)

7.3.14 Find constants A, B, C, and D such that

$$\frac{2x^3-x^2+2x-2}{(x^2+1)^2} = \frac{Ax+B}{x^2+1} + \frac{Cx+D}{(x^2+1)^2}$$

for all real numbers x. (*Hint*: Clear denominators.)

7.3.15 Find constants A, B, and C such that

$$\frac{5x^2-9x+19}{(x-2)(x^2+3)} = \frac{A}{x-2} + \frac{Bx+C}{x^2+3}$$

for all real numbers x different from 2. (*Hint*: Clear denominators.)

7.4 Summary

In this chapter, we studied rational functions, their properties, and their graphs. In particular, we saw that the sum, difference, product, and quotient (with nonzero denominator) of any two rational functions are again rational functions. Moreover, the composition of any two rational functions is again a rational function. We also saw that the set of all rational functions contains the set of all polynomial functions and that we operate on rational functions by the usual rules of high school algebra. We then singled out an important class of rational functions, known as linear rational functions, and studied the properties and graphs of these functions. We saw that every linear rational function which is not a constant is both one-to-one and onto and, hence, has an inverse. This inverse itself is again a linear rational function. The graph of a linear rational function is always a straight line (if the linear rational function happens to be a linear polynomial function) or a hyperbola (a special kind of a conic section).

In the latter case, the graph has exactly one horizontal asymptote and exactly one vertical asymptote. We then directed our attention to the study of arbitrary rational functions, with heavy emphasis on sketching their graphs and locating their asymptotes. We saw that, generally speaking, not all the asymptotes of the graphs of rational functions are horizontal or vertical asymptotes. Moreover, we gave numerous examples to show how a knowledge of interesting concepts such as intercepts, symmetry, and asymptotes greatly facilitates the sketching of the graphs of rational functions. Finally, we introduced the concepts of even and odd functions along with conditions for symmetry of a graph with respect to the *y* axis or origin.

8 Exponential and Logarithmic Functions

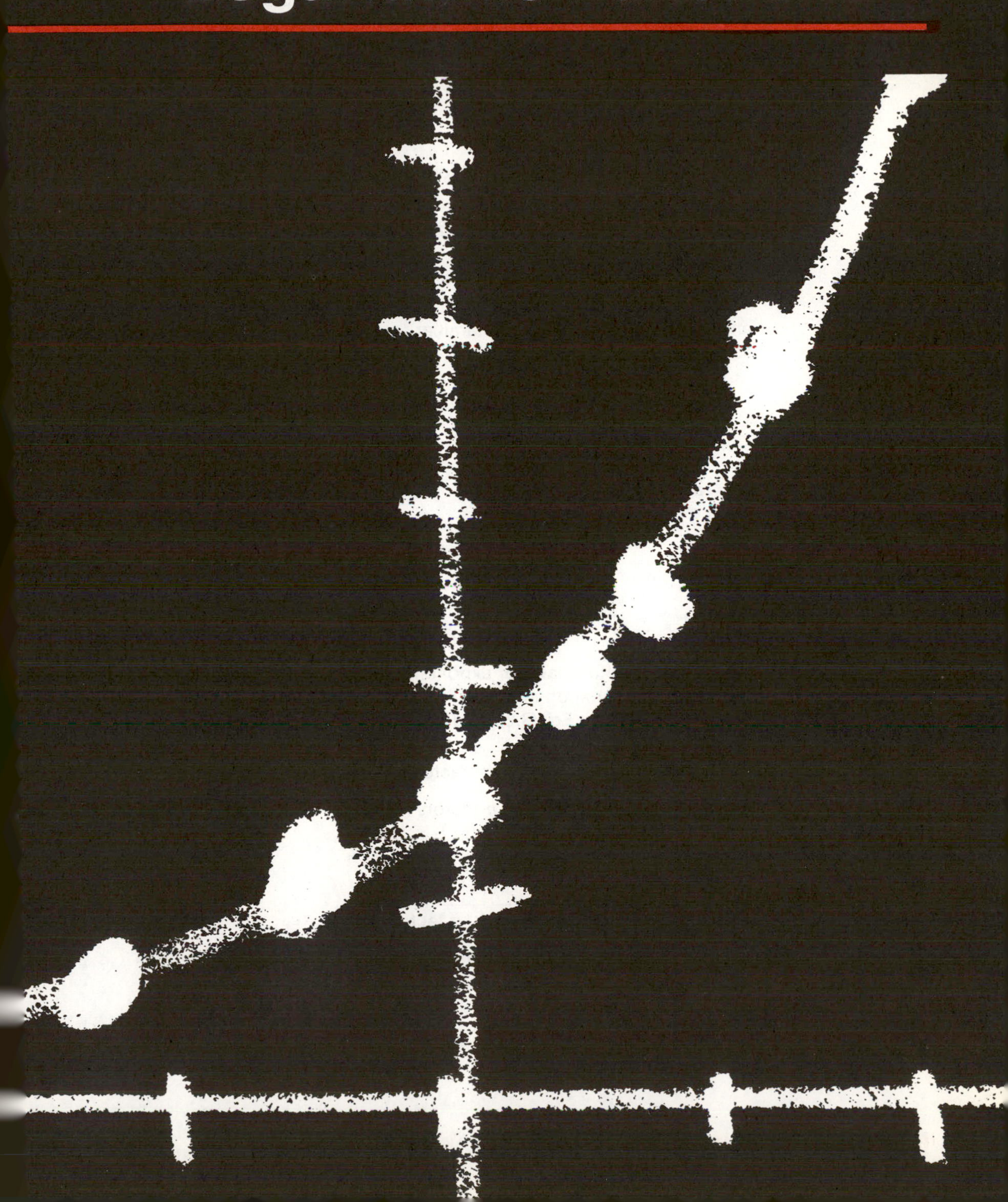

8.1 Exponents and Their Properties

In this section, we study exponents and their properties. The student is undoubtedly familiar with exponents in a^2, a^3, and, more generally, the exponent n in a^n, for n any positive integer. These are indeed the simplest types of exponents, and they will, in fact, be the starting point in our development.

Consider for a moment any nonzero real number a and any positive integer n. We recall from high school algebra that the nth *power* of a, denoted by a^n, was defined as follows:

$$a^n = a \cdot a \cdot a \cdots a \qquad n \text{ factors} \tag{8.1}$$

The number a is called the *base*, and the integer n is called the *exponent*. Thus,

$$a^1 = a \qquad a^2 = a \cdot a \qquad a^3 = a \cdot a \cdot a \qquad a^4 = a \cdot a \cdot a \cdot a$$

and so on. We also recall the familiar *laws of exponents* which can be stated as follows:

For any nonzero real number a and any positive integers m and n, we have

$$a^m \cdot a^n = a^{m+n} \tag{8.2}$$

$$\frac{a^m}{a^n} = a^{m-n} \qquad \text{if } m > n \tag{8.3}$$

$$\frac{a^m}{a^n} = \frac{1}{a^{n-m}} \qquad \text{if } n > m \tag{8.4}$$

$$(a^m)^n = a^{mn} \tag{8.5}$$

The validity of these laws of exponents can be intuitively seen by simply using the definition of exponents given in Eq. (8.1). Consider, for example, the law of exponents given in Eq. (8.2). By (8.1) we know that

$$a^m = a \cdot a \cdot a \cdots a \qquad m \text{ factors}$$

$$a^n = a \cdot a \cdot a \cdots a \qquad n \text{ factors}$$

and hence

$$\begin{aligned} a^m \cdot a^n &= \underbrace{(a \cdot a \cdot a \cdots a)}_{m \text{ factors}}\underbrace{(a \cdot a \cdot a \cdots a)}_{n \text{ factors}} \\ &= a \cdot a \cdot a \cdots a \qquad m + n \text{ factors} \\ &= a^{m+n} \end{aligned}$$

Thus, $a^m \cdot a^n = a^{m+n}$. In this same way we can intuitively justify the remaining laws (8.3) to (8.5).

It would be very restrictive indeed to always have the exponents m and n in these laws as positive integers. We are thus led to the question of whether we can extend these laws to zero and negative exponents. It turns out that it is quite feasible to make such an extension, as we now proceed to show. In attempting to develop this desired extension, our guiding principle will always be the following: *No matter how we define a^n, where n is any integer (positive, negative, or zero), we wish to retain the validity of the laws of exponents given in Eqs. (8.2) to (8.5).* It turns out that, with the aid of this general guiding principle, we are able to give plausible definitions for a^n when n is zero or when n is any negative integer.

Let us start by attempting to give a reasonable definition for a^0, following the general principle just stated. Thus, suppose that $a \neq 0$ and n is any positive integer. Then, on the one hand,

$$\frac{a^n}{a^n} = 1 \qquad \text{(since any } \textit{nonzero} \text{ number divided by itself is equal to 1)} \tag{8.6}$$

On the other hand, by Eq. (8.3), we would certainly like to have

$$\frac{a^n}{a^n} = a^{n-n} = a^0 \tag{8.7}$$

Now, in order for both (8.6) and (8.7) to hold, we are led to the following *definition*:

$$a^0 = 1 \qquad \text{for any nonzero real number } a \tag{8.8}$$

Note that 0^0 is *not* defined.

Next, we seek a plausible definition for a^{-n}, where $-n$ is any *negative* integer and $a \neq 0$. In view of (8.2), we would like to have

$$a^n \cdot a^{-n} = a^{n+(-n)} = a^0 = 1 \qquad \text{by Eq. (8.8)}$$

that is, $a^n \cdot a^{-n} = 1$. We are thus led to the following *definition*:

$$a^{-n} = \frac{1}{a^n} \qquad \text{for any negative integer } -n \text{ and nonzero real number } a \tag{8.9}$$

It turns out that if we define negative and zero exponents as indicated in (8.8) and (8.9), then all the laws of exponents given in Eqs. (8.2) to (8.5) are valid for all integers (positive, negative, or zero). Thus, our attempt to extend the above laws of exponents to *all* integral exponents proved to be quite successful. We shall continue to follow this principle in our next stage of development, where we will be attempting to give plausible definitions for rational exponents.

Suppose now that a is any *positive* real number, and n is any positive integer. Using one of the axioms for real numbers, known as the *Completeness axiom*, we can show that there exists exactly one positive real number t such that $t^n = a$. This unique number t is called the nth *root* of a, and it is denoted by $\sqrt[n]{a}$. In other words, we have the following:

$$\text{If } a > 0, \text{ then } \sqrt[n]{a} \text{ is that unique positive real number which when raised to the } n\text{th power gives } a. \tag{8.10}$$

For example,

$$\sqrt[2]{4} = 2 \qquad \text{(because } 2^2 = 4\text{)}$$
$$\sqrt[3]{64} = 4 \qquad \text{(because } 4^3 = 64\text{)}$$
$$\sqrt[4]{81} = 3 \qquad \text{(because } 3^4 = 81\text{)}$$

and so on. Thus, in view of (8.10), we always have

$$(\sqrt[n]{a})^n = a \qquad a > 0,\ \sqrt[n]{a} > 0 \tag{8.11}$$

In order to define arbitrary rational exponents, we first seek a plausible definition for $a^{1/n}$, where a is any positive real number and n any positive integer. Now, in view of (8.5), we would certainly like to have

$$(a^{1/n})^n = a^{(1/n)\cdot n} = a^1 = a$$

that is, $(a^{1/n})^n = a$. Hence, keeping (8.10) and (8.11) in mind, we are led to the following *definition*:

$$a^{1/n} = \sqrt[n]{a} \qquad \text{for any positive real number } a \text{ and any positive integer } n \tag{8.12}$$

Now, suppose that m and n are positive integers which have no proper common factors. Keeping in mind (8.1) and (8.12), we *define*

$$a^{m/n} = (\sqrt[n]{a})^m \qquad a > 0,\ m,\ n \text{ positive integers} \tag{8.13}$$

Moreover, if $-m/n$ is a negative rational number, then, keeping in mind (8.9) and (8.13), we *define*

$$a^{-m/n} = \frac{1}{a^{m/n}} = \frac{1}{(\sqrt[n]{a})^m} \qquad a > 0,\ m,\ n \text{ positive integers} \tag{8.14}$$

So far, we have defined a^r for all rational exponents r and all positive real numbers a. It turns out that *if we define rational exponents as in Eqs. (8.12) to (8.14), then the laws of exponents in Eqs. (8.2) to (8.5) are indeed valid for all rational numbers, assuming* $a > 0$.

How about irrational exponents? Roughly speaking, we obtain approximate values of irrational exponents by taking suitable *rational* exponent approximations. Let us illustrate this by an example. Suppose that we wish to find the value of $2^{\sqrt{2}}$ correct to three decimals. We proceed by first taking *rational* approximations of the exponent $\sqrt{2}$. (See Table A.) We have

$$\begin{aligned} 1.4 &< \sqrt{2} < 1.5 \\ 1.41 &< \sqrt{2} < 1.42 \\ 1.414 &< \sqrt{2} < 1.415 \\ 1.4142 &< \sqrt{2} < 1.4143 \end{aligned} \tag{8.15}$$

Of course, we can continue the above process of getting closer and closer approximations of $\sqrt{2}$ indefinitely. Now, in view of (8.15), we see intuitively that

$$\begin{aligned} 2^{1.4} &< 2^{\sqrt{2}} < 2^{1.5} \\ 2^{1.41} &< 2^{\sqrt{2}} < 2^{1.42} \\ 2^{1.414} &< 2^{\sqrt{2}} < 2^{1.415} \\ 2^{1.4142} &< 2^{\sqrt{2}} < 2^{1.4143} \end{aligned} \tag{8.16}$$

Furthermore, using logarithms (a topic which we shall discuss later in this chapter), we obtain

$$2.665 < 2^{1.4142} < 2^{1.4143} < 2.666 \tag{8.17}$$

Combining (8.16) and (8.17), we get

$$2.665 < 2^{\sqrt{2}} < 2.666$$

and hence the value of $2^{\sqrt{2}}$ correct to three decimals is 2.665.

It should be pointed out that a rigorous treatment of irrational exponents requires a sophisticated application of the Completeness axiom. Nevertheless, the above example gives an intuitive account of what happens in the case of irrational exponents. Thus, we shall *assume from now on* that a^r is defined for all positive real numbers a and for all real numbers r. Moreover, we shall also assume that the laws of exponents (8.2) to (8.5) hold for all a^r, where a is any *positive* real number and r is any real number.

In addition to the laws of exponents given in Eqs. (8.2) to (8.5), we further assume the following additional laws:

$$(ab)^n = a^n b^n \qquad \text{for all positive real } a, b \text{ and all real } n \tag{8.18}$$

$$\left(\frac{a}{b}\right)^n = \frac{a^n}{b^n} \qquad \text{for all positive real } a, b \text{ and all real } n \tag{8.19}$$

We also recall introducing earlier the concept of an nth root of a positive real number a, denoted by $\sqrt[n]{a}$, where n is a positive integer. This positive real number $\sqrt[n]{a}$ is also called a *radical.* Keeping in mind (8.5), (8.12), and (8.13), we easily see that Eqs. (8.18) and (8.19) imply the following useful properties of radicals: For any positive real numbers a, b and for any positive integer n, we have

$$\sqrt[n]{ab} = \sqrt[n]{a}\sqrt[n]{b} \tag{8.20}$$

$$\sqrt[n]{\frac{a}{b}} = \frac{\sqrt[n]{a}}{\sqrt[n]{b}} \tag{8.21}$$

$$\sqrt[n]{a^m} = (\sqrt[n]{a})^m \tag{8.22}$$

$$\sqrt[n]{\sqrt[m]{a}} = \sqrt[m]{\sqrt[n]{a}} = \sqrt[mn]{a} \tag{8.23}$$

So far, we have confined ourselves to radicals of the form $\sqrt[n]{a}$, where a is a *positive* real number and n is *any* positive integer. Now, if we restrict our positive integer n to being *odd*, then we can define $\sqrt[n]{a}$ for *any* real number a. Indeed, using the Completeness axiom for real numbers again, it can be shown that

> If a is any negative real number and n is any odd positive integer, then there exists exactly one negative real number t such that $t^n = a$. This unique number t is denoted by $\sqrt[n]{a}$. Moreover, we define the rational exponent, $a^{m/n}$, where $a < 0$ and n is any odd positive integer as in (8.12) to (8.14).

Thus, $\sqrt[3]{-8} = -2$ because $(-2)^3 = -8$, $\sqrt[5]{-1} = -1$ because $(-1)^5 = -1$, $\sqrt[3]{-64} = -4$ because $(-4)^3 = -64$, and so on. Furthermore, it is easy to verify that the properties of radicals given in (8.20) to (8.23) are valid when a and b are negative real numbers, *provided that the index of each radical involved (namely, m and n) is an odd positive integer.*

We now pause to give some examples.

Example 8.1 Simplify the following expression.

$$\frac{a^3(bc^{-1})^2}{a^{-4}b^{-5}c^3}$$

Solution Using the properties of exponents discussed in this section, we get

$$\frac{a^3(bc^{-1})^2}{a^{-4}b^{-5}c^3} = \frac{a^3b^2c^{-2}}{a^{-4}b^{-5}c^3}$$

$$= \frac{a^{3-(-4)}b^{2-(-5)}}{c^{3-(-2)}}$$

$$= \frac{a^7b^7}{c^5}$$

Thus,

$$\frac{a^3(bc^{-1})^2}{a^{-4}b^{-5}c^3} = \frac{a^7b^7}{c^5}$$

Example 8.2 Simplify the following expression.

$$\sqrt[3]{8} + \left(\frac{1}{2}\right)^{-3} + \sqrt[3]{-27} - (\sqrt[7]{5})^0 - \left(\frac{4}{9}\right)^{-3/2}$$

Solution Using the properties of exponents discussed in this section, we get

$$\sqrt[3]{8} = 2 \qquad \left(\frac{1}{2}\right)^{-3} = \left[\left(\frac{1}{2}\right)^{-1}\right]^3 = 2^3 = 8$$

$$\sqrt[3]{-27} = -3 \qquad (\sqrt[7]{5})^0 = 1$$

$$\left(\frac{4}{9}\right)^{-3/2} = \left[\left(\frac{4}{9}\right)^{-1}\right]^{3/2} = \left(\frac{9}{4}\right)^{3/2} = \left(\sqrt[2]{\frac{9}{4}}\right)^3 = \left(\frac{3}{2}\right)^3 = \frac{27}{8}$$

Hence,

$$\sqrt[3]{8} + \left(\frac{1}{2}\right)^{-3} + \sqrt[3]{-27} - (\sqrt[7]{5})^0 - \left(\frac{4}{9}\right)^{-3/2} = 2 + 8 - 3 - 1 - \frac{27}{8} = \frac{21}{8}$$

Example 8.3 Simplify $\sqrt{8} - \sqrt{50} + \sqrt{45} + \sqrt{18}$.

Solution Using the properties of radicals discussed in this section, we get

$$\sqrt{8} = \sqrt{(4)(2)} = \sqrt{4}\sqrt{2} = 2\sqrt{2}$$
$$\sqrt{50} = \sqrt{(25)(2)} = \sqrt{25}\sqrt{2} = 5\sqrt{2}$$
$$\sqrt{45} = \sqrt{(9)(5)} = \sqrt{9}\sqrt{5} = 3\sqrt{5}$$
$$\sqrt{18} = \sqrt{(9)(2)} = \sqrt{9}\sqrt{2} = 3\sqrt{2}$$

Hence,

$$\sqrt{8} - \sqrt{50} + \sqrt{45} + \sqrt{18} = 2\sqrt{2} - 5\sqrt{2} + 3\sqrt{5} + 3\sqrt{2} = 3\sqrt{5}$$

Example 8.4 Arrange the following real numbers in order of magnitude:

$$3^{-2/3} \qquad (9^{-1/5})(27^{-1/15}) \qquad \left(\frac{1}{3}\right)^{7/10} \qquad (3^{-1/4})^3$$

Solution Using the properties of exponents discussed in this section, we get

$$3^{-2/3} = \frac{1}{3^{2/3}} \qquad (9^{-1/5})(27^{-1/15}) = (3^{-2/5})(3^{-3/15}) = 3^{-3/5} = \frac{1}{3^{3/5}}$$

$$\left(\frac{1}{3}\right)^{7/10} = \frac{1}{3^{7/10}} \qquad (3^{-1/4})^3 = 3^{-3/4} = \frac{1}{3^{3/4}}$$

Since, $3^{3/5} < 3^{2/3} < 3^{7/10} < 3^{3/4}$, it follows that $1/3^{3/5} > 1/3^{2/3} > 1/3^{7/10} > 1/3^{3/4}$, and thus $(9^{-1/5})(27^{-1/15}) > 3^{-2/3} > (1/3)^{7/10} > (3^{-1/4})^3$.

Example 8.5 Find the values of m that make each equation in (*a*) to (*g*) true.

(*a*) $2^m = 8^{-3}$

(*b*) $\left(\frac{1}{3}\right)^m = 9^5$

(*c*) $(2^3)^4 = 16^{1/m}$

(*d*) $2^{(3^4)} = 16^{1/m}$

(*e*) $16^m = 64\sqrt{2}$

(*f*) $\left(\frac{1}{\sqrt{2}}\right)^m = \left(\frac{\sqrt[3]{2}}{4}\right)^{-5}$

(*g*) $(\sqrt{50})^m = 6250\sqrt{8}$

Solution

(*a*) Since $8 = 2^3$, the equation $2^m = 8^{-3}$ is equivalent to $2^m = (2^3)^{-3} = 2^{-9}$, and hence $m = -9$.

(*b*) Since $1/3 = 3^{-1}$ and $9 = 3^2$, the equation $(1/3)^m = 9^5$ is equivalent to $(3^{-1})^m = (3^2)^5$, or $3^{-m} = 3^{10}$, and hence $m = -10$.

(*c*) Since $16 = 2^4$, the equation $(2^3)^4 = 16^{1/m}$ is equivalent to $2^{12} = (2^4)^{1/m} = 2^{4/m}$, and hence $12 = 4/m$. Thus $m = 1/3$.

(*d*) Since $16 = 2^4$ and $3^4 = 81$, the equation $2^{(3^4)} = 16^{1/m}$ is equivalent to $2^{81} = (2^4)^{1/m} = 2^{4/m}$, and hence $4/m = 81$. Thus, $m = 4/81$.

(*e*) Since $16 = 2^4$, $64 = 2^6$, and $\sqrt{2} = 2^{1/2}$, the equation $16^m = 64\sqrt{2}$ is equivalent to $(2^4)^m = 2^6(2^{1/2})$, or $2^{4m} = 2^{13/2}$. Hence $4m = 13/2$, and thus $m = 13/8$.

(*f*) Since $\sqrt{2} = 2^{1/2}$, $\sqrt[3]{2} = 2^{1/3}$, and $4 = 2^2$,

$$\left(\frac{1}{\sqrt{2}}\right)^m = \left(\frac{\sqrt[3]{2}}{4}\right)^{-5} \quad \text{is equivalent to} \quad \left(\frac{1}{2^{1/2}}\right)^m = \left(\frac{2^{1/3}}{2^2}\right)^{-5}$$

The last equation is, by the properties of exponents we discussed in this section, equivalent to $2^{-m/2} = 2^{25/3}$ and hence $m = -50/3$.

(*g*) Since $\sqrt{50} = \sqrt{(25)(2)} = 5\sqrt{2}$ and $\sqrt{8} = \sqrt{(4)(2)} = 2\sqrt{2}$, the equation $(\sqrt{50})^m = 6250\sqrt{8}$ is equivalent to

$$\begin{aligned}(5\sqrt{2})^m &= 6250(2\sqrt{2}) = (5^4)(10)(2\sqrt{2})\\ &= 5^5 \cdot 2^2 \cdot \sqrt{2} = 5^5(\sqrt{2})^5 = (5\sqrt{2})^5\end{aligned}$$

Hence $(5\sqrt{2})^m = (5\sqrt{2})^5$, and thus $m = 5$.

Example 8.6

Find all numbers x such that

$$\sqrt{x+2} - \sqrt{x-3} = 1 \tag{8.24}$$

Solution

Suppose that we try to solve for one of the radicals in Eq. (8.24); we get

$$\sqrt{x+2} = \sqrt{x-3} + 1 \tag{8.25}$$

Now, if we square both sides of (8.25), we obtain

$$\begin{aligned}x + 2 &= (\sqrt{x-3} + 1)^2\\ &= x - 3 + 1 + 2\sqrt{x-3} = x - 2 + 2\sqrt{x-3}\end{aligned}$$

Solving for the only remaining radical above, we get $2\sqrt{x-3} = 4$, and hence $\sqrt{x-3} = 2$. Squaring both sides of this last equation, we get $x - 3 = 4$, or $x = 7$. We have thus shown that *if* Eq. (8.24) has a solution, *then* this solution is 7, so *we must still check to see whether 7 satisfies Eq. (8.24)*. Substituting $x = 7$ into Eq. (8.24), we see that 7 *is* indeed a solution.

Example 8.7

Find all numbers x such that

$$\sqrt{2x-1} - 2\sqrt{x-1} = -1 \tag{8.26}$$

Solution Solving for the first radical, we obtain

$$\sqrt{2x-1} = 2\sqrt{x-1} - 1$$

Now, squaring both sides of this equation, we get

$$2x - 1 = 4(x-1) + 1 - 4\sqrt{x-1}$$

Next, we solve in the last equation for the only radical left; we obtain

$$-4\sqrt{x-1} = 2x - 1 - 4x + 4 - 1$$
$$= -2x + 2$$

that is, $-4\sqrt{x-1} = -2x + 2$, or equivalently $-2\sqrt{x-1} = -x + 1$. Hence, squaring both sides of this final equation, we get $4(x-1) = x^2 - 2x + 1$, or $x^2 - 6x + 5 = 0$. Therefore $x = 1$ or $x = 5$. We have thus shown that *if* Eq. (8.26) has any solutions, *then* these solutions are 1 and/or 5. *Hence, we must still check each of these candidates for possible solutions.* Observe that 5 *is* a solution of (8.26) but 1 is *not* a solution. This follows since $\sqrt{9} - 2\sqrt{4} = -1$, but $\sqrt{1} - 2\sqrt{0} \neq -1$.

Remark The reason that we *must* always check each candidate for a possible solution is that the equations we get by the process described in the last two examples are, in general, *not* equivalent (i.e., we cannot reverse all the steps involved). It is interesting to observe that 1 *is* indeed a solution of the equation

$$\sqrt{2x-1} - 2\sqrt{x-1} = 1 \tag{8.27}$$

In fact, if we start with Eq. (8.27) and apply exactly the same method we used above, we obtain $2\sqrt{x-1} = -x + 1$ (instead of $-2\sqrt{x-1} = -x + 1$). However, when we square both sides again, we get $4(x-1) = x^2 - 2x + 1$, *which is precisely the same equation we obtained in solving (8.26)!* And so we see that we must always verify whether or not each value of x we obtain by this method gives a solution. Note also that by squaring both sides of an equation, we often introduce extra solutions. For example, 5 is the only solution to $4x = 20$. However, $(4x)^2 = 20^2$ has two solutions, namely 5 and -5.

Example 8.8 Express the fraction given in (8.28)

$$\frac{7 - 2\sqrt{5}}{13 - 4\sqrt{5}} \tag{8.28}$$

in a form which does *not* involve any radicals in the denominator.

Solution Let us multiply both numerator and denominator of (8.28) by $13 + 4\sqrt{5}$. For, if we do, the denominator becomes

$$(13 - 4\sqrt{5})(13 + 4\sqrt{5}) = 13^2 - (4\sqrt{5})^2$$
$$= 89$$

(Here we used the difference-of-two-squares formula.) Thus,

$$\frac{7-2\sqrt{5}}{13-4\sqrt{5}}=\frac{7-2\sqrt{5}}{13-4\sqrt{5}}\cdot\frac{13+4\sqrt{5}}{13+4\sqrt{5}}$$

$$=\frac{91+28\sqrt{5}-26\sqrt{5}-40}{13^2-(4\sqrt{5})^2}$$

$$=\frac{51+2\sqrt{5}}{89}$$

Hence,

$$\frac{7-2\sqrt{5}}{13-4\sqrt{5}}=\frac{51+2\sqrt{5}}{89} \tag{8.29}$$

and the latter fraction does *not* involve any radicals in the denominator.

Remark 1 Observe that it is easier to evaluate the fraction on the right side of Eq. (8.29) than the fraction on the left.

Remark 2 The above technique is known as *rationalizing the denominator.* In general, in order to rationalize the denominator of the fraction

$$\frac{a+b\sqrt{k}}{c+d\sqrt{k}}$$

where a, b, c, d, and k are rational numbers and k is not a perfect square, we simply multiply both numerator and denominator of this fraction by $c-d\sqrt{k}$, and then we apply the difference-of-two-squares formula in simplifying the *new* denominator. This new denominator is, in fact, equal to c^2-kd^2, which does *not* involve any radicals. Incidentally, the expression $c-d\sqrt{k}$ is usually called the *conjugate* of the expression $c+d\sqrt{k}$. It should be pointed out that this technique works even if c and d are products of rational numbers and square roots, as the following example illustrates.

Example 8.9 Express the fraction given in (8.30)

$$\frac{5\sqrt{3}-3\sqrt{7}}{4\sqrt{35}+9\sqrt{2}} \tag{8.30}$$

in a form which does not involve any radicals in the denominator.

Solution The conjugate of $4\sqrt{35}+9\sqrt{2}$ is $4\sqrt{35}-9\sqrt{2}$. Thus, we multiply both numerator and denominator of (8.30) by $4\sqrt{35}-9\sqrt{2}$, to get

$$\frac{5\sqrt{3}-3\sqrt{7}}{4\sqrt{35}+9\sqrt{2}}=\frac{5\sqrt{3}-3\sqrt{7}}{4\sqrt{35}+9\sqrt{2}}\cdot\frac{4\sqrt{35}-9\sqrt{2}}{4\sqrt{35}-9\sqrt{2}}$$

$$=\frac{20\sqrt{105}-45\sqrt{6}-12\sqrt{7(35)}+27\sqrt{14}}{(4\sqrt{35})^2-(9\sqrt{2})^2}$$

$$=\frac{20\sqrt{105}-45\sqrt{6}-84\sqrt{5}+27\sqrt{14}}{398}$$

We conclude this section by giving examples which illustrate the use of the properties of radicals given in Eqs. (8.20) to (8.23).

Example 8.10 Simplify each of the following. Express each answer in a form which does *not* involve any radicals in the denominator.

(*a*) $\sqrt{25/3}$ (*b*) $\sqrt{5/18}$ (*c*) $\sqrt[3]{189/40}$ (*d*) $\sqrt[5]{32^6}$ (*e*) $\sqrt[3]{\sqrt[2]{27}}$

Solution (*a*) In order to clear the radical in the denominator, we need to have 3^2 (instead of 3). In view of this, we multiply both numerator and denominator of $25/3$ by 3 first, as indicated below. (Observe that multiplying a number by $3/3$ does not change the value of the number.)

$$\sqrt{\frac{25}{3}}=\sqrt{\frac{25}{3}\cdot\frac{3}{3}}=\frac{\sqrt{25\cdot 3}}{\sqrt{3^2}}=\frac{\sqrt{25}\sqrt{3}}{3}=\frac{5\sqrt{3}}{3}$$

Thus $\sqrt{25/3}=5\sqrt{3}/3$.

(*b*) Observe that the denominator $18=3^2\cdot 2$. In order to clear the radical in the denominator, we need to have $3^2\cdot 2^2$ (instead of $3^2\cdot 2$). In view of this, we multiply both numerator and denominator by 2 first, as indicated below.

$$\sqrt{\frac{5}{18}}=\sqrt{\frac{5}{18}\cdot\frac{2}{2}}=\sqrt{\frac{10}{36}}=\frac{\sqrt{10}}{\sqrt{36}}=\frac{\sqrt{10}}{6}$$

Thus, $\sqrt{5/18}=\sqrt{10}/6$.

(*c*) Factoring both numerator and denominator of $189/40$, we obtain

$$\frac{189}{40}=\frac{(27)(7)}{(8)(5)}=\frac{3^3\cdot 7}{2^3\cdot 5}$$

Hence, in order to clear the radical in the denominator, we need to have $2^3\cdot 5^3$ (instead of $2^3\cdot 5$). In view of this, we multiply both numerator and denominator of $189/40$ by 5^2 first, as indicated below.

$$\sqrt[3]{\frac{189}{40}}=\sqrt[3]{\frac{3^3\cdot 7}{2^3\cdot 5}}=\sqrt[3]{\frac{3^3\cdot 7}{2^3\cdot 5}\cdot\frac{5^2}{5^2}}=\frac{\sqrt[3]{3^3\cdot(7\cdot 5^2)}}{\sqrt[3]{2^3\cdot 5^3}}$$

$$=\frac{\sqrt[3]{3^3}\sqrt[3]{7\cdot 5^2}}{\sqrt[3]{2^3}\sqrt[3]{5^3}}=\frac{3\sqrt[3]{175}}{10}$$

Thus, $\sqrt[3]{189/40}=3\sqrt[3]{175}/10$.

(*d*) Observe that $32=2^5$, and hence

$$\sqrt[5]{32^6}=(\sqrt[5]{32})^6=(\sqrt[5]{2^5})^6=2^6=64$$

Thus, $\sqrt[5]{32^6}=64$. Incidentally, it would have been more tedious to compute 32^6 first and then take the fifth root.

(*e*) Observe that $27=3^3$, and hence

$$\sqrt[3]{\sqrt[2]{27}}=\sqrt[2]{\sqrt[3]{27}}=\sqrt[2]{\sqrt[3]{3^3}}=\sqrt[2]{3}$$

Thus, $\sqrt[3]{\sqrt[2]{27}} = \sqrt{3}$. To first get an *approximate* value of $\sqrt[2]{27}$ and then to find an *approximate* value of the cube root of the result would have been more tedious (and perhaps a bit less accurate).

Problem Set 8.1

8.1.1 Simplify.

(*a*) $3^2 \cdot 3^{-1} \cdot 3^4$ (*c*) $12^2 \cdot 3^{-4} \cdot 6^{-2} \cdot 9^2$

(*b*) $\left(\frac{1}{5}\right)^{-2} \cdot 5^{-3}$ (*d*) $16^{-2} \cdot 4^3 \cdot 8^2 \cdot \left(\frac{1}{2}\right)^{-5}$

8.1.2 Simplify.

(*a*) $(a^2a^{-5}a^4)^{-3}$ (*c*) $\dfrac{a^{-2}(b^3c^5)^{-3}}{(a^{-5}b^{-4})^2c^{-10}}$

(*b*) $(a^{1/2}a^{-2/3}a^{1/6})^{-12}$ (*d*) $\dfrac{1}{(a^3a^{-5}a^2)^{100}}$

8.1.3 Simplify.

(*a*) $\sqrt[3]{\frac{8}{27}} - \left(\frac{4}{9}\right)^{-1/2} + (\sqrt[3]{7} - \sqrt[5]{11})^0$ (*c*) $\sqrt[3]{16} + 4\sqrt[3]{54} - 7\sqrt[3]{2}$

(*b*) $\sqrt{45} - \sqrt{20} + \sqrt{80}$ (*d*) $\left(\frac{4}{25}\right)^{-1/2} - \left(\frac{1}{8}\right)^{2/3} + \left(\frac{1}{7}\right)^{-2}$

8.1.4 For (*a*) through (*d*) arrange the following real numbers in order of magnitude.

(*a*) $4\sqrt{3}, 5\sqrt{2}, 3\sqrt{5}$ (*c*) $8^{2/3}, \left(\frac{1}{15}\right)^{-1/2}, \frac{2^{1/2}}{3^{-1}}$

(*b*) $18^{1/2}, \left(\frac{1}{2}\right)^{-2}, 2\sqrt{5}$ (*d*) $2\sqrt[3]{3}, \left(\frac{1}{3}\right)^{-1}, \left(\frac{1}{8}\right)^{-5/9}$

8.1.5 Find the values of m that make the equations in (*a*) through (*d*) true.

(*a*) $3^m = 9^{-1}$ (*c*) $\left(\frac{1}{\sqrt{2}}\right)^m = (\sqrt[3]{8})^{-1/5}$

(*b*) $\left(\frac{1}{2}\right)^{1/m} = 4^{2/3}$ (*d*) $16^m = 8\sqrt[3]{2}$

8.1.6 In (*a*) to (*d*), find all real numbers x (if any exist) that make the equalities true.

(*a*) $\sqrt{x-1} - 1 = \sqrt{x+1}$ (*c*) $2\sqrt{x+4} = \sqrt{x+11} + 2$

(*b*) $\sqrt{x+3} - \sqrt{x+8} = -1$ (*d*) $\sqrt{2x+7} - \sqrt{x-8} = 4$

8.1.7 Rationalize the denominators in the following expressions.

(*a*) $\dfrac{1+\sqrt{2}}{1-\sqrt{2}}$ (*c*) $\dfrac{3\sqrt{5}-2\sqrt{7}}{4\sqrt{10}-3\sqrt{35}}$

(*b*) $\dfrac{\sqrt{3}-\sqrt{2}}{\sqrt{3}+\sqrt{2}}$ (*d*) $\dfrac{4\sqrt{15}+3\sqrt{6}}{4\sqrt{5}-5\sqrt{3}}$

8.1.8 Express each of the following in a form which does *not* involve any radical in the denominator: (*a*) $\sqrt[3]{1/4}$; (*b*) $\sqrt{8/75}$; (*c*) $\sqrt[3]{-1/54}$; (*d*) $5\sqrt{1/98} + 3\sqrt{49/50} - 4\sqrt{25/18}$.

8.2 Exponential and Logarithmic Functions

In this section we will define exponential and logarithmic functions and study their properties. In particular, we shall see that these functions are inverses of each other. We shall also show that the exponential and logarithmic functions are *not* polynomial functions in disguise. Finally, we shall illustrate the use of logarithms in computational problems.

Suppose that b is any fixed positive real number, and suppose R is the set of all real numbers. Now, define a function $f: R \to R$ by

$$f(x) = b^x \qquad \text{for all real numbers } x \text{ (in } R) \tag{8.31}$$

This function $f: R \to R$ is called the *exponential function with base b*, and it is sometimes denoted by $\exp_b$.

Example 8.11 Sketch the graph of the exponential function $\exp_2$.

Solution Our present problem is concerned with sketching the graph of the function

$$f: x \to 2^x \qquad x \text{ any real number}$$

or, equivalently, the graph of the equation $y = f(x) = 2^x$. Let us make a brief table of values, as indicated below.

x	0	1	2	3	-1	-2	-3	$1/2$	$1/3$
$y = f(x) = 2^x$	1	2	4	8	$1/2$	$1/4$	$1/8$	$\sqrt{2}$	$\sqrt[3]{2}$

The graph of the exponential function f given by $f(x) = 2^x$ is sketched in Fig. 8.1. We now list several important facts regarding this function and its graph.

1. As x becomes very large and positive, $f(x)$ also becomes very large and positive. We describe this phenomenon by saying that $f(x)$ grows without bound as x grows without bound.
2. As x becomes very large and negative, $f(x)$ approaches zero. We describe this phenomenon by saying that $f(x)$ approaches zero as x grows numerically large in the negative direction.
3. The x axis is a horizontal asymptote of the graph *in the negative direction only* [in the sense that (2) is true; also see (1)].
4. The graph has no x intercept point. Moreover, the y intercept point is $(0,1)$.
5. The domain of f is the set of *all* real numbers, while the range of f is precisely the set of all *positive* real numbers.
6. f is a one-to-one function.

Example 8.12 Sketch the graph of the exponential function $\exp_{1/2}$.

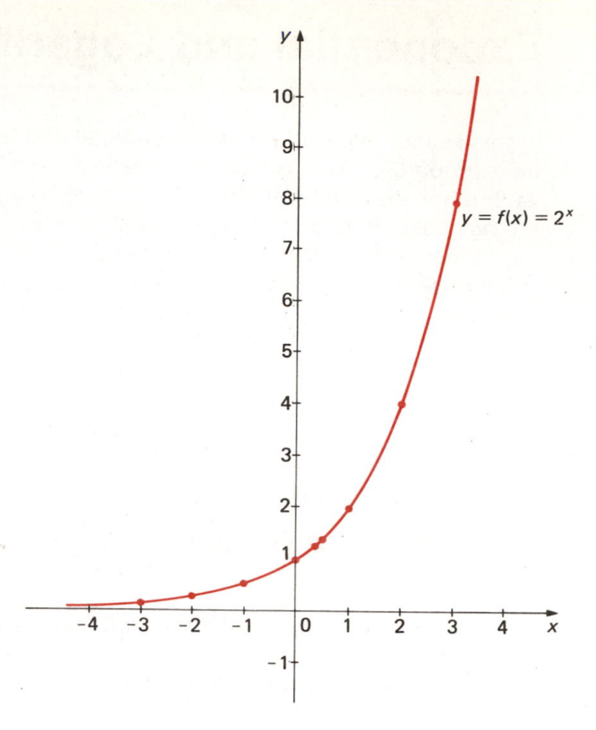

Fig. 8.1 Graph of $y = f(x) = 2^x$

Solution We are now concerned with sketching the graph of the function

$$f: x \to \left(\frac{1}{2}\right)^x \qquad x \text{ any real number}$$

or, equivalently, the graph of the equation $y = f(x) = (1/2)^x$. We first proceed to make a brief table of values, as indicated below.

x	0	1	2	3	−1	−2	−3
$y = f(x) = \left(\frac{1}{2}\right)^x$	1	1/2	1/4	1/8	2	4	8

The graph of the exponential function f given by $f(x) = (1/2)^x$ is sketched in Fig. 8.2. Let us list some of the important facts regarding this function f and its graph.

1. As x becomes very large and positive, $f(x)$ becomes very close to zero.
2. As x becomes very large and negative, $f(x)$ becomes very large and positive.
3. The x axis is a horizontal asymptote of the graph *in the positive direction only* [in the sense that (1) is true: also see (2)].

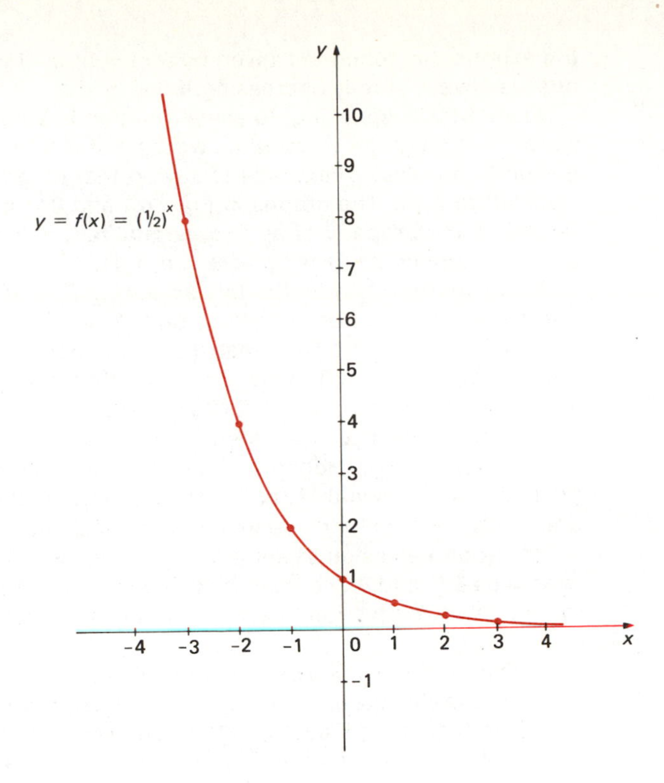

Fig. 8.2 Graph of $y = f(x) = (1/2)^x$

4. The graph has no x intercept point. Moreover, the y intercept point is (0, 1).
5. The domain of f is the set of *all* real numbers, while the range of f is precisely the set of all *positive* real numbers.
6. f is a one-to-one function.

It is interesting to observe that the graphs in Figs. 8.1 and 8.2 are the reflections of each other with respect to the y axis. For, observe that, for all real numbers x, $(x, 2^x)$ lies on the graph of $y = 2^x$, while $(-x, (1/2)^{-x})$ lies on the graph of $y = (1/2)^x$. Now, since $(1/2)^{-x} = 2^x$, we see that $(-x, 2^x)$ lies on the graph of $y = (1/2)^x$, while $(x, 2^x)$ lies on the graph of $y = 2^x$. This proves the assertion made.

The graphs in Figs. 8.1 and 8.2 also motivate the following definition:

> *A function f is said to be* **strictly increasing** *if, for all x_1 and x_2 in the domain of f, $x_1 < x_2$ always implies that $f(x_1) < f(x_2)$. Moreover, f is said to be* **strictly decreasing** *if, for all x_1 and x_2 in the domain of f, $x_1 < x_2$ always implies that $f(x_1) > f(x_2)$.*

Now, a glance at the graphs in Figs. 8.1 and 8.2 shows that the exponential function f given by $f(x) = 2^x$ is strictly increasing, while the exponential function f given by $f(x) = (1/2)^x$ is strictly decreasing. As a matter of fact, it turns out that the exponential function f given by $f(x) = b^x$ is always strictly increasing if $b > 1$, but f is always strictly decreasing if $0 < b < 1$.

Geometrically speaking, to say a function is *strictly increasing* amounts to saying that its graph is *rising* as we look at it from left to right. To say a function is *strictly decreasing* amounts to saying that its graph is *falling* as we look at it from left to right. The graphs in Figs. 8.1 and 8.2 support the claims just made regarding the property of an exponential function being strictly increasing (if $b > 1$) or strictly decreasing (if $0 < b < 1$).

Strictly increasing and strictly decreasing functions are necessarily one-to-one functions. The proof of this fact is quite easy. Suppose, for example, that f is a function which is strictly increasing, and suppose x_1 and x_2 are in the domain of f such that $f(x_1) = f(x_2)$. Now, if $x_1 < x_2$, then $f(x_1) < f(x_2)$, which is impossible since, by hypothesis, $f(x_1) = f(x_2)$. Consequently, we *cannot* possibly have $x_1 < x_2$. Similarly, if $x_2 < x_1$, then $f(x_2) < f(x_1)$, which is again a contradiction [since $f(x_1) = f(x_2)$]. Hence, we *cannot* have $x_2 < x_1$ either. Therefore $x_1 = x_2$. We have thus shown that $f(x_1) = f(x_2)$ always implies $x_1 = x_2$, and hence f is one-to-one. If f is strictly decreasing, the argument is very similar.

The properties listed in items 1 to 6 above, which describe the graphs in both Examples 8.11 and 8.12 are actually typical properties of the graph of $y = f(x) = b^x$, for *any* positive real number b, $b \neq 1$. Indeed, it turns out that

> The graph of $y = f(x) = b^x$ looks like the graph in Fig. 8.1 if $b > 1$, and, in this case, the function f and its graph have the properties 1 to 6 given in Example 8.11. Moreover, the graph of $y = f(x) = b^x$ looks like the graph in Fig. 8.2 if $0 < b < 1$, and, in this case, the function f and its graph have the properties 1 to 6 given in Example 8.12. (8.32)

A rigorous proof of the above facts is beyond the scope of this text. Nevertheless, the above graphs (in Figs. 8.1 and 8.2) give an intuitive account of what happens in the general case (depending on whether $b > 1$ or $0 < b < 1$). Incidentally, the case in which $b = 1$ is not very interesting because we obtain a constant function, given by $f(x) = b^x = 1^x = 1$. This naturally gives rise to the following question: Can an exponential function reduce to a constant, or, for that matter, to a polynomial function for other values of b? The answer is no. In fact, we shall now prove the following result:

> If b is any positive real number such that $b \neq 1$, then the exponential function f given by $f(x) = b^x$ is not equal to any polynomial function.

The proof is as follows. Suppose that

$$b^x = g(x) \qquad \text{where } g(x) \text{ is a polynomial} \tag{8.33}$$

In the first place, g cannot be a constant function, since

$$g(0) = b^0 = 1 \qquad g(1) = b^1 = b$$

and hence $g(1) \neq g(0)$ (since $b \neq 1$). Thus, the polynomial $g(x)$ has a *positive* degree, say, n. Now, squaring both sides of (8.33), we obtain

$$b^{2x} = [g(x)]^2 \tag{8.34}$$

But, by (8.33), we have

$$b^{2x} = g(2x) \tag{8.35}$$

and hence by Eqs. (8.34) and (8.35) we conclude that

$$g(2x) = [g(x)]^2 \tag{8.36}$$

Moreover, since the degree of $g(x)$ is equal to n, the degree of $g(2x)$ is also equal to n, and, moreover, the degree of $[g(x)]^2$ is equal to $2n$. Thus, in view of (8.36), we must have

$$n = \text{degree of } g(2x) = \text{degree of } [g(x)]^2 = 2n$$

and hence $n = 2n$. Thus $n = 0$, a contradiction, since $g(x)$ is of *positive* degree n. This contradiction proves that b^x cannot be a polynomial.

Now, keeping in mind that both (5) and (6) in Examples 8.11 and 8.12 hold for *all* exponential functions, we have the following fact:

> If R is the set of all real numbers, and if P is the set of all positive real numbers, then, for any positive *real* number b such that $b \neq 1$, the exponential function $f: R \to P$ given by
>
> $$f(x) = b^x \qquad \text{for all real numbers } x \text{ (in } R) \tag{8.37}$$
>
> is both one-to-one and onto. Hence, as shown in Chap. 1 and Sec. 7.2, this function f has an inverse.

In view of (8.37), if b is any fixed positive real number, $b \neq 1$, then the exponential function

$$f: R \to P \qquad f: x \to b^x \qquad b > 0,\ b \neq 1$$

has an inverse

$$g: P \to R \qquad g: b^x \to x \qquad b > 0,\ b \neq 1$$

This function g is called the *logarithm function with base b*, and is denoted by "$\log_b$." A consequence of this inverse relationship is the following very useful *definition:*

> Suppose b is any fixed positive real number such that $b \neq 1$. Then $x = \log_b y$ if and only if $y = b^x$. (8.38)

As an immediate corollary of (8.38), we have

$$y = b^{\log_b y} \qquad \text{and} \qquad x = \log_b b^x \tag{8.39}$$

The first equality in (8.39) says that logarithms are exponents. *Indeed* $\log_b y$ *is simply that exponent to which we should raise the base b in order to obtain the positive number y.* Incidentally, it is customary to write $\log_b y$ [instead of $\log_b (y)$]. Also, "$\log_b y$" is read as "logarithm of y with base b." *The important thing to keep in mind always is that, if* $y > 0$, $b > 0$, *and* $b \neq 1$, *then*

$$\log_b y = x \quad \textit{is equivalent to} \quad y = b^x \tag{8.40}$$

Thus, $\log_{10} 100 = 2$ is equivalent to $100 = 10^2$, which, of course, is true. We shall find (8.40) to be extremely useful throughout the rest of this section.

Now, since the functions f and g given by $y = f(x) = b^x$ and $x = g(y) = \log_b y$ are inverses of each other, we know (see Prob. 2.1.15) that *the graph of* $y = \log_b x$ *is simply the reflection of the graph of* $y = b^x$ *with respect to the line* $y = x$, as indicated in Fig. 8.3 for the case in which $b > 1$. (Recall that the graphs of a function and its inverse are reflections of each other with respect to the line $y = x$.)

With an eye on the graph of $y = \log_b x$, given in Fig. 8.3, we observe the following facts:

> Let R denote the set of all real numbers, and let P denote the set of all positive real numbers. Let b be any fixed real number, $b > 1$. Then the function $g: P \to R$ defined by $g(x) = \log_b x$ is a strictly

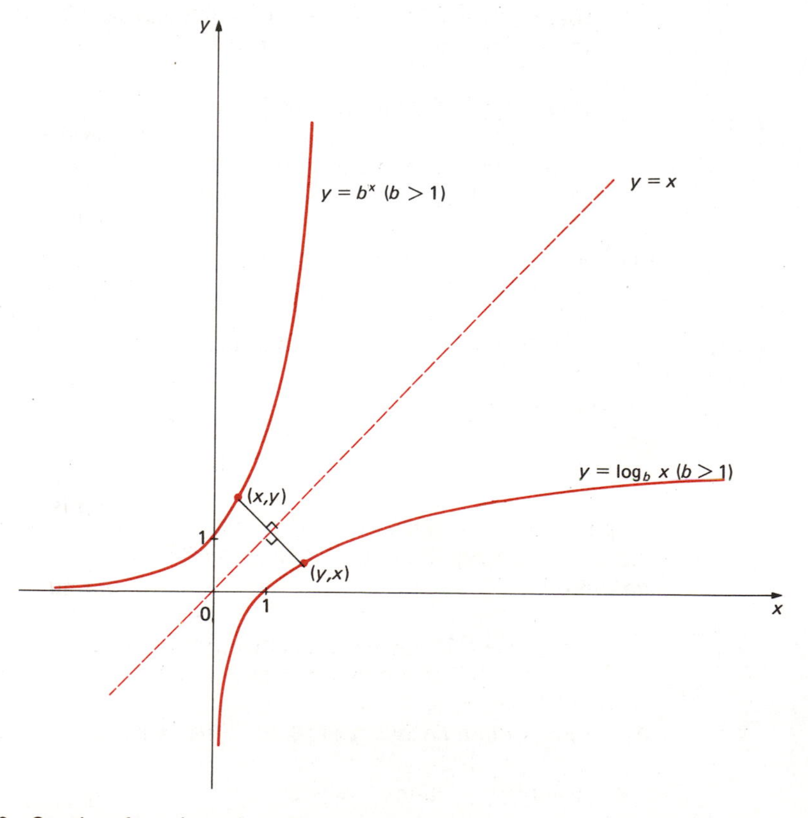

Fig. 8.3 Graphs of $y = b^x$ and $y = \log_b x$ when $b > 1$

increasing function which grows large without bound as x grows large without bound. Moreover, the y axis taken in the negative (but not the positive) direction is an asymptote of the graph of $y = \log_b x$. This graph has no y intercept point, and the point (1,0) is the only x intercept point. Finally, g is a one-to-one function whose domain is the set P of all positive real numbers, and whose range is the set of all real numbers. (8.41)

What can we say about the properties of the function $g: P \to R$ given by $g(x) = \log_b x$ when $0 < b < 1$? The student should sketch a graph to accommodate this situation, keeping in mind Figs. 8.2 and 8.3. It turns out that, when $0 < b < 1$, then g is a strictly decreasing function which grows numerically large in the negative direction as x grows without bound. Moreover, the y axis taken in the positive (but not the negative) direction is an asymptote of the graph of $y = \log_b x$. The rest of the properties given in (8.41) are still true when $0 < b < 1$.

We now list some additional properties of the logarithm function. These properties are extremely important, and are also very useful in computations.

Suppose that y_1 and y_2 are any positive real numbers, and suppose z is any real number. Suppose, further, that b is any fixed positive real number, $b \neq 1$. Then

$$\log_b (y_1 y_2) = \log_b y_1 + \log_b y_2 \tag{8.42}$$

$$\log_b \frac{y_1}{y_2} = \log_b y_1 - \log_b y_2 \tag{8.43}$$

$$\log_b y_1^z = z \log_b y_1 \tag{8.44}$$

To prove these results, suppose that we set

$$\log_b y_1 = x_1 \qquad \log_b y_2 = x_2 \tag{8.45}$$

Then, by (8.40), we know that (8.45) is equivalent to

$$y_1 = b^{x_1} \qquad y_2 = b^{x_2} \tag{8.46}$$

Hence, $y_1 y_2 = b^{x_1} b^{x_2} = b^{x_1 + x_2}$, and thus

$$y_1 y_2 = b^{x_1 + x_2} \tag{8.47}$$

Now, by (8.40), we have $\log_b (y_1 y_2) = x_1 + x_2 = \log_b y_1 + \log_b y_2$, by (8.45). Thus, (8.42) is true. Next we prove (8.43). By (8.46), we have

$$\frac{y_1}{y_2} = b^{x_1 - x_2} \tag{8.48}$$

Hence, by (8.40) we obtain $\log_b (y_1/y_2) = x_1 - x_2$, and therefore by (8.45),

$$\log_b \frac{y_1}{y_2} = \log_b y_1 - \log_b y_2 \tag{8.49}$$

Thus, (8.43) is true.

Finally we prove (8.44). Using (8.46), we get $y_1^z = b^{zx_1}$, and hence by (8.40),

$$\log_b y_1^z = zx_1 \tag{8.50}$$

Therefore by (8.45)

$$\log_b y_1^z = z \log_b y_1 \tag{8.51}$$

Thus, (8.44) is true, and the proof is complete.

We may summarize Eqs. (8.42) to (8.44) as follows:

> The logarithm of a product is equal to the sum of the logarithms of the factors, while the logarithm of a quotient is equal to the difference of the logarithms. Finally, in taking the logarithm of a number k^t, the exponent t becomes a factor. (8.52)

In addition to Eqs. (8.42) to (8.44), an extremely useful result (especially in the application of logarithms to computational problems) is obtained by setting $x = 0$, and $x = 1$, in the last equality in (8.39). The result is

$$\log_b 1 = 0 \qquad \log_b b = 1 \tag{8.53}$$

As an interesting application of (8.44) let us now prove that the logarithm function is *not* a polynomial function in disguise. Thus, suppose that b is any fixed positive real number, $b \neq 1$, and suppose

$$f(x) = \log_b x \qquad \text{for all positive real numbers } x$$

where $f(x)$ is a *polynomial* of degree n. We claim that $n \neq 0$, since if $n = 0$, then $f(x)$ would be a constant polynomial. But this is impossible, since

$$f(1) = \log_b 1 = 0 \qquad \text{and} \qquad f(b) = \log_b b = 1$$

Hence $f(x)$ is a polynomial of *positive* degree n. Now, since $f(x) = \log_b x$, we obtain

$$f(x^2) = \log_b x^2 = 2 \log_b x \qquad \text{by (8.44)}$$
$$= 2f(x)$$

Thus, $f(x^2) = 2f(x)$. Now, $f(x^2)$ is a polynomial of degree $2n$ [since the degree of $f(x)$ is equal to n], while $2f(x)$ is a polynomial of degree n. Hence $2n = n$, which is false since $n > 0$. This contradiction proves that *the logarithm function is not equal to any polynomial function.*

Example 8.13 Evaluate (*a*) $\log_2 8$; (*b*) $\log_2 \frac{1}{4}$; (*c*) $\log_2 2^{\sqrt{2}}$; (*d*) $\log_2 \sqrt[3]{1/32}$; (*e*) $\log_2 16^{2/3}$; (*f*) $\log_2 (\frac{1}{4})^{-4/5}$; (*g*) $\log_2 (\sqrt[3]{4}/\sqrt[5]{16})$.

Solution (*a*) Using (8.44) and (8.53), we obtain

$$\log_2 8 = \log_2 2^3 = 3 \log_2 2 = 3$$

(*b*) Using Eqs. (8.43), (8.44), and (8.53), we get

$$\log_2 \tfrac{1}{4} = \log_2 1 - \log_2 4 = \log_2 1 - \log_2 2^2 = \log_2 1 - 2 \log_2 2$$
$$= 0 - 2 = -2$$

Thus, $\log_2 \frac{1}{4} = -2$.

(*c*) By (8.44) and (8.53), we obtain

$$\log_2 2^{\sqrt{2}} = \sqrt{2} \log_2 2 = \sqrt{2}$$

(*d*) By (8.43), (8.44), and (8.53), we get

$$\log_2 \sqrt[3]{\frac{1}{32}} = \log_2\left(\frac{1}{2^5}\right)^{1/3} = \frac{1}{3}(\log_2 1 - \log_2 2^5)$$
$$= \frac{1}{3}(0 - 5 \log_2 2) = -\frac{5}{3}$$

(*e*) By (8.44) and (8.53), we obtain

$$\log_2 16^{2/3} = \log_2 (2^4)^{2/3} = \log_2 2^{8/3} = \tfrac{8}{3} \log_2 2 = \tfrac{8}{3}$$

(*f*) Again by (8.44) and (8.53), we get

$$\log_2 (\tfrac{1}{4})^{-4/5} = \log_2 (2^{-2})^{-4/5} = \log_2 2^{8/5} = \tfrac{8}{5} \log_2 2 = \tfrac{8}{5}$$

(*g*) By (8.44) and (8.53), we obtain

$$\log_2 \frac{\sqrt[3]{4}}{\sqrt[5]{16}} = \log_2 \frac{2^{2/3}}{2^{4/5}} = \log_2 2^{-2/15} = -\tfrac{2}{15} \log_2 2 = -\tfrac{2}{15}$$

Example 8.14 Express the following logarithms in terms of $\log_{10} 2$ and $\log_{10} 3$: (*a*) $\log_{10} 4$; (*b*) $\log_{10} 5$; (*c*) $\log_{10} 6$; (*d*) $\log_{10} 8$; (*e*) $\log_{10} 9$; (*f*) $\log_{10} 12$; (*g*) $\log_{10} 15$.

Solution Using the properties of logarithms given in Eqs. (8.42) to (8.44) and (8.53), we easily verify the following:

(*a*) $\log_{10} 4 = \log_{10} 2^2 = 2 \log_{10} 2$

(*b*) $\log_{10} 5 = \log_{10} \dfrac{10}{2} = \log_{10} 10 - \log_{10} 2 = 1 - \log_{10} 2$

(*c*) $\log_{10} 6 = \log_{10} (2 \cdot 3) = \log_{10} 2 + \log_{10} 3$

(*d*) $\log_{10} 8 = \log_{10} 2^3 = 3 \log_{10} 2$

(*e*) $\log_{10} 9 = \log_{10} 3^2 = 2 \log_{10} 3$

(*f*) $\log_{10} 12 = \log_{10} (2^2 \cdot 3) = \log_{10} 2^2 + \log_{10} 3 = 2 \log_{10} 2 + \log_{10} 3$

$$(g)\quad \log_{10} 15 = \log_{10} \frac{30}{2} = \log_{10} 30 - \log_{10} 2 = \log_{10} (10 \cdot 3) - \log_{10} 2$$
$$= \log_{10} 10 + \log_{10} 3 - \log_{10} 2 = 1 + \log_{10} 3 - \log_{10} 2$$

In the above two examples, we have used the bases 2 and 10. This naturally gives rise to the following question: Is there a formula by means of which we can express a logarithm with a given base in terms of other logarithms with different bases? The answer is furnished in the following useful result:

> Suppose that a, b, and c are any positive real numbers such that $a \neq 1$ and $b \neq 1$. Then
>
> $$\log_a c = \frac{\log_b c}{\log_b a} \tag{8.54}$$

To prove this, suppose that we set

$$\log_b a = x \qquad \log_a c = y \qquad \log_b c = z \tag{8.55}$$

Then, by (8.40), the equalities in (8.55) are equivalent to

$$b^x = a \qquad a^y = c \qquad b^z = c$$

and hence,

$$b^z = c = a^y = (b^x)^y = b^{xy}$$

Thus, $b^z = b^{xy}$, and hence $z = xy$, since the exponential function is a one-to-one function. Now, in view of (8.55) and the fact that $z = xy$, we get

$$\log_b c = (\log_b a)(\log_a c) \tag{8.56}$$

Moreover, since $a \neq 1$, $\log_b a \neq 0$, and hence we can divide both sides of (8.56) by (the *nonzero* number) $\log_b a$. When we do this, we obtain (8.54).

In defining $\log_b x$, the only restriction we imposed on the base b is that $b > 0$ and $b \neq 1$. However, in practice, the two choices for the base b which are, by far, most frequently used are $b = 10$ and $b = e$, where e is a special irrational number which is described below. In view of our decimal number system, the choice $b = 10$ for a base of logarithms has excellent advantages from the computational viewpoint. Incidentally, for every positive real number x, $\log_{10} x$ is called the *common logarithm of* x, and is usually written as $\log x$ (since the base is understood to be 10). On the other hand, for every positive real number x, $\log_e x$ is called the *natural logarithm of* x, and is usually written as $\ln x$. Natural logarithms play an important role in both pure and applied mathematics, and we now pause to give an intuitive account of the number e, which is the base used in natural logarithms.

Consider for a moment the numbers

$$\left(1 + \frac{1}{n}\right)^n \qquad n = 1, 2, 3, 4, \ldots \tag{8.57}$$

Let us evaluate a few of the numbers described in (8.57). For example, let us take $n = 1, 2, 3, 4$. The results, correct to two decimal places, are

$$\left(1+\frac{1}{1}\right)^1 = 2 \qquad \left(1+\frac{1}{2}\right)^2 = 2.25 \qquad \left(1+\frac{1}{3}\right)^3 = 2.37 \qquad \left(1+\frac{1}{4}\right)^4 = 2.44$$

It turns out that, *when n becomes very large,* $(1 + 1/n)^n$ *becomes very close to a specific number which we denote by e.* (A proof of this fact is well beyond the scope of this text.) This number e is an irrational number, which, correct to six places, is

$$e = 2.71828 \ldots \tag{8.58}$$

In view of (8.54), we can, if we wish, express natural logarithms in terms of common logarithms, and, conversely, express common logarithms in terms of natural logarithms. Indeed, taking $a = e$ and $b = 10$ in (8.54), we get

$$\log_e c = \frac{\log_{10} c}{\log_{10} e} \qquad \text{or} \qquad \ln c = \frac{\log c}{\log e} \tag{8.59}$$

which is the formula to use if we wish to express natural logarithms in terms of common logarithms. Moreover, taking $a = 10$ and $b = e$ in (8.54), we obtain

$$\log_{10} c = \frac{\log_e c}{\log_e 10} \qquad \text{or} \qquad \log c = \frac{\ln c}{\ln 10} \tag{8.60}$$

which is the formula to use if we wish to express common logarithms in terms of natural logarithms.

Problem Set 8.2

8.2.1 Sketch the graph of each of the functions f given by

(*a*) $f(x) = 3^x$ (*d*) $f(x) = e^{-x}$

(*b*) $f(x) = 3^{-x}$ (*e*) $f(x) = \left(\frac{1}{3}\right)^x$

(*c*) $f(x) = e^x$ (*f*) $f(x) = \left(\frac{1}{3}\right)^{-x}$

8.2.2 Sketch, using the same coordinate axes of Prob. 8.2.1, the graphs of the inverses of all the functions given in Prob. 8.2.1.

8.2.3 (*a*) Sketch the graph of the function f given by

$$f(x) = \frac{e^x + e^{-x}}{2}$$

Is this function even or odd? (See Sec. 7.3 for definition of even and odd functions.)

(*b*) Sketch the graph of the function f given by

$$f(x) = \frac{e^x - e^{-x}}{2}$$

Is this function even or odd?

Remark: These functions are examples of the class of *hyperbolic functions.*

8.2.4 Sketch the graphs of

(*a*) $y = f(x) = \ln x$ (*c*) $y = f(x) = \log x$

(*b*) $y = f(x) = \ln \frac{1}{x}$ (*d*) $y = f(x) = \log \frac{1}{x}$

Use the tables in the back of this text in preparing your table of values for each function.

8.2.5 Sketch, using the same coordinate axes of Prob. 8.2.4, the graphs of the inverses of all functions given in Prob. 8.2.4.

8.2.6 Suppose that a and b are any two positive real numbers such that $a \neq 1$ and $b \neq 1$. Prove that:

$$\log_a b = \frac{1}{\log_b a}$$

8.2.7 Given that $\log e = 0.4343$, find $\ln 10$.

8.2.8 Simplify.

(*a*) $\log_2 \sqrt[3]{2}$ (*d*) $\log_9 3^{4/5}$

(*b*) $\log_{1/2} \sqrt[5]{2}$ (*e*) $\log_{1/4} (1/8)^{-3/5}$

(*c*) $\log_3 3\sqrt{3}$ (*f*) $\log_{1/9} \sqrt[3]{1/27}$

8.2.9 Express the following logarithms in terms of log 2, log 3, and log 7.

(*a*) $\log 14$ (*d*) $\log 56$

(*b*) $\log 35$ (*e*) $\log \sqrt[3]{7/5}$

(*c*) $\log 1/15$ (*f*) $\log (4/63)^{1/5}$

8.2.10 Find each real number x such that:

(*a*) $\log_3 (x + 1) + \log_3 (x - 1) = 1$

(*b*) $\log_2 (x + 1) + \log_2 (x + 2) = \log_2 12$

(*c*) $\frac{1}{2} \log_b x - \log_b 4 = \log_b 3$

(*d*) $2^{2x+1} - 10(2^x) = -12$ [*Hint*: Let $y = 2^x$.]

8.3 Mantissa and Characteristic of Common Logarithms

We shall devote this section to the study of common logarithms and their application to computational problems. To begin with, we know that every positive real number z can be written, using the so-called *scientific notation*, in the form

$$z = a \times 10^n \qquad n \text{ integer, } a \text{ real, } 1 \leq a < 10 \qquad (8.61)$$

For example,

$$517 = 5.17 \times 10^2 \qquad 51.7 = 5.17 \times 10^1 \qquad 0.0517 = 5.17 \times 10^{-2}$$

and so on. Now, in view of (8.61) and (8.42), we have

$$\log z = \log a + \log 10^n = (\log a) + n$$

that is,

$$\log z = \log a + n \qquad 1 \leq a < 10,\ n \text{ integer} \tag{8.62}$$

Moreover, by (8.61) again, $1 \leq a < 10$, and hence by (8.53)

$$0 \leq \log a < 1 \tag{8.63}$$

Combining (8.61) to (8.63), we conclude that

> Logarithm of any positive real number z can be written as a sum of a nonnegative real number less than 1 (namely, $\log a$) and an integer (namely, n). (8.64)

The nonnegative real number $\log a$ in (8.62) is called the *mantissa* of $\log z$, while the integer n in (8.62) is called the *characteristic* of $\log z$. Observe that, by (8.63), the *mantissa is always a nonnegative real number which is strictly less than* 1. Thus, (8.62) can now be written as follows:

$$\log z = \log a + n = \text{mantissa} + \text{characteristic} \tag{8.65}$$

Example 8.15 Write $\log z$ as a sum of its mantissa and characteristic, if (*a*) $z = 517$; (*b*) $z = 51.7$; (*c*) $z = 0.0517$.

Solution First, we express each of the given numbers z in scientific notation, as indicated below:

(*a*) $517 = 5.17 \times 10^2$
(*b*) $51.7 = 5.17 \times 10^1$
(*c*) $0.0517 = 5.17 \times 10^{-2}$

Hence,

(*a*) $\log 517 = \log 5.17 + 2$
(*b*) $\log 51.7 = \log 5.17 + 1$
(*c*) $\log 0.0517 = \log 5.17 - 2$

Thus, the mantissas of all the above logarithms are equal to $\log 5.17$, while the characteristics of $\log z$ in (*a*), (*b*), and (*c*), are 2, 1, and -2, respectively.

The above example shows that the mantissa of the logarithm of a positive real number is always a nonnegative real number strictly less than 1 which does not depend on the position of the decimal point. Moreover, the position of the decimal point determines (and is determined by) the characteristic of the logarithm.

A table of common logarithms (Table C) appears in the back of this text. We use this table in order to find the mantissas of the logarithms under consideration, while the positions of the decimal points tell us the characteristics of the logarithms under consideration (see Example 8.15). We illustrate this in Example 8.16. *Observe that* $\approx$ *is used to indicate "is approximately equal to."*

Example 8.16 Find (*a*) log 517; (*b*) log 51.7; (*c*) log 0.0517.

Solution As we have seen in Example 8.15, we know that

$$\log 517 = \log 5.17 + 2$$

$$\log 51.7 = \log 5.17 + 1$$

$$\log 0.0517 = \log 5.17 - 2$$

Now, from Table C (see the back of this text), we find that

$$\log 5.17 \approx 0.7135$$

Substituting this result in the above equalities, we obtain

$$\log 517 \approx 2.7135$$

$$\log 51.7 \approx 1.7135$$

$$\log 0.0517 \approx \overline{2}.7135$$

The notation in this last equality needs some explanation. *The number* $\overline{2}.7135$ *is to be understood here to denote the sum* $0.7135 + (-2)$; that is,

$$\overline{2}.7135 = 0.7135 + (-2)$$

We prefer *not* to write $0.7135 + (-2)$ as $-(1.2865)$. For, if we were to do so, we would not know (without further calculation) what the mantissa and characteristic of the logarithm are. In other words, if we were to write

$$\log 0.0517 \approx -(1.2865) = -0.2865 - 1$$

then we would *not* have the desired form for log 0.0517 which is prescribed in Eqs. (8.64) and (8.65). Incidentally, in certain computational problems, it may be desirable to write

$$\overline{2}.7135 = 8.7135 - 10$$

(See Example 8.18 below.)

We now give some examples to indicate the use of logarithms in computational problems. For convenience, *from now on we shall always write* $\log x$ *to mean* $\log_{10} x$.

Example 8.17 Use logarithms to find an approximate value of (493) (51.4) (138).

Solution Let $x = (493)(51.4)(138)$. Then (see Table C)

$$\begin{aligned}\log x &= \log 493 + \log 51.4 + \log 138\\ &\approx 2.6928 + 1.7110 + 2.1399\\ &= 6.5437\end{aligned}$$

Thus,

$$\log x \approx 6.5437 = 0.5437 + 6 \tag{8.66}$$

Now, to find x, we try to locate, in the body of the table of logarithms, the number 0.5437. Unfortunately, this number does *not* appear in the body of the table. However, we find from Table C that

$$\log 3.49 \approx 0.5428$$

$$\log 3.50 \approx 0.5441$$

Since the mantissa of $\log x$ is approximately equal to 0.5437, which lies between 0.5428 and 0.5441, we use the *linear interpolation method* to get an approximate value of x. This we do as follows:

$$\text{Difference} = 0.01 \left[\begin{aligned}\log 3.49 &\approx 0.5428\\ \log\ (3.49 + d) &\approx 0.5437\\ \log 3.50 &\approx 0.5441\end{aligned}\right] \text{difference} = 0.0013$$

Since $0.5437 - 0.5428 = 0.0009$, we set the proportion $d/0.01 = 0.0009/0.0013$, and hence

$$d = \frac{0.0009}{0.0013}(0.01) \approx 0.007$$

Thus,

$$\log 3.497 \approx 0.5437$$

and hence by (8.66), $x \approx (3.497)(10^6)$. Therefore,

$$(493)(51.4)(138) \approx (3.497)(10^6)$$

Example 8.18 Use logarithms to find an approximate value of 4.83/93.5.

Solution Let $x = 4.83/93.5$. Then, using Table C,

$$\begin{aligned}\log x &= \log 4.83 - \log 93.5\\ &\approx 0.6839 - 1.9708\\ &= 0.6839 + (8.0292 - 10)\\ &= 8.7131 - 10\end{aligned}$$

Hence,

$$\log x \approx \bar{2}.7131 \tag{8.67}$$

or,

$$\log x \approx 0.7131 + (-2)$$

Now, we find in Table C that

$$\log 5.16 \approx 0.7126 \qquad \log 5.17 \approx 0.7135$$

Hence, interpolating as in the previous example, we obtain

$$\log 5.166 \approx 0.7131$$

Thus, by (8.67), we get $\log x \approx 0.7131 + (-2) \approx \log 5.166 + \log (10^{-2})$. Hence, $x \approx 5.166 \times 10^{-2}$, that is,

$$x \approx 0.05166$$

Therefore,

$$\frac{4.83}{93.5} \approx 0.05166$$

Example 8.19 Use logarithms to find an approximate value of $\sqrt[5]{93.4}$.

Solution Let $x = \sqrt[5]{93.4}$. Then, recalling the definition of fractional exponents, we have

$$x = (93.4)^{1/5}$$

Hence (see Table C),

$$\begin{aligned} \log x &= \tfrac{1}{5} \log 93.4 \\ &\approx \tfrac{1}{5}(1.9703) \\ &= 0.39406 \end{aligned}$$

Thus,

$$\log x \approx 0.39406 \tag{8.68}$$

Now, we find from Table C that

$$\log 2.47 \approx 0.3927 \qquad \log 2.48 \approx 0.3945$$

Interpolating as indicated in Example 8.17, we obtain

$$\log 2.478 \approx 0.39406$$

Hence, by (8.68), we get

$$x \approx 2.478$$

Thus,

$$\sqrt[5]{93.4} \approx 2.478$$

Example 8.20 Use logarithms to find an approximate solution of the exponential equation

$$3^x = 119 \tag{8.69}$$

Solution Taking logarithms of both sides of (8.69), we get

$$x \log 3 = \log 119$$

Hence (from Table C),

$$x = \frac{\log 119}{\log 3} \approx \frac{2.0755}{0.4771} \approx 4.35$$

Thus

$$x \approx 4.35$$

Problem Set 8.3

8.3.1 Find an approximate value of x such that $2^x = 3^{2x-1}$.

8.3.2 Find approximate values of the following by using logarithms.

(*a*) $(198)(293)(174)$

(*b*) $\dfrac{(413)(915)^2}{(31.5)^3}$

(*c*) $(21.3)^{-4/5}(937)^{1/5}$

(*d*) $\sqrt[7]{\dfrac{(983)^3(145)^2}{10.8}}$

(*e*) $\dfrac{(492)^{-10}(38.1)^5}{(981)^{-8}}$

(*f*) $\sqrt[5]{15.3\sqrt[3]{984}}$

8.3.3 Find all real numbers b and x such that $b^x = 1$.

8.3.4 Find approximate values of the following by using logarithms: (*a*) 2^{30}; (*b*) 3^{20}; (*c*) $(1.01)^{10}$.

8.3.5 Using logarithms, find an approximate value of x such that (*a*) $x^2 = 89.3$; (*b*) $x^3 = 3.71$; (*c*) $e^x = 10^{1-x}$.

8.3.6 Without using any tables of logarithms, find x such that

(*a*) $10^{3 \log x} = e^{30}$

(*b*) $\log 100^x - \log\left(\dfrac{1}{10}\right)^{x+1} = 0$

(*c*) $\log (x^2 - 3) = 0$

(*d*) $2 \log x = \log (3x - 2)$

8.4 Applications

In this section, we discuss some of the applications of exponential functions in describing population growth, radioactive decay, and compound interest.

Biologists have observed that the amount of time needed for certain bacteria to double in number does not depend on the original number of bacteria present at the beginning of the experiment. It is important to understand this phenomenon because it arises in many situations in various disciplines. Let us take a closer look at what happens to the number of bacteria at an arbitrary time t. Thus, suppose that

$$N_0 = \text{number of bacteria at beginning of experiment (i.e., at time } t = 0)$$

$$N(t) = \text{number of bacteria at time } t$$

$$k = \text{amount of time needed for bacteria to double its number}$$

Since $N(0) = N_0$ by hypothesis, it follows from the definition of k that

$$N(k) = 2N_0$$

$$N(2k) = 2N(k) = 2^2 N_0$$

$$N(3k) = 2N(2k) = 2^3 N_0$$

and, in general,

$$N(rk) = 2^r N_0 \qquad \text{for all positive integers } r \tag{8.70}$$

In order to extend (8.70) to *all real numbers* t, it seems plausible to set $rk = t$ in (8.70), and hence $r = t/k$, to obtain

$$N(t) = 2^{t/k} N_0 \qquad \text{for all real numbers } t \tag{8.71}$$

Accordingly, we *assume* that (8.71) gives the law for population growth of certain bacteria.

Example 8.21 A biologist observed, in growing a certain type of bacteria, that the number of bacteria 1 hr after the experiment started was 100, while 2 hr after the experiment started, the number was 500. Assuming that the amount of time needed for the bacteria to double does not depend on the original number of bacteria present at the beginning of the experiment, find the number of bacteria at the beginning of the experiment and 3 hr after the experiment started. How much time is needed for this bacteria to double?

Solution We are told that

$$N(1) = 100 \qquad N(2) = 500$$

Hence, using Eq. (8.71), we obtain

$$100 = N(1) = 2^{1/k}N_0 \tag{8.72}$$

and

$$500 = N(2) = 2^{2/k}N_0 \tag{8.73}$$

Now, dividing (8.72) into (8.73), we obtain

$$\frac{500}{100} = \frac{2^{2/k}N_0}{2^{1/k}N_0} = 2^{2/k-1/k} = 2^{1/k}$$

Hence,

$$2^{1/k} = 5 \tag{8.74}$$

and thus (8.71) becomes

$$N(t) = 2^{t/k}N_0 = (2^{1/k})^t N_0 = 5^t N_0$$

that is,

$$N(t) = 5^t N_0 \tag{8.75}$$

Now, combining (8.72) and (8.75), we obtain

$$100 = N(1) = 5^1 N_0 = 5N_0$$

and hence $N_0 = 20$. Substituting this in (8.75), we get

$$N(t) = 5^t(20) \tag{8.76}$$

Thus, (8.76) gives the law of growth for the bacteria under consideration. In particular, setting $t = 3$ in (8.76), we obtain

$$N(3) = 5^3(20) = 2500$$

In other words, the number of bacteria present at time $t = 3$ is 2500. Finally, to find k, we take the logarithm of both sides of (8.74):

$$\frac{1}{k}\log 2 = \log 5$$

Hence (see Table C),

$$k = \frac{\log 2}{\log 5} \approx \frac{0.3010}{0.6990} \approx 0.43$$

Thus, the bacteria under consideration double approximately every 0.43 hr.

Let us now consider another type of problem. Physicists observed that the time required for one-half of a radioactive substance to disintegrate does not depend on the quantity of the substance present. This amount of time (required for one-half of such a substance to disintegrate) is called the *half-life* of the substance. For example, the half-life of radioactive bismuth (radium E) is 5 days. Hence, at the end of 10 days, three-fourths of a given amount will have disintegrated, leaving one-fourth of the initial amount. At the end of 15 days, an additional one-eighth of the initial amount will have disintegrated, leaving one-eighth of such initial amount. Similarly, at the end of 20 days, only one-sixteenth of the original amount will be present, and so on. We may tabulate these results as follows:

Time in days	0	5	10	15	20
Radioactive amount present	W_0	$\frac{1}{2}W_0$	$(\frac{1}{2})^2W_0$	$(\frac{1}{2})^3W_0$	$(\frac{1}{2})^4W_0$

This table suggests that if

$$W(t) = \text{radioactive amount present at time } t$$

$$W_0 = \text{radioactive amount present initially, i.e., at time } t = 0$$

$$k = \text{half-life of substance}$$

then

$$W(t) = W_0(\tfrac{1}{2})^{t/k} = 2^{-t/k}W_0 \tag{8.77}$$

This formula is quite plausible, and is indeed consistent with the above table (with $k = 5$). We shall thus *assume* that (8.77) gives the law for any phenomenon involving radioactive decay. It is noteworthy to observe the great similarity of formulas (8.71) and (8.77). Let us illustrate Eq. (8.77) by some examples.

Example 8.22 Suppose that radium decomposes in such a way that at the end of 1620 yr one-half of the original amount remains. What fraction of this radium remains after 810 yr?

Solution By hypothesis, the half-life of radium is equal to 1620 yr. Our problem, then, is to find the ratio $W(810)/W_0$, given that $k = 1620$. Now, using Eq. (8.77), we obtain

$$W(810) = 2^{-810/1620}W_0$$

and hence (see Table A)

$$\frac{W(180)}{W_0} = 2^{-1/2} = \frac{1}{\sqrt{2}} = \frac{\sqrt{2}}{2} \approx \frac{1.414}{2} = 0.707$$

Thus, the fraction remaining after 810 yr is approximately equal to 0.707 of the original amount.

Example 8.23 Find the half-life of uranium if one-third of the substance disintegrates in 0.26 billion yr.

Solution By hypothesis, we are told that

$$W(0.26) = \tfrac{2}{3}W_0$$

Substituting this in (8.77), we obtain

$$\tfrac{2}{3}W_0 = W(0.26) = 2^{-0.26/k}W_0$$

Hence,

$$2^{-0.26/k} = \tfrac{2}{3} \tag{8.78}$$

Now, taking logarithms of both sides of (8.78), we get

$$-\frac{0.26}{k}\log 2 = \log 2 - \log 3$$

Hence (by Table C)

$$\begin{aligned} k &= \frac{(0.26)\log 2}{\log 3 - \log 2} \\ &\approx \frac{(0.26)(0.3010)}{0.4771 - 0.3010} \\ &\approx 0.45 \end{aligned}$$

Thus, the half-life of uranium is approximately equal to 0.45 billion yr.

As a final application of exponential functions, we now direct our attention to compound-interest considerations.

Suppose that \$100 is deposited in a bank, and suppose the bank pays an annual interest rate of 5 percent compounded quarterly. How much will the balance be at the end of 2 yr? To solve this problem, we first find the rate of interest for one interest period (that is, one-fourth of a year). Clearly, we have

$$\text{Rate of interest for one } \textit{interest period} = \frac{5}{4} \text{ percent}$$

Now, since there are four interest periods in 1 yr, the total number of interest periods in 2 yr is equal to eight. We tabulate below the amounts to which the original principal of \$100 accumulates at the end of each one of these eight interest periods.

Interest Period	Amount P_i accumulated at the end of i interest periods
1	$P_1 = 100\left(1 + \frac{5/4}{100}\right)$
2	$P_2 = 100\left(1 + \frac{5/4}{100}\right)^2 \quad \left[= P_1\left(1 + \frac{5/4}{100}\right)\right]$
3	$P_3 = 100\left(1 + \frac{5/4}{100}\right)^3 \quad \left[= P_2\left(1 + \frac{5/4}{100}\right)\right]$
4	$P_4 = 100\left(1 + \frac{5/4}{100}\right)^4 \quad \left[= P_3\left(1 + \frac{5/4}{100}\right)\right]$
5	$P_5 = 100\left(1 + \frac{5/4}{100}\right)^5 \quad \left[= P_4\left(1 + \frac{5/4}{100}\right)\right]$
6	$P_6 = 100\left(1 + \frac{5/4}{100}\right)^6 \quad \left[= P_5\left(1 + \frac{5/4}{100}\right)\right]$
7	$P_7 = 100\left(1 + \frac{5/4}{100}\right)^7 \quad \left[= P_6\left(1 + \frac{5/4}{100}\right)\right]$
8	$P_8 = 100\left(1 + \frac{5/4}{100}\right)^8 \quad \left[= P_7\left(1 + \frac{5/4}{100}\right)\right]$

Thus, we see that at the end of eight interest periods (that is, 2 yr), the original principal of $100 accumulates to

$$100\left(1 + \frac{5/4}{100}\right)^8 \tag{8.79}$$

Let us generalize this situation and ask the following question instead: If $$P$ is deposited in a bank which pays an annual interest rate of r percent compounded k times a year, how much will the balance be at the end of m yr? Arguing exactly as we did above, we can easily check that [see (8.79)]

$$\text{Balance at end of } m \text{ yr} = P\left(1 + \frac{r/k}{100}\right)^{km} \tag{8.80}$$

In verifying (8.80), observe that

$$\text{Rate of interest for one } \textit{interest period} = \frac{r}{k} \text{ percent}$$

$$\text{Number of interest periods} = km$$

Now, suppose that we increase the number of interest periods without bound. Thus, to help our imagination, we may first think of the interest as being compounded monthly, then weekly, then daily, then hourly, and so on. In the theoretical situation in which we visualize the interest to be compounded instantly (from one moment to the next), we say that the interest is *compounded continuously.* This is an extremely interesting situation. Let us take a closer look at what happens to Eq. (8.80) in the event that the interest is compounded *continuously*. Now, certainly Eq. (8.80) is valid no matter how large is the number of interest periods in 1 yr (namely, k). Observe that when we are dealing with *continuous* interest, k becomes arbitrarily large (or, as we say, k is large without bound). In view of (8.80), it is reasonable to expect that

> Balance of \$$P$ at end of m yr with annual rate of interest equal to r percent compounded continuously is equal to the number which
>
> $$P\left(1+\frac{r/k}{100}\right)^{km} \tag{8.81}$$
>
> approaches as k becomes arbitrarily large.

So we are now confronted with the problem of finding what number (if any such number exists) the expression in (8.81) approaches (or becomes close to) as k increases beyond all bounds. We have an idea how to tackle this problem if we recall (see Sec. 8.2) that

$$\left(1+\frac{1}{n}\right)^n \text{ approaches } e \text{ as } n \text{ grows without bound} \tag{8.82}$$

This certainly suggests that we set, in (8.81),

$$\frac{r/k}{100}=\frac{1}{n} \qquad \text{and hence} \qquad \frac{r}{100k}=\frac{1}{n} \qquad k=\frac{nr}{100} \tag{8.83}$$

Thus,

$$km=\frac{nmr}{100} \tag{8.84}$$

Hence, substituting (8.83) and (8.84) in (8.81), we get

$$\begin{aligned} P\left(1+\frac{r/k}{100}\right)^{km} &= P\left(1+\frac{1}{n}\right)^{nmr/100} \\ &= P\left[\left(1+\frac{1}{n}\right)^n\right]^{mr/100} \end{aligned}$$

Thus,

$$P\left(1+\frac{r/k}{100}\right)^{km} = P\left[\left(1+\frac{1}{n}\right)^n\right]^{mr/100} \tag{8.85}$$

Now, let us imagine that k becomes arbitrarily large. In view of the last equation in (8.83), we have $n=100k/r$, and hence n, too, becomes arbitrarily large as k becomes arbitrarily large (since r is a fixed finite number). The net result, then, is that as k becomes arbitrarily large, n does also, and hence, by (8.82),

$$\left(1+\frac{1}{n}\right)^n \text{ approaches } e \text{ as } k \text{ grows without bound} \tag{8.86}$$

Thus, combining (8.85) and (8.86), we see, intuitively, that

$$P\left(1+\frac{r/k}{100}\right)^{km} \text{ approaches } P(e^{mr/100}) \text{ as } k \text{ grows without bound}$$

We can summarize our results as follows:

> If \$$P$ is deposited in a bank which pays an annual rate of interest of r percent compounded continuously, then the balance at the end of m yr is given by (8.87)
>
> $$\text{Balance} = P(e^{mr/100}) \qquad (e \approx 2.71828\ldots)$$

Let us illustrate (8.87) by some examples.

Example 8.24 Find the balance to which \$100 accumulates after 5 yr, if the annual rate of interest is 6 percent compounded continuously.

Solution Applying (8.87) and using Table B in the back of this text, we obtain

$$\text{Balance} = 100e^{5(6)/100} = 100e^{0.3} \approx 134.99$$

Thus, the balance to the nearest cent is \$134.99.

Example 8.25 How much money should be deposited at an annual rate of interest of 6 percent compounded continuously in order for the balance after 10 yr to be \$1000?

Solution Suppose that P denotes the principal to be deposited. Then by (8.87) we have

$$1000 = Pe^{10(6)/100} = Pe^{0.6}$$

Thus,

$$P = 1000e^{-0.6}$$

Now, using Table B, we get

$$P \approx 548.81$$

Thus, the principal which must be deposited is approximately equal to \$548.81.

Example 8.26 At what annual rate of interest should \$100 be deposited in order for it to doubl in 10 yr if (*a*) interest is compounded continuously; (*b*) interest is compoundec quarterly?

Solution (*a*) Let r be the annual rate of interest. Since interest is compounded *continuously*, we apply (8.87) to get

$$200 = 100e^{10r/100}$$

and hence

$$e^{0.1r} = 2 \tag{8.88}$$

Therefore, taking *natural* logarithms (with base e) of both sides of (8.88), we get $(0.1r)\ \ln e = \ln 2$, or $0.1r = \ln 2$. Hence, using Table D, we obtain

$$r = \frac{\ln 2}{0.1} \approx 6.93$$

Thus, the annual rate of interest is approximately equal to 6.93 percent.

(*b*) Let r be the annual rate of interest. Since interest is compounded quarterly, we apply Eq. (8.80) to get

$$200 = 100\left(1 + \frac{r/4}{100}\right)^{4(10)}$$

and hence,

$$\left(1 + \frac{r}{400}\right)^{40} = 2$$

Now, taking logarithms of both sides, we get

$$40 \log\left(1 + \frac{r}{100}\right) = \log 2 \approx 0.3010$$

Thus,

$$\log\left(1 + \frac{r}{400}\right) \approx 0.0075$$

Hence

$$1 + \frac{r}{400} \approx 1.0174 \qquad (\text{because } \log 1.0174 \approx 0.0075)$$

and thus $r \approx 6.96$. Therefore, the annual rate of interest approximately equals 6.96 percent. Observe that the difference between the two rates of interest in (*a*) and (*b*) is only 0.03 percent = 0.0003, which is relatively small.

xample 8.27 Find the approximate amount of time needed in order for \$100 to triple if the annual rate of interest is 5 percent, provided that (*a*) interest is compounded continuously; (*b*) interest is compounded semiannually.

lution (*a*) Let m denote the number of years needed. Since interest is compounded *continuously*, we apply (8.87) to get

$$300 = 100e^{m(5)/100}$$

and hence

$$e^{m/20} = 3$$

Now, taking *natural* logarithms of both sides, we obtain $(m/20) \ln e = \ln 3$, and hence (see Table D)

$$m = 20 \ln 3 \approx 20(1.0986) \approx 22$$

Thus, the time needed approximately equals 22 yr.

(*b*) Let m denote the number of years needed. Since interest is compounded semiannually, we apply Eq. (8.80) to get

$$300 = 100\left(1 + \frac{5/2}{100}\right)^{2m}$$

and hence

$$\left(1 + \frac{5}{200}\right)^{2m} = 3$$

Now, taking logarithms of both sides, we obtain

$$2m \log \left(1 + \frac{5}{200}\right) = \log 3$$

Hence (using Table C),

$$m = \frac{\log 3}{2 \log (1.025)} \approx \frac{0.4771}{2(0.0107)}$$
$$\approx 22.3$$

Thus, the time needed is approximately equal to 22.3 yr. Observe that the difference between the two times in (*a*) and (*b*) is only 0.3 yr on a 22-yr period.

Example 8.28 Do Example 8.24 with interest compounded annually.

Solution Applying Eq. (8.80), we get

$$\text{Balance} = 100\left(1 + \frac{6/1}{100}\right)^{1(5)} = 100(1.06)^5$$
$$\approx 133.82$$

Thus the balance is approximately equal to \$133.82. It is perhaps a little surprising that the difference between this balance and the balance when interest was compounded *continuously* is only about \$1.17!

Problem Set 8.4

8.4.1 At the end of 10 min, $1/15$ of a sample of polonium remains. What is the half-life?

8.4.2 The half-life of radon is 3.85 days. What fraction of a given sample of radon remains at the end of (*a*) 1 wk; (*b*) 30 days?

8.4.3 Assume that the time needed for a colony of bacteria to double its number does not depend on the number of bacteria present at the beginning of the experiment, and assume that at the end of 2 hr the number of bacteria is 250, while at the end of 3 hr it is 400. Find the number of bacteria at the beginning of the experiment. Also, find the number of bacteria at the end of 6 hr. How long does it take for this type of bacteria to double its number?

8.4.4 A colony of bacteria doubles its number every $1/2$ hr. If the number of bacteria present 3 hr after the experiment started is 96,000, find the number of bacteria at the beginning of the experiment. What is the number of bacteria present after 5 hr?

8.4.5 Find the balance after 10 yr of $500 invested in a bank which gives 4 percent interest, if (*a*) interest is compounded quarterly; (*b*) interest is compounded continuously.

8.4.6 How much money should be deposited in a bank which gives 5 percent interest so that 12 yr later the balance will be $1000, if (*a*) interest is compounded semiannually; (*b*) interest is compounded continuously?

8.4.7 At what rate of interest should we invest $500 so that in 10 yr it will double, if (*a*) interest is compounded annually; (*b*) interest is compounded continuously?

8.4.8 How long does it take for $200 to triple at an annual rate of interest of 6 percent, if (*a*) interest is compounded semiannually; (*b*) interest is compounded continuously?

8.4.9 Suppose that, when $\$P$ is deposited at an annual interest rate of 4 percent compounded continuously, the balance after m yr is $1000, while the balance after m yr at 6 percent annual rate of interest compounded continuously is $2000. Find the approximate values of P and m.

8.4.10 Which yields a bigger balance: $100 invested for 10 yr at 5 percent annual rate of interest compounded continuously, or $100 invested for 10 yr at 6 percent annual rate of interest compounded annually? Determine the amount of the difference.

.5 Summary

In this chapter we studied exponents and their properties. In particular, we saw that both rational and irrational exponents can be defined such that the validity of the familiar fundamental laws of exponents is retained. Moreover, we discussed exponential functions, and we proved that these are *not* polynomial functions in disguise. We also saw that exponential functions do have *inverse* functions, known as logarithmic functions. Thus, $y = \log_b x$ is equivalent to $x = b^y$. The most frequently used values of the base b are $b = 10$ and $b = e$. Furthermore, we studied the properties of both exponential and logarithmic functions. We learned, for example, that $\log_b (x_1 x_2) = \log_b x_1 + \log_b x_2$, $\log_b (x_1/x_2) = \log_b x_1 - \log_b x_2$, and $\log_b x^t = t \log_b x$. Also, we saw that logarithms are extremely useful for computational purposes. Finally, we gave some applications of exponential functions and logarithms in the areas of population growth, radioactive decay, and compound interest.

9 Trigonometric Functions

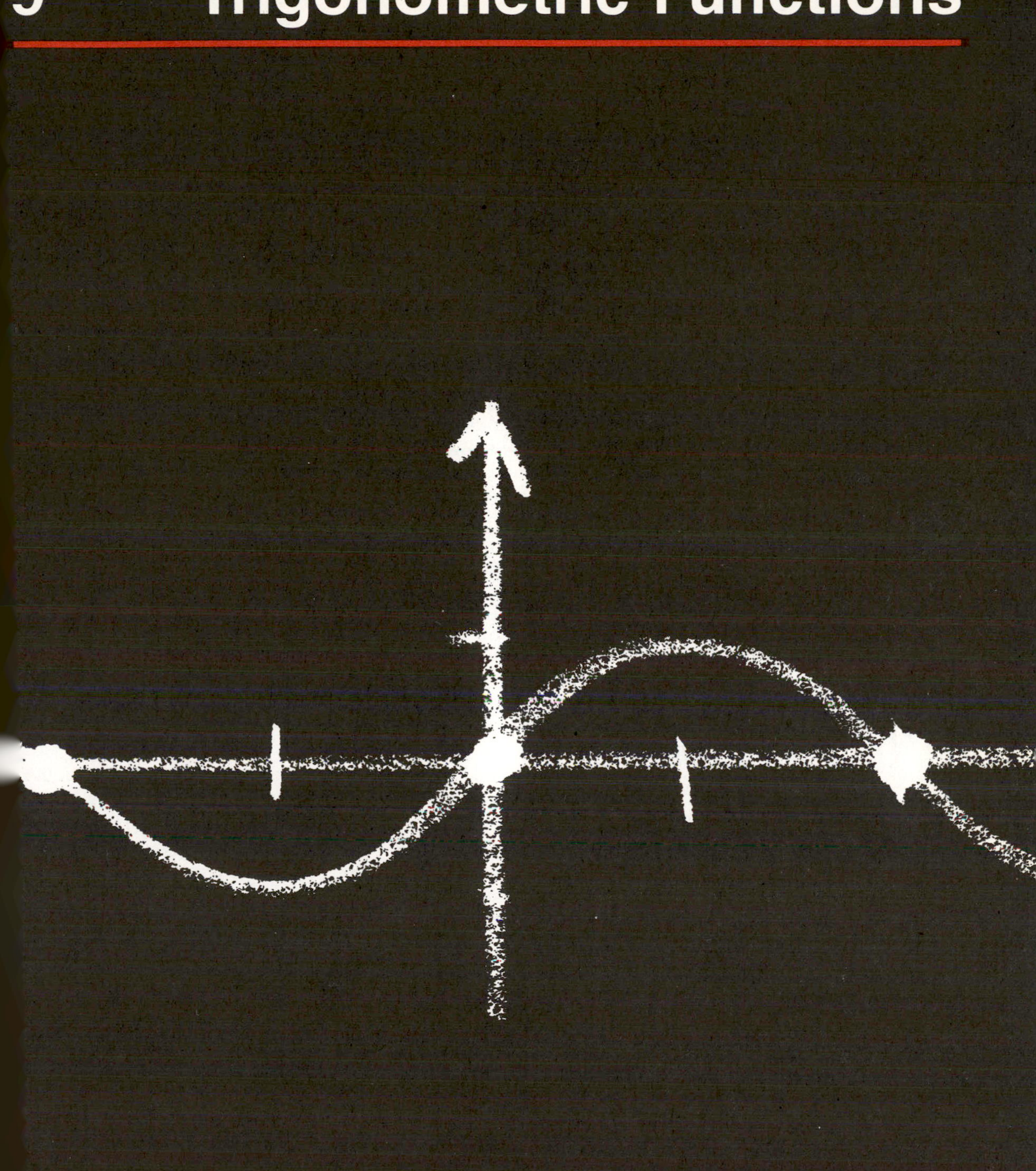

9.1 Definition of Trigonometric Functions

In this section, we will define the trigonometric functions. We will also discuss the two most frequently used measures for angles, known as degree measure and radian measure.

We recall from high school geometry that a *ray* (or half-line) is that part of a line which extends in one direction only from a point, called the endpoint. An *angle* is obtained from two rays which have the same endpoint, as indicated in Fig. 9.1. These two rays are called the *sides* of the angle, while the common endpoint is called the *vertex* of the angle.

If the order of the sides of an angle is important, we talk about a *directed angle.* Thus, if AB is chosen as the *initial side* and AC as the *terminal side* of an angle θ, we denote this angle by BAC and draw one single arrow *from* the initial side *to* the terminal side, as indicated in Fig. 9.2.

Observe that angle θ in Fig. 9.2 denotes angle BAC and *not* angle CAB. We further agree that if a directed angle is measured in the *counterclockwise* direction, it is to be considered as a *positive* angle, while if it is measured in the *clockwise* direction, it is a *negative* angle, as indicated in Fig. 9.2. (Note that in measuring an angle, we always start with the initial side.)

Next, we define what is meant by an angle in standard position. A directed angle θ is said to be in *standard position* in a coordinate plane if both of the following conditions are satisfied:

1. The vertex of angle θ is the origin point $\mathcal{O}$.
2. The initial side of angle θ is the x axis in the positive direction.

Thus in Fig. 9.3, angle θ is a positive directed angle in standard position, while angle α is a negative directed angle in standard position. *Every* directed angle in the plane can be put into a unique standard position, and for this reason, it suffices to define the trigonometric functions for only directed angles *in standard position.*

Suppose that θ is the directed angle in standard position indicated in Fig. 9.4, and suppose $A:(u,v)$ is any point, except the origin, on the terminal side of angle θ. Suppose, further, that AB is the line segment which is perpendicular to the x axis and which meets the x axis at B. Moreover, suppose that the positive length of the line segment $\mathcal{O}A$ is r (see Fig. 9.4). Observe that triangle $\mathcal{O}BA$ is a right triangle with sides $\mathcal{O}B$, BA, and with $\mathcal{O}A$ as hypotenuse. Hence, by the Pythagorean theorem, we have $u^2 + v^2 = r^2$, and thus

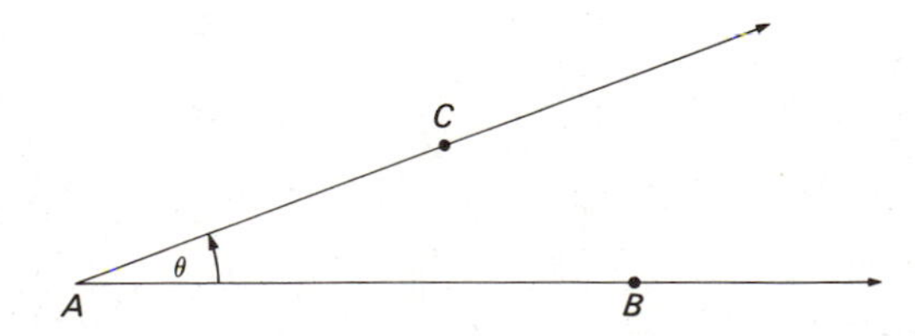

Fig. 9.1 Rays and angle

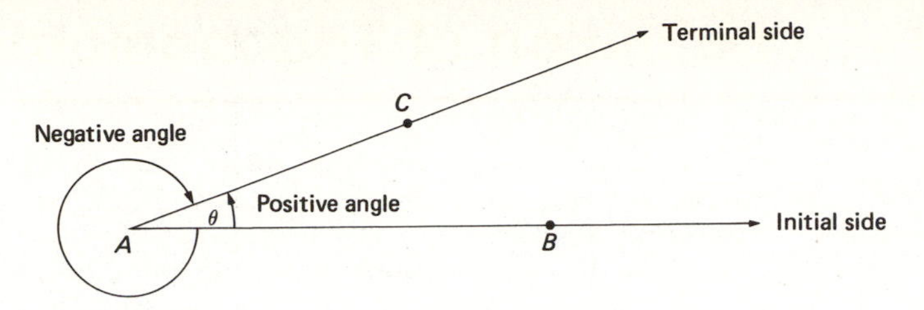

Fig. 9.2 Directed angle

$$r = \sqrt{u^2 + v^2} \qquad r > 0$$

Now, we define six trigonometric functions known as sine, cosine, tangent, cotangent, secant, and cosecant, which are abbreviated sin, cos, tan, cot, sec, and csc, in the following fashion (keep in mind Fig. 9.4):

$$\sin \theta = \frac{v}{r} = \frac{v}{\sqrt{u^2 + v^2}} \tag{9.1}$$

$$\cos \theta = \frac{u}{r} = \frac{u}{\sqrt{u^2 + v^2}} \tag{9.2}$$

$$\tan \theta = \frac{v}{u} \qquad \text{provided that } u \neq 0 \tag{9.3}$$

$$\cot \theta = \frac{u}{v} \qquad \text{provided that } v \neq 0 \tag{9.4}$$

$$\sec \theta = \frac{r}{u} = \frac{\sqrt{u^2 + v^2}}{u} \qquad \text{provided that } u \neq 0 \tag{9.5}$$

$$\csc \theta = \frac{r}{v} = \frac{\sqrt{u^2 + v^2}}{v} \qquad \text{provided that } v \neq 0 \tag{9.6}$$

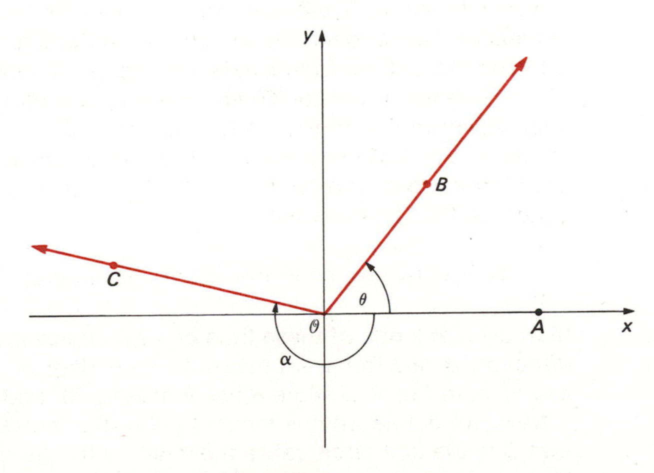

Fig. 9.3 Directed angles in standard position

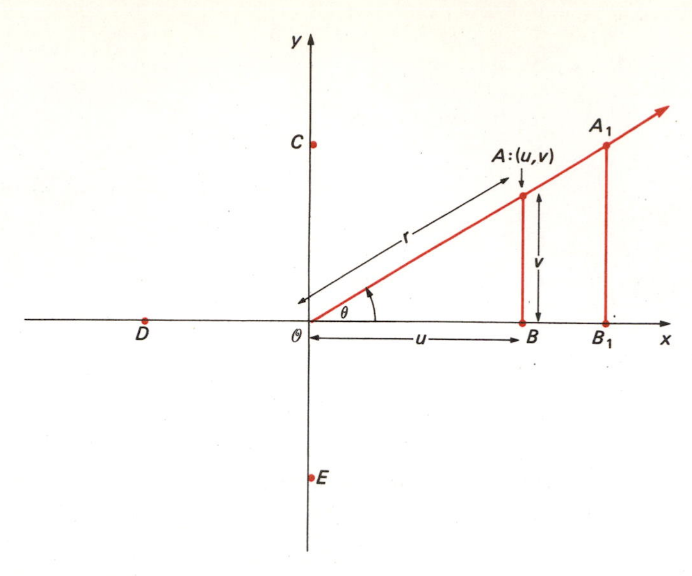

Fig. 9.4 Definition of trigonometric functions. Observe that $r = \sqrt{u^2 + v^2}$, $r > 0$.

It should be emphasized:

> The definitions of trigonometric functions are independent of the choice of point A on the terminal side of the angle θ. (9.7)

Indeed, if in Fig. 9.4 we choose any other point A_1 on the terminal side of θ different from the origin, then, by properties of similar triangles, we easily see that we obtain the *same values* for each of the trigonometric functions given in Eqs. (9.1) to (9.6) whether we use triangle $\mathcal{O}BA$ or triangle $\mathcal{O}B_1A_1$ (since these two triangles are similar by the way they are constructed). In other words, the *above definitions given in (9.1) to (9.6) depend only on the directed angle* θ.

In measuring angles, the most frequently used measures are known as *degree measure* and *radian measure.* In order to define these two measures, we first construct a circle C with the origin $\mathcal{O}$ as center and with radius 1 (for convenience). This circle is known as the *unit circle.* Let $A:(1,0)$ be the point of intersection of C and the x axis. Now, starting with this point A, we divide the circumference of C into 360 equal parts, and then join each of these points of subdivision to the origin $\mathcal{O}$. We thus obtain 360 little central angles of equal measure. Each of these little central angles, *provided that it is measured in the counterclockwise direction,* is said to have a measure of *one degree* (abbreviated as 1°). Thus, we see

> 360° corresponds to the entire circumference of C (9.8)

Moreover, any one of these little central angles *measured in the clockwise direction* is said to have a measure of -1 degree (abbreviated as $-1°$). For example, in Fig. 9.5, angle α has measure 30° and angle β has measure $-40°$.

Next, we define what is meant by a radian measure. Thus, suppose again that C is the unit circle (of radius 1 and with the origin $\mathcal{O}$ as center). We say that

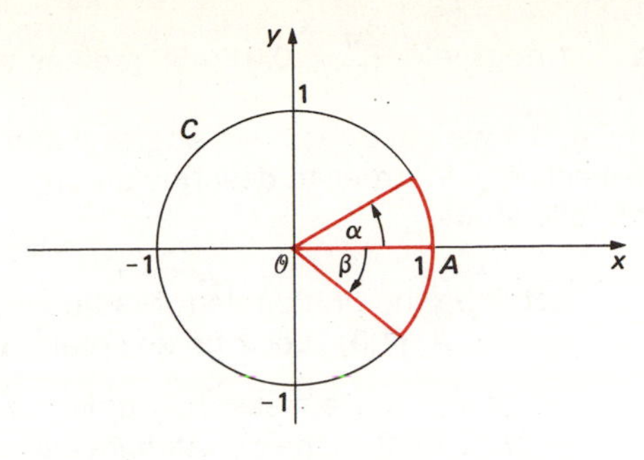

Fig. 9.5 Degree measure. Angle α has measure 30° and angle β has measure −40°.

A directed angle in standard position, measured in the counterclockwise direction, has radian measure 1 if the arc it subtends along the circumference of the unit circle C is equal to the radius 1 of C. Thus, angle θ in Fig. 9.6 has radian measure 1. On the other hand, if such a measurement takes place in the clockwise direction, we say that the directed angle has radian measure −1. Thus, angle α in Fig. 9.6 has radian measure −1.

Now, since the circumference of a circle of radius r is equal to $2\pi r$, it is clear that the circumference of the unit circle C is equal to 2π, and hence

$$2\pi \text{ radians corresponds to the entire circumference of } C \tag{9.9}$$

Comparing (9.8) and (9.9), we see that

$$2\pi \text{ radians} = 360° \tag{9.10}$$

In particular, (9.10) readily gives

$$1 \text{ radian} = \left(\frac{360}{2\pi}\right)^{\circ} = 57.296° \text{ approximately} \tag{9.11}$$

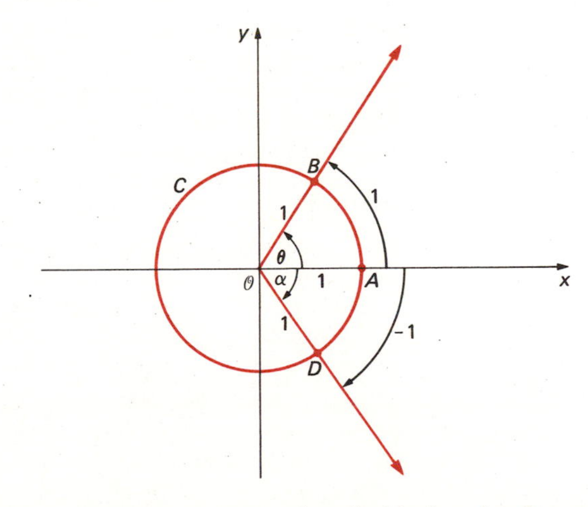

Fig. 9.6 Radian measure. Radian measure of angle θ is 1, and radian measure of angle α is −1.

$$1 \text{ degree} = \frac{2\pi}{360} = 0.017453 \text{ radians approximately} \tag{9.12}$$

In Fig. 9.6 we described two angles θ and α of radian measures 1 and -1, respectively. In order to describe an angle of *arbitrary* radian measure x, we do the following:

> Let C be the unit circle (of radius 1 and with origin $\mathcal{O}$ as center), and let $A:(1,0)$. Let x be any real number.
>
> (1) If $x > 0$, we measure x units (of length) along the circumference of C in the *counterclockwise* direction, starting this measurement from the point (1,0). This way we obtain *the* point P_x on C. The angle $A\mathcal{O}P_x$ is said to have a *positive radian measure* x (see Fig. 9.7).
>
> (2) If $x < 0$, we measure $|x|$ units (of length) along the circumference of C in the *clockwise* direction, again starting this measurement from the point (1,0). This way we obtain *the* point N_x on C. The angle $A\mathcal{O}N_x$ is said to have a *negative radian measure* x (see Fig. 9.7).
>
> (3) If $x = 0$, then the angle $A\mathcal{O}A$ is said to have a *zero radian measure.*

In the above definition of radian measure, it should be emphasized that the unit (of length) used in the measurement along the circumference of the unit circle C is the same as the unit (of length) used in the measurement along the coordinate axes.

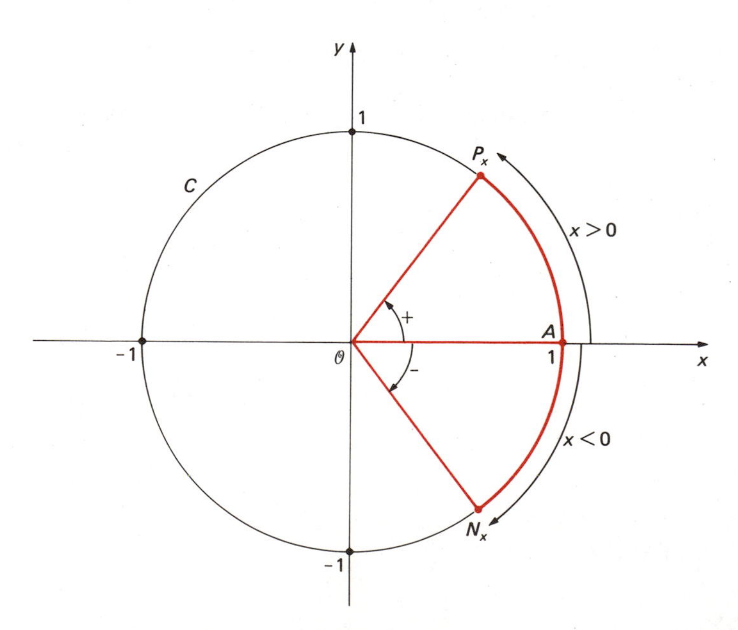

Fig. 9.7 Radian measure. $\angle A\mathcal{O}P_x$ has a positive radian measure x, while $\angle A\mathcal{O}N_x$ has a negative radian measure x.

Remark In order to find the degree measure, radian measure, or any of the trigonometric functions of an angle θ which is *not* in standard position, we first put θ into standard position, and then we apply the definitions given in this section (referring to angles in standard position).

Example 9.1 Give the radian and degree measures of certain angles in order to illustrate (9.11) and (9.12).

Solution Using (9.11) and (9.12), we can easily verify the following facts:

Radian measure	0	$\frac{\pi}{6}$	$\frac{\pi}{4}$	$\frac{\pi}{3}$	$\frac{\pi}{2}$	$\frac{2\pi}{3}$	$\frac{3\pi}{4}$	$\frac{5\pi}{6}$	π	$\frac{7\pi}{6}$	$\frac{5\pi}{4}$	$\frac{4\pi}{3}$	$\frac{3\pi}{2}$
Degree measure	0	30	45	60	90	120	135	150	180	210	225	240	270

Radian measure	$\frac{5\pi}{3}$	$\frac{7\pi}{4}$	$\frac{11\pi}{6}$	2π
Degree measure	300	315	330	360

Example 9.2 Find the values of the trigonometric functions of 45, 135, 225, and 315°.

Solution In Fig. 9.8, we draw the unit circle, and display on the circumference the points A, B, C, D, corresponding to 45, 135, 225, and 315°. The coordinates of these

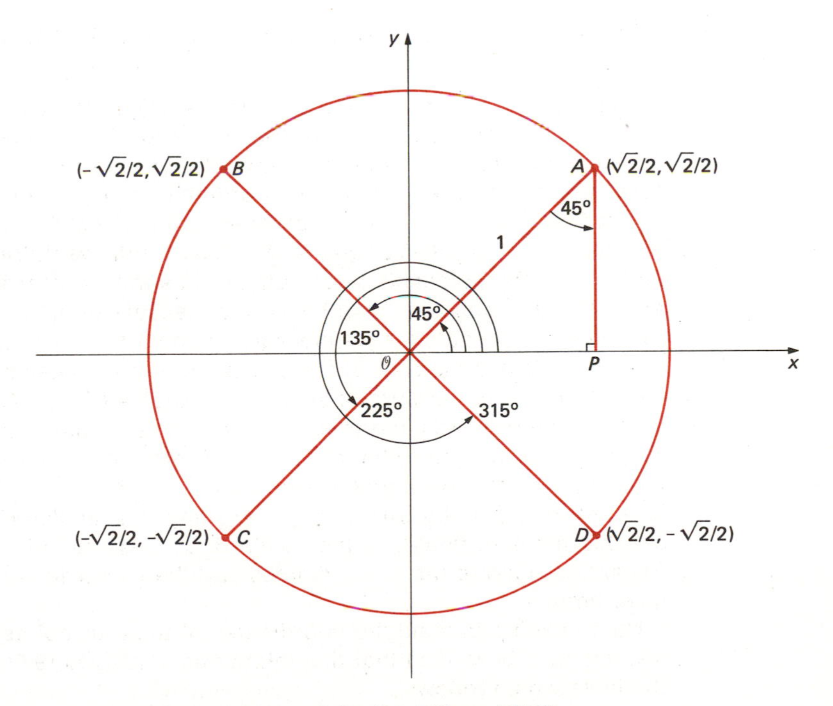

Fig. 9.8 Trigonometric functions of 45, 135, 225, and 315°

four points will enable us to read off the functional values of the trigonometric functions of these angles. In order to find the coordinates of the point A in Fig. 9.8, we draw a line AP perpendicular to the x axis and meeting the x axis at P. Now (see Remark on p. 219) since the measure of angle $\mathcal{O}PA = 90°$, the measure of angle $P\mathcal{O}A = 45°$, and since the sum of the measures of the angles of a triangle is equal to 180°, it follows that the measure of angle $\mathcal{O}AP = 45°$. Thus, the triangle $\mathcal{O}PA$ is *both* a right triangle and an isosceles triangle. Now, suppose that the length of each of the sides of triangle $\mathcal{O}PA$ is equal to a. Then, by the Pythagorean theorem, we have $a^2 + a^2 = 1$, and thus $a = \sqrt{1/2} = \sqrt{2}/2$. Hence the coordinates of the point A are $(\sqrt{2}/2, \sqrt{2}/2)$. Arguing in a similar way, we easily verify that the coordinates of B, C, D are as indicated in Fig. 9.8 (recall that the x coordinates are positive in the first and fourth quadrants but negative in the other quadrants, while the y coordinates are positive in the first and second quadrants but negative in the other quadrants).

Now, keeping in mind the coordinates of the four points A, B, C, D in Fig. 9.8, it is easily seen that the definitions in (9.1) to (9.6) yield the following facts which, for convenience, we state in a tabular form.

θ	$\sin\theta$	$\cos\theta$	$\tan\theta$	$\cot\theta$	$\sec\theta$	$\csc\theta$
45°	$\sqrt{2}/2$	$\sqrt{2}/2$	1	1	$2/\sqrt{2}$	$2/\sqrt{2}$
135°	$\sqrt{2}/2$	$-\sqrt{2}/2$	−1	−1	$-2/\sqrt{2}$	$2/\sqrt{2}$
225°	$-\sqrt{2}/2$	$-\sqrt{2}/2$	1	1	$-2/\sqrt{2}$	$-2/\sqrt{2}$
315°	$-\sqrt{2}/2$	$\sqrt{2}/2$	−1	−1	$2/\sqrt{2}$	$-2/\sqrt{2}$

Example 9.3 Find the values of the trigonometric functions of 30, 150, 210, and 330°.

Solution In Fig. 9.9, we draw the unit circle and display on the circumference the points A, B, C, D, corresponding to 30, 150, 210, and 330°. These four points will enable us to read off the functional values of the trigonometric functions of these angles. In order to find the coordinates of point A in Fig. 9.9, we join A and D and let P be the point of intersection of AD and the x axis. Now, since measure of the positive angle $P\mathcal{O}D = 330°$, it follows that measure of angle $D\mathcal{O}P = 30°$. By high school geometry, it thus follows that the triangles $\mathcal{O}PA$ and $\mathcal{O}PD$ are congruent, and hence the measure of angle $AP\mathcal{O} = 90°$. Moreover, the measure of angle $\mathcal{O}AP =$ measure of angle $PD\mathcal{O}$. But the measure of angle $\mathcal{O}AP = 60°$ (since the sum of the measures of the angles of a triangle equals 180°). Hence, the measure of each of the angles $PD\mathcal{O}$, $\mathcal{O}AP$, and $D\mathcal{O}A$ is equal to 60°. In other words, the triangle $\mathcal{O}DA$ is an equilateral triangle, and hence the length of the line segment $PA = 1/2$. Therefore, by the Pythagorean theorem, the length of the line segment $\mathcal{O}P = \sqrt{1^2 - (1/2)^2} = \sqrt{3}/2$. Thus, the coordinates of the point A are $(\sqrt{3}/2, 1/2)$. Arguing in a similar way, we can easily see that the coordinates of the points B, C, D are as indicated in Fig. 9.9 (recall the signs of the x coordinates and the y coordinates in the various quadrants).

Now, keeping in mind the coordinates of the four points A, B, C, D in Fig. 9.9, it is readily verified that the definitions in (9.1) to (9.6) yield the facts in the table which follows.

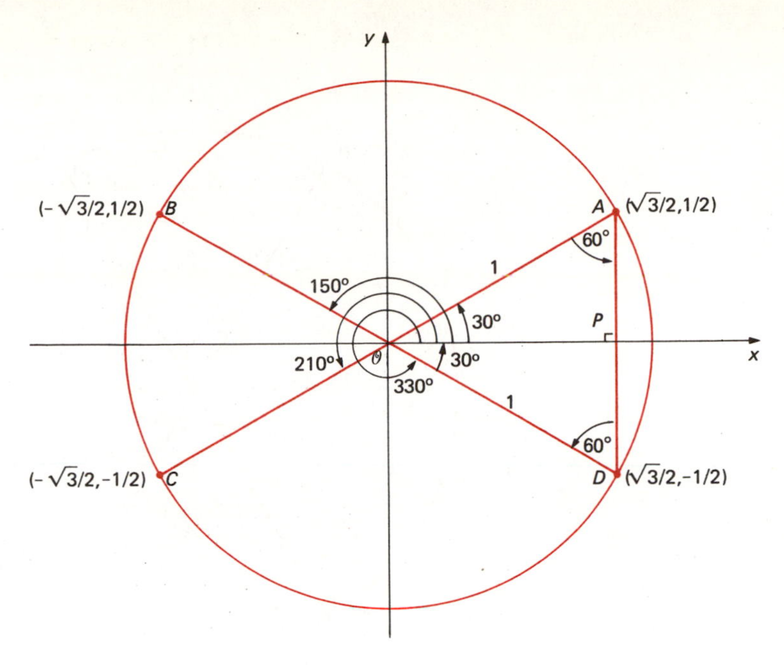

Fig. 9.9 Trigonometric functions of 30, 150, 210, and 330°

θ	$\sin\theta$	$\cos\theta$	$\tan\theta$	$\cot\theta$	$\sec\theta$	$\csc\theta$
30°	$\frac{1}{2}$	$\frac{\sqrt{3}}{2}$	$\frac{1}{\sqrt{3}}$	$\sqrt{3}$	$\frac{2}{\sqrt{3}}$	2
150°	$\frac{1}{2}$	$-\frac{\sqrt{3}}{2}$	$-\frac{1}{\sqrt{3}}$	$-\sqrt{3}$	$-\frac{2}{\sqrt{3}}$	2
210°	$-\frac{1}{2}$	$-\frac{\sqrt{3}}{2}$	$\frac{1}{\sqrt{3}}$	$\sqrt{3}$	$-\frac{2}{\sqrt{3}}$	-2
330°	$-\frac{1}{2}$	$\frac{\sqrt{3}}{2}$	$-\frac{1}{\sqrt{3}}$	$-\sqrt{3}$	$\frac{2}{\sqrt{3}}$	-2

xample 9.4 Find the values of the trigonometric functions of 60, 120, 240, and 300°.

olution In Fig. 9.10 we draw the unit circle and display on the circumference the points *A*, *B*, *C*, *D* corresponding to 60, 120, 240, and 300°. These four points will enable us to read off the functional values of the trigonometric functions of these angles. Now, to find the coordinates of point *A* in Fig. 9.10, we should continue to keep in mind the triangle $\mathcal{O}PA$ in Fig. 9.9. Indeed, if *AQ* is a line segment perpendicular to the *x* axis meeting the *x* axis at *Q*, then, triangle $\mathcal{O}QA$ in Fig. 9.10 is congruent to triangle $\mathcal{O}PA$ in Fig. 9.9. In view of this, we see that the coordinates of *A* are as indicated in Fig. 9.10. Arguing in a similar way, we readily verify that the coordinates of the points *B*, *C*, *D* are also as indicated in Fig. 9.10 (recall the signs of the *x* coordinates and *y* coordinates in the various quadrants).

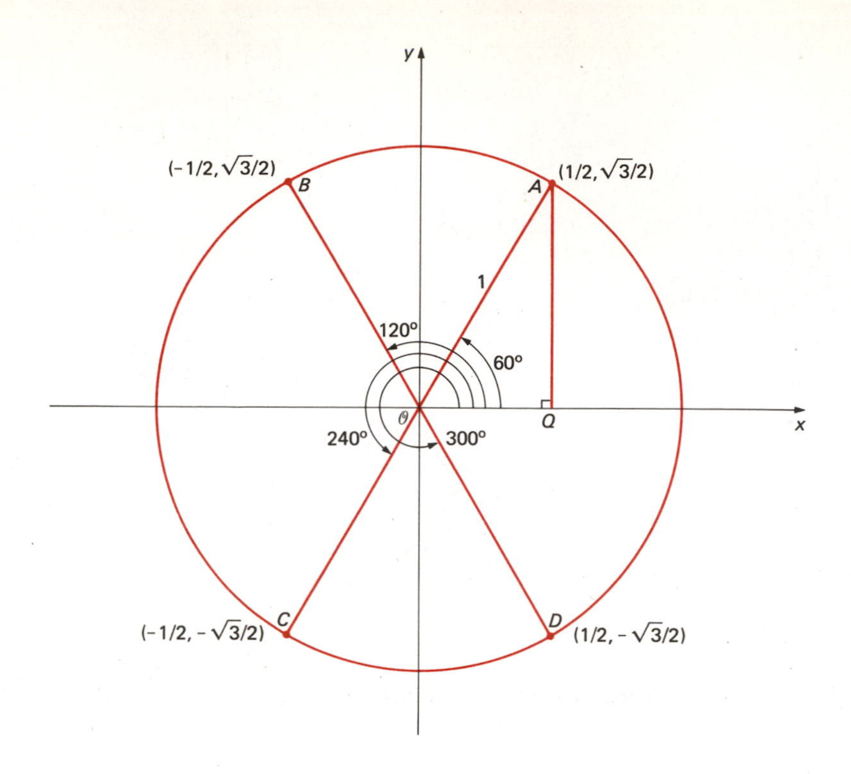

Fig. 9.10 Trigonometric functions of 60, 120, 240, 300°

Now, keeping in mind the coordinates of the four points A, B, C, D in Fig. 9.10, it is easily seen that the definitions in (9.1) to (9.6) yield the facts displayed in the table below.

θ	$\sin\theta$	$\cos\theta$	$\tan\theta$	$\cot\theta$	$\sec\theta$	csc
60°	$\frac{\sqrt{3}}{2}$	$\frac{1}{2}$	$\sqrt{3}$	$\frac{1}{\sqrt{3}}$	2	$\frac{2}{\sqrt{3}}$
120°	$\frac{\sqrt{3}}{2}$	$-\frac{1}{2}$	$-\sqrt{3}$	$-\frac{1}{\sqrt{3}}$	-2	$\frac{2}{\sqrt{3}}$
240°	$-\frac{\sqrt{3}}{2}$	$-\frac{1}{2}$	$\sqrt{3}$	$\frac{1}{\sqrt{3}}$	-2	$-\frac{2}{\sqrt{\;}}$
300°	$-\frac{\sqrt{3}}{2}$	$\frac{1}{2}$	$-\sqrt{3}$	$-\frac{1}{\sqrt{3}}$	2	$-\frac{2}{\sqrt{\;}}$

Let us tabulate, for convenience, the values of the trigonometric functions we obtained in the last three examples.

θ in radian measure	θ in degree measure	$\sin\theta$	$\cos\theta$	$\tan\theta$	$\cot\theta$	$\sec\theta$	$\csc\theta$
$\frac{\pi}{6}$	30	$\frac{1}{2}$	$\frac{\sqrt{3}}{2}$	$\frac{1}{\sqrt{3}}$	$\sqrt{3}$	$\frac{2}{\sqrt{3}}$	2
$\frac{\pi}{4}$	45	$\frac{\sqrt{2}}{2}$	$\frac{\sqrt{2}}{2}$	1	1	$\frac{2}{\sqrt{2}}$	$\frac{2}{\sqrt{2}}$
$\frac{\pi}{3}$	60	$\frac{\sqrt{3}}{2}$	$\frac{1}{2}$	$\sqrt{3}$	$\frac{1}{\sqrt{3}}$	2	$\frac{2}{\sqrt{3}}$
$\frac{2\pi}{3}$	120	$\frac{\sqrt{3}}{2}$	$-\frac{1}{2}$	$-\sqrt{3}$	$-\frac{1}{\sqrt{3}}$	-2	$\frac{2}{\sqrt{3}}$
$\frac{3\pi}{4}$	135	$\frac{\sqrt{2}}{2}$	$-\frac{\sqrt{2}}{2}$	-1	-1	$-\frac{2}{\sqrt{2}}$	$\frac{2}{\sqrt{2}}$
$\frac{5\pi}{6}$	150	$\frac{1}{2}$	$-\frac{\sqrt{3}}{2}$	$-\frac{1}{\sqrt{3}}$	$-\sqrt{3}$	$-\frac{2}{\sqrt{3}}$	2
$\frac{7\pi}{6}$	210	$-\frac{1}{2}$	$-\frac{\sqrt{3}}{2}$	$\frac{1}{\sqrt{3}}$	$\sqrt{3}$	$-\frac{2}{\sqrt{3}}$	-2
$\frac{5\pi}{4}$	225	$-\frac{\sqrt{2}}{2}$	$-\frac{\sqrt{2}}{2}$	1	1	$-\frac{2}{\sqrt{2}}$	$-\frac{2}{\sqrt{2}}$
$\frac{4\pi}{3}$	240	$-\frac{\sqrt{3}}{2}$	$-\frac{1}{2}$	$\sqrt{3}$	$\frac{1}{\sqrt{3}}$	-2	$-\frac{2}{\sqrt{3}}$
$\frac{5\pi}{3}$	300	$-\frac{\sqrt{3}}{2}$	$\frac{1}{2}$	$-\sqrt{3}$	$-\frac{1}{\sqrt{3}}$	2	$-\frac{2}{\sqrt{3}}$
$\frac{7\pi}{4}$	315	$-\frac{\sqrt{2}}{2}$	$\frac{\sqrt{2}}{2}$	-1	-1	$\frac{2}{\sqrt{2}}$	$-\frac{2}{\sqrt{2}}$
$\frac{11\pi}{6}$	330	$-\frac{1}{2}$	$\frac{\sqrt{3}}{2}$	$-\frac{1}{\sqrt{3}}$	$-\sqrt{3}$	$\frac{2}{\sqrt{3}}$	-2

In the next section, we shall study the properties of trigonometric functions and the identities they give rise to.

Problem Set 9.1

9.1.1 Express the following radian measures as degree measures: $\pi/5$, $2\pi/15$, $28\pi/45$, $-\pi/12$, $-7\pi/30$.

9.1.2 Express the following degree measures as radian measures: 10, 75, 162, -300, -340.

9.1.3 Find the values of all the trigonometric functions of the following angles: -30, -45, -60, -120, -135, and $-150°$.

9.1.4 Find the values of all the trigonometric functions of the angles with radian measure $-7\pi/6$, $-5\pi/4$, $-4\pi/3$, $-5\pi/3$, $-7\pi/4$, $-11\pi/6$.

9.1.5 Compare the values of the trigonometric functions of 60 and 150°. Explain.

9.1.6 Compare the values of the trigonometric functions of 60 and 30°. Explain.

9.1.7 Compare the values of the trigonometric functions of 45 and 225°. Explain.

9.1.8 Compare the values of the trigonometric functions of 60 and 330°. Explain.

9.1.9 Compare the values of the trigonometric functions of 30 and 210°. Explain.

9.1.10 Compare the values of the trigonometric functions of 45 and −45°. Explain.

9.1.11 Compare the values of the trigonometric functions of 30 and −330°. Explain.

9.1.12 In each of the following parts, the coordinates of a point B on the terminal side of a directed angle θ in standard position are given. Find the values of all of the six trigonometric functions of θ. (Be careful. None of these points is on the unit circle.)

(*a*) $(2,2)$ (*d*) $(3,4)$ (*g*) $(-3,-4)$,
(*b*) $(1,\sqrt{3})$ (*e*) $(-3,4)$ (*h*) $(-2,2)$
(*c*) $(\sqrt{3},1)$ (*f*) $(4,-3)$

9.1.13 In each of the following parts, the value of one of the six trigonometric functions of a directed angle θ in standard position and in the *first quadrant* is given. Find the values of the remaining five trigonometric functions of θ.

(*a*) $\sin\theta=\frac{1}{2}$ (*d*) $\cot\theta=3$
(*e*) $\sec\theta=\sqrt{3}$
(*b*) $\cos\theta=\frac{1}{4}$ (*f*) $\csc\theta=\sqrt{5}$
(*c*) $\tan\theta=\frac{3}{4}$

9.2 Trigonometric Identities

In Sec. 9.1, we defined trigonometric functions as *functions of angles.* In many situations, however, it is more advantageous to view trigonometric functions as *functions of real numbers,* and we now direct our attention to introducing trigonometric functions from this point of view. We also study in this section some of the basic properties of the trigonometric functions and the relationships which exist among these functions. Such a study will lead to a substantia number of trigonometric identities.

Suppose that x is a real number, and suppose (see Fig. 9.7 and the discussic preceding that figure)

$$\theta_x \text{ is an angle in standard position with } \textit{radian} \text{ measure } x \qquad (9.1$$

as indicated in Fig. 9.11. We now define six trigonometric functions as *functions of real numbers* in the following fashion:

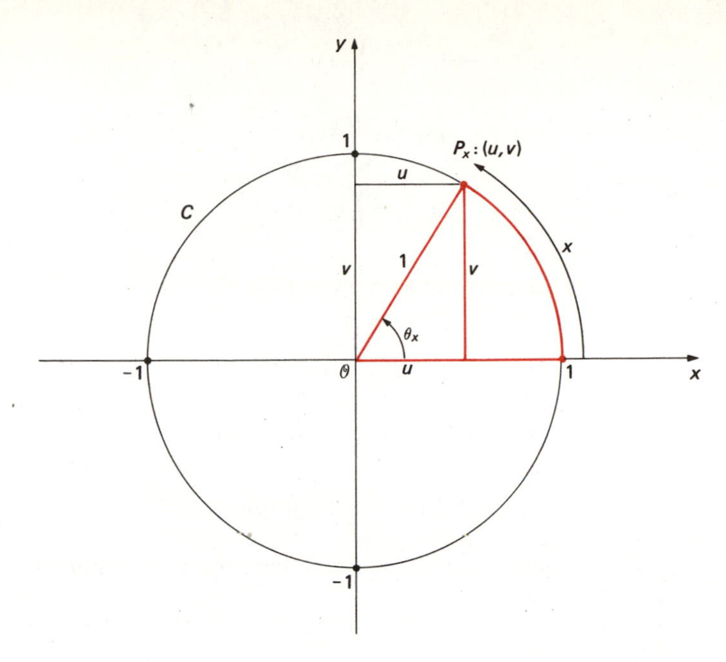

Fig. 9.11 Definition of trigonometric functions as functions of real numbers

$$\begin{aligned}
&\sin x = \sin \theta_x \\
&\cos x = \cos \theta_x \\
&\tan x = \tan \theta_x \quad \text{when } \tan \theta_x \text{ is defined} \\
&\cot x = \cot \theta_x \quad \text{when } \cot \theta_x \text{ is defined} \\
&\sec x = \sec \theta_x \quad \text{when } \sec \theta_x \text{ is defined} \\
&\csc x = \csc \theta_x \quad \text{when } \csc \theta_x \text{ is defined}
\end{aligned} \tag{9.14}$$

In other words,

> Sine of the real number x is now identified with sine of the angle θ_x of radian measure x. Similarly, cosine of the real number x is now identified with cosine of the angle θ_x of radian measure x, and so on. It should be emphasized that the measure of angle θ_x here and in (9.14) is x radians and not x degrees. (9.15)

It should also be pointed out that, strictly speaking, we should use different symbols on the left-hand sides of the definitions given in (9.14). We do not do this, however, since the functional values are equal (that is, $\sin x = \sin \theta_x$, etc.).

In (9.14) and (9.15), we have seen that *the trigonometric functions may now be viewed as functions of real numbers.* Thus recalling the definitions of trigonometric functions given in (9.1) to (9.6) [also see (9.7)] and keeping in mind Fig. 9.11 with the *unit circle,* we see that

$$\sin x = \sin \theta_x = \frac{v}{1} = v \tag{9.16}$$

$$\cos x = \cos \theta_x = \frac{u}{1} = u \tag{9.17}$$

$$\tan x = \tan \theta_x = \frac{v}{u} \qquad \text{if } u \neq 0 \tag{9.18}$$

$$\cot x = \cot \theta_x = \frac{u}{v} \qquad \text{if } v \neq 0 \tag{9.19}$$

$$\sec x = \sec \theta_x = \frac{1}{u} \qquad \text{if } u \neq 0 \tag{9.20}$$

$$\csc x = \csc \theta_x = \frac{1}{v} \qquad \text{if } v \neq 0 \tag{9.21}$$

In view of (9.16) and (9.17), it follows that

sin x and cos x are defined for all real numbers x (9.22)

Moreover, by (9.18) and (9.20), we note that (see Fig. 9.11)

tan x and sec x are defined for all real numbers x except when $x = \pm\pi/2, \pm 3\pi/2, \pm 5\pi/2, \ldots$ (9.23)

Furthermore, by (9.19) and (9.21), we observe that (see Fig. 9.11)

cot x and csc x are defined for all real numbers x except when $x = 0, \pm\pi, \pm 2\pi, \ldots$ (9.24)

Thus, if R denotes the set of all real numbers and if

$$R_1 = \{x \mid x \text{ real}, x \neq k \cdot \frac{\pi}{2}, k \text{ any } \textit{odd} \text{ integer}\} \tag{9.25}$$

$$R_2 = \{x \mid x \text{ real}, x \neq k\pi, k \textit{ any} \text{ integer}\} \tag{9.26}$$

then

$$\begin{array}{lll} \sin : R \to R & \cos : R \to R & \tan : R_1 \to R \\ \cot : R_2 \to R & \sec : R_1 \to R & \csc : R_2 \to R \end{array} \tag{9.27}$$

Example 9.5 Find the values of the trigonometric functions of 0, $\pi/2$, π, and $3\pi/2$.

Solution In Fig. 9.12 we draw the unit circle and display on its circumference the points $P_0, P_{\pi/2}, P_\pi, P_{3\pi/2}$ corresponding to 0, $\pi/2$, π, $3\pi/2$. These four points will enable us to read off the functional values of the trigonometric functions of 0, $\pi/2$, π, and $3\pi/2$, respectively. Indeed, recalling the definitions given in (9.16) to (9.21) and Fig. 9.11, we can easily read off the trigonometric functional values of 0. In fact, keeping in mind the point P_0: (1,0) in Fig. 9.12, we see that the

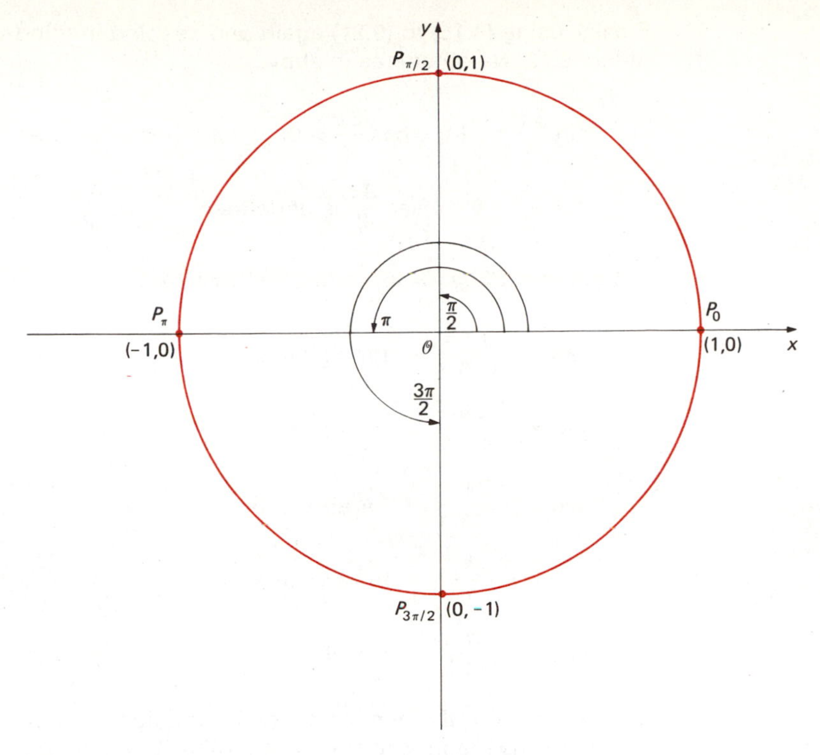

Fig. 9.12 Trigonometric functions of 0, $\pi/2$, π, and $3\pi/2$

values of u and v in (9.16) to (9.21) are given by $u = 1$, $v = 0$, and thus Eqs. (9.16) to (9.21) give

$$\sin 0 = 0 \qquad \cos 0 = 1 \qquad \tan 0 = 0$$

$$\cot 0 \text{ is undefined} \qquad \sec 0 = \frac{1}{1} = 1 \qquad \csc 0 \text{ is undefined}$$

Similarly, using (9.16) to (9.21) again and keeping in mind the point $P_{\pi/2}: (0,1)$ in Fig. 9.12, we see that the values of u and v in Eqs. (9.16) to (9.21) are now given by $u = 0$, $v = 1$, and thus Eqs. (9.16) to (9.21) now give

$$\sin \frac{\pi}{2} = 1 \qquad \cos \frac{\pi}{2} = 0 \qquad \tan \frac{\pi}{2} \text{ is undefined}$$

$$\cot \frac{\pi}{2} = 0 \qquad \sec \frac{\pi}{2} \text{ is undefined} \qquad \csc \frac{\pi}{2} = \frac{1}{1} = 1$$

Again, using (9.16) to (9.21) and keeping in mind the point $P_\pi: (-1,0)$ in Fig. 9.12, we get (arguing as above)

$$\sin \pi = 0 \qquad \cos \pi = -1 \qquad \tan \pi = 0$$

$$\cot \pi \text{ is undefined} \qquad \sec \pi = \frac{1}{-1} = -1 \qquad \csc \pi \text{ is undefined}$$

Finally, using (9.16) to (9.21) again and keeping in mind the point $P_{3\pi/2}: (0,-1)$ in Fig. 9.12, we obtain (as in above)

$$\sin \frac{3\pi}{2} = -1 \qquad \cos \frac{3\pi}{2} = 0 \qquad \tan \frac{3\pi}{2} \text{ is undefined}$$

$$\cot \frac{3\pi}{2} = 0 \qquad \sec \frac{3\pi}{2} \text{ is undefined} \qquad \csc \frac{3\pi}{2} = \frac{1}{-1} = -1$$

A glance at Eqs. (9.16) to (9.21) shows that

$$\tan x = \frac{\sin x}{\cos x} \qquad \text{for all } x \text{ in } R_1 \tag{9.28}$$

$$\cot x = \frac{\cos x}{\sin x} \qquad \text{for all } x \text{ in } R_2 \tag{9.29}$$

$$\sec x = \frac{1}{\cos x} \qquad \text{for all } x \text{ in } R_1 \tag{9.30}$$

$$\csc x = \frac{1}{\sin x} \qquad \text{for all } x \text{ in } R_2 \tag{9.31}$$

$$\cot x = \frac{1}{\tan x} \qquad \text{for all } x \text{ in } R_1 \cap R_2 \tag{9.32}$$

Note that as a further simplification in notation, we have been writing $\sin x$ to mean $\sin(x)$, and similarly for the other trigonometric functions. Moreover, we shall write $\sin^2 x$ to mean $(\sin x)^2$, and similarly for $\cos^2 x$, $\tan^3 x$, and so on.

Next, suppose we square and add both sides of (9.16) and (9.17), we obtain

$$\sin^2 x + \cos^2 x = v^2 + u^2 = u^2 + v^2 = 1$$

since (u,v) lies on the unit circle C (see Fig. 9.11). Thus,

$$\sin^2 x + \cos^2 x = 1 \qquad \text{for all real numbers } x \tag{9.33}$$

Now, if we were to divide both sides of (9.33) first by $\cos^2 x$ and then by $\sin^2 x$, we would get

$$\frac{\sin^2 x}{\cos^2 x} + 1 = \frac{1}{\cos^2 x} \qquad \text{if } \cos x \neq 0$$

$$1 + \frac{\cos^2 x}{\sin^2 x} = \frac{1}{\sin^2 x} \qquad \text{if } \sin x \neq 0$$

In view of (9.28) to (9.31), the above two identities become

$$\tan^2 x + 1 = \sec^2 x \qquad \text{for all } x \text{ in } R_1 \text{ [see (9.27)]} \qquad (9.3$$

$$1 + \cot^2 x = \csc^2 x \qquad \text{for all } x \text{ in } R_2 \text{ [see (9.27)]} \qquad (9.3$$

Moreover, using (9.16) and keeping in mind Fig. 9.11, we see that $|\sin x| = |v| \leq 1$. Similarly, using (9.17), we get $|\cos x| = |u| \leq 1$ (see Fig. 9.11). We have thus shown that

$$-1 \le \sin x \le 1 \qquad -1 \le \cos x \le 1 \qquad \text{for all real numbers } x \tag{9.36}$$

Now, since

$$\sec x = \frac{1}{\cos x} \qquad \text{and} \qquad \csc x = \frac{1}{\sin x} \qquad \text{[see (9.30) and (9.31)]}$$

(9.36) readily implies that

$$|\sec x| \ge 1 \qquad \text{for all } x \text{ in } R_1 \tag{9.37}$$

$$|\csc x| \ge 1 \qquad \text{for all } x \text{ in } R_2 \tag{9.38}$$

It is instructive to inquire about the signs of the values of the various trigonometric functions in the four quadrants. Thus, suppose θ is any directed angle in standard position whose terminal side lies in the first quadrant (for simplicity's sake, we say that θ lies in the first quadrant), and suppose that $P: (u,v)$ is any point on the terminal side of angle θ, other than the origin, as indicated in Fig. 9.13. Using Eqs. (9.1) to (9.6) we easily see that, for θ in the *first quadrant*,

$$\sin \theta = \frac{v}{r} > 0 \qquad \cos \theta = \frac{u}{r} > 0 \qquad \tan \theta = \frac{v}{u} > 0$$

$$\cot \theta = \frac{u}{v} > 0 \qquad \sec \theta = \frac{r}{u} > 0 \qquad \csc \theta = \frac{r}{v} > 0$$

Thus, *all* the trigonometric functions of θ are *positive* in the *first quadrant.* In other words, we have

The values of *all* of the six trigonometric functions of θ are *positive* if θ lies in the *first quadrant*. (9.39)

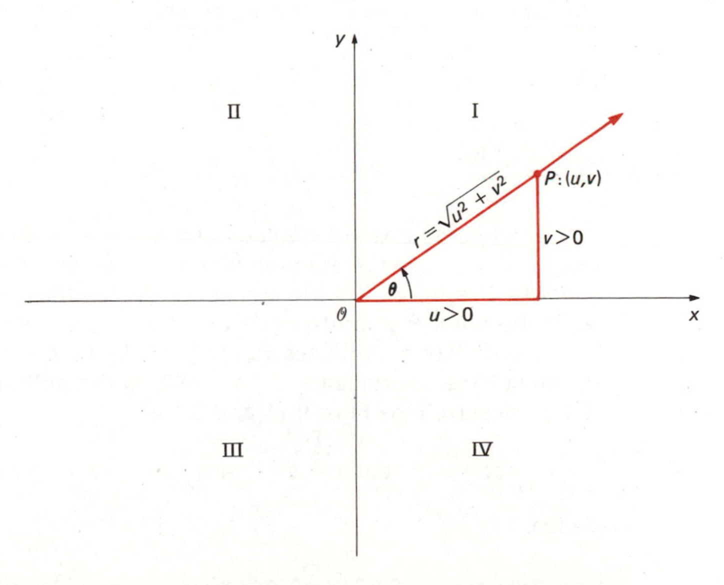

Fig. 9.13 Signs of values of trigonometric functions in the first quadrant

Arguing in the same way, and keeping in mind that the x coordinate of any point P in the *second quadrant* is negative while the y coordinate of P is positive, we easily see that

$$\begin{array}{lll} \sin\theta > 0 & \cos\theta < 0 & \\ \tan\theta < 0 & \cot\theta < 0 & \text{if } \theta \text{ lies in the second quadrant} \\ \sec\theta < 0 & \csc\theta > 0 & \end{array} \qquad (9.40$$

Similarly, since both x coordinate and y coordinate of any point in the *third quadrant* are negative, we easily verify, using the same sort of argument as above, that

$$\begin{array}{lll} \sin\theta < 0 & \cos\theta < 0 & \\ \tan\theta > 0 & \cot\theta > 0 & \text{if } \theta \text{ lies in the third quadrant} \\ \sec\theta < 0 & \csc\theta < 0 & \end{array} \qquad (9.4$$

Finally, since the x coordinate of any point P in the *fourth quadrant* is positive while the y coordinate of P is negative, we can easily check, arguing again as above, that

$$\begin{array}{lll} \sin\theta < 0 & \cos\theta > 0 & \\ \tan\theta < 0 & \cot\theta < 0 & \text{if } \theta \text{ lies in the fourth quadrant} \\ \sec\theta > 0 & \csc\theta < 0 & \end{array} \qquad (9.4$$

We summarize the results we obtained thus far in the following table:

Quadrant where θ lies	$\sin\theta$ and $\csc\theta$	$\cos\theta$ and $\sec\theta$	$\tan\theta$ and co[t]
I	+	+	+
II	+	−	−
III	−	−	+
IV	−	+	−

Now let us compare the values of the trigonometric functions of $-x$ with the values of the trigonometric functions of x, where x is any real number. Thus, suppose C is the unit circle and suppose that the point P_x corresponds to x, while the point P_{-x} corresponds to $-x$, as indicated in Fig. 9.14. Suppose that the point P_x has coordinates (u,v). Then, by high school geometry, the point P_{-x} must have coordinates $(u,-v)$. But, by the definitions given in (9.16) and (9.17), we have (see Figs. 9.11 and 9.14)

$$\sin x = v \qquad \cos x = u \qquad \sin(-x) = -v \qquad \cos(-x) = u$$

Hence,

$$\sin(-x) = -\sin x \qquad \cos(-x) = \cos x \qquad \text{for all } x \text{ in } R \qquad (9.$$

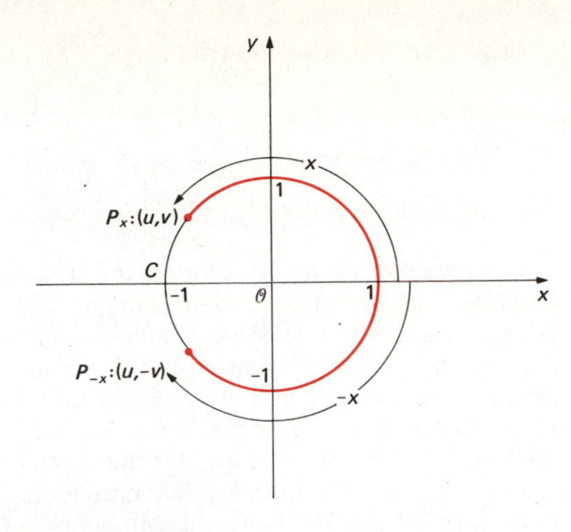

Fig. 9.14 Trigonometric functions of $-x$

Therefore, combining (9.28) and (9.43), we obtain

$$\tan(-x) = \frac{\sin(-x)}{\cos(-x)} = \frac{-\sin x}{\cos x} = -\tan x \qquad (x \in R_1)$$

that is,

$$\tan(-x) = -\tan x \qquad \text{for all } x \text{ in } R_1 \tag{9.44}$$

Similarly, combining (9.29) and (9.43), we get

$$\cot(-x) = \frac{\cos(-x)}{\sin(-x)} = \frac{\cos x}{-\sin x} = -\cot x \qquad (x \in R_2)$$

that is,

$$\cot(-x) = -\cot x \qquad \text{for all } x \text{ in } R_2 \tag{9.45}$$

Now, combining (9.30), (9.31), and (9.43), we obtain

$$\sec(-x) = \frac{1}{\cos(-x)} = \frac{1}{\cos x} = \sec x \qquad (x \in R_1)$$

$$\csc(-x) = \frac{1}{\sin(-x)} = \frac{1}{-\sin x} = -\csc x \qquad (x \in R_2)$$

Thus,

$$\sec(-x) = \sec x \qquad \text{for all } x \text{ in } R_1 \tag{9.46}$$

$$\csc(-x) = -\csc x \qquad \text{for all } x \text{ in } R_2 \tag{9.47}$$

To sum up, we have shown that

$$\begin{array}{lll}\sin(-x) = -\sin x & \cos(-x) = \cos x & (x \in R)\\ \tan(-x) = -\tan x & \sec(-x) = \sec x & (x \in R_1)\\ \cot(-x) = -\cot x & \csc(-x) = -\csc x & (x \in R_2)\end{array} \tag{9.48}$$

It is quite natural to inquire how the trigonometric functions of a sum (or a difference) of two real numbers α and β compare with the trigonometric functions of α and those of β. We shall now study this situation. In preparation for the next result, we first establish the following fact, known as the distance formula. Thus, suppose that $P_1:(x_1,y_1)$ and $P_2:(x_2,y_2)$ are any two points in the xy coordinate plane, as indicated in Fig. 9.15. What is the positive length of the line segment P_1P_2? To answer this question, we first draw a line P_1Q parallel to the x axis and another line P_2Q parallel to the y axis (see Fig. 9.15). Then triangle P_1QP_2 is a right triangle with sides P_1Q and QP_2, and with hypotenuse P_1P_2. Moreover, as is easily seen, side P_1Q has length $x_2 - x_1$ and side QP_2 has length $y_2 - y_1$. Let r denote the positive length of the line segment P_1P_2. Then, by the Pythagorean theorem, we know that

$$r^2 = (x_2 - x_1)^2 + (y_2 - y_1)^2$$

We have thus shown the

Distance formula: If $P_1:(x_1,y_1)$ and $P_2:(x_2,y_2)$, then the length of $P_1P_2 = r = \sqrt{(x_2 - x_1)^2 + (y_2 - y_1)^2}$. (9.49)

We shall use this formula presently in deriving the so-called *Addition theorems* for trigonometric functions.

Now, suppose that α and β are any real numbers. Can we compute $\cos(\alpha + \beta)$ in terms of the trigonometric functions of α and β? To answer this question, suppose that C is the unit circle and A is the point $(1,0)$. Suppose, further, that as indicated in Fig. 9.16,

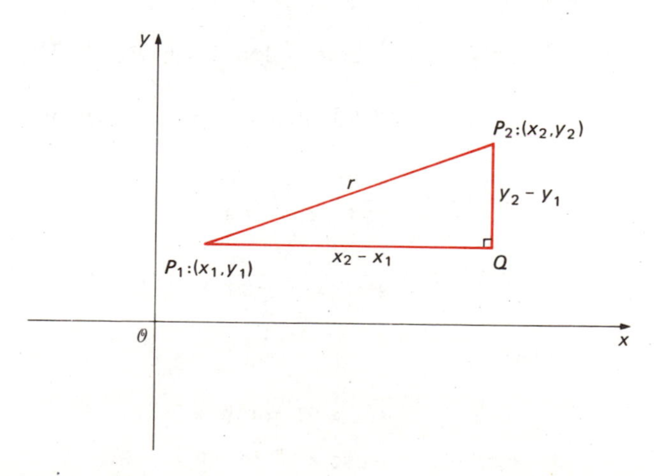

Fig. 9.15 Distance formula

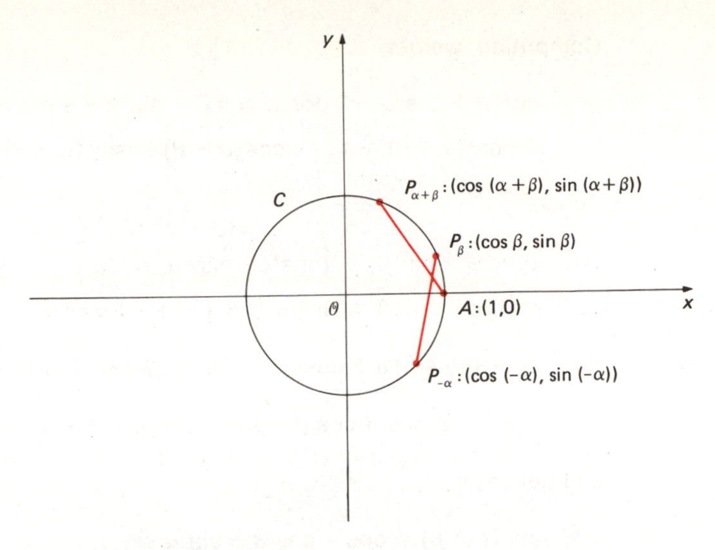

Fig. 9.16 Addition theorem for $\cos(\alpha+\beta)$

P_β is the point on C such that the arc AP_β is equal to β

$P_{\alpha+\beta}$ is the point on C such that the arc $AP_{\alpha+\beta}$ is equal to $\alpha+\beta$

$P_{-\alpha}$ is the point on C such that the arc $AP_{-\alpha}$ is equal to $-\alpha$

Now, since arc $P_{-\alpha}P_\beta$ and arc $AP_{\alpha+\beta}$ (both taken in the counterclockwise direction) are of equal length (namely, $\alpha+\beta$), it follows by high school geometry that

$$\text{Length of chord } P_{-\alpha}P_\beta = \text{length of chord } AP_{\alpha+\beta} \tag{9.50}$$

Moreover, since C is the unit circle, it follows from the definitions of cosine and sine given in (9.16) and (9.17) that (see Figs. 9.11 and 9.16)

coordinates of $P_{-\alpha}$ are $(\cos(-\alpha), \sin(-\alpha)) = (\cos\alpha, -\sin\alpha)$ [see (9.43)]

coordinates of P_β are $(\cos\beta, \sin\beta)$

coordinates of $P_{\alpha+\beta}$ are $(\cos(\alpha+\beta), \sin(\alpha+\beta))$

Now, keeping these facts in mind and recalling that the coordinates of A are $(1,0)$, we obtain, upon applying the distance formula (9.49),

$$\text{Length of } P_{-\alpha}P_\beta = \sqrt{(\cos\beta-\cos\alpha)^2+[\sin\beta-(-\sin\alpha)]^2} \tag{9.51}$$

$$\text{Length of } AP_{\alpha+\beta} = \sqrt{[\cos(\alpha+\beta)-1]^2+[\sin(\alpha+\beta)-0]^2} \tag{9.52}$$

Hence, combining (9.50) to (9.52) and squaring both sides to get rid of the radicals, we obtain

$$(\cos\beta-\cos\alpha)^2+(\sin\beta+\sin\alpha)^2=[\cos(\alpha+\beta)-1]^2+\sin^2(\alpha+\beta)$$

Computing, we get

$$\cos^2\beta + \cos^2\alpha - 2\cos\alpha\cos\beta + \sin^2\beta + \sin^2\alpha + 2\sin\alpha\sin\beta$$
$$= \cos^2(\alpha+\beta) + 1 - 2\cos(\alpha+\beta) + \sin^2(\alpha+\beta)$$

Hence,

$$(\cos^2\beta + \sin^2\beta) + (\cos^2\alpha + \sin^2\alpha) - 2(\cos\alpha\cos\beta - \sin\alpha\sin\beta)$$
$$= [\cos^2(\alpha+\beta) + \sin^2(\alpha+\beta)] + 1 - 2\cos(\alpha+\beta)$$

Now, applying $\sin^2\theta + \cos^2\theta = 1$ [see (9.33)] to the above identity, we obtain

$$1 + 1 - 2(\cos\alpha\cos\beta - \sin\alpha\sin\beta) = 1 + 1 - 2\cos(\alpha+\beta)$$

and hence,

$$\cos(\alpha+\beta) = \cos\alpha\cos\beta - \sin\alpha\sin\beta$$

We have thus established the following important *Addition theorem*:

$$\cos(\alpha+\beta) = \cos\alpha\cos\beta - \sin\alpha\sin\beta \qquad \text{for any real numbers } \alpha, \beta \qquad (9.53)$$

How about a formula for $\cos(\alpha-\beta)$, where α, β are any real numbers? This is easily achieved by combining (9.53) and (9.43), as indicated below:

$$\begin{aligned}\cos(\alpha-\beta) &= \cos[\alpha + (-\beta)]\\ &= \cos\alpha\cos(-\beta) - \sin\alpha\sin(-\beta) \qquad \text{by (9.53)}\\ &= \cos\alpha\cos\beta - \sin\alpha(-\sin\beta) \qquad \text{by (9.43)}\\ &= \cos\alpha\cos\beta + \sin\alpha\sin\beta\end{aligned}$$

We thus have the following important *Subtraction theorem*:

$$\cos(\alpha-\beta) = \cos\alpha\cos\beta + \sin\alpha\sin\beta \qquad \text{for any real numbers } \alpha, \beta \qquad (9.54)$$

Armed with the above fundamental identities, we are now able to derive some important corollaries by specializing the values of α and β. Indeed, setting $\alpha = \pi/2$ in (9.54) and recalling that $\cos(\pi/2) = 0$ and $\sin(\pi/2) = 1$ (see Example 9.5), we get

$$\begin{aligned}\cos\left(\frac{\pi}{2} - \beta\right) &= \cos\frac{\pi}{2}\cos\beta + \sin\frac{\pi}{2}\sin\beta\\ &= 0\cos\beta + 1\sin\beta\\ &= \sin\beta\end{aligned}$$

Thus,

$$\cos\left(\frac{\pi}{2} - \beta\right) = \sin\beta \qquad \text{for all real numbers } \beta \qquad (9.55)$$

Now, if we set, in (9.55), $\beta = \pi/2 - \theta$, we obtain

$$\cos\left[\frac{\pi}{2} - \left(\frac{\pi}{2} - \theta\right)\right] = \sin\left(\frac{\pi}{2} - \theta\right)$$

that is, $\cos\theta = \sin(\pi/2 - \theta)$. We have thus shown that

$$\sin\left(\frac{\pi}{2} - \theta\right) = \cos\theta \qquad \text{for all real numbers } \theta \tag{9.56}$$

It is quite natural now to inquire about formulas for $\sin(\alpha + \beta)$ and $\sin(\alpha - \beta)$, where α, β are any real numbers. With an eye on Eq. (9.55), we proceed as follows:

$$\begin{aligned}
\sin(\alpha + \beta) &= \cos\left[\frac{\pi}{2} - (\alpha + \beta)\right] && \text{by (9.55)} \\
&= \cos\left[\left(\frac{\pi}{2} - \alpha\right) - \beta\right] \\
&= \cos\left(\frac{\pi}{2} - \alpha\right)\cos\beta + \sin\left(\frac{\pi}{2} - \alpha\right)\sin\beta && \text{by (9.54)} \\
&= \sin\alpha\cos\beta + \cos\alpha\sin\beta && \text{by (9.55), (9.56)}
\end{aligned}$$

We have thus established the following important *Addition theorem*:

$$\sin(\alpha + \beta) = \sin\alpha\cos\beta + \cos\alpha\sin\beta \qquad \text{for all real numbers } \alpha, \beta \tag{9.57}$$

Next, we consider $\sin(\alpha - \beta)$. Keeping in mind (9.57) and (9.43), we obtain

$$\begin{aligned}
\sin(\alpha - \beta) &= \sin[\alpha + (-\beta)] \\
&= \sin\alpha\cos(-\beta) + \cos\alpha\sin(-\beta) && \text{by (9.57)} \\
&= \sin\alpha\cos\beta + \cos\alpha(-\sin\beta) && \text{by (9.43)} \\
&= \sin\alpha\cos\beta - \cos\alpha\sin\beta
\end{aligned}$$

We have thus shown the following important *Subtraction theorem*:

$$\sin(\alpha - \beta) = \sin\alpha\cos\beta - \cos\alpha\sin\beta \qquad \text{for all real numbers } \alpha, \beta \tag{9.58}$$

So far, we have obtained formulas for $\sin(\alpha + \beta)$, $\sin(\alpha - \beta)$, $\cos(\alpha + \beta)$, and $\cos(\alpha - \beta)$, where α, β are any real numbers; it is now natural to seek formulas for $\tan(\alpha + \beta)$, $\tan(\alpha - \beta)$, and so on. Fortunately, such formulas are very easy to achieve now, in view of the identities (9.28) to (9.31) which express $\tan x$, $\cot x$, $\sec x$, and $\csc x$ in terms of $\sin x$ and $\cos x$. To illustrate the style of proof we have in mind, let us consider $\tan(\alpha + \beta)$, where α, β are real numbers. Indeed, we have

$$\begin{aligned}
\tan(\alpha + \beta) &= \frac{\sin(\alpha + \beta)}{\cos(\alpha + \beta)} \qquad \text{if } \cos(\alpha + \beta) \neq 0 && \text{by (9.28)} \\
&= \frac{\sin\alpha\cos\beta + \cos\alpha\sin\beta}{\cos\alpha\cos\beta - \sin\alpha\sin\beta} && \text{by (9.57), (9.53)}
\end{aligned}$$

We have thus shown that

$$\tan(\alpha+\beta) = \frac{\sin\alpha\cos\beta + \cos\alpha\sin\beta}{\cos\alpha\cos\beta - \sin\alpha\sin\beta} \qquad \text{if } \cos(\alpha+\beta) \neq 0 \tag{9.59}$$

If we wish to express the right side of the above identity in terms of tan α and tan β, all we need to do is to divide both numerator and denominator of (9.59) by cos α cos β (assuming, of course, that cos α cos $\beta \neq 0$). If we do this, (9.59) becomes

$$\begin{aligned}\tan(\alpha+\beta) &= \left(\frac{\sin\alpha\cos\beta}{\cos\alpha\cos\beta} + \frac{\cos\alpha\sin\beta}{\cos\alpha\cos\beta}\right)\Big/\left(\frac{\cos\alpha\cos\beta}{\cos\alpha\cos\beta} - \frac{\sin\alpha\sin\beta}{\cos\alpha\cos\beta}\right)\\ &= \frac{\tan\alpha + \tan\beta}{1-\tan\alpha\tan\beta} \qquad \text{by (9.28)}\end{aligned}$$

We have thus shown that

$$\tan(\alpha+\beta) = \frac{\tan\alpha+\tan\beta}{1-\tan\alpha\tan b} \qquad \begin{array}{l}\text{if } \cos\alpha \neq 0,\ \cos\beta \neq 0,\\ \cos(\alpha+\beta) \neq 0\end{array} \tag{9.60}$$

Next, consider $\tan(\alpha-\beta)$. Combining (9.60) and (9.44), we obtain

$$\begin{aligned}\tan(\alpha-\beta) &= \tan[\alpha+(-\beta)]\\ &= \frac{\tan\alpha+\tan(-\beta)}{1-\tan\alpha\tan(-\beta)} \qquad \text{by (9.60)}\\ &= \frac{\tan\alpha-\tan\beta}{1+\tan\alpha\tan\beta} \qquad \text{by (9.44)}\end{aligned}$$

We thus have

$$\tan(\alpha-\beta) = \frac{\tan\alpha-\tan\beta}{1+\tan\alpha\tan\beta} \qquad \begin{array}{l}\text{if } \cos\alpha \neq 0,\ \cos\beta \neq 0,\\ \cos(\alpha-\beta) \neq 0\end{array} \tag{9.61}$$

Now, if we wish to obtain a formula for $\cot(\alpha+\beta)$ in terms of cot α and cot β, we argue as follows:

$$\begin{aligned}\cot(\alpha+\beta) &= \frac{\cos(\alpha+\beta)}{\sin(\alpha+\beta)} \qquad \text{if } \sin(\alpha+\beta) \neq 0, \text{ by (9.29)}\\ &= \frac{\cos\alpha\cos\beta-\sin\alpha\sin\beta}{\sin\alpha\cos\beta+\cos\alpha\sin\beta} \qquad \text{by (9.53), (9.57)}\\ &= \left(\frac{\cos\alpha\cos\beta}{\sin\alpha\sin\beta} - \frac{\sin\alpha\sin\beta}{\sin\alpha\sin\beta}\right)\Big/\left(\frac{\sin\alpha\cos\beta}{\sin\alpha\sin\beta} + \frac{\cos\alpha\sin\beta}{\sin\alpha\sin\beta}\right)\\ &= \frac{\cot\alpha\cot\beta-1}{\cot\beta+\cot\alpha} \qquad \text{by (9.29)}\end{aligned}$$

if sin $\alpha \neq 0$, sin $\beta \neq 0$, and $\sin(\alpha+\beta) \neq 0$. We have thus shown that

$$\cot(\alpha+\beta) = \frac{\cot\alpha\cot\beta-1}{\cot\alpha+\cot\beta} \qquad \begin{array}{l}\text{if } \sin\alpha \neq 0,\ \sin\beta \neq 0,\\ \sin(\alpha+\beta) \neq 0\end{array} \tag{9.62}$$

Finally, in order to obtain a formula for $\cot(\alpha - \beta)$, we argue as follows:

$$\begin{aligned}\cot(\alpha-\beta) &= \cot[\alpha + (-\beta)] \\ &= \frac{\cot\alpha\cot(-\beta) - 1}{\cot\alpha + \cot(-\beta)} && \text{by (9.62)} \\ &= \frac{-\cot\alpha\cot\beta - 1}{\cot\alpha - \cot\beta} && \text{by (9.45)}\end{aligned}$$

Hence,

$$\cot(\alpha-\beta) = \frac{\cot\alpha\cot\beta + 1}{\cot\beta - \cot\alpha} \qquad \begin{array}{l}\text{if } \sin\alpha \neq 0,\ \sin\beta \neq 0, \\ \sin(\alpha-\beta) \neq 0\end{array} \tag{9.63}$$

Some extremely important trigonometric identities arise by specializing $\alpha = \beta$ in formulas (9.57), (9.53), (9.60), and (9.62). Indeed, by setting $\alpha = \beta$ in these formulas, we obtain

$$\sin 2\alpha = 2\sin\alpha\cos\alpha \qquad \text{for all real numbers } \alpha \tag{9.64}$$

$$\cos 2\alpha = \cos^2\alpha - \sin^2\alpha \qquad \text{for all real numbers } \alpha \tag{9.65}$$

$$\tan 2\alpha = \frac{2\tan\alpha}{1 - \tan^2\alpha} \qquad \text{if } \cos\alpha \neq 0,\ \cos 2\alpha \neq 0 \tag{9.66}$$

$$\cot 2\alpha = \frac{\cot^2\alpha - 1}{2\cot\alpha} \qquad \text{if } \sin\alpha \neq 0,\ \sin 2\alpha \neq 0 \tag{9.67}$$

It is also interesting to observe that since, by (9.33) $\sin^2\alpha + \cos^2\alpha = 1$, Eq. (9.65) can be written in either of the following forms:

$$\cos 2\alpha = (1 - \sin^2\alpha) - \sin^2\alpha = 1 - 2\sin^2\alpha$$

or

$$\cos 2\alpha = \cos^2\alpha - (1 - \cos^2\alpha) = 2\cos^2\alpha - 1$$

We thus have [in addition to (9.65) above]

$$\cos 2\alpha = 1 - 2\sin^2\alpha \qquad \text{for all real numbers } \alpha \tag{9.68}$$

$$\cos 2\alpha = 2\cos^2\alpha - 1 \qquad \text{for all real numbers } \alpha \tag{9.69}$$

Now, if we set $\alpha = \theta/2$ in (9.68) and (9.69) and solve for $\sin\alpha$ and $\cos\alpha$, respectively, we obtain

$$\sin\frac{\theta}{2} = \pm\sqrt{\frac{1 - \cos\theta}{2}} \qquad \text{for all real numbers } \theta \tag{9.70}$$

$$\cos\frac{\theta}{2} = \pm\sqrt{\frac{1 + \cos\theta}{2}} \qquad \text{for all real numbers } \theta \tag{9.71}$$

The remaining trigonometric functions of $\theta/2$ may be obtained from (9.70) and (9.71), upon using (9.28) to (9.31). The ambiguity in sign in both (9.70) and

(9.71) will be removed as soon as we know the quadrant in which $\theta/2$ lies. [See the table following (9.42).]

Equations (9.53) to (9.71) are extremely useful in finding the trigonometric functions of certain real numbers (or angles) which we were unable to obtain before. We illustrate this in the following examples.

Example 9.6 Find the values of the trigonometric functions of $\pi/12$ and $7\pi/12$.

Solution Let us first express each of the numbers given above as a sum or difference of two numbers whose trigonometric functions are already known. This done, we can then use the above addition (or subtraction) theorems. In view of this, let us write

$$\frac{\pi}{12}=\frac{\pi}{3}-\frac{\pi}{4} \qquad \frac{7\pi}{12}=\frac{\pi}{3}+\frac{\pi}{4} \tag{9.72}$$

Then,

$$\begin{aligned}\sin\frac{\pi}{12}&=\sin\left(\frac{\pi}{3}-\frac{\pi}{4}\right)\\&=\sin\frac{\pi}{3}\cos\frac{\pi}{4}-\cos\frac{\pi}{3}\sin\frac{\pi}{4} && \text{by (9.58)}\\&=\frac{\sqrt{3}}{2}\cdot\frac{\sqrt{2}}{2}-\frac{1}{2}\cdot\frac{\sqrt{2}}{2} && \text{(see table in Sec. 9.1)}\\&=\frac{\sqrt{6}-\sqrt{2}}{4}\end{aligned}$$

Thus,

$$\sin\frac{\pi}{12}=\frac{\sqrt{6}-\sqrt{2}}{4} \tag{9.73}$$

Moreover,

$$\begin{aligned}\cos\frac{\pi}{12}&=\cos\left(\frac{\pi}{3}-\frac{\pi}{4}\right)\\&=\cos\frac{\pi}{3}\cos\frac{\pi}{4}+\sin\frac{\pi}{3}\sin\frac{\pi}{4} && \text{by (9.54)}\\&=\frac{1}{2}\cdot\frac{\sqrt{2}}{2}+\frac{\sqrt{3}}{2}\cdot\frac{\sqrt{2}}{2} && \text{(see table in Sec. 9.1)}\\&=\frac{\sqrt{2}+\sqrt{6}}{4}\end{aligned}$$

Thus,

$$\cos\frac{\pi}{12}=\frac{\sqrt{2}+\sqrt{6}}{4} \tag{9.74}$$

Using Eqs. (9.73) and (9.74) together with (9.28) to (9.31), the student should verify that

$$\tan\frac{\pi}{12}=\frac{\sqrt{6}-\sqrt{2}}{\sqrt{2}+\sqrt{6}} \qquad \cot\frac{\pi}{12}=\frac{\sqrt{2}+\sqrt{6}}{\sqrt{6}-\sqrt{2}}$$
$$\sec\frac{\pi}{12}=\frac{4}{\sqrt{2}+\sqrt{6}} \qquad \csc\frac{\pi}{12}=\frac{4}{\sqrt{6}-\sqrt{2}} \tag{9.75}$$

Next, consider the trigonometric functions of $7\pi/12$. We have

$$\begin{aligned}\sin\frac{7\pi}{12}&=\sin\left(\frac{\pi}{3}+\frac{\pi}{4}\right)\\ &=\sin\frac{\pi}{3}\cos\frac{\pi}{4}+\cos\frac{\pi}{3}\sin\frac{\pi}{4} \qquad \text{by (9.57)}\\ &=\frac{\sqrt{6}+\sqrt{2}}{4}\end{aligned}$$

and hence,

$$\sin\frac{7\pi}{12}=\frac{\sqrt{6}+\sqrt{2}}{4} \tag{9.76}$$

Moreover,

$$\begin{aligned}\cos\frac{7\pi}{12}&=\cos\left(\frac{\pi}{3}+\frac{\pi}{4}\right)\\ &=\cos\frac{\pi}{3}\cos\frac{\pi}{4}-\sin\frac{\pi}{3}\sin\frac{\pi}{4} \qquad \text{by (9.53)}\\ &=\frac{\sqrt{2}-\sqrt{6}}{4}\end{aligned}$$

and hence,

$$\cos\frac{7\pi}{12}=\frac{\sqrt{2}-\sqrt{6}}{4} \tag{9.77}$$

Using Eqs. (9.76) and (9.77) together with (9.28) to (9.31), the student should verify that

$$\tan\frac{7\pi}{12}=\frac{\sqrt{6}+\sqrt{2}}{\sqrt{2}-\sqrt{6}} \qquad \cot\frac{7\pi}{12}=\frac{\sqrt{2}-\sqrt{6}}{\sqrt{6}+\sqrt{2}}$$
$$\sec\frac{7\pi}{12}=\frac{4}{\sqrt{2}-\sqrt{6}} \qquad \csc\frac{7\pi}{12}=\frac{4}{\sqrt{6}+\sqrt{2}} \tag{9.78}$$

The above example agrees with the table [following (9.42)] which gives the signs of the values of the trigonometric functions in the various quadrants. We also observe from the above calculations that

$$\sin\frac{7\pi}{12} = \cos\frac{\pi}{12} \qquad \cos\frac{7\pi}{12} = -\sin\frac{\pi}{12} \qquad \text{etc.}$$

Is this a coincidence? The next example provides the answer to this question.

Example 9.7 Show that, for any real number α,

$$\sin\left(\frac{\pi}{2} + \alpha\right) = \cos\alpha \qquad \cos\left(\frac{\pi}{2} + \alpha\right) = -\sin\alpha \tag{9.79}$$

(Note that $7\pi/12 = \pi/2 + \pi/12$.)

Solution By (9.57), we have

$$\begin{aligned}\sin\left(\frac{\pi}{2} + \alpha\right) &= \sin\frac{\pi}{2}\cos\alpha + \cos\frac{\pi}{2}\sin\alpha \\ &= 1\cdot\cos\alpha + 0\cdot\sin\alpha \qquad \text{(see Example 9.5)} \\ &= \cos\alpha\end{aligned}$$

Thus,

$$\sin\left(\frac{\pi}{2} + \alpha\right) = \cos\alpha \qquad \text{for all real numbers } \alpha$$

Moreover, by (9.53) we have

$$\begin{aligned}\cos\left(\frac{\pi}{2} + \alpha\right) &= \cos\frac{\pi}{2}\cos\alpha - \sin\frac{\pi}{2}\sin\alpha \\ &= 0\cdot\cos\alpha - 1\cdot\sin\alpha = -\sin\alpha\end{aligned}$$

Hence,

$$\cos\left(\frac{\pi}{2} + \alpha\right) = -\sin\alpha \qquad \text{for all real numbers } \alpha$$

So the results that $\sin(7\pi/12) = \cos(\pi/12)$ and $\cos(7\pi/12) = -\sin(\pi/12)$ were *not* a coincidence.

Remark Using the results of Example 9.7, together with Eqs. (9.28) to (9.31), we can easily obtain identities for $\tan(\pi/2 + \alpha)$, $\cot(\pi/2 + \alpha)$, etc., in terms of the trigonometric functions of α.

Example 9.8 Find the values of the trigonometric functions of $\pi/24$.

Solution Since $\pi/24 = \frac{1}{2}\cdot\pi/12$, and since we already know the trigonometric function values of $\pi/12$ (see Example 9.6), we use the identities (9.70) and (9.71). Indeed, setting $\theta = \pi/12$ in (9.70) and (9.71), we get

$$\sin\frac{\pi}{24} = \pm\sqrt{\frac{1-\cos(\pi/12)}{2}}$$

$$\cos\frac{\pi}{24} = \pm\sqrt{\frac{1+\cos(\pi/12)}{2}}$$

Now, since $\pi/24$ lies in the first quadrant, both $\sin(\pi/24)$ and $\cos(\pi/24)$ are *positive.* Using this fact, together with the results of Example 9.6, we obtain

$$\sin\frac{\pi}{24} = \sqrt{\frac{1-(\sqrt{2}+\sqrt{6})/4}{2}} = \sqrt{\frac{4-\sqrt{2}-\sqrt{6}}{8}} \tag{9.80}$$

$$\cos\frac{\pi}{24} = \sqrt{\frac{1+(\sqrt{2}+\sqrt{6})/4}{2}} = \sqrt{\frac{4+\sqrt{2}+\sqrt{6}}{8}} \tag{9.81}$$

Using these results, together with Eqs. (9.28) to (9.31), we can easily find the remaining trigonometric function values of $\pi/24$.

Remark It should be emphasized that the formulas we established in this section are also valid for angles. Indeed, in view of (9.15), the above identities are true whether we view trigonometric functions as functions of real numbers or as functions of angles.

The equalities embodied in Eqs. (9.28) to (9.35), in Eqs. (9.43) to (9.48), and (9.53) to (9.71) as well are examples of trigonometric identities. In general, a *trigonometric identity* is an equality $F = G$, involving trigonometric functions, which is true for all values for which both F and G are defined. The trigonometric identities we established in this section can be used to establish further trigonometric identities, as the following examples show.

Example 9.9 Prove that for all real numbers x

$$2\cos^4 x + \cos^2 x\sin^2 x - \sin^4 x = 2\cos^2 x - \sin^2 x \tag{9.82}$$

Solution We first rewrite the left side of (9.82) as indicated in the first step below, and then proceed as shown in the following computation.

$$\begin{aligned} 2\cos^4 x + \cos^2 x\sin^2 x - \sin^4 x &= (\cos^4 x + \cos^2 x\sin^2 x) + (\cos^4 x - \sin^4 x) \\ &= \cos^2 x\,(\cos^2 x + \sin^2 x) \\ &\quad + (\cos^2 x - \sin^2 x)(\cos^2 x + \sin^2 x) \\ &= (\cos^2 x)(1) + (\cos^2 x - \sin^2 x)(1) \qquad \text{by (9.33)} \\ &= 2\cos^2 x - \sin^2 x \end{aligned}$$

which is the right side of (9.82). This establishes the identity (9.82).

Example 9.10 Prove that, for all real numbers x, y such that all the functions involved are defined and each denominator is nonzero,

$$\frac{\sin(x+y)}{\cos x\cos y} = \tan x + \tan y \tag{9.83}$$

Solution

Let us try to use some of the trigonometric identities established in this section in order to express the right side of (9.83) in terms of the functions "sin" and "cos." In this way, we hope to obtain the left side of (9.83). We indicate below the computation involved, together with the trigonometric identities used.

$$\tan x + \tan y = \frac{\sin x}{\cos x} + \frac{\sin y}{\cos y} \qquad \text{by (9.28)}$$

$$= \frac{\sin x \cos y + \cos x \sin y}{\cos x \cos y}$$

$$= \frac{\sin(x+y)}{\cos x \cos y} \qquad \text{by (9.57)}$$

which is the left side of (9.83). This establishes the identity (9.83).

Remark

In trigonometric identities, we always assume that all the real numbers involved are chosen so that each trigonometric function involved is defined and, moreover, each denominator is not equal to zero. Thus, in (9.83) we tacitly assume that $\cos x \neq 0$ and $\cos y \neq 0$.

Example 9.11

Prove that, for all real numbers x such that all the functions involved are defined and each denominator is nonzero,

$$\frac{1+\cos x}{\sin x} = \cot \frac{x}{2} \tag{9.84}$$

Solution

In view of the nature of the right side of (9.84), let us try to express the left side in terms of the trigonometric functions of $x/2$, as indicated below. Now, if we let $\alpha = x/2$ in (9.69) and (9.64), we obtain $1 + \cos x = 2\cos^2(x/2)$ and $\sin x = 2\sin(x/2)\cos(x/2)$. Hence,

$$\frac{1+\cos x}{\sin x} = \frac{2\cos^2(x/2)}{2\sin(x/2)\cos(x/2)}$$

$$= \frac{\cos(x/2)}{\sin(x/2)}$$

$$= \cot\frac{x}{2} \qquad \text{by (9.29)}$$

which is the right side of (9.84). This establishes the identity (9.84).

Example 9.12

Prove that, for all real numbers x such that all the functions involved are defined and each denominator is nonzero,

$$\sqrt{\frac{1-\sin x}{1+\sin x}} = \left|\frac{\tan(x/2)-1}{\tan(x/2)+1}\right| \tag{9.85}$$

Solution

In view of the right side of (9.85), we express the left side in terms of the trigonometric functions of $x/2$, and then we try (hopefully) to remove the square root symbol. Now, as we have remarked above, $\sin x = 2 \sin(x/2) \cos(x/2)$, and hence $1 - \sin x = 1 - 2 \sin(x/2) \cos(x/2) = \sin^2(x/2) + \cos^2(x/2) - 2 \sin(x/2) \cos(x/2)$, since [as we have seen in Eq. (9.33)] $\sin^2(x/2) + \cos^2(x/2) = 1$. Thus,

$$1 - \sin x = \sin^2\frac{x}{2} - 2 \sin\frac{x}{2} \cos\frac{x}{2} + \cos^2\frac{x}{2} = \left(\sin\frac{x}{2} - \cos\frac{x}{2}\right)^2 \tag{9.86}$$

Similarly,

$$1 + \sin x = \sin^2\frac{x}{2} + 2 \sin\frac{x}{2} \cos\frac{x}{2} + \cos^2\frac{x}{2} = \left(\sin\frac{x}{2} + \cos\frac{x}{2}\right)^2 \tag{9.87}$$

Hence, dividing (9.86) by (9.87), we obtain

$$\frac{1 - \sin x}{1 + \sin x} = \left[\frac{\sin(x/2) - \cos(x/2)}{\sin(x/2) + \cos(x/2)}\right]^2 \tag{9.88}$$

Now, if we divide both parts of the fraction contained in brackets on the right side of (9.88) by $\cos(x/2)$, we get

$$\frac{\sin(x/2) - \cos(x/2)}{\sin(x/2) + \cos(x/2)} = \frac{\tan(x/2) - 1}{\tan(x/2) + 1} \tag{9.89}$$

since by (9.28), $\sin(x/2)/\cos(x/2) = \tan(x/2)$. Hence, by (9.88) and (9.89) we have

$$\frac{1 - \sin x}{1 + \sin x} = \left[\frac{\tan(x/2) - 1}{\tan(x/2) + 1}\right]^2 \tag{9.90}$$

Now, taking the *positive* square roots of both sides of (9.90), we obtain (9.85).

In the next section, we consider the graphs of the trigonometric functions. We shall see that the trigonometric functions are examples of *periodic functions.*

Problem Set 9.2

9.2.1 Use the identities of this section to find the values of all the trigonometric functions of $5\pi/12$.

9.2.2 Find all the values of the trigonometric functions of $\pi/8$. [*Hint*: Use formulas (9.70) and (9.71).]

9.2.3 Find all the values of the trigonometric functions of $\pi/24 = \pi/6 - \pi/8$. Compare your results with those obtained in Example 9.8.

9.2.4 Prove that $\sin 3\alpha = 3 \sin \alpha - 4 \sin^3\alpha$, for all real numbers α. [*Hint*: $\sin 3\alpha = \sin(\alpha + 2\alpha)$.]

9.2.5 Prove that $\cos 3\alpha = 4\cos^3\alpha - 3\cos\alpha$, for all real numbers α. [*Hint*: $\cos 3\alpha = \cos(\alpha + 2\alpha)$.]

9.2.6 Prove the following identities, assuming that α is any real number such that all functions involved are defined:

(*a*) $\tan(\pi/2 + \alpha) = -\cot\alpha$

(*b*) $\cot(\pi/2 + \alpha) = -\tan\alpha$

(*c*) $\csc(\pi/2 + \alpha) = \sec\alpha$

(*d*) $\sec(\pi/2 + \alpha) = -\csc\alpha$

9.2.7 Use the addition and subtraction theorems for trigonometric functions to show the following:

(*a*) $\sin(\pi - \alpha) = \sin\alpha$

(*b*) $\cos(\pi - \alpha) = -\cos\alpha$

(*c*) $\sin(\pi + \alpha) = -\sin\alpha$

(*d*) $\cos(\pi + \alpha) = -\cos\alpha$

9.2.8 Prove the following identities, assuming that x is any real number such that all functions involved are defined.

(*a*) $\cot x = \tan\left(\frac{\pi}{2} - x\right) = -\tan\left(x - \frac{\pi}{2}\right)$

(*b*) $\csc x = \sec\left(\frac{\pi}{2} - x\right) = \sec\left(x - \frac{\pi}{2}\right)$

9.2.9 Prove the following identities, assuming that x is any real number such that all functions involved are defined and each denominator is nonzero.

(*a*) $\sin x = \dfrac{2\tan(x/2)}{1 + \tan^2(x/2)}$

(*b*) $\cos x = \dfrac{1 - \tan^2(x/2)}{1 + \tan^2(x/2)}$

(*c*) $\tan x = \dfrac{2\tan(x/2)}{1 - \tan^2(x/2)}$

9.2.10 Prove the following identities, assuming that x is any real number such that all functions involved are defined and each denominator is nonzero.

(*a*) $2\sin^2 x = \tan x \sin 2x$

(*b*) $\sin 4x = 4\sin x\cos x\cos 2x$

(*c*) $\dfrac{\cos x}{1 - \sin x} = \dfrac{1 + \sin x}{\cos x}$

(*d*) $\dfrac{\sin x}{1 - \cos x} = \dfrac{1 + \cos x}{\sin x}$

(*e*) $\dfrac{\sin 2x}{\cos 2x + 1} = \tan x$

(*f*) $1 - 2\cos^2 x + \cos^4 x = \sin^4 x$

(*g*) $\sin^4 x - \cos^4 x = 2\sin^2 x - 1$

(*h*) $\cos^4\frac{x}{2} - \sin^4\frac{x}{2} = \cos x$

(*i*) $\dfrac{1 - \cos x}{\sin x} = \tan\dfrac{x}{2}$

(*j*) $(\sin x + \cos x)^2 = 1 + \sin 2x$

(*k*) $\sin^2 x - \cos^2 x = \dfrac{1 - \cot^2 x}{1 + \cot^2 x}$

(*l*) $\tan x + \cot x = \sec x \csc x$

(*m*) $\dfrac{\cot x - \tan x}{\cot x + \tan x} = \cos 2x$

(*n*) $(\tan x - \sec x)^2 = \dfrac{1 - \sin x}{1 + \sin x}$

(*o*) $(\cot x - \csc x)^2 = \dfrac{1 - \cos x}{1 + \cos x}$

(*p*) $\dfrac{\cot x}{1 - \tan x} + \dfrac{\tan x}{1 - \cot x} = 1 + \tan x + \cot x$

(*q*) $\dfrac{\sin x}{1 + \cos x} = \csc x - \cot x = \tan \dfrac{x}{2}$

(*r*) $\left(\dfrac{1 - \tan x}{1 + \tan x}\right)^2 = \dfrac{1 - \sin 2x}{1 + \sin 2x}$

(*s*) $\dfrac{1 + \cot x}{\tan x} = \dfrac{\sin x + \cos x}{\sec x - \cos x}$

(*t*) $\dfrac{\cos x - \sec x - \tan x}{\tan x + \sec x} = -\sin x$

(*u*) $\dfrac{2 \cos x(\cos x + 1)}{1 - \sin x + \cos x} = 1 + \sin x + \cos x$

(*v*) $\dfrac{\tan x + \sin x}{2 \tan x} = \cos^2 \dfrac{x}{2}$

9.2.11 Prove the following identities.

(*a*) $\sin(x + y) + \sin(x - y) = 2 \sin x \cos y$

(*b*) $\sin(x + y) - \sin(x - y) = 2 \cos x \sin y$

(*c*) $\cos(x + y) + \cos(x - y) = 2 \cos x \cos y$

(*d*) $\cos(x + y) - \cos(x - y) = -2 \sin x \sin y$

9.2.12 By setting $x + y = \alpha$, $x - y = \beta$, and solving for x and y in terms of α and β, verify that the identities in Prob. 9.2.11 yield the following identities, where α, β are any real numbers.

(*a*) $\sin \alpha + \sin \beta = 2 \sin \dfrac{\alpha + \beta}{2} \cos \dfrac{\alpha - \beta}{2}$

(*b*) $\sin \alpha - \sin \beta = 2 \cos \dfrac{\alpha + \beta}{2} \sin \dfrac{\alpha - \beta}{2}$

(*c*) $\cos \alpha + \cos \beta = 2 \cos \dfrac{\alpha + \beta}{2} \cos \dfrac{\alpha - \beta}{2}$

(*d*) $\cos \alpha - \cos \beta = -2 \sin \dfrac{\alpha + \beta}{2} \sin \dfrac{\alpha - \beta}{2}$

9.2.13 Find $\sin 15°$ and $\cos 15°$.

9.2.14 Find the maximum and minimum values of each of the functions $f: R \to R$ (R is the set of real numbers) given below. For what values of x are these maximum and minimum values achieved?

(*a*) $f(x) = \sin x \cos x$ (*b*) $f(x) = \cos^2 x - \sin^2 x$

[*Hint*: Use the formulas for the trigonometric functions of $2x$.]

9.3 Periodic Functions and Graphs of Trigonometric Functions

In this section, we discuss the graphs of trigonometric functions. We shall see that all the trigonometric functions are periodic functions.

Now, suppose x is any real number, and suppose that C is the unit circle (see Fig. 9.17). Then the perimeter of C is equal to 2π, and hence, by the *definitions of trigonometric functions* given in Eqs. (9.16) to (9.21), we have

$$f(x) = f(x + 2\pi) \qquad \text{for all real numbers } x \text{ for which } f(x) \text{ is defined} \tag{9.91}$$

where f is any one of the functions sin, cos, tan, cot, sec, csc. In fact, since the perimeter of C is equal to 2π, the point P_x representing x and the point $P_{x+2\pi}$ representing $x + 2\pi$ are precisely the same point (see Fig. 9.17), and hence by (9.16) to (9.21), x and $x + 2\pi$ have precisely the same trigonometric function values (keep in mind Fig. 9.11).

With (9.91) as motivation, we now introduce the concept of a periodic function. Thus, suppose that R denotes the set of real numbers, and suppose D is a subset of R. Suppose that $f: D \to R$, and suppose a is a nonzero real number. We say that f is a *periodic function* if both of the following conditions are satisfied:

1. Whenever x is in D, so are both $x - a$ and $x + a$.
2. $f(x + a) = f(x)$ for all x in D.

We also call the nonzero real number a a *period* of f. The *least positive* real number a (if such exists) which satisfies both (1) and (2) is called the *fundamental period* of f. In view of the above discussion, it is easy to see that each of the six trigonometric functions is a periodic function of period 2π. Moreover, using the definitions given in (9.18) and (9.19), we readily verify that

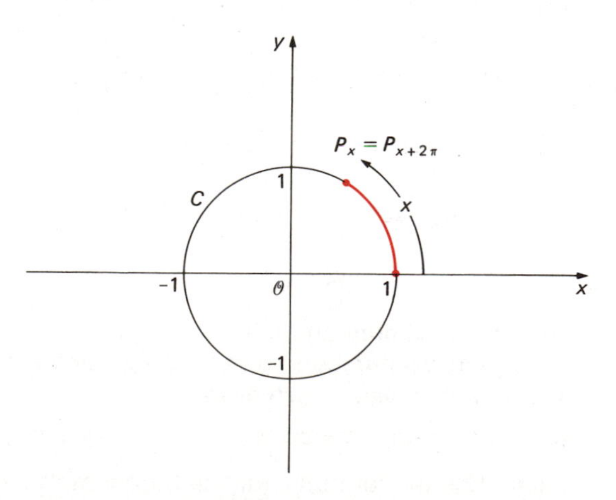

Fig. 9.17 Points P_x and $P_{x+2\pi}$, representing x and $x + 2\pi$, respectively

$$\tan(x+\pi) = \tan x \qquad \text{for all } x \text{ for which } \tan x \text{ is defined} \tag{9.92}$$

$$\cot(x+\pi) = \cot x \qquad \text{for all } x \text{ for which } \cot x \text{ is defined} \tag{9.93}$$

Hence, "tan" and "cot" are periodic functions of period π (as well as of period 2π). In fact, it turns out (see the exercises in Problem Set 9.3) that the *fundamental* period of "sin," "cos," "sec," and "csc" is equal to 2π, while the *fundamental* period of "tan" and "cot" is equal to π. These facts will be confirmed soon, when we sketch the graphs of the trigonometric functions. Incidentally, not all functions are periodic. For example, any nonconstant polynomial function is *not* a periodic function. Moreover, the exponential and logarithmic functions we discussed in Chap. 8 are *not* periodic functions. It should also be pointed out that, even though most periodic functions have a fundamental period, there are some periodic functions which do *not* have a fundamental period (see Problem Set 9.3).

We now turn our attention to sketching the graphs of the six trigonometric functions. Let us begin with the graph of the function "sin."

Recalling the definition of the function sin given in terms of the unit circle C in (9.16) (also keep in mind Fig. 9.11), it is easy to see that

$$\sin x = 0 \quad \text{if and only if} \quad x = k\pi \qquad \text{where } k \text{ is } \textit{any} \text{ integer}$$

Hence, the x intercept points of the graph of the function sin are all those, and only those, points of the form $(k\pi,0)$, where $k = 0, \pm1, \pm2, \pm3, \ldots$. Also, we easily see that $(0,\sin 0)$, that is, $(0,0)$, is the one and only y intercept point of the graph of the function sin. Moreover, $\sin(-x) = -\sin x$ [see Eq. (9.43)] for all real numbers x, and hence the graph of the function sin is symmetric with respect to the origin (but not with respect to the y axis). Furthermore, $-1 \leq \sin x \leq 1$ [see (9.36)] for all real numbers x, and hence the entire graph of the function sin is contained in the "horizontal strip" bounded by the lines $y=-1$ and $y=1$. Because the function sin is a periodic function of period 2π, we only need to observe the graph between $x=-\pi$ and $x=\pi$ to determine the entire graph, since the graph repeats itself in exacly the same pattern for the remaining values of x. Indeed, $\sin x = \sin(x+2\pi) = \sin(x+4\pi) = \sin(x+6\pi)$, and so on; also, $\sin x = \sin(x-2\pi) = \sin(x-4\pi) = \sin(x-6\pi)$, and so on. Keeping all the above facts in mind, and using the brief table of values for the function sin indicated below (these are obtained from the table at the end of Sec. 9.1 and Example 9.5), we now sketch the graph of $y=\sin x$ (see Fig. 9.18). Observe that, since the graph is symmetric with respect to the origin [i.e., $\sin(-x) = -\sin x$], we need not tabulate the values of $\sin x$ when x is negative.

x	0	$\frac{\pi}{6}$	$\frac{\pi}{4}$	$\frac{\pi}{3}$	$\frac{\pi}{2}$	$\frac{2\pi}{3}$	$\frac{3\pi}{4}$	$\frac{5\pi}{6}$	π
$y=\sin x$	0	$\frac{1}{2}$	$\frac{\sqrt{2}}{2}$	$\frac{\sqrt{3}}{2}$	1	$\frac{\sqrt{3}}{2}$	$\frac{\sqrt{2}}{2}$	$\frac{1}{2}$	0

Next, we consider the graph of the function "cos." Recalling the definition of the function cos given in terms of the unit circle C in (9.17) (also keep in mind Fig. 9.11), it is easily seen that

$$\cos x = 0 \quad \text{if and only if} \quad x = \frac{\pi}{2} + k\pi$$

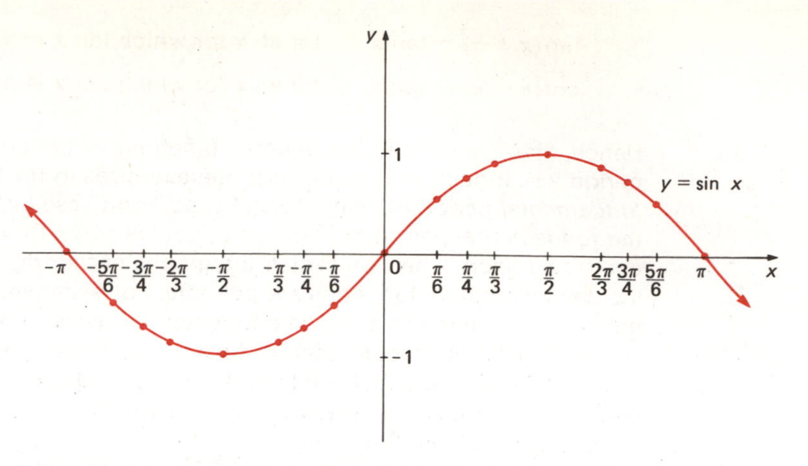

Fig. 9.18 Graph of $y = \sin x$. Observe that since the function "sin" is periodic of period 2π, the rest of the graph follows exactly the same pattern as the part indicated for $-\pi \leq x \leq \pi$.

where k is *any* integer. Hence, the x intercept points of the graph of cos are all those, and only those, points of the form $(\pi/2 + k\pi, 0)$, where $k = 0, \pm 1, \pm 2, \pm 3, \ldots$. Also, we easily see that $(0, \cos 0)$ that is, $(0,1)$, is the one and only y intercept point of the graph of cos. Moreover, since $\cos(-x) = \cos x$ [by Eq. (9.43)] for all real numbers x, the graph of the function cos is symmetric with respect to the y axis (but not with respect to the origin). Furthermore, $-1 \leq \cos x \leq 1$ [see (9.36)] for all real numbers x, and hence the entire graph of cos is contained in the horizontal strip bounded by the lines $y = -1$ and $y = 1$. Because cos is a periodic function of period 2π, we only need to observe the graph between $x = -\pi$ and $x = \pi$ to determine the entire graph, since the graph repeats itself in exactly the same pattern for the remaining values of x.

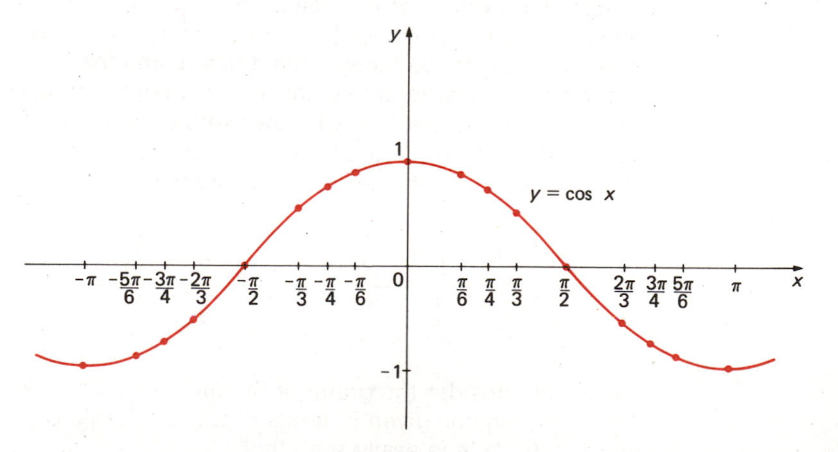

Fig. 9.19 Graph of $y = \cos x$. Observe that since the function "cos" is periodic of period 2π, the rest of the graph follows exactly the same pattern as the part indicated for $-\pi \leq x \leq \pi$.

A brief table of values, together with the graph of $y = \cos x$, are now given (see Fig. 9.19).

x	0	$\pm\frac{\pi}{6}$	$\pm\frac{\pi}{4}$	$\pm\frac{\pi}{3}$	$\pm\frac{\pi}{2}$	$\pm\frac{2\pi}{3}$	$\pm\frac{3\pi}{4}$	$\pm\frac{5\pi}{6}$	$\pm\pi$
$y = \cos x$	1	$\frac{\sqrt{3}}{2}$	$\frac{\sqrt{2}}{2}$	$\frac{1}{2}$	0	$-\frac{1}{2}$	$-\frac{\sqrt{2}}{2}$	$-\frac{\sqrt{3}}{2}$	-1

We now direct our attention to the graph of the function "tan." Recalling the definition of tan given in terms of the unit circle C in (9.18) (also keeping in mind Fig. 9.11), it is easily seen that

$$\tan x = 0 \quad \text{if and only if} \quad x = k\pi \qquad \text{where } k \text{ is } any \text{ integer}$$

Hence, the x intercept points of the graph of tan are all those, and only those, points of the form $(k\pi,0)$, where $k = 0, \pm1, \pm2, \pm3, \ldots$. We recognize that these are precisely the same as the x intercept points of the graph of sin. Of course, this is not surprising, since $\tan x = \sin x/\cos x$ (if $\cos x \neq 0$). Also, $(0,\tan 0)$, that is, $(0,0)$, is the one and only y intercept point of the graph of tan. Moreover, since $\tan(-x) = -\tan x$ [see Eq. (9.44)], for all real numbers x for which $\tan x$ is defined, the graph of tan is symmetric with respect to the origin (but not with respect to the y axis). Furthermore, as x becomes very close to $\pi/2$ (from either side), $\tan x$ becomes very large numerically, since when x is very close to $\pi/2$, $\sin x$ is very close to 1 *but* $\cos x$ is very close to zero. Thus $\tan x = \sin x/\cos x$ is numerically very large. This shows that $x = \pi/2$ is a *vertical asymptote* of the graph of $y = \tan x$. A similar argument shows that

$$x = \frac{\pi}{2} + k\pi \qquad k = \pm1, \pm2, \pm3, \ldots$$

are also asymptotes of the graph of $y = \tan x$. Moreover, since $\tan(x + \pi) = \tan x$, the function tan is a periodic function of period π. Hence we only need to observe the graph between $x = -\pi/2$ and $x = \pi/2$ to determine the entire graph, since the graph repeats itself in exactly the same pattern for the remaining values of x for which $\tan x$ is defined. Keeping all the above facts in mind, and using the brief table of values for tan indicated below, we now sketch the graph of $y = \tan x$ (see Fig. 9.20). Observe that, since the graph is symmetric with respect to the origin [that is, $\tan(-x) = -\tan x$], we need not tabulate the values of $\tan x$ when x is negative.

x	0	$\frac{\pi}{6}$	$\frac{\pi}{4}$	$\frac{\pi}{3}$
$y = \tan x$	0	$\frac{1}{\sqrt{3}}$	1	$\sqrt{3}$

Arguments similar to those for tan give the following facts for the function cot.

1. The x intercept points of the graph of the function cot are precisely the same as the x intercept points of the graph of the function cos, namely, all points of the form $(\pi/2 + k\pi,0)$, where k is *any* integer. Moreover, there is no y intercept point, since $\cot 0$ is *not* defined.

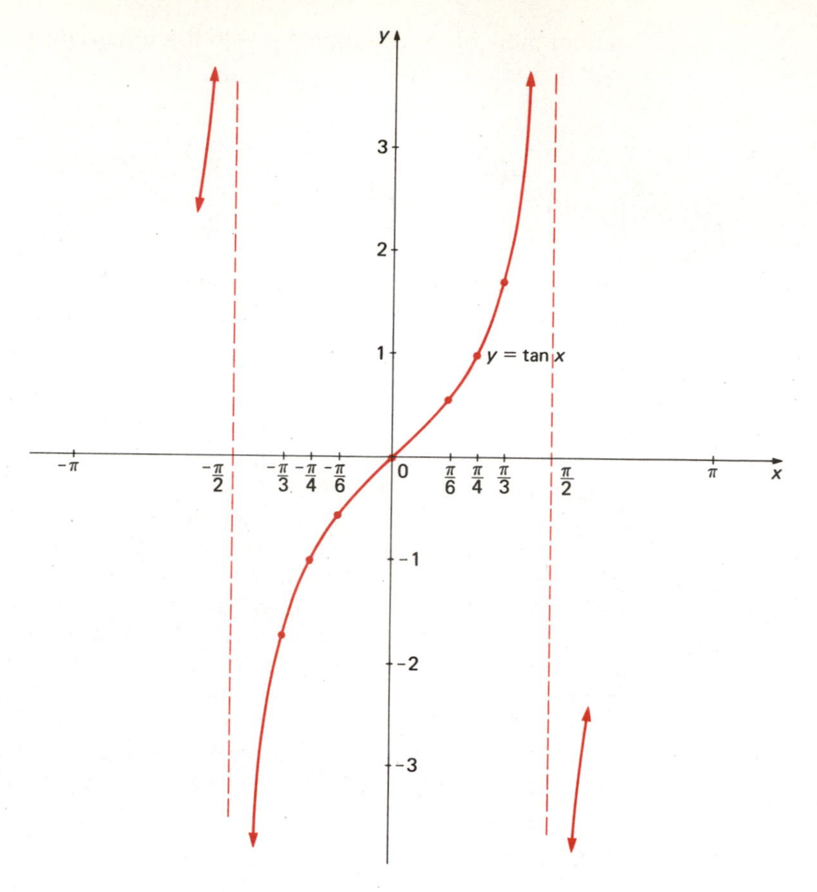

Fig. 9.20 Graph of $y = \tan x$. Since the function "tan" is periodic of period π, the rest of the graph follows exactly the same pattern as the part indicated for $-\pi/2 < x < \pi/2$.

2. The graph of the function cot is symmetric with respect to the origin (but not with respect to the y axis).
3. The lines $x = 0, \pm\pi, \pm 2\pi, \pm 3\pi, \ldots$ are vertical asymptotes of the graph of $y = \cot x$.
4. The function cot is periodic of period π.

Using these facts, together with the brief table of values indicated below, we now sketch the graph of cot, as indicated in Fig. 9.21. Observe that since the graph is symmetric with respect to the origin [that is, $\cot(-x) = -(\cot x)$], we need not tabulate the values of $\cot x$ when x is negative.

x	$\frac{\pi}{6}$	$\frac{\pi}{4}$	$\frac{\pi}{3}$	$\frac{\pi}{2}$
$y = \cot x$	$\sqrt{3}$	1	$\frac{1}{\sqrt{3}}$	0

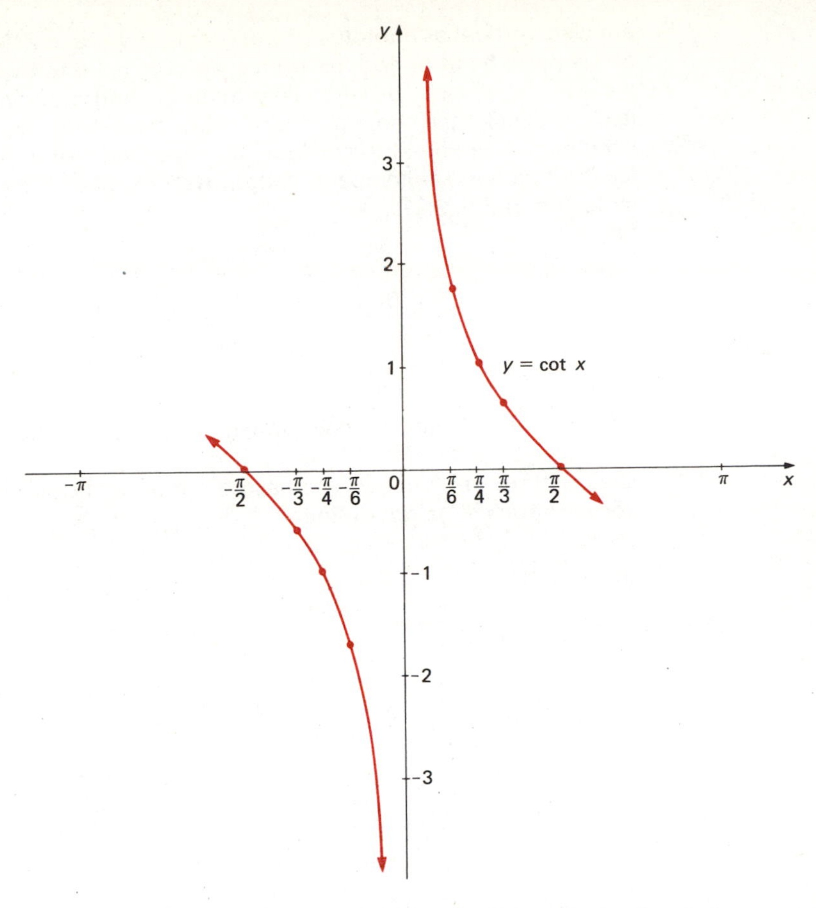

Fig. 9.21 Graph of $y = \cot x$. Since the function "cot" is periodic of period π, the rest of the graph follows exactly the same pattern as the part indicated for $-\pi/2 \leq x \leq \pi/2$ ($x \neq 0$).

We now direct our attention to sketching the graph of the function "sec." Since $\sec x = 1/\cos x$ (if $\cos x \neq 0$), it is clear that the graph of the function sec has *no* x intercept point (since $1/\cos x = 0$ has no solutions). In fact, we know that $|\sec x| \geq 1$ [see (9.37)], for all x for which $\sec x$ is defined. Hence the graph of the function sec lies above (or on) the line $y = 1$ or below (or on) the line $y = -1$. Moreover, the point (0, sec 0), that is, (0,1), is the *only* y intercept point. Furthermore, since $\sec(-x) = \sec x$ [see Eq. (9.46)], for all x for which $\sec x$ is defined, the graph of $y = \sec x$ is symmetric with respect to the y axis (but not with respect to the origin). Also, when x is very close to $\pi/2$ (on either side), $\cos x$ is very close to zero, and hence $\sec x = 1/\cos x$ becomes very large numerically. Thus, the line $x = \pi/2$ is a vertical asymptote of the graph of $y = \sec x$. A similar argument shows that the lines

$$x = \frac{\pi}{2} + k\pi \qquad k = \pm 1, \pm 2, \pm 3, \ldots$$

are also vertical asymptotes of the graph of $y = \sec x$. Moreover, the function sec is periodic of period 2π. Hence we only need to observe the graph between $x = -\pi$ and $x = \pi$ in order to determine the entire graph, since the graph repeats itself in exactly the same manner for the remaining values of x for which $\sec x$ is defined. Keeping all these facts in mind, and using the brief table of values for the function sec indicated below, we now sketch the graph of $y = \sec x$, as indicated in Fig. 9.22.

x	0	$\pm\frac{\pi}{6}$	$\pm\frac{\pi}{4}$	$\pm\frac{\pi}{3}$	$\pm\frac{2\pi}{3}$	$\pm\frac{3\pi}{4}$	$\pm\frac{5\pi}{6}$	$\pm\pi$
$y = \sec x$	1	$\frac{2}{\sqrt{3}}$	$\sqrt{2}$	2	-2	$-\sqrt{2}$	$-\frac{2}{\sqrt{3}}$	-1

Arguments similar to those for sec give the following facts for the function csc.

1. The graph of the function csc has no x intercept point and no y intercept point (recall that csc 0 is not defined).

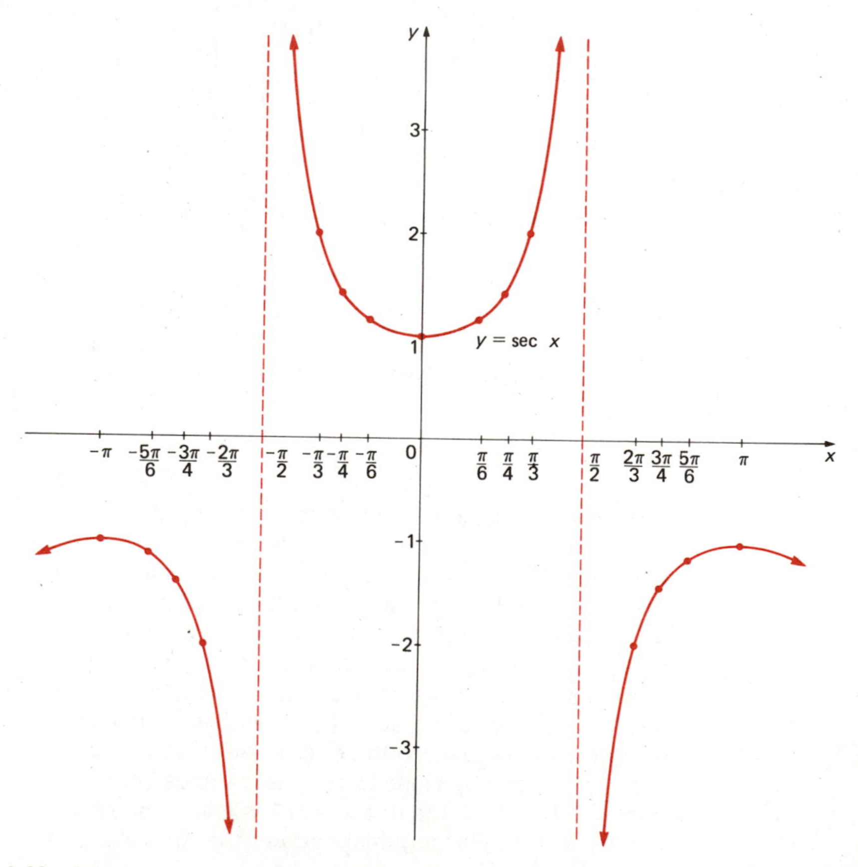

Fig. 9.22 Graph of $y = \sec x$. Since the function "sec" is periodic of period 2π, the rest of the graph follows exactly the same pattern as indicated for $-\pi \leq x \leq \pi$ ($x \neq -\pi/2$, $x \neq \pi/2$).

2. Since $\csc(-x) = -\csc x$ [see Eq. (9.47)], for all real numbers x for which $\csc x$ is defined, the graph of csc is symmetric with respect to the origin (but not with respect to the y axis).

3. Since $|\csc x| \geq 1$ [see (9.38)], for all x for which $\csc x$ is defined, the graph of the function csc lies above (or on) the line $y = 1$ or below (or on) the line $y = -1$.

4. When x is very close to zero (on either side), $\sin x$ is very close to zero, and hence $\csc x = 1/\sin x$ becomes very large numerically. Thus, the line $x = 0$ is a vertical asymptote of the graph of $y = \csc x$. Similarly, the lines $x = k\pi$, for $k = \pm 1, \pm 2, \pm 3, \ldots$, are also vertical asymptotes of the graph of $y = \csc x$.

5. The function csc is periodic of period 2π.

Keeping all the above facts in mind, and using the brief table of values for csc indicated below, we now sketch the graph of $y = \csc x$, as indicated in Fig. 9.23. Observe that since the graph is symmetric with respect to the origin [that is, $\csc(-x) = -\csc x$], we need not tabulate the values of $\csc x$ when x is negative.

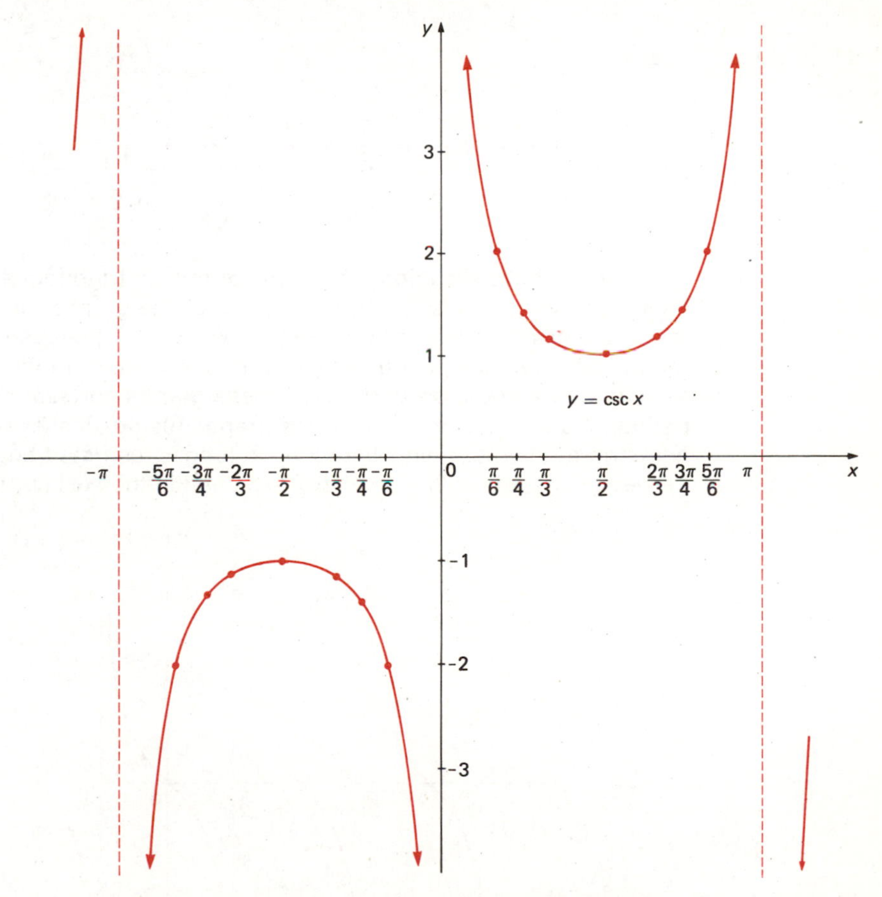

Fig. 9.23 Graph of $y = \csc x$. Since the function "csc" is periodic of period 2π, the rest of the graph follows exactly the same pattern as indicated for $-\pi < x < \pi$ $(x \neq 0)$.

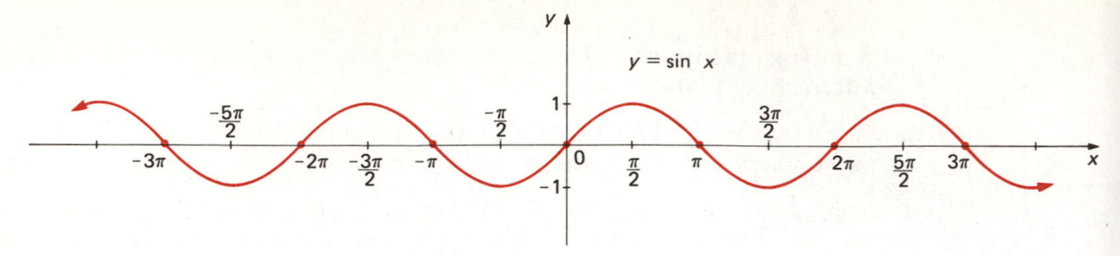

Fig. 9.24 Graph of $y = \sin x$

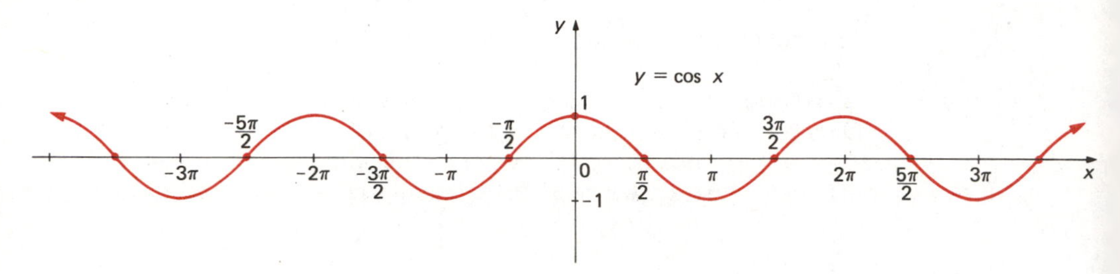

Fig. 9.25 Graph of $y = \cos x$

x	$\frac{\pi}{6}$	$\frac{\pi}{4}$	$\frac{\pi}{3}$	$\frac{\pi}{2}$	$\frac{2\pi}{3}$	$\frac{3\pi}{4}$	$\frac{5\pi}{6}$
$y = \csc x$	2	$\sqrt{2}$	$\frac{2}{\sqrt{3}}$	1	$\frac{2}{\sqrt{3}}$	$\sqrt{2}$	2

So far, we have sketched the graphs of the six trigonometric functions, with heavy emphasis on displaying those parts of the graphs for which $-\pi \leq x \leq \pi$ (or, in the cases of tan and cot, $-\pi/2 \leq x \leq \pi/2$). The reason is that since the trigonometric functions are periodic functions, their graphs are fully determined as soon as we know the portions of these graphs corresponding to one full period. It is instructive to sketch the graphs (using smaller scales) of the trigonometric functions again where x is no longer restricted between $-\pi$ and π (or between $-\pi/2$ and $\pi/2$). The graphs sketched in the next six figures (9.24 to 9.29)

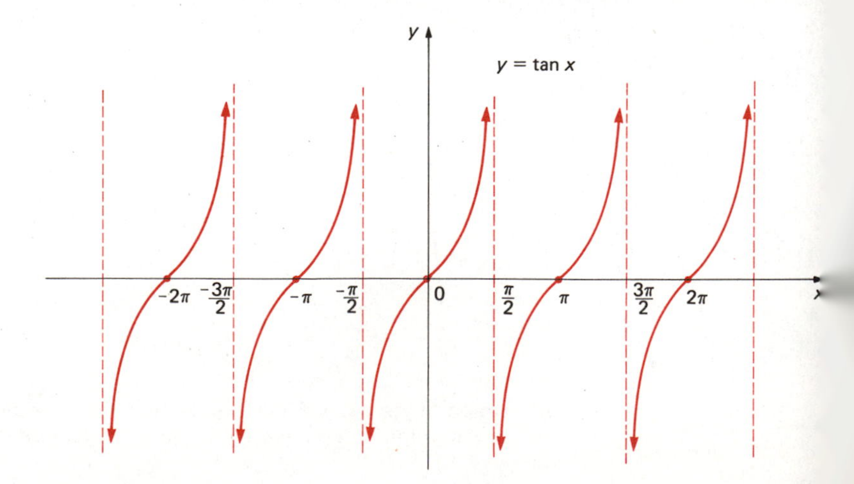

Fig. 9.26 Graph of $y = \tan x$

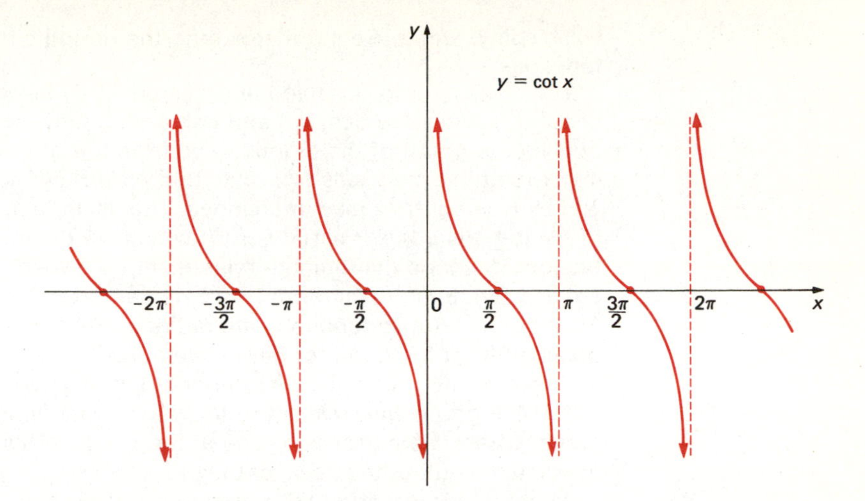

Fig. 9.27 Graph of $y = \cot x$

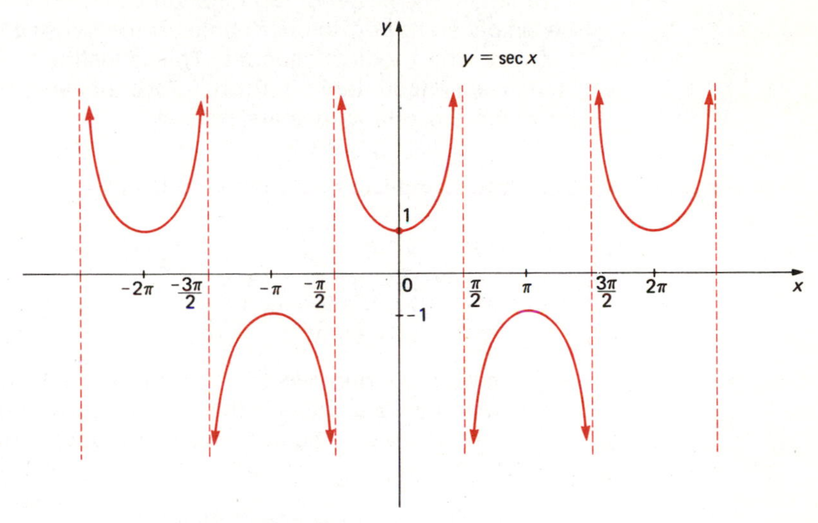

Fig. 9.28 Graph of $y = \sec x$

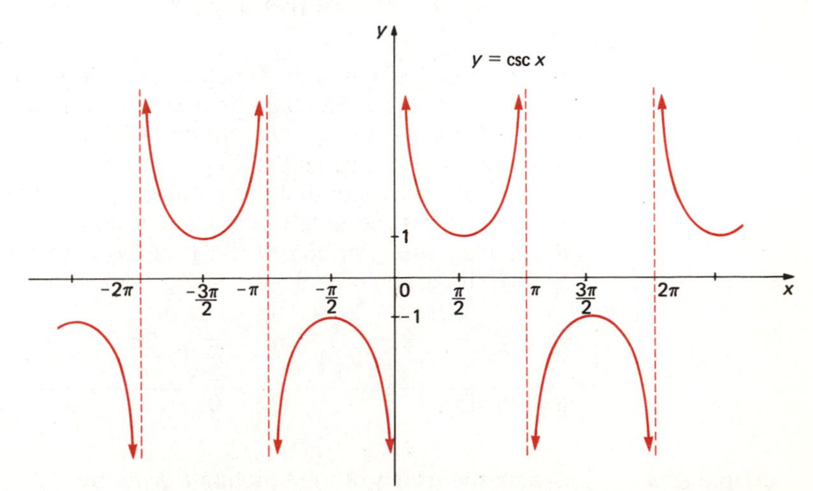

Fig. 9.29 Graph of $y = \csc x$

will display, and make more apparent, the periodic nature of the trigonometric functions.

It is interesting to observe the great similarity between the graphs of the pairs of functions sin and cos, tan and cot, and, finally, sec and csc. In fact, one obtains the graph of the function cos from the graph of the function sin by simply shifting the y axis $\pi/2$ units to the *right* (without changing its direction) and by leaving the x axis unchanged. This is, in fact, the geometric interpretation of the identity $\cos x = \sin(x + \pi/2)$, which we proved in Example 9.7. Similarly, one can describe the intimate relationship between the graphs of the functions cot and tan. First, recall from Prob. 9.2.6*a* that $\cot x = -\tan(x + \pi/2)$ for all x for which these two functions are defined. Hence, the graph of cot can be obtained from the graph of tan in two stages: First, shift the y axis $\pi/2$ units to the *right* (without changing its direction), and leave the x axis unchanged. Next, reflect the graph with respect to the x axis; that is, find the mirror image of the graph, where the mirror is placed at the x axis. (This explains the meaning of the minus sign in the above identity.)

To obtain the graph of the function csc from the graph of the function sec, we simply shift the y axis $\pi/2$ units to the *left* (without changing its direction), and leave the x axis unchanged. This is justified in view of a trigonometric identity we encountered in Prob. 9.2.8*b*, namely, $\csc x = \sec(x - \pi/2)$ for all x for which these functions are defined.

Example 9.13 Sketch the graph of the function f given by

$$f(x) = -3 \sin 2x \tag{9.94}$$

Find a period for the function f.

Solution Since, for all real numbers x, $-1 \leq \sin 2x \leq 1$ [see (9.36)], we see that $f(x)$ is bounded by -3 and 3. In other words, the entire graph of the function f in (9.94) is contained in the horizontal strip bounded by $y = -3$ and $y = 3$. Moreover, since

$$f(x + \pi) = -3 \sin[2(x + \pi)] = -3 \sin(2x + 2\pi) = -3 \sin 2x = f(x)$$

we see that f is a periodic function of period π. In fact, it can be verified from the graph of f (see Fig. 9.30) that π is the *fundamental* period of f. Observe also that (0,0) is the only y intercept point. Moreover, the points $(k \cdot \pi/2, 0)$, where k is *any* integer, give all the x intercept points of the graph (why?). Also, it seems reasonable to expect that the graph of the function f in (9.94) has essentially the same shape as the graph of the sine function. Of course, as we remarked above the *fundamental* period of f is π (not 2π), and f is bounded by -3 and 3 (not -1 and 1). The graph of f appears in Fig. 9.30. A brief table of values is indicated here

x	0	$\frac{\pi}{4}$	$\frac{\pi}{2}$	$\frac{3\pi}{4}$	π	$-\frac{\pi}{4}$	$-\frac{\pi}{2}$	$-\frac{3\pi}{4}$	$-\pi$
$y = -3 \sin 2x$	0	-3	0	3	0	3	0	-3	0

Example 9.14 Sketch the graph of the function f given by

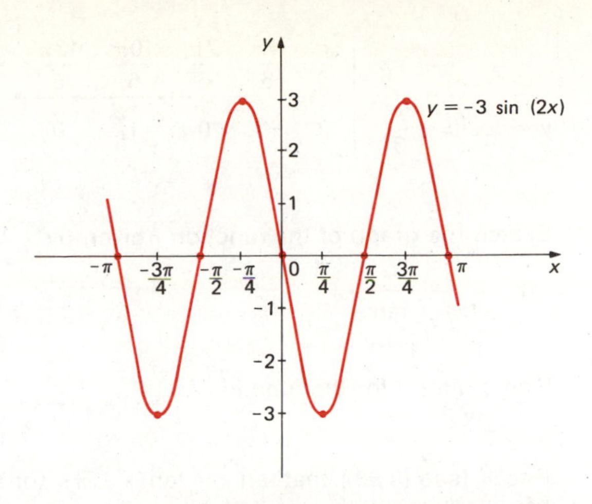

Fig. 9.30 Graph of $y = f(x) = -3\sin(2x)$. Observe that, since the function f is periodic, of period π, the rest of the graph follows exactly the same pattern as the part indicated for $-\pi/2 \le x \le \pi/2$.

$$f(x) = \cos\left(x + \frac{\pi}{3}\right) \tag{9.95}$$

Solution Since, for all real numbers x, $-1 \le \cos(x + \pi/3) \le 1$ [by (9.36)], we see that $f(x)$ is bounded by -1 and 1. In other words, the entire graph of f in (9.95) is contained in the horizontal strip bounded by $y = -1$ and $y = 1$. Moreover, since

$$f(x + 2\pi) = \cos\left[(x + 2\pi) + \frac{\pi}{3}\right] = \cos\left[\left(x + \frac{\pi}{3}\right) + 2\pi\right] = \cos\left(x + \frac{\pi}{3}\right) = f(x)$$

we see that f is a periodic function of period 2π. In fact, it can be verified from the graph of f (see Fig. 9.31) that 2π is the fundamental period of f. Moreover, it seems reasonable to expect that the graph of the function f in (9.95) has essentially the same shape as the graph of the cosine function. A brief table of values is given now. f is graphed in Fig. 9.31.

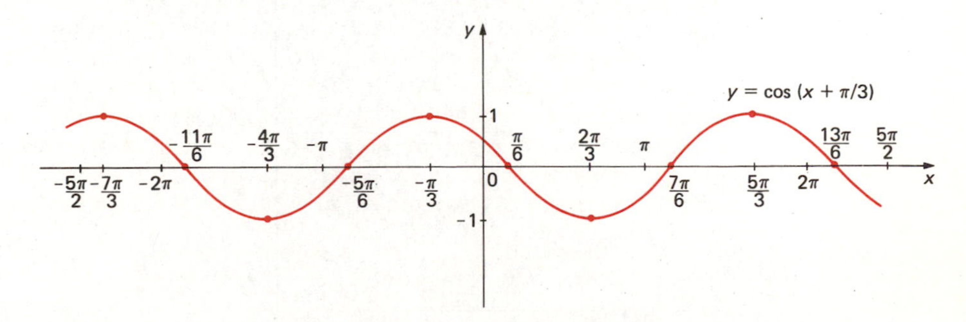

Fig. 9.31 Graph of $y = f(x) = \cos(x + \pi/3)$. Observe that the function f is periodic, of period 2π, and thus the rest of the graph of f follows exactly the same pattern as the part indicated for $-11\pi/6 \le x \le \pi/6$ (or for $\pi/6 \le x \le 13\pi/6$, etc.).

x	$\frac{\pi}{6}$	$\frac{4\pi}{6}$	$\frac{7\pi}{6}$	$\frac{10\pi}{6}$	$\frac{13\pi}{6}$	$\frac{-2\pi}{6}$	$\frac{-5\pi}{6}$	$\frac{-8\pi}{6}$	$\frac{-11\pi}{6}$	$\frac{-14\pi}{6}$
$y = \cos\left(x + \frac{\pi}{3}\right)$	0	−1	0	1	0	1	0	−1	0	1

Example 9.15 Sketch the graph of the function f given by

$$f(x) = \tan \frac{x}{2} \tag{9.96}$$

Find a period for the function f.

Solution Recall [see (9.92)] that $\tan x = \tan(x + \pi)$, for all real numbers x for which $\tan x$ is defined. Hence, if $f(x)$ is defined, then

$$f(x + 2\pi) = \tan\left(\frac{x + 2\pi}{2}\right) = \tan\left(\frac{x}{2} + \pi\right) = \tan \frac{x}{2} = f(x)$$

Thus, f is periodic, of period 2π. In fact, it can be verified from the graph (see Fig. 9.32) that the fundamental period of f is 2π. The graph of $y = \tan(x/2)$ is similar to the graph of $y = \tan x$ (sketched earlier). Of course, the fundamental period of $y = \tan(x/2)$ is 2π (not π). We indicate below a brief table of values. The graph of $y = \tan(x/2)$ appears in Fig. 9.32. Observe that the lines $x = k\pi$, where $k = \pm 1, \pm 3, \pm 5, \ldots$, are vertical asymptotes of the graph of f.

x	0	$\frac{\pi}{2}$	$\frac{2\pi}{3}$	$\frac{3\pi}{2}$	2π	$\frac{-\pi}{2}$	$\frac{-2\pi}{3}$	$\frac{-4\pi}{3}$	$\frac{-3\pi}{2}$
$y = \tan \frac{x}{2}$	0	1	$\sqrt{3}$	−1	0	−1	$-\sqrt{3}$	$\sqrt{3}$	1

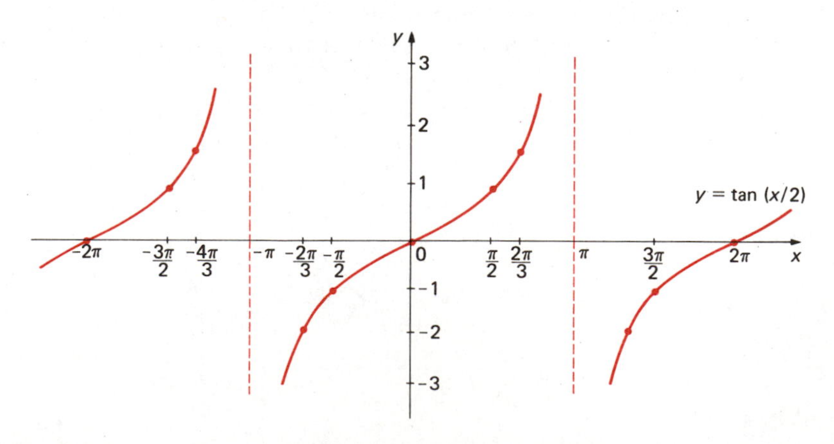

Fig. 9.32 Graph of $y = f(x) = \tan(x/2)$. The function f is periodic, of period 2π, and hence the graph of f follows exactly the same pattern as the part indicated for $-\pi < x < \pi$.

Example 9.16 Find (if possible) real numbers A and β such that

$$\sin 3x + \sqrt{3} \cos 3x = A \sin(3x + \beta) \tag{9.97}$$

Solution We know that [see Eq. (9.57)]

$$\sin(3x + \beta) = \sin 3x \cos \beta + \cos 3x \sin \beta$$

and hence

$$A \sin(3x + \beta) = (A \cos \beta) \sin 3x + (A \sin \beta) \cos 3x \tag{9.98}$$

Combining (9.97) and (9.98), we obtain

$$\sin 3x + \sqrt{3} \cos 3x = (A \cos \beta) \sin 3x + (A \sin \beta) \cos 3x$$

This suggests that we choose (if possible) A and β such that

$$A \cos \beta = 1 \qquad \text{and} \qquad A \sin \beta = \sqrt{3} \tag{9.99}$$

Squaring and adding both sides of the equations in (9.99), we get

$$A^2(\cos^2\beta + \sin^2\beta) = 1^2 + (\sqrt{3})^2 = 4 \tag{9.100}$$

But [see Eq. (9.33)] $\cos^2\beta + \sin^2\beta = 1$, and hence (9.100) reduces to $A^2 = 4$. In view of this, we may choose

$$A = 2 \tag{9.101}$$

Combining (9.99) and (9.101), we get

$$\cos \beta = \frac{1}{2} \qquad \sin \beta = \frac{\sqrt{3}}{2} \tag{9.102}$$

Hence, we may choose

$$\beta = \frac{\pi}{3} \tag{9.103}$$

Substituting (9.101) and (9.103) in (9.97), we obtain

$$\sin 3x + \sqrt{3} \cos 3x = 2 \sin\left(3x + \frac{\pi}{3}\right) \tag{9.104}$$

Remark Using the technique of Example 9.16, it can be shown that, for *all* real numbers a, b, θ, the expression

$$a \sin \theta + b \cos \theta \qquad \text{not both } a \text{ and } b \text{ are zero}$$

can *always* be expressed in the form

$$A \sin(\theta + \beta)$$

for some real numbers A and β. Furthermore, A *may* always be chosen to be positive (for convenience), although this is not required.

Example 9.17 Sketch the graph of the function f given by

$$f(x) = \sin 3x + \sqrt{3} \cos 3x \tag{9.105}$$

Solution By (9.104), we know that (9.105) is equivalent to

$$f(x) = 2 \sin\left(3x + \frac{\pi}{3}\right) \tag{9.106}$$

However, it is easier to sketch the graph of the function f if we view it in the form (9.106) rather than in the form (9.105). Indeed, a discussion similar to the one given in Example 9.13 shows that the graph of the function f given in (9.106) has essentially the same shape as the graph of the sine function. Of course, f is now bounded by -2 and 2 (not -1 and 1), and its fundamental period is $2\pi/3$ (not 2π). A brief table of values is indicated below, and the graph of f appears in Fig. 9.33.

x	$\frac{\pi}{18}$	$\frac{4\pi}{18}$	$\frac{7\pi}{18}$	$\frac{10\pi}{18}$	$\frac{13\pi}{18}$	$\frac{16\pi}{18}$	$\frac{-2\pi}{18}$	$\frac{-5\pi}{18}$	$\frac{-8\pi}{18}$	$\frac{-11\pi}{18}$	$\frac{-14\pi}{18}$	$\frac{-17\pi}{18}$
$y = f(x)$	2	0	−2	0	2	0	0	−2	0	2	0	−2

In the next chapter, we will be dealing with triangles. In that chapter, it will be more natural to view trigonometric functions *as functions of angles* (see Sec. 9.1). We shall so view them.

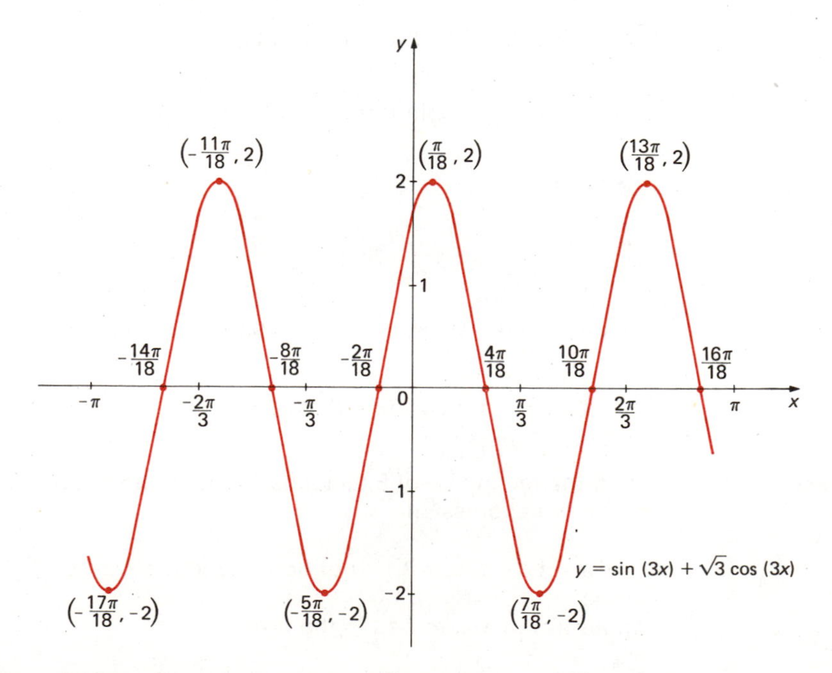

Fig. 9.33 Graph of $y = f(x) = \sin(3x) + \sqrt{3}\cos(3x)$

Problem Set 9.3

9.3.1 Prove that the fundamental period of the function sin is equal to 2π. [*Hint*: Let a be the fundamental period of the function sin. Then $0 < a \leq 2\pi$ (Why?). Thus, $\sin 0 = \sin a$, and hence $\sin a = 0$. Now, if $a \neq 2\pi$, then $a = \pi$ (Why?). But, then $\sin(\pi/2) = \sin(\pi/2 + a) = \sin(\pi/2 + \pi)$, and hence $1 = -1$, a contradiction. Therefore $a = 2\pi$.]

9.3.2 Use the ideas given in the hint in Prob. 9.3.1 to show that the fundamental period of each of the functions cos, sec, and csc is equal to 2π, while the fundamental period of the functions tan and cot is equal to π.

9.3.3 Prove that every constant function f is periodic, and that every nonzero real number is a period of f. Show that f has *no* fundamental period.

9.3.4 Sketch the graphs of the functions f given below. Use the same coordinate axes.

(*a*) $f(x) = 3 \cos x$ (*c*) $f(x) = \frac{1}{2} \cos x$

(*b*) $f(x) = -2 \cos x$ (*d*) $f(x) = -\frac{1}{3} \cos x$

9.3.5 Sketch the graphs of the functions f given below. Use the same coordinate axes.

(*a*) $f(x) = \tan 2x$ (*c*) $f(x) = \tan \dfrac{x}{2}$

(*b*) $f(x) = \tan(-3x)$

9.3.6 Find the fundamental period of each function given in Probs. 9.3.4 and 9.3.5.

9.3.7 Sketch the graphs of the functions f given below.

(*a*) $f(x) = 3 \sin 2x$ (*c*) $f(x) = -4 \sin \dfrac{x}{2}$

(*b*) $f(x) = -4 \sin \dfrac{3x}{2}$ (*d*) $f(x) = 3 \cos\left(\dfrac{-x}{2}\right)$

9.3.8 Find the fundamental period of each function given in Prob. 9.3.7.

9.3.9 Sketch the graph of each of the functions f given below.

(*a*) $f(x) = \sin\left(x - \dfrac{\pi}{4}\right)$ (*c*) $f(x) = 3 \sin\left(2x + \dfrac{\pi}{3}\right)$

(*b*) $f(x) = \sin\left(x + \dfrac{\pi}{4}\right)$ (*d*) $f(x) = -2 \sin\left(3x - \dfrac{\pi}{2}\right)$

9.3.10 Find the fundamental period of each function given in Prob. 9.3.9.

9.3.11 Sketch the graphs of the functions f given below.

(*a*) $f(x) = |\sin x|$ (*c*) $f(x) = |\tan x|$

(*b*) $f(x) = |\cos x|$

9.3.12 Find the fundamental period of each of the functions given in Prob. 9.3.11.

9.3.13 Find the fundamental period of each of the functions f given below.

(*a*) $f(x) = \sin^2 x$ (*b*) $f(x) = \cos^2 x$

9.3.14 Sketch the graph of each of the functions given in Prob. 9.3.13.

9.3.15 Use the method described in Example 9.17 to sketch the graphs of the functions given by

(*a*) $f(x) = \sin 3x + \cos 3x$ (*b*) $f(x) = -\sin 2x + \sqrt{3} \cos 2x$

9.3.16 Use the method described in Example 9.16 to show that, for all real numbers x,

(*a*) $|\sin x + \cos x| \leq \sqrt{2}$ (*b*) $|\sqrt{3} \sin x - \cos x| \leq 2$

9.4 Summary

In this chapter, we viewed trigonometric functions in two ways: as functions of angles, and as functions of real numbers. The measures of angles most frequently used are *degree measure* and *radian measure.* These two measures are related as follows: 2π radians $= 360°$. Numerous trigonometric identities were established. For example, we learned that, for any real numbers α, β,

$$\sin^2\alpha + \cos^2\alpha = 1 \qquad \cos(\alpha \pm \beta) = \cos\alpha \cos\beta \mp \sin\alpha \sin\beta$$

$$\sin(\alpha \pm \beta) = \sin\alpha \cos\beta \pm \cos\alpha \sin\beta \qquad \sin 2\alpha = 2 \sin\alpha \cos\alpha$$

$$\cos 2\alpha = \cos^2\alpha - \sin^2\alpha = 2\cos^2\alpha - 1 = 1 - 2\sin^2\alpha$$

and so on. Furthermore, *all these (and other) trigonometric identities are valid for angles* (as well as for real numbers).

We also saw that if f is any one of the six trigonometric functions sin, cos, tan, cot, sec, or csc, then

$$f(x) = f(x + 2\pi) \qquad \text{for all } x \text{ for which } f(x) \text{ is defined}$$

For this reason, the trigonometric functions are called *periodic* functions. In terms of graphs, this means that the entire graph of a trigonometric function is an exact repetition of that portion of the graph corresponding to $-\pi \leq x \leq \pi$ (or to $-\pi/2 \leq x \leq \pi/2$ in the cases of tan and cot). Moreover, the functions sin, tan, cot, and csc are *odd* functions in the sense that $\sin(-x) = -\sin x$, $\tan(-x) = -\tan x$, etc. However, the functions cos and sec are *even* functions in the sense that $\cos(-x) = \cos x$ and $\sec(-x) = \sec x$.

10 Further Properties of Trigonometric Functions

10.1 Solution of Triangles: Law of Sines and Law of Cosines

In this section, we consider certain properties of triangles, such as the laws of sines and cosines. We also apply these laws in order to find the measurements of the sides and angles of triangles. Our starting point is the study of right triangles. Later in this chapter, we discuss arbitrary triangles.

Suppose that ABC is a right triangle, where the measure of angle ACB is 90° (see Fig. 10.1). Suppose, further, that the two acute angles of the triangle ABC have measures $\alpha°$ and $\beta°$ while the two *legs* are of lengths a and b, as indicated in Fig. 10.1. Moreover, suppose that the length of the hypotenuse AB is c. Then, by the Pythagorean theorem, we have

$$a^2 + b^2 = c^2 \tag{10.1}$$

We now introduce a rectangular coordinate system as follows: Let us agree to choose the positive direction of the x axis to be along BC with the origin at point B. This choice is quite attractive, since the coordinates of the vertices A, B, C now take a simple form, as indicated below (see Fig. 10.1):

$$A:(a,b) \qquad B:(0,0) \qquad C:(a,0)$$

Moreover, since angle CBA $[=\beta]$ is a directed angle in standard position and $A:(a,b)$ is on the terminal side of β, it follows from the definitions of trigonometric functions (as *functions of angles*) given in Sec. 9.1 that

$$\begin{aligned} \sin\beta &= \frac{b}{c} & \cot\beta &= \frac{a}{b} \\ \cos\beta &= \frac{a}{c} & \sec\beta &= \frac{c}{a} \\ \tan\beta &= \frac{b}{a} & \csc\beta &= \frac{c}{b} \end{aligned} \tag{10.2}$$

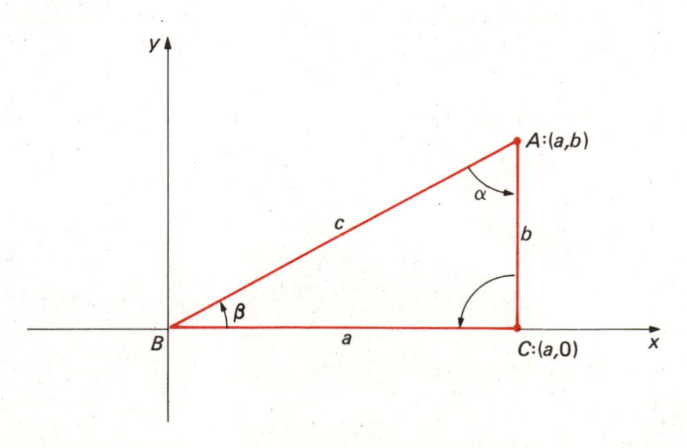

Fig. 10.1 Right triangle

We also have, of course,

$$\alpha + \beta = 90° \tag{10.3}$$

The relations given in Eqs. (10.1) to (10.3) enable us to find all the sides and angles of triangle ABC as soon as we know any of the following: (1) two legs, (2) one leg and the hypotenuse, (3) one leg and one acute angle, or (4) the hypotenuse and one acute angle. The following examples illustrate the procedure to be used.

Example 10.1 Suppose that in the right triangle ABC given in Fig. 10.1, we have

$$a = 30 \qquad b = 40$$

Find c, α, and β.

Solution By (10.1) we have

$$c^2 = a^2 + b^2 = 30^2 + 40^2 = 2500$$

and hence

$$c = 50$$

Now, by (10.2), we have

$$\tan \beta = \frac{b}{a} = \frac{40}{30} \approx 1.3333$$

Hence, using Table E in the back of this text, we find

$$\beta \approx 53°$$

and thus

$$\alpha = 90° - \beta \approx 37°$$

Example 10.2 Suppose that in the right triangle ABC given in Fig. 10.1, we have $b = 20$ and $c = 40$. Find a, α, and β.

Solution Using (10.1), we obtain

$$a^2 = c^2 - b^2 = 40^2 - 20^2 = 1200$$

and hence

$$a = \sqrt{1200} = 20\sqrt{3}$$

Now, by (10.2) we have

$$\sin \beta = \frac{b}{c} = \frac{20}{40} = \frac{1}{2}$$

Hence,

$$\beta = 30° \qquad \alpha = 90° - 30° = 60°$$

Example 10.3 Suppose that in the right triangle ABC given in Fig. (10.1), we have $a = 10$ and $\beta = 40°$. Find b, c, and α.

Solution First, observe that $\alpha = 90° - \beta = 50°$. Now, by (10.2) we have

$$\sec \beta = \frac{c}{a} = \frac{c}{10}$$

and hence

$$c = 10 \sec \beta = 10 \sec 40° \approx 10(1.305)$$

Thus,

$$c \approx 13.05$$

Moreover,

$$\tan \beta = \frac{b}{a} = \frac{b}{10}$$

and hence, using Table E, we get

$$b = 10 \tan \beta = 10 \tan 40° \approx 10(0.8391)$$

Thus,

$$b \approx 8.39$$

Example 10.4 Suppose that in the right triangle ABC given in Fig. 10.1, we have $c = 100$ and $\alpha = 25°$. Find a, b, and β.

Solution First, observe that $\beta = 90° - \alpha = 65°$. Now, by (10.2), we have

$$\sin \beta = \frac{b}{c} = \frac{b}{100}$$

and hence, using Table E, we have

$$b = 100 \sin \beta = 100 \sin 65° \approx 100(0.9063)$$

Thus,

$$b \approx 90.63$$

Moreover, using (10.2) again, we find

$$\cos\beta = \frac{a}{c} = \frac{a}{100}$$

and hence, using Table E, we obtain

$$a = 100 \cos\beta = 100 \cos 65° \approx 100(0.4226)$$

Thus,

$$a \approx 42.26$$

In preparation for the next result, known as the law of sines, we first establish the following fact:

If ABC is any triangle with sides a, b, c and angles α, β, γ, as indicated in Fig. 10.2, then

$$\text{Area of triangle } ABC = \tfrac{1}{2}bc \sin\alpha$$
$$\text{Area of triangle } ABC = \tfrac{1}{2}ca \sin\beta \qquad (10.4)$$
$$\text{Area of triangle } ABC = \tfrac{1}{2}ab \sin\gamma$$

To prove this, we first introduce a rectangular coordinate system. For convenience, we choose the positive direction of the x axis to be along one of the sides of triangle ABC, say AB, with the origin at point A, as indicated in Fig. 10.2. Let CD be the line segment which is perpendicular to the x axis and which meets the x axis at D. Suppose, further, that the coordinates of C are (u,v). Then, since angle BAC $[=\alpha]$ is a directed angle in standard position, and since $C:(u,v)$ is on the terminal side of α, it follows from the definition of trigonometric functions given in Sec. 9.1 that

$$\sin\alpha = \frac{v}{b}$$

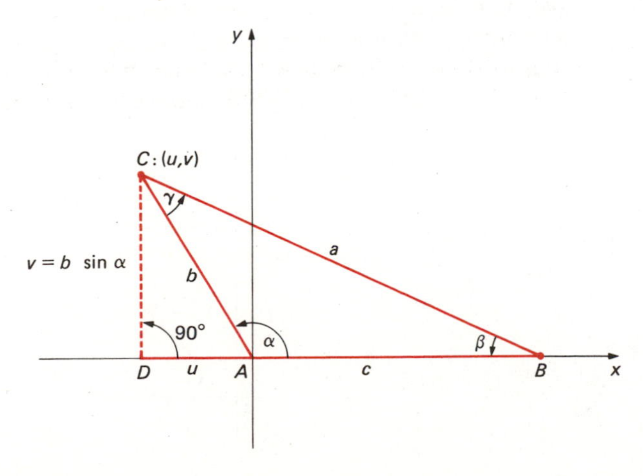

Fig. 10.2 Law of sines

and hence

$$v = b \sin \alpha \tag{10.5}$$

Now, we recall that the area of a triangle is equal to one-half the product of the base and height, and hence (see Fig. 10.2)

$$\text{Area of triangle } ABC = \tfrac{1}{2}cv = \tfrac{1}{2}cb \sin \alpha$$

by (10.5). We have thus shown that

$$\text{Area of triangle } ABC = \tfrac{1}{2}bc \sin \alpha$$

The other formulas in (10.4) are proved in a similar way (place the x axis along different sides of the triangle).

Now, by (10.4), we have

$$\tfrac{1}{2}bc \sin \alpha = \tfrac{1}{2}ca \sin \beta = \tfrac{1}{2}ab \sin \gamma \tag{10.6}$$

Hence by dividing (10.6) by $\tfrac{1}{2}abc$, we obtain

$$\frac{\sin \alpha}{a} = \frac{\sin \beta}{b} = \frac{\sin \gamma}{c} \tag{10.7}$$

We have thus proved the following important result, known as the law of sines.

Law of Sines

If ABC is any triangle with sides a, b, c and angles α, β, γ, as indicated in Fig. 10.2, then

$$\frac{\sin \alpha}{a} = \frac{\sin \beta}{b} = \frac{\sin \gamma}{c}$$

We shall soon illustrate the law of sines by some examples. But first let us establish another important result, known as the *law of cosines.* Now, suppose that ABC is any triangle with sides a, b, c and angles α, β, γ, as indicated in Fig. 10.3. Once again, let us choose the side AB to be along the positive

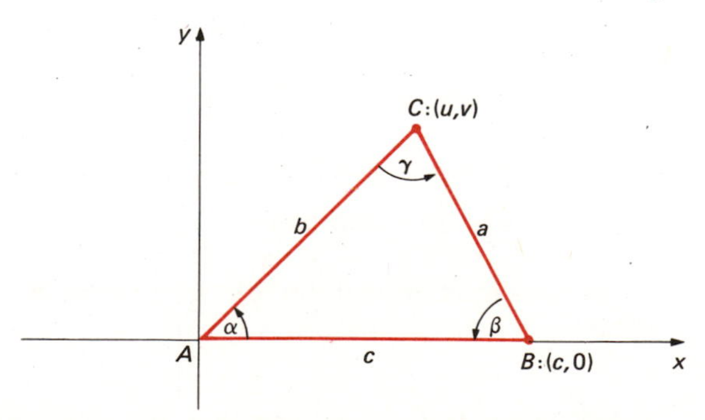

Fig. 10.3 Law of cosines

direction of the x axis, with A at the origin. Then, the coordinates of A and B are

$$A:(0,0) \qquad B:(c,0)$$

Now, suppose that the coordinates of C are (u,v). Observe that angle BAC $[=\alpha]$ is a directed angle in standard position, and $C:(u,v)$ is on the terminal side of α. Therefore, recalling the definition of trigonometric functions given in Sec. 9.1 and imagining that we have drawn a line CD perpendicular to and meeting the x axis at D, we obtain

$$\sin \alpha = \frac{v}{b} \qquad \cos \alpha = \frac{u}{b} \tag{10.8}$$

Hence, by (10.8),

$$v = b \sin \alpha \qquad u = b \cos \alpha \tag{10.9}$$

Now, as indicated in Fig. 10.3, we see that

$$\text{Distance between } B:(c,0) \text{ and } C:(u,v) \text{ is } a \tag{10.10}$$

Applying the distance formula established in Eq. (9.49), we obtain

$$\begin{aligned}
\text{Distance from } B:(c,0) \text{ to } C:(u,v) &= \sqrt{(u-c)^2 + (v-0)^2} \\
&= \sqrt{(b \cos \alpha - c)^2 + (b \sin \alpha)^2} \qquad \text{by (10.9)} \\
&= \sqrt{b^2\cos^2\alpha + c^2 - 2(b \cos \alpha)c + b^2\sin^2\alpha} \\
&= \sqrt{b^2(\cos^2\alpha + \sin^2\alpha) + c^2 - 2bc \cos \alpha} \\
&= \sqrt{b^2 + c^2 - 2bc \cos \alpha} \qquad \text{by Eq. (9.33)}
\end{aligned}$$

We have thus shown that

$$\text{Distance between } (c,0) \text{ and } (u,v) = \sqrt{b^2 + c^2 - 2bc \cos \alpha} \tag{10.11}$$

Comparing (10.10) and (10.11) and squaring, we obtain

$$a^2 = b^2 + c^2 - 2bc \cos \alpha \tag{10.12}$$

Similarly, we can show that

$$b^2 = c^2 + a^2 - 2ca \cos \beta \tag{10.13}$$

$$c^2 = a^2 + b^2 - 2ab \cos \gamma \tag{10.14}$$

We have thus proved the following important result, known as the law of cosines.

Law of Cosines If ABC is any triangle with sides a, b, c and angles α, β, γ, as indicated in Fig. 10.3, then

$$a^2 = b^2 + c^2 - 2bc \cos \alpha$$

$$b^2 = c^2 + a^2 - 2ca \cos \beta$$

$$c^2 = a^2 + b^2 - 2ab \cos \gamma$$

Remark If the triangle ABC is a right triangle, say $\alpha = 90°$, then one of the forms of the above law of cosines reduces to $a^2 = b^2 + c^2$, which is the Pythagorean theorem. This follows since $\cos \alpha = \cos 90° = 0$.

We now pause to give some examples which illustrate the use of the law of sines and the law of cosines.

Example 10.5 In triangle ABC, we have $a = 10$, $\beta = 45°$, and $\gamma = 60°$. Find b, c, and α.

Solution By high school geometry, we know that

$$\alpha = 180° - \beta - \gamma = 75°$$

Now, by the law of sines, we have

$$\frac{\sin \alpha}{a} = \frac{\sin \beta}{b} = \frac{\sin \gamma}{c}$$

and hence, by substitution,

$$\frac{\sin 75°}{10} = \frac{\sin 45°}{b} = \frac{\sin 60°}{c} \tag{10.15}$$

Using Table E and the first equality in (10.15), we obtain

$$b = \frac{10 \sin 45°}{\sin 75°} \approx \frac{10(0.7071)}{(0.9659)} \approx 7.3$$

$$b \approx 7.3$$

Finally, to find c, we have by (10.15),

$$\frac{\sin 75°}{10} = \frac{\sin 60°}{c}$$

Using Table E, we get

$$c = \frac{10 \sin 60°}{\sin 75°} \approx \frac{10(0.8660)}{(0.9659)} \approx 9.0$$

Thus $c \approx 9.0$.

Example 10.6 In triangle ABC, we have $a = 10$, $b = 15$, and $c = 20$. Find α, β, and γ.

Solution Using the law of cosines, we easily verify that

$$\cos \alpha = \frac{b^2 + c^2 - a^2}{2bc} = \frac{15^2 + 20^2 - 10^2}{2(15)(20)} = \frac{525}{600} = 0.8750$$

Using Table E, we observe that $\alpha \approx 29°$. Similarly, using the law of cosines again, we get

$$\cos\beta = \frac{c^2 + a^2 - b^2}{2ca} = \frac{20^2 + 10^2 - 15^2}{2(20)(10)} = \frac{275}{400} = 0.6875$$

Using Table E, we obtain $\beta \approx 46.5°$. Finally, to obtain γ, we use the law of cosines again to get

$$\cos\gamma = \frac{a^2 + b^2 - c^2}{2ab} = \frac{10^2 + 15^2 - 20^2}{2(10)(15)} = -\frac{75}{300} = -0.2500 \tag{10.16}$$

Now, the fact that $\cos\gamma$ is *negative* tells us that the angle γ is obtuse (that is, $\gamma > 90°$). In view of this, let us write

$$\gamma = 90° + \delta \qquad \text{where } \delta \text{ is an acute angle} \tag{10.17}$$

Combining (10.16) and (10.17), we get

$$\cos(90° + \delta) = -0.2500 \tag{10.18}$$

But, by Example 9.7, we have

$$\cos(90° + \delta) = -\sin\delta \tag{10.19}$$

Hence, by (10.18) and (10.19), we obtain

$$\sin\delta = 0.2500$$

Now, using Table E, we find $\delta \approx 14.5°$, and hence by (10.17), $\gamma = 90° + \delta \approx 104.5°$.

Remark We may check the correctness of our work by observing that

$$\alpha + \beta + \gamma \approx 29° + 46.5° + 104.5° = 180°$$

Example 10.7 In triangle ABC, we have $a = 10$, $b = 5$, and $\alpha = 60°$. Find c, β, and γ.

Solution By the law of sines, we have

$$\frac{\sin\alpha}{a} = \frac{\sin\beta}{b} = \frac{\sin\gamma}{c}$$

and hence,

$$\frac{\sin 60°}{10} = \frac{\sin\beta}{5} = \frac{\sin\gamma}{c} \tag{10.20}$$

Thus,

$$\sin\beta = \frac{5\sin 60°}{10} = \frac{\sqrt{3}}{4} \approx 0.433 \tag{10.21}$$

Now, since the function sin is positive in *both* the first and second quadrants, Eq. (10.21) allows two *possible* values of β, one acute (<90°) and the other obtuse (>90°). Indeed, recalling the result given in Prob. 9.2.7*a*, we have

$$\sin(180° - \beta) = \sin \beta \tag{10.22}$$

Thus, the two *possible* angles β which satisfy (10.21) are, in view of (10.22), supplementary angles (i.e., their sum is 180°). Now, using Table E, we see that a solution of (10.21) is given by $\beta \approx 25.5°$. Hence, in view of (10.22), the two *possible* solutions β of (10.21) are $\beta \approx 25.5°$ and $\beta \approx 154.5°$. Now, since $\alpha = 60°$, we *cannot* possibly have $\beta \approx 154.5°$ (since the sum of the angles of a triangle is equal to 180°). Thus the only feasible value of β is

$$\beta \approx 25.5° \tag{10.23}$$

Hence, $\gamma = 180° - \alpha - \beta \approx 94.5°$; that is,

$$\gamma \approx 94.5° \tag{10.24}$$

Finally, to find *c*, we use (10.20) again, to get

$$\frac{\sin 60°}{10} = \frac{\sin \gamma}{c} \approx \frac{\sin 94.5°}{c} \tag{10.25}$$

Therefore, recalling Example 9.7, Eq. (10.25) yields (see Table E)

$$c \approx \frac{10 \sin 94.5°}{\sin 60°} = \frac{10 \cos 4.5°}{\sin 60°}$$

$$\approx \frac{10(0.9969)}{0.8660} \approx 11.5$$

Thus,

$$c \approx 11.5 \tag{10.26}$$

Example 10.8 In triangle *ABC*, we have $a = 4.5$, $b = 5$, and $\alpha = 60°$. Find c, β, and γ.

Solution By the law of sines we have

$$\frac{\sin \alpha}{a} = \frac{\sin \beta}{b} = \frac{\sin \gamma}{c} \tag{10.27}$$

and hence

$$\frac{\sin 60°}{4.5} = \frac{\sin \beta}{5} = \frac{\sin \gamma}{c}$$

Thus,

$$\sin \beta = \frac{5 \sin 60°}{4.5} \approx \frac{5(0.8660)}{4.5} \approx 0.9622 \tag{10.28}$$

Hence, using Table E and keeping (10.22) in mind, we obtain the following possible values of β:

$$\beta \approx 74° \qquad \text{or} \qquad \beta \approx 180° - 74° = 106° \tag{10.29}$$

Since $\gamma = 180° - \alpha - \beta$, Eq. (10.29) yields two values of γ, namely,

$$\gamma \approx 46° \qquad \text{or} \qquad \gamma \approx 14° \tag{10.30}$$

We are thus led to distinguishing two cases.

Case 1 $\gamma = 46°$

In this case, (10.27) yields

$$\frac{\sin 60°}{4.5} = \frac{\sin \gamma}{c} \approx \frac{\sin 46°}{c}$$

and hence (see Table E)

$$c \approx \frac{4.5 \sin 46°}{\sin 60°} \approx \frac{4.5(0.7193)}{(0.8660)} \approx 3.7 \tag{10.31}$$

Case 2 $\gamma \approx 14°$

In this case, (10.27) yields

$$\frac{\sin 60°}{4.5} = \frac{\sin \gamma}{c} \approx \frac{\sin 14°}{c}$$

and hence (see Table E)

$$c \approx \frac{4.5 \sin 14°}{\sin 60°} \approx \frac{4.5(0.2419)}{(0.8660)} \approx 1.3 \tag{10.32}$$

We summarize our results as follows: There are two solutions of the problem posed in this example. These two solutions are:

Case 1: $\beta \approx 74° \qquad \gamma \approx 46° \qquad c \approx 3.7$

Case 2: $\beta \approx 106° \qquad \gamma \approx 14° \qquad c \approx 1.3$

The two triangles corresponding to the above two cases are indicated in Fig. 10.4 as triangles ACB_1 and ACB_2.

Example 10.9 Given that ABC is a triangle with $a = 4$, $b = 5$, and $\alpha = 60°$, find c, β, and γ.

Solution By the law of sines, we have

$$\frac{\sin \alpha}{a} = \frac{\sin \beta}{b} = \frac{\sin \gamma}{c}$$

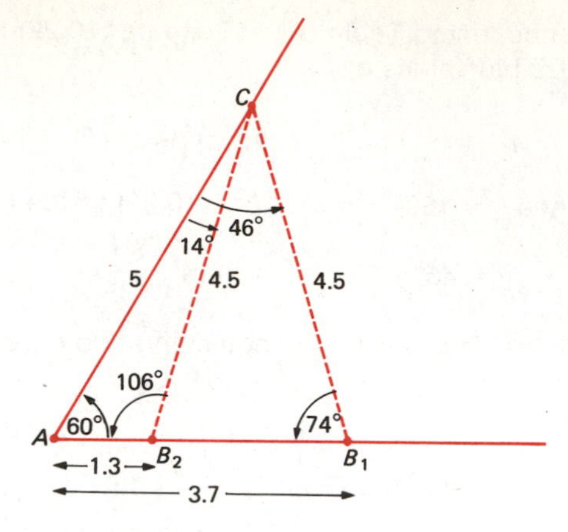

Fig. 10.4 "Solutions" of a triangle

and hence

$$\frac{\sin 60°}{4} = \frac{\sin \beta}{5} = \frac{\sin \gamma}{c} \tag{10.33}$$

Now, the first equality in (10.33) gives

$$\sin \beta = \frac{5 \sin 60°}{4} = \frac{5\sqrt{3}}{8} = \frac{\sqrt{75}}{\sqrt{64}} > 1$$

and hence

$$\sin \beta > 1 \tag{10.34}$$

But, by (9.36), $-1 \leq \sin \beta \leq 1$ *for all values of* β, and hence (10.34) is *never* true no matter what value we choose for β. Thus, *there exists no triangle satisfying the hypotheses of the present example.* So it is absurd to try to determine c, β, and γ.

Remark It is instructive to examine the present situation geometrically. Indeed, if CD denotes the line which is perpendicular to the line AX and which meets AX at D, as indicated in Fig. 10.5, then

$$\text{Length of } CD = 5 \sin 60° = \frac{5\sqrt{3}}{2} > 4$$

Thus, length of $CD > 4$. Hence, for *any* point B on the line AX, we must have

$$\text{Length of } BC \;\; [= a] \geq \text{length of } CD > 4$$

by high school geometry. This implies that $a > 4$, which contradicts the hypothesis that $a = 4$. Hence *no* triangle satisfying the conditions given in this example can possibly exist.

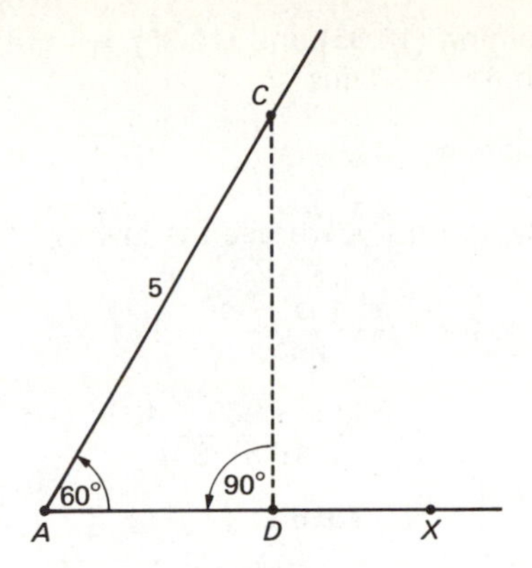

Fig. 10.5 Application of law of sines to the nonexistence of triangles with certain conditions

Remark The last three examples show that if we are given two sides of a triangle together with an angle *which is not the angle included by these two given sides,* then we may have any one of the following three possibilities:

1. There is *exactly one solution* (as in Example 10.7).
2. There are *exactly two solutions* (as in Example 10.8).
3. There are *no solutions* (as in Example 10.9).

For this reason, this situation (involving two sides and a nonincluded angle) is often described as the *ambiguous case.* Incidentally, if the given angle were *included* by the two given sides, then we would have, of course, a *unique* solution (see the following example and recall Side-Angle-Side).

Example 10.10 In triangle ABC, we have $b = 30$, $c = 20$, and $\alpha = 40°$. Find a, β, and γ.

Solution By the law of cosines, we have (see Table E)

$$\begin{aligned} a^2 &= b^2 + c^2 - 2bc \cos \alpha \\ &= 30^2 + 20^2 - 2(30)(20) \cos 40° \\ &\approx 1300 - 1200(0.7660) \\ &\approx 380.8 \end{aligned}$$

Thus $a \approx \sqrt{380.8} \approx 19.5$. Now, using the law of cosines again, we get

$$\cos \beta = \frac{c^2 + a^2 - b^2}{2ca} \approx \frac{400 + 380.8 - 900}{2(20)(19.5)} \approx -0.1528 \tag{10.35}$$

Hence β is obtuse (that is, $\beta > 90°$). Let $\beta = 90° + \delta$. Then, by Example 9.7 we have

$$\cos \beta = \cos(90° + \delta) = -\sin \delta \tag{10.36}$$

Combining (10.35) and (10.36), we get $\sin \delta \approx 0.1528$. Hence, using Table E, we obtain $\delta \approx 9°$. Thus

$$\beta = 90° + \delta \approx 99° \tag{10.37}$$

Finally, to find γ, we use the law of cosines again to obtain

$$\cos \gamma = \frac{a^2 + b^2 - c^2}{2ab}$$

$$\approx \frac{380.8 + 900 - 400}{2(19.5)(30)}$$

$$\approx 0.7528$$

Using Table E, we obtain $\gamma \approx 41°$.

Remark We may check the correctness of our work by observing that

$$\alpha + \beta + \gamma \approx 40° + 99° + 41° = 180°$$

In the next section, we shall consider a very important trigonometric method for representing complex numbers, known as the *polar form*. This will be particularly useful in solving certain types of polynomial equations.

Problem Set 10.1

10.1.1 Find the legs a and b, the hypotenuse c, and the acute angles α and β of a right-angled triangle ABC, given

(a) $\alpha = 60°$, $c = 4$ (d) $\alpha = 50°$, $b = 5$
(b) $\beta = 30°$, $c = 2$ (e) $a = 10$, $b = 30$
(c) $\alpha = 40°$, $a = 3$ (f) $a = 5$, $c = 15$

10.1.2 Find the sides a, b, c and the angles α, β, γ of a triangle ABC, given

(a) $a = 2$, $b = 3$, $c = 4$ (c) $a = 10$, $\alpha = 80°$, $\beta = 35°$
(b) $a = 5$, $b = 10$, $\gamma = 35°$ (d) $a = 20$, $\beta = 40°$, $\gamma = 120°$

10.1.3 Find the area of a triangle ABC with sides a, b, c and angles α, β, γ, given

(a) $a = 10$, $b = 20$, $\gamma = 50°$ (c) $a = 10$, $\beta = 20°$, $\gamma = 110°$
(b) $a = 5$, $\alpha = 85°$, $\beta = 40°$ (d) $a = 8$, $b = 10$, $c = 15$

10.1.4 Find all possible triangles ABC (if any) with sides a, b, c and angles α, β, γ, given

(a) $b = 6$, $c = 4$, $\gamma = 30°$ (c) $a = 5$, $c = 2$, $\gamma = 20°$
(b) $a = 9$, $b = 10$, $\alpha = 50°$

10.1.5 Use the law of sines to determine whether or not there is a triangle ABC with sides a, b, c and angles α, β, γ which satisfies each condition below. If there is such a triangle, find all its sides and angles.

(a) $a = 5$, $b = 2$, $\beta = 40°$ (c) $c = 2$, $a = 4$, $\gamma = 31°$
(b) $b = 5$, $c = 10$, $\gamma = 100°$

10.1.6 The radius of a circle is 10 in. and the length of a chord *AB* is 5 in. Find the central angle which is subtended by *AB*.

10.1.7 Find the perimeter of a regular pentagon which can be inscribed in a circle of radius 10 in.

10.1.8 Prove that if *ABC* is a triangle with sides a, b, c and angles α, β, γ, then

$$\frac{\sin \alpha - \sin \beta}{\sin \alpha + \sin \beta} = \frac{a - b}{a + b}$$

[*Hint*: Use the law of sines.]

10.1.9 Suppose that *ABC* is a triangle with sides a, b, c and angles α, β, γ. Prove the following:

(*a*) $$\frac{1 + \cos \alpha}{2} = \frac{(b + c + a)(b + c - a)}{4bc}$$

(*b*) $$\frac{1 - \cos \alpha}{2} = \frac{(a + b - c)(a - b + c)}{4bc}$$

(*c*) $$\cos \frac{\alpha}{2} = \sqrt{\frac{(b + c + a)(b + c - a)}{4bc}}$$

(*d*) $$\sin \frac{\alpha}{2} = \sqrt{\frac{(a + b - c)(a - b + c)}{4bc}}$$

[*Hint*: Use the law of cosines.]

10.1.10 A parallelogram has two adjacent sides of lengths 5 and 10 in. and an included angle 100°. Find the length of each of the diagonals. Also, find the area of the parallelogram.

10.1.11 At a point 500 ft from the base of a building, the angle between the horizontal and the line to the top of the building is 40°. Find the height of the building.

10.1.12 A piece of wire 40 in. long is stretched from level ground to the top of a pole 20 in. high. Find the angle between the pole and the wire.

10.1.13 A man standing 100 ft from the foot of a flagpole, which is at his eye level, observes that the angle of elevation of the top of the flagpole is 40°, as indicated in the figure. Find the height of the pole.

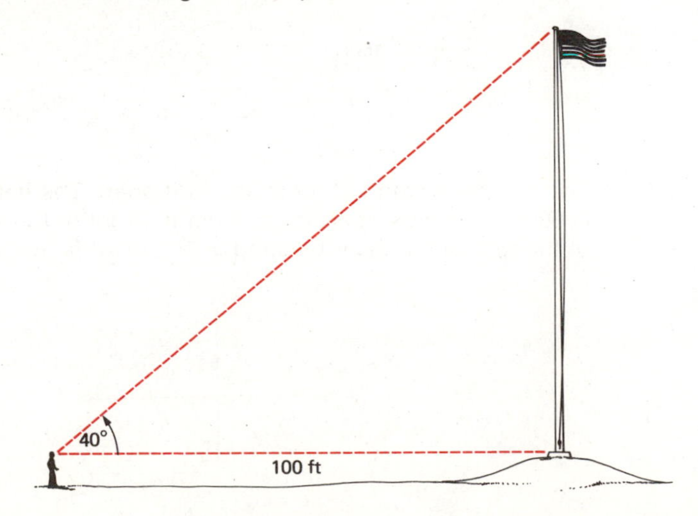

10.1.14 When a man 6 ft tall looks towards the top of a ladder, his line of vision makes an angle of 25° with the horizontal. However, when a child 4 ft tall looks at the top of the ladder from the same spot, the child's line of vision makes an angle of 40° with the horizontal. How tall is the ladder?

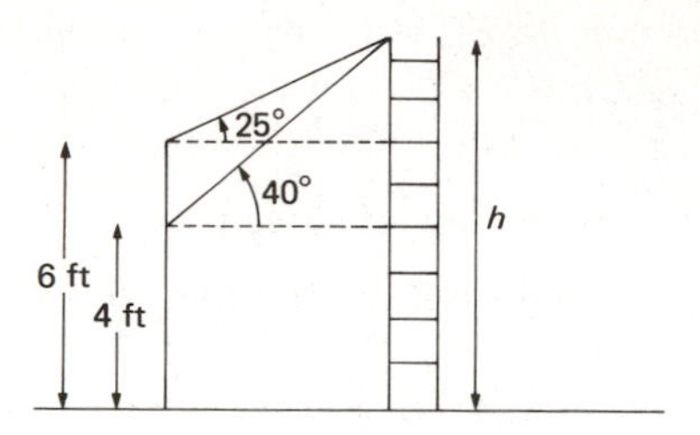

10.1.15 The angle subtended to the ground from the top of a building at a distance x ft from its base is 50°. The angle subtended at a distance 100 ft further away is 35°. Find the height of the building.

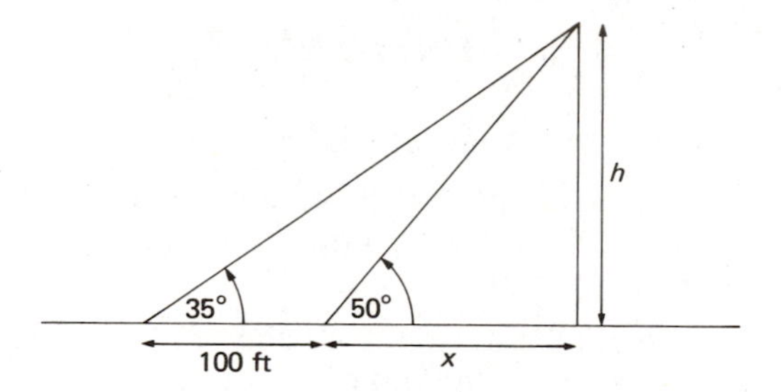

10.1.16 From the top of a lighthouse, 150 ft above the ocean, the angle of depression of a boat is 20°. How far is the boat from the lighthouse?

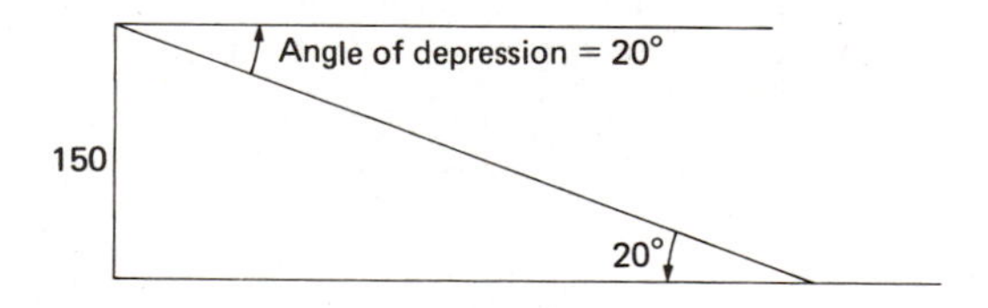

10.1.17 Two buildings with flat roofs are 50 ft apart. The height of the shorter building is 30 ft. The angle of elevation from the roof of the shorter building to the edge of the roof of the taller building is 35°. What is the height of the taller building?

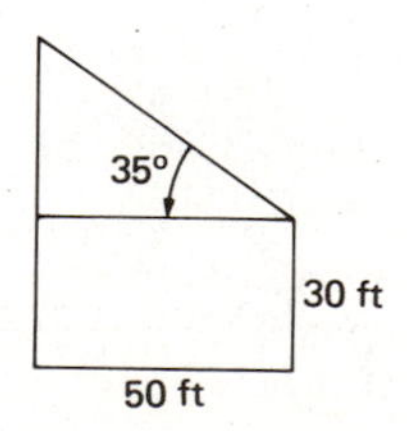

10.1.18 A ladder is 30 ft long and has its base in the street. The ladder makes an angle of 25° with the street when its top rests on a building on one side of the street, and it makes an angle of 35° with the street when its top rests on a building on the other side of the street. How wide is the street?

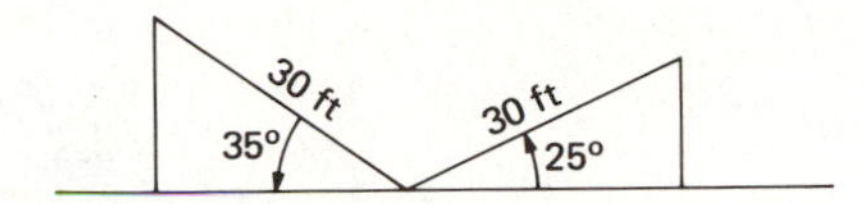

10.1.19 From a point A on level ground, the angles of elevation of the top B and the bottom C of a flagpole situated on top of a hill are 50 and 40°, respectively. The height of the flagpole is 100 ft. What is the height of the hill above level ground?

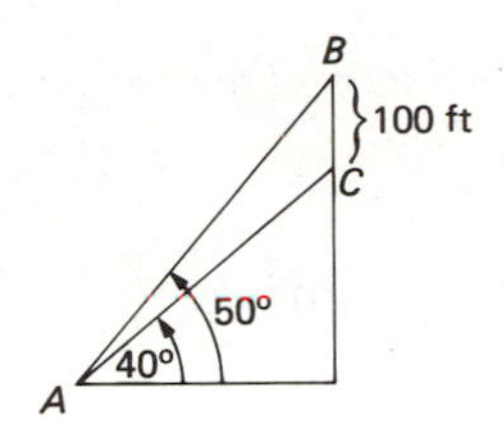

10.2 Polar Form

In this section, we use trigonometric functions to express complex numbers in what is known as the *polar form.* We also discuss the applications of complex numbers in polar form to finding solutions of certain types of polynomial equations.

In Chap. 4, we defined a complex number to be a number z of the form

$$z = x + yi \qquad x, y \text{ real}, \ i^2 = -1 \tag{10.38}$$

We also made the following identification of the complex number z given in (10.38)

The point (x,y), x and y real, represents the complex number z in (10.38). (10.39)

Now, suppose that $P:(x,y)$ represents the complex number $z = x + yi$ (x, y real), as indicated in Fig. 10.6. Suppose, further, that θ is the directed angle in standard position whose terminal side is $\mathcal{O}P$ (where $\mathcal{O}$ is the origin), and suppose that r denotes the nonnegative length of the line segment $\mathcal{O}P$, as indicated in Fig. 10.6. Finally, suppose that PQ is the line segment which is perpendicular to the x axis and which meets the x axis at Q. Then, using the definitions of trigonometric functions given in Sec. 9.1, we have (see Fig. 10.6)

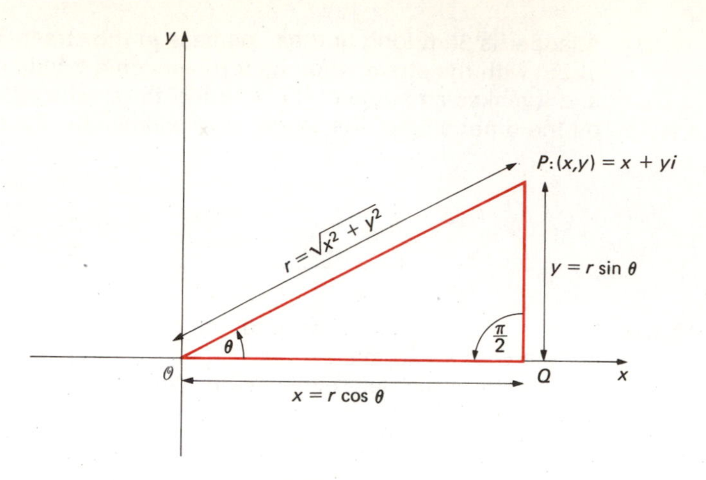

Fig. 10.6 Polar form of the complex number $z = x + yi$ (x, y real) is $z = r(\cos\theta + i\sin\theta)$.

$$\cos\theta = \frac{x}{r} \qquad \sin\theta = \frac{y}{r}$$

and hence

$$x = r\cos\theta \qquad y = r\sin\theta \tag{10.40}$$

Moreover, by the Pythagorean theorem, we have $r^2 = x^2 + y^2$, and hence (since $r \geq 0$)

$$r = \sqrt{x^2 + y^2} \tag{10.41}$$

Now, combining Eqs. (10.40) and (10.41), we easily see that

$$\cos\theta = \frac{x}{\sqrt{x^2 + y^2}} \qquad \sin\theta = \frac{y}{\sqrt{x^2 + y^2}} \tag{10.42}$$

Note that if both $\cos\theta$ and $\sin\theta$ are known *and* $0 \leq \theta < 2\pi$, then θ is uniquely determined. The numbers r and θ described in (10.41) and (10.42) are called the *polar coordinates* of the point P (see Fig. 10.6). Again, suppose $z = x + yi$ (x, y real) is any complex number. Then, in view of (10.40), we have

$$z = x + yi = (r\cos\theta) + (r\sin\theta)i = r(\cos\theta + i\sin\theta) \tag{10.43}$$

We summarize our results as follows:

> If $z = x + yi$ (x, y real) is any complex number, then we may rewrite z in the form $z = r(\cos\theta + i\sin\theta)$, where r is a nonnegative real number given by $r = \sqrt{x^2 + y^2}$ and θ is a real number satisfying (10.42). The form $z = r(\cos\theta + i\sin\theta)$ is called the *polar form* of z. Moreover, r is called the *modulus* of z, while θ is called the *argument* of z. Observe that (see Chap. 4) the modulus of z is equal to $|z|$.

We should state that in polar form, θ is usually expressed in *radian measure*. Let us illustrate these concepts by some examples.

Example 10.11 Find the polar form of the complex number $z = 1 + i$.

Solution Here $x = 1$, $y = 1$, and hence by (10.41) and (10.42) we have

$$r = \sqrt{1^2 + 1^2} = \sqrt{2} \tag{10.44}$$

$$\cos \theta = \frac{1}{\sqrt{2}} \qquad \sin \theta = \frac{1}{\sqrt{2}} \tag{10.45}$$

In view of (10.45), we may choose

$$\theta = \frac{\pi}{4} \tag{10.46}$$

Now, by definition, the polar form of $z = x + yi$ (x, y real) is $z = r(\cos \theta + i \sin \theta)$, and hence the polar form of $z = 1 + i$ is [see Eqs. (10.44) and (10.46)]

$$z = \sqrt{2}\left(\cos \frac{\pi}{4} + i \sin \frac{\pi}{4}\right)$$

as illustrated in Fig. 10.7.

Remark Since $\cos(\theta + 2\pi k) = \cos \theta$ and $\sin(\theta + 2\pi k) = \sin \theta$ for all integers k, we may also take the polar form of $z = x + yi$ (x, y real) to be $z = r[\cos(\theta + 2\pi k) + i \sin(\theta + 2\pi k)]$, where r and θ are as described in (10.41) and (10.42) and k is *any* integer. For example, other polar forms of $z = 1 + i$ in the preceding example are

$$z = \sqrt{2}\left(\cos \frac{9\pi}{4} + i \sin \frac{9\pi}{4}\right) = \sqrt{2}\left(\cos \frac{17\pi}{4} + i \sin \frac{17\pi}{4}\right)$$

Example 10.12 Find the polar form of the complex number $z = -\sqrt{3} + i$.

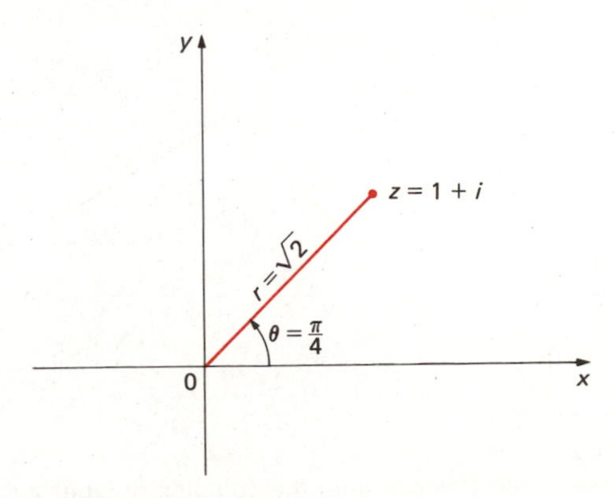

Fig. 10.7 Polar form of the complex number $z = 1 + i$ is $z = \sqrt{2}[\cos(\pi/4) + i \sin(\pi/4)]$.

Solution Here $x = -\sqrt{3}$, $y = 1$, and hence by (10.41) and (10.42) we have

$$r = \sqrt{(-\sqrt{3})^2 + 1^2} = 2 \tag{10.47}$$

$$\cos\theta = \frac{-\sqrt{3}}{2} \qquad \sin\theta = \frac{1}{2} \tag{10.48}$$

In view of (10.48), we may choose

$$\theta = \frac{5\pi}{6} \tag{10.49}$$

Hence, the polar form of $z = -\sqrt{3} + i$ is [see (10.47) and (10.49)]

$$z = 2\left(\cos\frac{5\pi}{6} + i\sin\frac{5\pi}{6}\right)$$

as illustrated in Fig. 10.8.

In Chap. 4 we were able to give geometric interpretations of the sum and difference of two complex numbers, and this led to the so-called parallelogram rule. However, we were unable at that time to give geometric interpretation of the product or quotient of two complex numbers. Using the polar form of a complex number, we are now able to give geometric interpretations of products and quotients of complex numbers.

Suppose that z_1 and z_2 are complex numbers in polar form, say,

$$z_1 = r_1(\cos\theta_1 + i\sin\theta_1) \tag{10.50}$$

$$z_2 = r_2(\cos\theta_2 + i\sin\theta_2) \tag{10.51}$$

where $r_1 \geq 0$, $r_2 \geq 0$, and θ_1, θ_2 both real. Then, recalling that $i^2 = -1$, and using Eqs. (9.53) and (9.57), we get

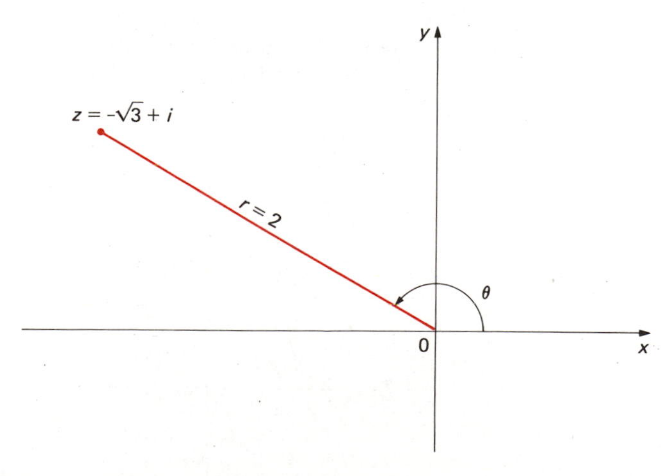

Fig. 10.8 Polar form of the complex number $z = -\sqrt{3} + i$ is $z = 2[\cos(5\pi/6) + i\sin(5\pi/6)]$

$$\begin{aligned} z_1 z_2 &= r_1 r_2 (\cos\theta_1 + i\sin\theta_1)(\cos\theta_2 + i\sin\theta_2) \\ &= r_1 r_2 [(\cos\theta_1\cos\theta_2 - \sin\theta_1\sin\theta_2) + i(\cos\theta_1\sin\theta_2 + \sin\theta_1\cos\theta_2)] \\ &= r_1 r_2 [\cos(\theta_1+\theta_2) + i\sin(\theta_1+\theta_2)] \end{aligned}$$

We have thus shown that

$$\begin{aligned} &\textit{If} && z_1 = r_1(\cos\theta_1 + i\sin\theta_1) \\ &\textit{and} && z_2 = r_2(\cos\theta_2 + i\sin\theta_2) \\ &\textit{then} && z_1 z_2 = r_1 r_2[\cos(\theta_1+\theta_2) + i\sin(\theta_1+\theta_2)] \end{aligned} \tag{10.52}$$

In other words, in order *to multiply two complex numbers in polar form, simply multiply the two moduli and add the two arguments*, as indicated in (10.52) and as illustrated in Fig. 10.9.

Next, consider the quotient z_1/z_2, where z_1 and z_2 are as in (10.50) and (10.51) and $z_2 \neq 0 + 0i$. Now, recalling that $i^2 = -1$, and using Eqs. (9.33), (9.54), and (9.58), we readily verify that

$$\begin{aligned} \frac{z_1}{z_2} &= \frac{r_1(\cos\theta_1 + i\sin\theta_1)}{r_2(\cos\theta_2 + i\sin\theta_2)} \\ &= \frac{r_1(\cos\theta_1 + i\sin\theta_1)}{r_2(\cos\theta_2 + i\sin\theta_2)} \cdot \frac{\cos\theta_2 - i\sin\theta_2}{\cos\theta_2 - i\sin\theta_2} \\ &= \frac{r_1[(\cos\theta_1\cos\theta_2 + \sin\theta_1\sin\theta_2) + i(\sin\theta_1\cos\theta_2 - \cos\theta_1\sin\theta_2)]}{r_2(\cos^2\theta_2 + \sin^2\theta_2)} \\ &= \frac{r_1}{r_2}[\cos(\theta_1-\theta_2) + i\sin(\theta_1-\theta_2)] \end{aligned}$$

We have thus shown that

$$\begin{aligned} &\textit{If} && z_1 = r_1(\cos\theta_1 + i\sin\theta_1) \\ &\textit{and} && z_2 = r_2(\cos\theta_2 + i\sin\theta_2) \qquad z_2 \neq (0+0i) \\ &\textit{then} && z_1/z_2 = r_1/r_2[\cos(\theta_1-\theta_2) + i\sin(\theta_1-\theta_2)] \end{aligned} \tag{10.53}$$

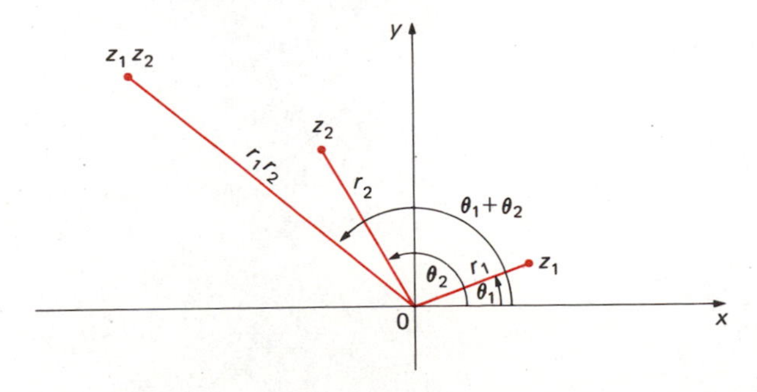

Fig. 10.9 Geometric interpretation of the product of two complex numbers $z_1 = r_1(\cos\theta_1 + i\sin\theta_1)$ and $z_2 = r_2(\cos\theta_2 + i\sin\theta_2)$

In other words, in order *to divide a complex number z_1 by a nonzero complex number z_2, simply divide the two moduli and subtract the two arguments,* as indicated in (10.53) and as illustrated in Fig. 10.10.

It is interesting to consider the special case of (10.52) in which we choose z_1 and z_2 to be *equal*, say,

$$z_2 = z_1 = r(\cos\theta + i\sin\theta)$$

In this important special case, (10.52) gives

$$[r(\cos\theta + i\sin\theta)]^2 = r^2(\cos 2\theta + i\sin 2\theta) \tag{10.54}$$

Similarly, if we multiply both sides of (10.54) by $r(\cos\theta + i\sin\theta)$, we obtain

$$\begin{aligned}[r(\cos\theta + i\sin\theta)]^3 &= [r^2(\cos 2\theta + i\sin 2\theta)][r(\cos\theta + i\sin\theta)] \\ &= r^3(\cos 3\theta + i\sin 3\theta) \qquad \text{by (10.52)}\end{aligned}$$

Thus,

$$[r(\cos\theta + i\sin\theta)]^3 = r^3(\cos 3\theta + i\sin 3\theta)$$

Continuing in this fashion, we see intuitively that

$$[r(\cos\theta + i\sin\theta)]^n = r^n(\cos n\theta + i\sin n\theta) \qquad \text{for all positive integers } n \tag{10.55}$$

Equation (10.55) is extremely useful, and is known as *De Moivre's theorem.* It turns out that (10.55) is true *without* the restriction that n be a positive integer. However, the formulation we gave of (10.55) is quite satisfactory for our purposes.

We now apply De Moivre's theorem to finding all the zeros (real or nonreal) of the polynomial

$$f(x) = x^n - 1 \qquad n \text{ a positive integer}$$

In this connection, we have the following important result:

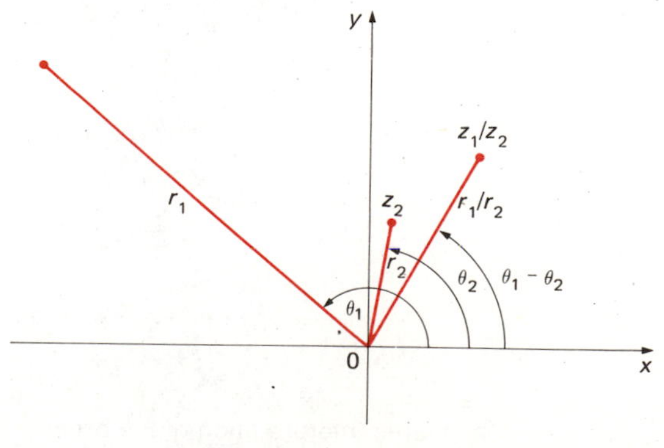

Fig. 10.10 Geometric interpretation of the quotient z_1/z_2 of the two complex numbers $z_1 = r_1(\cos\theta_1 + i\sin\theta_1)$ and $z_2 = r_2(\cos\theta_2 + i\sin\theta_2) \neq (0 + 0i)$

The solutions of the equation $x^n - 1 = 0$, where n is any positive integer, are given by

$$x = \cos\left(\frac{2\pi}{n} \cdot k\right) + i \sin\left(\frac{2\pi}{n} \cdot k\right) \qquad k = 0, 1, 2, \ldots, n-1 \tag{10.56}$$

The proof of this statement is easy. Indeed, raising both sides of (10.56) to the nth power and using De Moivre's theorem, we obtain

$$\begin{aligned} x^n &= \left[\cos\left(\frac{2\pi}{n} \cdot k\right) + i \sin\left(\frac{2\pi}{n} \cdot k\right)\right]^n \\ &= \cos\left[\left(\frac{2\pi}{n} \cdot k\right)n\right] + i \sin\left[\left(\frac{2\pi}{n} \cdot k\right)n\right] \\ &= \cos 2k\pi + i \sin 2k\pi \\ &= 1 + i(0) \qquad \text{(since } k \text{ is an } \textit{integer}\text{)} \\ &= 1 \end{aligned}$$

In other words, we have shown that if x is as in (10.56), then $x^n = 1$. It can also be shown (although we omit the proof) that if $k = 0, 1, 2, \ldots, n-1$, then the resulting solutions described in (10.56) are *distinct.* We have thus obtained *exactly n solutions* of the equation $x^n - 1 = 0$. Hence, by the general form of the Fundamental Theorem of Algebra, (10.56) gives *all* the solutions of $x^n - 1 = 0$. Indeed, these n solutions are obtained by simply taking, in (10.56), $k = 0$, then $k = 1$, then $k = 2$, etc., until finally we take $k = n - 1$. A curious question now arises: What happens if, in (10.56), we take $k = n$, or $k = n + 1$, or $k = n + 2$, and so on? In attempting to answer this question, let us first substitute $k = n$ in (10.56):

$$x = \cos\left(\frac{2\pi}{n} \cdot n\right) + i \sin\left(\frac{2\pi}{n} \cdot n\right) = 1 + i(0) = 1$$

This, however, is *not* a new solution, since if we set $k = 0$ in (10.56), we obtain

$$x = \cos 0 + i \sin 0 = 1 + i(0) = 1$$

In other words, *setting $k = n$ in (10.56) gives exactly the same solution (namely, 1) as setting $k = 0$ in (10.56) does.* Similarly, it can be shown that setting $k = n + 1$ in (10.56) yields exactly the same solution as setting $k = 1$ in (10.56) does, and so on. In fact, (10.56) gives *all* the solutions of $x^n - 1 = 0$.

xample 10.13 Find all complex numbers x such that

$$x^3 = 1 \tag{10.57}$$

Also, represent these solutions geometrically.

olution Here $n = 3$, and hence by (10.56), the solutions of (10.57) are given by

$$x = \cos\left(\frac{2\pi}{3} \cdot k\right) + i \sin\left(\frac{2\pi}{3} \cdot k\right) \qquad k = 0, 1, 2 \tag{10.58}$$

This gives us *exactly three* solutions, namely,

$$x = x_0 = \cos 0 + i \sin 0 = 1 \qquad \text{[by taking } k = 0 \text{ in (10.58)]}$$

$$x = x_1 = \cos\frac{2\pi}{3} + i\sin\frac{2\pi}{3} = -\frac{1}{2} + \frac{\sqrt{3}}{2}i \qquad \text{[by taking } k = 1 \text{ in (10.58)]}$$

$$x = x_2 = \cos\frac{4\pi}{3} + i\sin\frac{4\pi}{3} = -\frac{1}{2} - \frac{\sqrt{3}}{2}i \qquad \text{[by taking } k = 2 \text{ in (10.58)]}$$

These solutions all lie on the unit circle, and are sketched in Fig. 10.11. [Note that if we take $k = 5$, say, in (10.58), we obtain $x = \cos(10\pi/3) + i\sin(10\pi/3) = x_2$ (Why?). Thus, $k = 5$ does *not* give new solutions.]

Example 10.14 Find all complex numbers x such that

$$x^6 = 1 \tag{10.59}$$

Also, represent these solutions geometrically.

Solution Here $n = 6$, and hence by (10.56) the solutions of (10.59) are given by

$$x = \cos\left(\frac{2\pi}{6}\cdot k\right) + i\sin\left(\frac{2\pi}{6}\cdot k\right) \qquad k = 0, 1, 2, 3, 4, 5 \tag{10.60}$$

This gives us *exactly six* solutions, namely,

$$x = x_0 = \cos 0 + i\sin 0 = 1 \qquad \text{[by taking } k = 0 \text{ in (10.60)]}$$

$$x = x_1 = \cos\frac{\pi}{3} + i\sin\frac{\pi}{3} \qquad \text{[by taking } k = 1 \text{ in (10.60)]}$$

$$x = x_2 = \cos\frac{2\pi}{3} + i\sin\frac{2\pi}{3} \qquad \text{[by taking } k = 2 \text{ in (10.60)]}$$

$$x = x_3 = \cos\pi + i\sin\pi = -1 \qquad \text{[by taking } k = 3 \text{ in (10.60)]}$$

$$x = x_4 = \cos\frac{4\pi}{3} + i\sin\frac{4\pi}{3} \qquad \text{[by taking } k = 4 \text{ in (10.60)]}$$

$$x = x_5 = \cos\frac{5\pi}{3} + i\sin\frac{5\pi}{3} \qquad \text{[by taking } k = 5 \text{ in (10.60)]}$$

These solutions all lie on the unit circle, and are sketched in Fig. 10.12.

A glance at Figs. 10.11 and 10.12 shows that the solutions of $x^3 = 1$ and the solutions of $x^6 = 1$ are equally spaced on the circumference of the unit circle, and that 1 is always a solution of both of these equations (which, of course, is obvious). Furthermore, all the central angles above subtended by the arcs joining consecutive solutions of $x^3 = 1$ and of $x^6 = 1$ are equal. These facts (about the n solutions of $x^n = 1$) are always true regardless of the value of n. Incidentally, the solutions of the equation $x^n = 1$ (n is a positive integer) are called nth *roots of unity*. Thus, *(10.56) gives all the* nth *roots of unity* (there are exactly n of them), for every positive integer n.

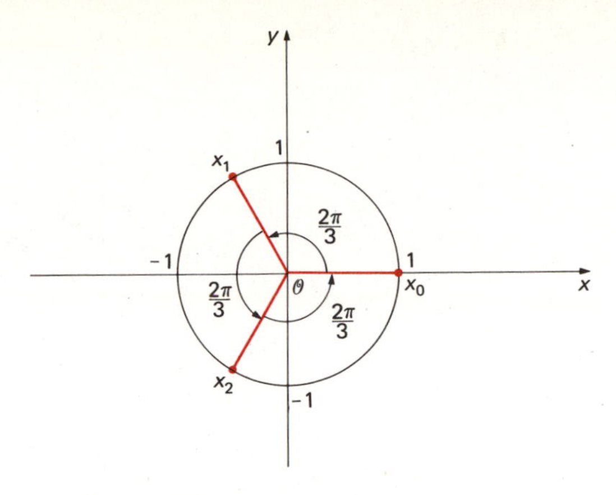

Fig. 10.11 Solutions of the equation $x^3 = 1$

We conclude this section by giving the following generalization of (10.56):

> Let $z_0 = r(\cos\theta + i\sin\theta)$, $r > 0$, θ real, be any nonzero complex number in polar form. Then the solutions of
>
> $x^n = z_0$
>
> are given by
>
> $$x = \sqrt[n]{r}\left(\cos\frac{\theta + 2\pi k}{n} + i\sin\frac{\theta + 2\pi k}{n}\right) \qquad k = 0, 1, 2, \ldots, n-1 \tag{10.61}$$

Once again, observe that (10.61) gives exactly n distinct solutions of $x^n = z_0$, where z_0 is any given nonzero complex number (expressed in polar form, as

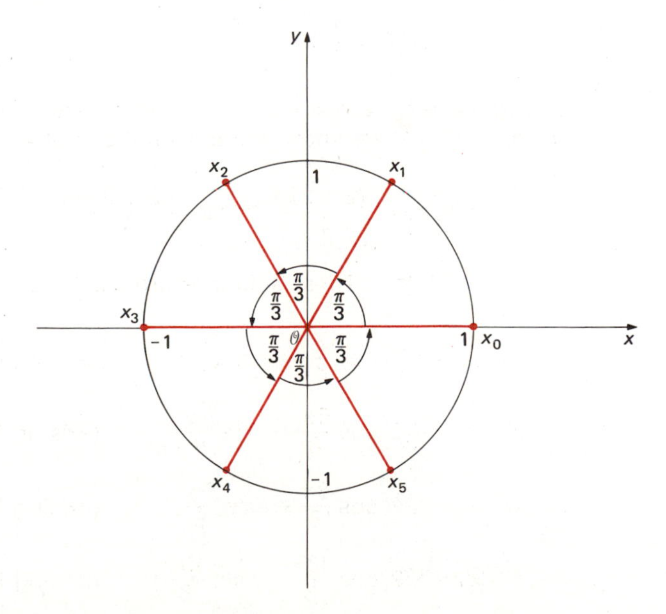

Fig. 10.12 Solutions of the equation $x^6 = 1$

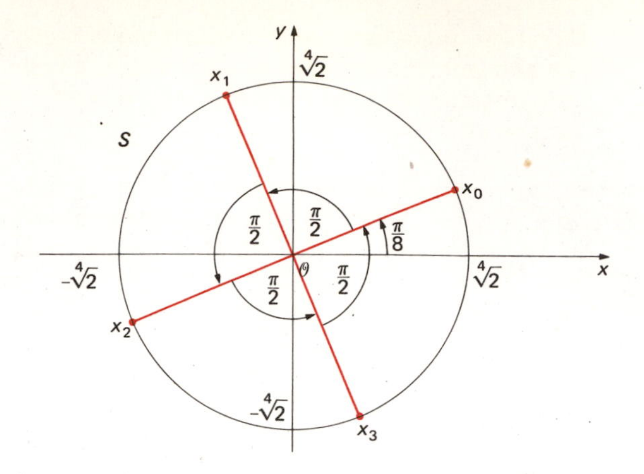

Fig. 10.13 Solution of the equation $x^4 = 2i$

indicated above). Moreover, the proof of (10.61) is quite similar to the proof of (10.56). Indeed, if we were to raise both sides of (10.61) to the nth power and use De Moivre's theorem, we would obtain at once $x^n = z_0$. We omit the proof that the solutions described in (10.61) (corresponding to $k = 0$, $k = 1$, $k = 2$, . . . , $k = n - 1$) are all *distinct.*

Let us illustrate this by an example.

Example 10.15 Find all solutions of $x^4 = 2i$. Also, represent these solutions geometrically.

Solution We first express the complex number $z_0 = 2i = 0 + 2i$ in polar form. It is readily verified that

$$z_0 = 2i = 2\left(\cos \frac{\pi}{2} + i \sin \frac{\pi}{2}\right)$$

Thus, we have $r = 2$, $\theta = \pi/2$, $n = 4$, and we are now ready to apply (10.61). Indeed, by (10.61) we know that the solutions of $x^4 = 2i$ are given by

$$x = \sqrt[4]{2}\left(\cos \frac{\pi/2 + 2\pi k}{4} + i \sin \frac{\pi/2 + 2\pi k}{4}\right)$$

where $k = 0, 1, 2, 3$. These solutions are easily seen to reduce to

$$x = x_0 = \sqrt[4]{2}\left(\cos \frac{\pi}{8} + i \sin \frac{\pi}{8}\right) \qquad \text{(arising from } k = 0\text{)}$$

$$x = x_1 = \sqrt[4]{2}\left(\cos \frac{5\pi}{8} + i \sin \frac{5\pi}{8}\right) \qquad \text{(arising from } k = 1\text{)}$$

$$x = x_2 = \sqrt[4]{2}\left(\cos \frac{9\pi}{8} + i \sin \frac{9\pi}{8}\right) \qquad \text{(arising from } k = 2\text{)}$$

$$x = x_3 = \sqrt[4]{2}\left(\cos \frac{13\pi}{8} + i \sin \frac{13\pi}{8}\right) \qquad \text{(arising from } k = 3\text{)}$$

Moreover, these solutions all lie on the circumference of the circle S with radius $\sqrt[4]{2}$ and with the origin as its center, and are sketched in Fig. 10.13. Observe that, once again, the solutions x_0, x_1, x_2, x_3 are equally spaced on the circumference of S.

In the next section, we discuss *inverse trigonometric functions*. Since the trigonometric functions are *not* one-to-one functions, it becomes absolutely essential to restrict their domains, if we expect to have inverses for these functions.

Problem Set 10.2

10.2.1 Express the following complex numbers in polar form, and represent the results geometrically.

(a) $1-i$ (g) $1+\sqrt{3}i$ (m) $-\frac{1}{2}-\frac{\sqrt{3}}{2}i$

(b) $-1+i$ (h) $-1+\sqrt{3}i$ (n) $\frac{1}{\sqrt{2}}+\frac{1}{\sqrt{2}}i$

(c) $\sqrt{3}-i$ (i) $-1-\sqrt{3}i$ (o) $\frac{1}{\sqrt{2}}-\frac{1}{\sqrt{2}}i$

(d) $\sqrt{3}+i$ (j) $\frac{1}{2}-\frac{\sqrt{3}}{2}i$ (p) $-\frac{1}{\sqrt{2}}+\frac{1}{\sqrt{2}}i$

(e) $-\sqrt{3}-i$ (k) $\frac{1}{2}+\frac{\sqrt{3}}{2}i$ (q) $-\frac{1}{\sqrt{2}}-\frac{1}{\sqrt{2}}i$

(f) $1-\sqrt{3}i$ (l) $-\frac{1}{2}+\frac{\sqrt{3}}{2}i$

10.2.2 Represent geometrically z_1z_2, z_1/z_2, and z_2/z_1 in each of the following.

(a) $z_1=3+4i$, $z_2=3-4i$ (c) $z_1=-1-\sqrt{3}i$, $z_2=-\sqrt{3}+i$

(b) $z_1=1-i$, $z_2=\sqrt{3}+i$ (d) $z_1=5+12i$, $z_2=\frac{1}{2}+\frac{\sqrt{3}}{2}i$

10.2.3 Use De Moivre's theorem to evaluate each of the following.

(a) $\left(\cos\frac{\pi}{4}+i\sin\frac{\pi}{4}\right)^{10}$

(b) $\left(\cos\frac{\pi}{4}-i\sin\frac{\pi}{4}\right)^{10}$ [*Hint*: $\cos\frac{\pi}{4}=\cos\left(\frac{-\pi}{4}\right)$ and $\sin\frac{\pi}{4}=-\sin\left(\frac{-\pi}{4}\right)$]

(c) $\left(\cos\frac{\pi}{3}+i\sin\frac{\pi}{3}\right)^{100}$

(d) $\left(\cos\frac{5\pi}{6}+i\sin\frac{5\pi}{6}\right)^{50}$

(e) $\left[2\left(\cos\frac{7\pi}{4}+i\sin\frac{7\pi}{4}\right)\right]^{10}$

(f) $\left[2\left(\cos\frac{7\pi}{4}-i\sin\frac{7\pi}{4}\right)\right]^{10}$ [*Hint*: See the hint in part *b*.]

10.2.4 Use the polar form and De Moivre's theorem to evaluate each of the following.

(a) $(-1+\sqrt{3}i)^7$ (d) $(2-2i)^{10}$

(b) $(-1-\sqrt{3}i)^7$ (e) $(2\sqrt{3}-2i)^8$

(c) $(2+2i)^{10}$ (f) $(2\sqrt{3}+2i)^8$

10.2.5 Find all nth roots of unity, where $n=4, 5, 7, 8, 9, 10$. In each case, represent these roots geometrically.

10.2.6 Express each of the following complex numbers z in polar form, and then find all the fifth roots of z in each case. Also, represent these roots geometrically.

(a) $z=1+i$ (e) $z=-1$

(b) $z=1-i$ (f) $z=-2$

(c) $z=-1+i$ (g) $z=4i$

(d) $z=-1-i$ (h) $z=-4i$

10.2.7 Suppose that n is any positive integer, and suppose $x_0=\cos(2\pi/n)+i\sin(2\pi/n)$. Prove that the nth roots of unity are precisely $x_0, x_0^2, x_0^3, \ldots, x_0^n$ $[=1]$. Represent these roots geometrically.

10.2.8 Suppose that n is any positive integer. Show that the polynomial $f(x)=x^n-1$ has at most two real zeros. For what values of n does $f(x)$ have exactly one real zero; exactly two real zeros? Explain.

10.2.9 Prove that if x is any nth root of unity (n is a positive integer), then $1/x$ is also an nth root of unity.

10.2.10 Show that if $\cos\theta+i\sin\theta\neq 0$, then $1/(\cos\theta+i\sin\theta)=\cos(-\theta)+i\sin(-\theta)$.

10.2.11 Use the result in Prob. 10.2.10 to show that De Moivre's theorem is true for all *negative* integers, provided, of course, that $\cos\theta+i\sin\theta\neq 0$.

10.2.12 Show that the result of Prob. 10.2.7 is still valid if we replace $x_0=\cos(2\pi/n)+i\sin(2\pi/n)$ by $x_0=\cos(2\pi/n)-i\sin(2\pi/n)$.

10.3 Inverse Trigonometric Functions

In this section, we consider the possibility of defining inverses for the trigonometric functions. We have to proceed cautiously here since, in view of the *periodic* nature of the trigonometric functions, none of the trigonometric functions is a one-to-one function. Thus, we cannot possibly hope to be able to define inverses of trigonometric functions unless we can somehow restore the one-to-one property to these functions. This we can do by *restricting the domains* in an appropriate fashion, as we now proceed to show.

Let us start with the function $y=\sin x$, the graph of which appeared earlier in Figs. 9.18 and 9.24. A glance at the graph of the function sin shows that if we restrict the domain of the function sin to the set of all real numbers x such that $-\pi/2\leq x\leq\pi/2$, then the restricted function is indeed a one-to-one function (see Fig. 10.14). Now, let

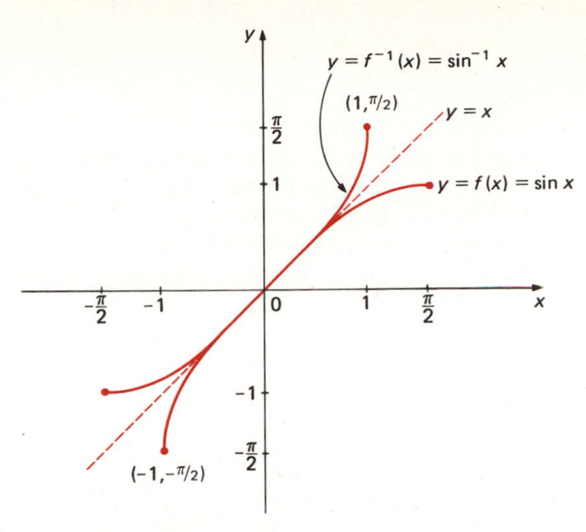

Fig. 10.14 Graph of the function "$\sin^{-1}$" (i.e., graph of $y = \sin^{-1} x$) as obtained from graph of the function "sin"

$$D = \text{set of all real numbers between } -\frac{\pi}{2} \text{ and } \frac{\pi}{2}, \text{ inclusive} \tag{10.62}$$

$$I = \text{set of all real numbers between } -1 \text{ and } 1, \text{ inclusive} \tag{10.63}$$

Then, the function $f: D \to I$ given by $f(x) = \sin x$ is (see Fig. 10.14) *both one-to-one and onto*, and hence, as we have shown in Chaps. 1 and 7, f has an inverse $f^{-1}: I \to D$ given by

$$f^{-1}(\sin x) = x$$

We also recall that the graph of the function f^{-1} [or, equivalently, the graph of the equation $y = f^{-1}(x)$] is simply the reflection (or mirror image) of the graph of f with respect to the line $y = x$, as indicated in Fig. 10.14 (see Prob. 2.1.15). This function f^{-1} is denoted by "$\sin^{-1}$" or "arcsin". We are thus led to the following definition:

> *Let x be any real number such that $-1 \leq x \leq 1$. We say that*
>
> $y = \sin^{-1} x$ (*read as: y equals inverse sine of x*)
>
> *if and only if both of the following conditions hold:*
>
> $x = \sin y$ *and* $-\frac{\pi}{2} \leq y \leq \frac{\pi}{2}$
>
> *This function* $\sin^{-1}$ *has domain I and range D, where* $I = \{x | -1 \leq x \leq 1\}$ *and* $D = \{y | -\pi/2 \leq y \leq \pi/2\}$.

The graph of $y = \sin^{-1} x$ appears in Fig. 10.14. Incidentally, we certainly could have chosen a set D_1 different from D such that the restricted function sin, *with domain D_1*, is a one-to-one function. Indeed, one such choice for D_1 could have been

$$D_1 = \text{set of all real numbers between } -\frac{3\pi}{2} \text{ and } -\frac{\pi}{2}, \text{ inclusive}$$

Such a choice, of course, would lead to another definition of $\sin^{-1}$. The main thing to keep in mind here is that whatever restriction we make on the domain of sin, this new (restricted) domain must be such that sin is a one-to-one function in this restricted domain.

Arguing in an analogous fashion, we define the function $\cos^{-1}$ (read inverse cosine) as follows (keep in mind the graph of the cosine function sketched in Figs. 9.19 and 9.25).

> *Let x be any real number such that $-1 \leq x \leq 1$. We say that*
>
> $y = \cos^{-1} x$ (*also written* $y = \arccos x$)
>
> *if and only if both of the following conditions hold:*
>
> $x = \cos y$ *and* $0 \leq y \leq \pi$
>
> *This function* $\cos^{-1}$ *has domain I and range D_2, where* $I = \{x \mid -1 \leq x \leq 1\}$ *and* $D_2 = \{y \mid 0 \leq y \leq \pi\}$.

Once again, the set D_2 given in this definition is *not* the only possible choice which makes the restricted function cos become a one-to-one function. We make a suitable choice (such as D_2) in order to facilitate defining the function $\cos^{-1}$. The graph of $\cos^{-1}$ appears in Fig. 10.15. This graph, as well as the graph given in Fig. 10.14, can of course be obtained by reflecting with respect to the line $y = x$ appropriate portions of the graphs of the corresponding trigonometric functions, as indicated in Fig. 10.14. The student should verify this fact in the following graphs.

Again, arguing as above, we define "$\tan^{-1}$" (read inverse tangent) as follows (keep in mind the graph of the tangent function sketched in Figs. 9.20 and 9.26).

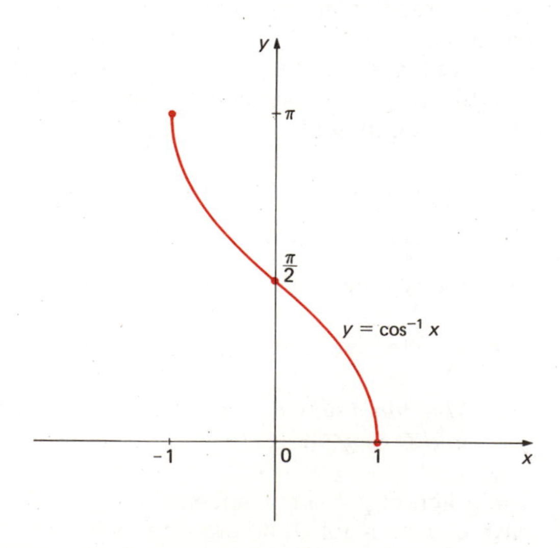

Fig. 10.15 Graph of "$\cos^{-1}$" (i.e., graph of $y = \cos^{-1} x$)

Let x be any real number. We say that

$y = \tan^{-1}x$ *(also written* $y = \arctan x$*)*

if and only if both of the following conditions hold:

$$x = \tan y \quad \text{and} \quad -\frac{\pi}{2} < y < \frac{\pi}{2}$$

This function $\tan^{-1}$ *has domain R (the set of all real numbers) and range* D_3*, where* $D_3 = \{y | -\pi/2 < y < \pi/2\}$.

Again, we could have made another appropriate choice for D_3 for which the restricted tan is a one-to-one function. However, we make a suitable choice (such as D_3) in order to facilitate defining the function $\tan^{-1}$. The graph of $\tan^{-1}$ appears in Fig. 10.16.

The definitions of the remaining inverse trigonometric functions are similar, and are given in the exercises.

Example 10.16 Evaluate the following: (*a*) $\sin^{-1} \frac{1}{2}$; (*b*) $\sin^{-1}(-\frac{1}{2})$; (*c*) $\cos^{-1}(\sqrt{2}/2)$; (*d*) $\cos^{-1}(-\sqrt{2}/2)$; (*e*) $\tan^{-1}0$.

Solution (*a*) let $y = \sin^{-1} \frac{1}{2}$. Then, by definition,

$$\sin y = \frac{1}{2} \quad \text{and} \quad -\frac{\pi}{2} \le y \le \frac{\pi}{2}$$

Now, since $\sin y$ is positive, y cannot possibly lie in the fourth quadrant, and hence y must lie in the first quadrant (recall that, by definition, $-\pi/2 \le y \le \pi/2$). Thus, $\sin y = \frac{1}{2}$ and y lies in the first quadrant. Hence $y = \pi/6$ (see the table at the end of Sec. 9.1). Thus, $\sin^{-1} \frac{1}{2} = \pi/6$.

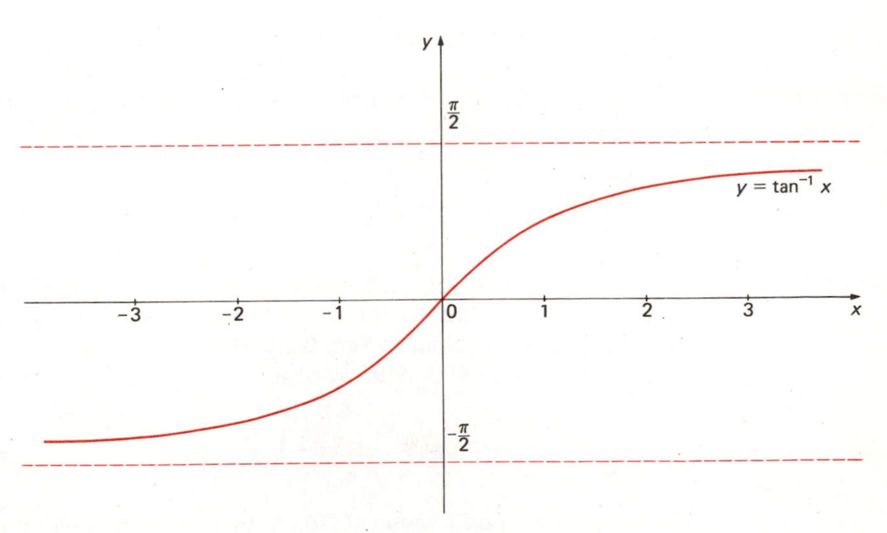

Fig. 10.16 Graph of "$\tan^{-1}$" (i.e., graph of $y = \tan^{-1} x$)

(*b*) let $y = \sin^{-1}(-1/2)$. Then, by definition,

$$\sin y = -\frac{1}{2} \qquad \text{and} \qquad -\frac{\pi}{2} \leq y \leq \frac{\pi}{2}$$

Now, since $\sin y$ is negative, y cannot possibly lie in the first quadrant, and hence $-\pi/2 \leq y \leq 0$ (recall that, by definition, $-\pi/2 \leq y \leq \pi/2$). Therefore, $\sin y = -1/2$ and $-\pi/2 \leq y \leq 0$. Hence, $y = -\pi/6$, and thus $\sin^{-1}(-1/2) = -\pi/6$.

(*c*) Let $y = \cos^{-1}(\sqrt{2}/2)$. Then, by definition,

$$\cos y = \frac{\sqrt{2}}{2} \qquad \text{and} \qquad 0 \leq y \leq \pi$$

Arguing as above, we see that y lies in the first quadrant and, moreover, $\cos y = \sqrt{2}/2$. Hence $y = \pi/4$, and thus $\cos^{-1}(\sqrt{2}/2) = \pi/4$.

(*d*) Let $y = \cos^{-1}(-\sqrt{2}/2)$. Then, by definition,

$$\cos y = -\frac{\sqrt{2}}{2} \qquad \text{and} \qquad 0 \leq y \leq \pi$$

and hence y lies in the second quadrant. Now, the only real number y which lies between $\pi/2$ and π with $\cos y = -\sqrt{2}/2$ is $3\pi/4$. Hence $y = 3\pi/4$. Thus

$$\cos^{-1}\left(-\frac{\sqrt{2}}{2}\right) = \frac{3\pi}{4}$$

(*e*) Let $y = \tan^{-1} 0$. Then, by definition,

$$\tan y = 0 \qquad \text{and} \qquad -\frac{\pi}{2} < y < \frac{\pi}{2}$$

Hence, $y = 0$, and thus $\tan^{-1} 0 = 0$.

Example 10.17 Find all real numbers x, where $0 \leq x < 2\pi$, which satisfy the trigonometric equation

$$\cos^2\frac{x}{2} + \cos 2x = -\frac{1}{2} \tag{10.64}$$

Solution Let us first express both of the trigonometric functions on the left side of (10.64) in terms of the trigonometric functions of x, using the fundamental identities we established in Sec. 9.2. Indeed, using Eqs. (9.69) and (9.71), we easily see that (10.64) is equivalent to

$$\frac{1 + \cos x}{2} + (2\cos^2 x - 1) = -\frac{1}{2} \tag{10.65}$$

Now, multiplying both sides of (10.65) by 2 and simplifying, we obtain

$$4\cos^2 x + \cos x = 0$$

or, $\cos x(4\cos x + 1) = 0$. Hence,

$$\cos x = 0 \qquad \text{or} \qquad \cos x = -1/4$$

Now, the only solutions of $\cos x = 0$, where $0 \le x < 2\pi$, are

$$x = \frac{\pi}{2} \qquad \text{and} \qquad x = \frac{3\pi}{2} \tag{10.66}$$

Moreover, the only solutions of $\cos x = -1/4$, where $0 \le x < 2\pi$, are

$$x = \cos^{-1}(-1/4) \qquad \text{and} \qquad x = \cos^{-1} 1/4 + \pi \tag{10.67}$$

This follows, since the function cos is negative in the *second* and *third* quadrants (observe that, *by definition*, $\pi/2 < \cos^{-1}(-1/4) < \pi$ while $0 < \cos^{-1} 1/4 < \pi/2$). Using Table E, we see that

$$\cos^{-1} 1/4 \approx 1.3177 \tag{10.68}$$

and hence

$$\cos^{-1}(-1/4) = \pi - \cos^{-1} 1/4 \approx \pi - 1.3177 \tag{10.69}$$

Thus, the solutions x of (10.64), where $0 \le x < 2\pi$, are [see Eqs. (10.66) to (10.69)]

$$x = \frac{\pi}{2} \qquad x = \frac{3\pi}{2} \qquad x \approx \pi - 1.3177 \qquad x \approx \pi + 1.3177$$

xample 10.18 Prove that, for all real numbers x such that $-1 \le x \le 1$,

$$\sin^{-1}x + \cos^{-1}x = \frac{\pi}{2} \tag{10.70}$$

olution Let

$$y = \cos^{-1}x \tag{10.71}$$

Then, by definition of the function $\cos^{-1}$,

$$x = \cos y \qquad \text{and} \qquad 0 \le y \le \pi \tag{10.72}$$

Now, by Eq. (9.56), we have

$$\sin\left(\frac{\pi}{2} - y\right) = \cos y \tag{10.73}$$

Hence, by (10.72) and (10.73), we get

$$\sin\left(\frac{\pi}{2} - y\right) = x \tag{10.74}$$

Moreover, since by (10.72), $0 \leq y \leq \pi$, it follows that

$$-\frac{\pi}{2} \leq \frac{\pi}{2} - y \leq \frac{\pi}{2} \tag{10.75}$$

Recalling the definition of $\sin^{-1}$, we easily see that (10.74) and (10.75) are equivalent to

$$\frac{\pi}{2} - y = \sin^{-1} x \tag{10.76}$$

Combining (10.76) and (10.71), we obtain (10.70).

We conclude this section by remarking that inverse trigonometric functions turn out to be quite useful in the study of calculus. They make their appearance in numerous ways in both pure and applied mathematics.

Problem Set 10.3

10.3.1 Evaluate:

(*a*) $\sin^{-1} \frac{\sqrt{3}}{2}$ (*d*) $\tan^{-1}(-\sqrt{3})$

(*b*) $\sin^{-1}\left(-\frac{\sqrt{3}}{2}\right)$ (*e*) $\cos^{-1} 1/2$

(*c*) $\tan^{-1}\sqrt{3}$ (*f*) $\cos^{-1}(-1/2)$

10.3.2 Evaluate:

(*a*) $\sin(\sin^{-1} x)$ where $-1 \leq x \leq 1$

(*b*) $\cos(\cos^{-1} x)$ where $-1 \leq x \leq 1$

(*c*) $\tan(\tan^{-1} x)$ where x is any real number

10.3.3 Evaluate, *without using tables*, the following.

(*a*) $\sin(\cos^{-1} 0.7)$ (*b*) $\cos[\sin^{-1}(-0.6)]$

(*c*) $\sin(2 \cos^{-1} 12/13)$ [*Hint*: Let $y = \cos^{-1} 12/13$ and use the formula for $\sin 2y$ in Sec. 9.2.]

10.3.4 (*a*) Sketch the graphs of $y = \tan x$ and $y = x$, using the same coordinate system.

(*b*) Find approximate solutions of the equation $\tan x = x$, where $0 \leq x \leq 2\pi$. [*Hint*: Consider the points of intersection of the graphs of $y = \tan x$ and $y = x$.]

10.3.5 Prove the following:

(*a*) $\tan^{-1} 4/3 - \tan^{-1} 1/7 = \pi/4$

(*b*) $\tan^{-1} 1/3 + \tan^{-1} 1/2 = \pi/4$

[*Hint*: For part *a*, let $\alpha = \tan^{-1} 4/3$, $\beta = \tan^{-1} 1/7$, and then evaluate $\tan(\alpha - \beta)$, as given in Sec. 9.2. Use a similar argument for part *b*.]

10.3.6 For each of the following trigonometric equations, find all real numbers x (if any) where $0 \leq x < 2\pi$, such that the equation is satisfied. Use any trigonometric identities that are appropriate.

(*a*) $\sin x = 1/2$

(*b*) $\cos x = -\sqrt{3}/2$

(*c*) $\tan x = -1$

(*d*) $\sin x + \cos x = 1$

(*e*) $\cos 2x = \sin x$

(*f*) $\cos 2x + \cos^2 x = 1/2$

(*g*) $\cos 2x - 2 \sin^2 \frac{x}{2} + 1 = 0$

(*h*) $\sin^2 x - \cos^2 x - \sin 2x = 0$

(*i*) $\cos 2x = \cos x$

(*j*) $\sin^2 x - 2 \cos x + 2 = 0$

(*k*) $\sin 3x \cos x + \cos 3x \sin x = 1$

(*l*) $\cos x + \sqrt{3} \sin x = 2$

(*m*) $\sin x - \cos x + 1 = 0$

(*n*) $\tan x + \sin 2x = 0$

(*o*) $\sin x + \cos 2x = 1$

(*p*) $2 \cos^2 x + \cos x - 1 = 0$

(*q*) $\sin x \cos x = -1/4$

(*r*) $\cos^2 x - \sin^2 x = -1/2$

(*s*) $\sin 2x = \cos x$

(*t*) $\cos 2x - \cos x + 1 = 0$

(*u*) $\sin^2 x + \cos^2 x = 2$

(*v*) $\sin x + \cos x = -3$

(*w*) $\sin x \cos x = 1$

(*x*) $\cos 2x = 2 \sin x + \frac{\sqrt{3}}{2}$

(*y*) $\frac{\tan x}{\cot x} = 0$

(*z*) $\sin x + 1 = 4 \cos x$

10.3.7 Prove that, for all real numbers x such that $-1 \leq x \leq 1$, $\sin(\cos^{-1} x) = \cos(\sin^{-1} x)$. [*Hint*: Use the result in Example 10.18.]

10.3.8 Do the following exist? If so, find their values. Explain.

(*a*) $\sin^{-1} 2$

(*b*) $\cos^{-1} 4$

(*c*) $\tan^{-1}(-1)$

(*d*) $\cot[\cos^{-1}(\pi/2)]$

(*e*) $\sin^{-1}(\tan \pi)$

(*f*) $\sin(\tan^{-1} \pi)$

(*g*) $\tan(\sin^{-1} \pi)$

(*h*) $\tan^{-1}(\sin \pi)$

(*i*) $\cos(\tan^{-1} 1)$

(*j*) $\tan(\cos^{-1} 1)$

10.3.9 Is the restriction that $-1 \leq x \leq 1$ necessary in Example 10.18? Explain.

10.3.10 Verify that the graphs of $\cos^{-1}$ and $\tan^{-1}$ can be obtained by reflecting with respect to the line $y = x$ appropriate portions of the graphs of cos and tan, as indicated in Fig. 10.14.

10.3.11 Let x be any real number. We say that $y = \cot^{-1} x$ if and only if $x = \cot y$ and $0 < y < \pi$. Sketch the graph of $y = \cot^{-1} x$.

10.3.12 Let x be any real number such that $|x| \geq 1$ (that is, $x \geq 1$ or $x \leq -1$). We say that $y = \sec^{-1} x$ if and only if $x = \sec y$ and $0 \leq y \leq \pi$, but $y \neq \pi/2$. Sketch the graph of $y = \sec^{-1} x$.

10.3.13 Let x be any real number such that $|x| \geq 1$. We say that $y = \csc^{-1} x$ if and only if $x = \csc y$ and $-\pi/2 \leq y \leq \pi/2$, but $y \neq 0$. Sketch the graph of $y = \csc^{-1} x$.

10.4 Summary

Viewing trigonometric functions as functions of angles was seen in this chapter to be convenient for "solving" triangles. In this connection, we established the following fundamental laws:

1. *Law of sines*: $\dfrac{\sin \alpha}{a} = \dfrac{\sin \beta}{b} = \dfrac{\sin \gamma}{c}$

2. *Law of cosines*: $a^2 = b^2 + c^2 - 2bc \cos \alpha$

For any complex number $z = x + yi$ (x, y real), the *polar form* of z is given by $z = r(\cos \theta + i \sin \theta)$. Here $r = \sqrt{x^2 + y^2}$ and θ satisfies the equations $\cos \theta = x/r$ *and* $\sin \theta = y/r$. In this connection, a very important theorem, known as De Moivre's theorem, asserts that

$$[r(\cos \theta + i \sin \theta)]^n = r^n(\cos n\theta + i \sin n\theta)$$

Using De Moivre's theorem, we were then able to show that the solutions of the polynomial equation $x^n - 1 = 0$ are given by

$$x = \cos \frac{2\pi k}{n} + i \sin \frac{2\pi k}{n} \qquad k = 0, 1, 2, \ldots, n-1$$

These solutions all lie on the unit circle C and are equally spaced on the circumference of C.

Finally, inverse trigonometric functions were defined by first restricting the domain of each of the trigonometric functions in such a way that these trigonometric functions become one-to-one functions. For example, the *inverse sine* function (denoted by $\sin^{-1}$ or by $\arcsin x$) was defined as follows:

> *For any real number* x *such that* $-1 \le x \le 1$, $y = \sin^{-1}x$ *if and only if* $x = \sin y$ *and* $-\pi/2 \le y \le \pi/2$.

Similar definitions were given for the other inverse trigonometric functions.

11 Analytic Geometry

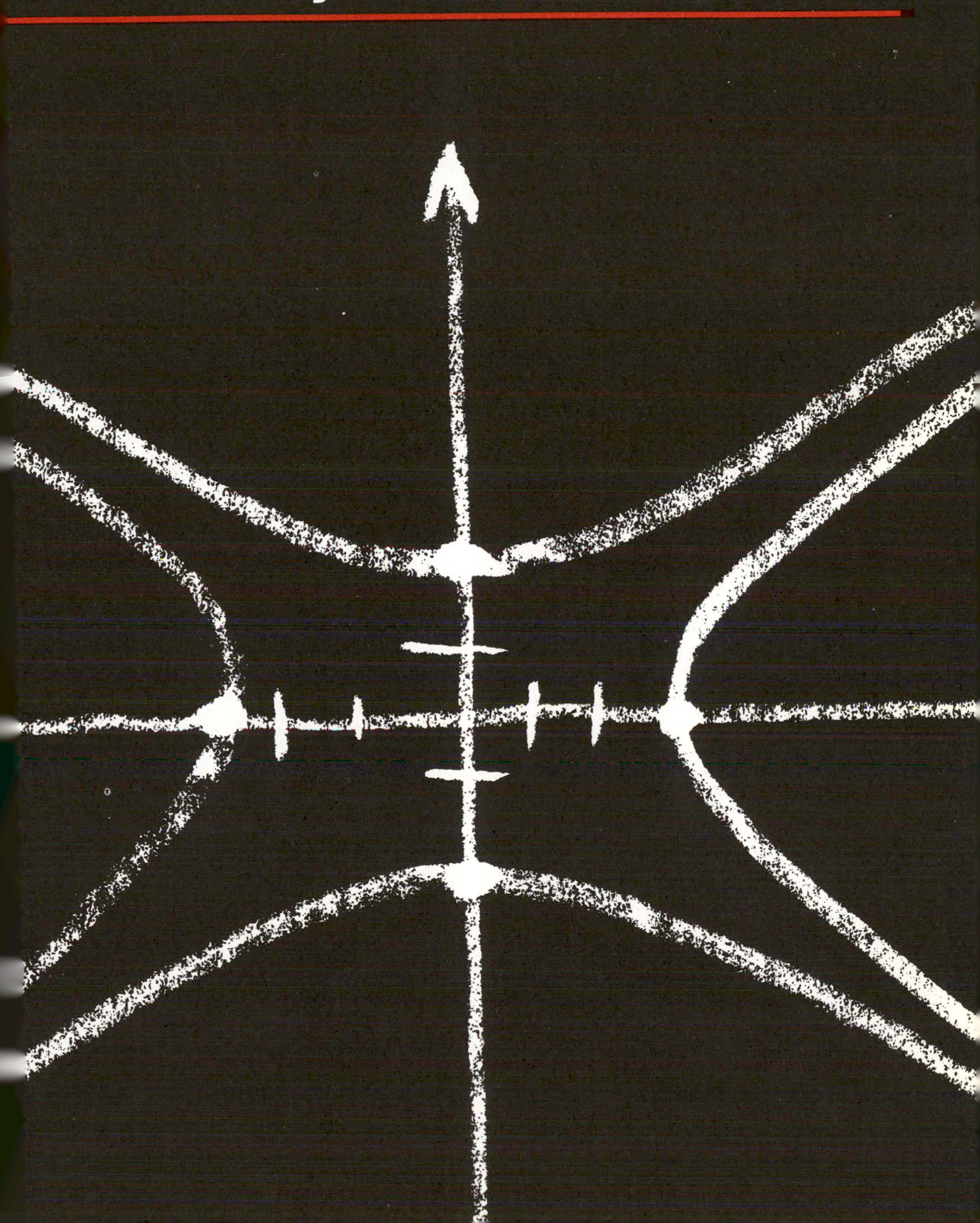

11.1 Straight Lines

Consider for a moment a straight line ℓ, as indicated in Fig. 11.1. If line ℓ does not pass through the origin $\mathcal{O}$, we draw a line ℓ' through $\mathcal{O}$ which is parallel to line ℓ. The directed angle $A\mathcal{O}B$ in standard position indicated in Fig. 11.1 is called the *inclination* of line ℓ. In other words, the *inclination of a straight line* ℓ is the smallest angle θ measured from the positive direction of the x axis to line ℓ in the counterclockwise direction. Thus, each of the two equal angles $A\mathcal{O}B$ and ADC in Fig. 11.1 gives the inclination of the line ℓ. Observe that if θ is the inclination of a given line ℓ, then, as is easily seen from the definition of inclination,

$$0 \leq \theta < \pi \tag{11.1}$$

In other words, the *inclination of any line is always a nonnegative real number which is strictly less than* π. In particular, the x axis and every line parallel to the x axis have inclination zero, while the y axis and every line parallel to the y axis have inclination $\pi/2$. Now, we define the *slope* of a line ℓ as follows (also see Sec. 2.1):

$$\text{Slope of } \ell = \tan\theta \qquad \text{where } \theta = \text{inclination of } \ell \qquad \text{if } \theta \neq \frac{\pi}{2} \tag{11.2}$$

However,

The slope of a line with inclination $\pi/2$ is *not* defined. (11.3)

In other words, *if a line* ℓ *is parallel to the y axis (or is the y axis itself), then the slope of* ℓ *is not defined; otherwise, the slope of* ℓ *is defined to be the tangent of the inclination* θ.

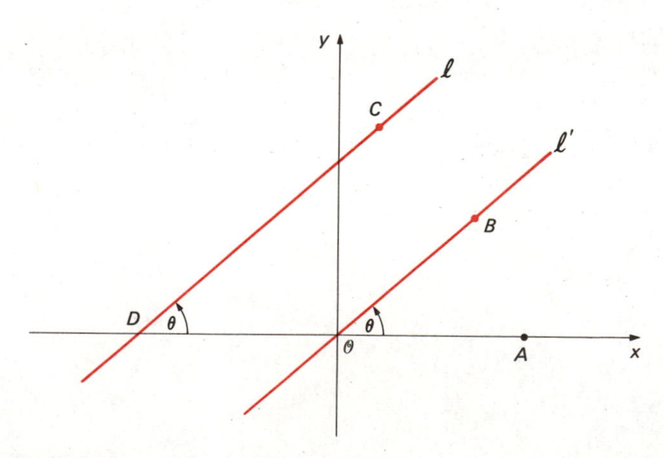

Fig. 11.1 Inclination and slope of a straight line. Line ℓ has inclination θ and slope "tan θ."

How does one find the slope of a line ℓ which is *not* vertical (that is, ℓ is not the y axis and ℓ is not parallel to the y axis)? The answer is as follows: simply find two distinct points $P_1:(x_1,y_1)$ and $P_2:(x_2,y_2)$ on ℓ, and then calculate the fraction $(y_2-y_1)/(x_2-x_1)$. This fraction is the slope of ℓ. To prove this, let us draw a line P_1Q which is parallel to the x axis, and another line P_2Q which is parallel to the y axis, as indicated in Fig. 11.2. Then, by high school geometry, we easily see that (see Fig. 11.2)

$$\text{Angle } QP_1P_2 = \text{angle } ABP_1 = \theta = \text{inclination of } \ell$$

Now, a glance at the right triangle P_1QP_2 shows that the length of $QP_2 = y_2 - y_1$ and the length of $P_1Q = x_2 - x_1$, and hence

$$\tan\theta = \frac{y_2 - y_1}{x_2 - x_1}$$

Since line ℓ has inclination θ, it follows that

$$\text{Slope of } \ell = \tan\theta = \frac{y_2 - y_1}{x_2 - x_1}$$

We have thus shown the following fact, known as the *slope formula* (also see Sec. 2.1):

> The slope of a nonvertical line ℓ passing through the distinct points $P_1:(x_1,y_1)$ and $P_2:(x_2,y_2)$ is given by (11.4)
>
> $$\text{Slope of } \ell = \frac{y_2 - y_1}{x_2 - x_1} \qquad (\textit{slope formula})$$

We now direct our attention to the problem of finding an equation of a line ℓ, assuming that we have enough information available to determine line ℓ. In general, *by an equation of a given line ℓ we mean an equation which is satisfied by the coordinates of all points on ℓ but by no others.*

To begin, suppose that a line ℓ passes through a point $P_1:(x_1,y_1)$ and has slope m. What is an equation for ℓ? To answer this question, we let $P:(x,y)$ be

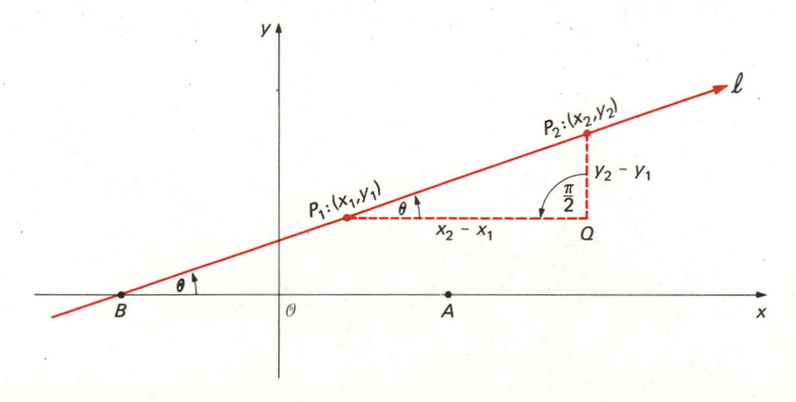

Fig. 11.2 Slope formula: slope of $\ell = (y_2 - y_1)/(x_2 - x_1)$

any point on line ℓ distinct from $P_1; (x_1,y_1)$, as indicated in Fig. 11.3. Then, by the slope formula, we know that

$$\text{Slope of } \ell = \frac{y - y_1}{x - x_1} \tag{11.5}$$

But, by hypothesis, we also have

$$\text{Slope of } \ell = m \tag{11.6}$$

Hence, by (11.5) and (11.6), $y - y_1 = m(x - x_1)$. Observe that this equation is also satisfied by the coordinates of P_1. Moreover, it is easy to see that the above steps are reversible. We thus have the following:

> **Point-slope form:** *An equation of the line ℓ which passes through the point $P_1: (x_1,y_1)$ and which has slope m is* (11.7)
>
> $$y - y_1 = m(x - x_1)$$

Next, suppose that a nonvertical line ℓ passes through two distinct points $P_1: (x_1,y_1)$ and $P_2: (x_2,y_2)$. We wish to find an equation for line ℓ. Thus, suppose that $P: (x,y)$ is any point on ℓ distinct from both P_1 and P_2, as indicated in Fig. 11.4. Then, by the slope formula (11.4), we have

$$\text{Slope of } \ell = \frac{y - y_1}{x - x_1} \qquad [\text{since } P_1: (x_1,y_1) \text{ and } P: (x,y) \text{ lie on } \ell]$$

Moreoever,

$$\text{Slope of } \ell = \frac{y_2 - y_1}{x_2 - x_1} \qquad [\text{since } P_1: (x_1,y_1) \text{ and } P_2: (x_2,y_2) \text{ lie on } \ell]$$

Hence,

$$\frac{y - y_1}{x - x_1} = \frac{y_2 - y_1}{x_2 - x_1}$$

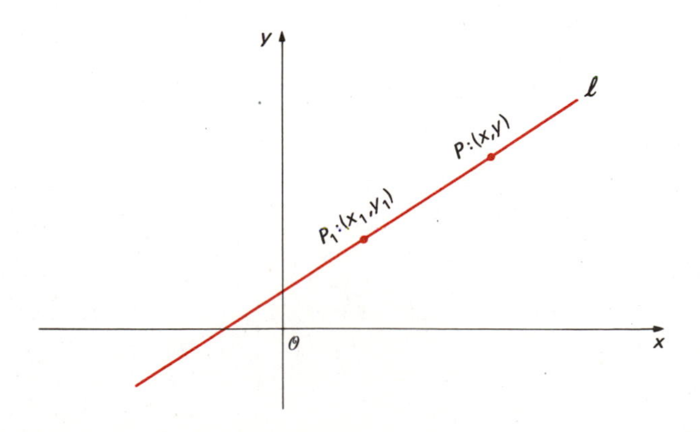

Fig. 11.3 Point-slope form: Equation of ℓ is $y - y_1 = m(x - x_1)$.

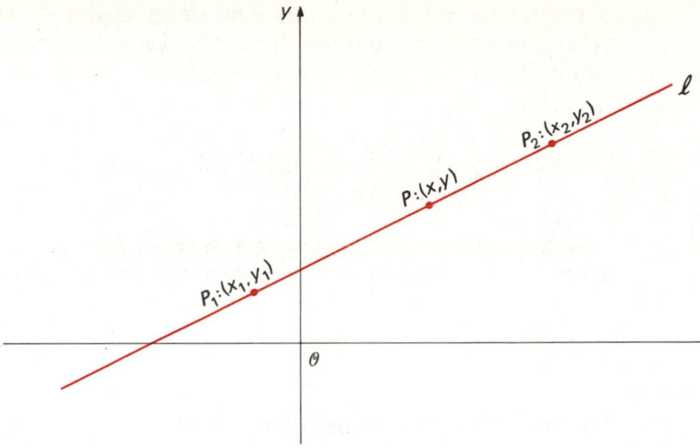

Fig. 11.4 Two-point form: $y - y_1 = \frac{y_2 - y_1}{x_2 - x_1}(x - x_1)$

and thus

$$y - y_1 = \left(\frac{y_2 - y_1}{x_2 - x_1}\right) \cdot (x - x_1)$$

Observe that this equation is also satisfied by the coordinates of both P_1 and P_2. Moreover, it is easy to see that the above steps are reversible. We have thus established the following:

> ***Two-point form:*** *An equation of the nonvertical line ℓ which passes through the two distinct points $P_1: (x_1,y_1)$ and $P_2: (x_2,y_2)$ is*
>
> $$y - y_1 = \left(\frac{y_2 - y_1}{x_2 - x_1}\right) \cdot (x - x_1)$$

emark

If the line ℓ above is vertical, then it is easy to see that $x_1 = x_2$, and that an equation of line ℓ is, in this case, $x = x_1$. (The above formula is not applicable in this case.)

Next, we consider another form, known as the *slope-intercept form.* Thus, suppose that ℓ is a line with slope m and with y intercept b. Then ℓ passes through the point $(0,b)$. Hence, by the point-slope form, an equation of ℓ is

$$y - b = m(x - 0)$$

and thus

$$y = mx + b$$

Hence, we have shown the following:

> ***Slope-intercept form:*** *An equation of a line ℓ with slope m and with y intercept b is*
>
> $$y = mx + b$$

Next, suppose a line ℓ has x intercept a and y intercept b, where $a \neq 0$ and $b \neq 0$. What is an equation for ℓ? Since line ℓ passes through the two distinct points $(a,0)$ and $(0,b)$, we may apply the two-point form we established earlier to get

$$y - 0 = \left(\frac{b-0}{0-a}\right) \cdot (x - a)$$

which simplifies to $bx + ay = ab$. Now, dividing both sides of the last equation by the *nonzero* number ab, we see that the above equation is equivalent to

$$\frac{x}{a} + \frac{y}{b} = 1$$

We have thus shown the following:

> ***Two-intercept form:*** *The equation of a line with x intercept a and with y intercept b, where $a \neq 0$ and $b \neq 0$, is*
>
> $$\frac{x}{a} + \frac{y}{b} = 1$$

A careful examination of the above forms shows that they are all expressible as follows:

$$Ax + By + C = 0 \qquad A, B, C \text{ real, } \textit{not both } A, B \textit{ zero} \tag{11.8}$$

This suggests that perhaps *every* line has an equation of the form (11.8). This is indeed the case, as we now proceed to show.

To begin with, if a line ℓ is a vertical line (that is, ℓ is the y axis or ℓ is parallel to the y axis), then the equation of ℓ is easily seen to be

$$x = k \qquad \text{where } k \text{ is a constant}$$

that is,

$$1 \cdot x + (-k) = 0 \tag{11.9}$$

Clearly, (11.9) is of the form (11.8) (take $A = 1$, $B = 0$, $C = -k$). Next, suppose that ℓ is *not* a vertical line. Then, ℓ has a slope m, say. Moreover, since ℓ is not parallel to the y axis, ℓ intersects the y axis at a point $(0,b)$, say. Hence, by the slope-intercept form, an equation of ℓ is $y = mx + b$, which can be written as

$$mx + (-1)y + b = 0 \tag{11.10}$$

which again is of the form (11.8) (take $A = m$, $B = -1$, $C = b$). Thus, in any case ℓ has an equation of the form (11.8). We have thus shown the following:

> *Every straight line has an equation of the form*
>
> $$Ax + By + C = 0 \qquad A, B, C \textit{ real, not both } A, B \textit{ zero} \tag{11.11}$$
>
> *This form is called the* **standard form**.

Conversely, the graph of equation (11.8) is always a straight line. To prove this, we distinguish two cases.

ase 1 $B = 0$

Then, by (11.8), $A \neq 0$, and hence $x = -C/A$. Clearly, the graph of the equation $x = -C/A$ is a vertical line passing through $(-C/A, 0)$.

ase 2 $B \neq 0$

Then, solving equation (11.8) for y, we get

$$y = \frac{-A}{B}x + \frac{-C}{B} \tag{11.12}$$

Now, recalling the slope-intercept form, we see that the graph of (11.12) is a straight line with slope $-A/B$ and with y intercept $-C/B$. We have thus shown the following:

> The graph of any linear equation
>
> $Ax + By + C = 0$ $\quad$ A, B, C real, not both A, B zero
>
> in standard form is always a straight line. This line has slope $-A/B$ and y intercept $-C/B$ if $B \neq 0$, and is a vertical line passing through $(-C/A, 0)$ if $B = 0$. (11.13)

We now pause to give some examples.

xample 11.1 Find the slope of the line ℓ joining (1,2) and (−3,5).

olution By the slope formula (11.4), we have

$$\text{Slope of } \ell = \frac{5-2}{-3-1} = -\frac{3}{4}$$

xample 11.2 Find an equation of the line ℓ joining (1,2) and (−3,5).

olution By the two-point form, we know that an equation of line ℓ is

$$y - 2 = \left(\frac{5-2}{-3-1}\right) \cdot (x-1)$$

which is equivalent to $-4(y-2) = 3(x-1)$, and $-4y + 8 = 3x - 3$. Simplifying, we have

$$3x + 4y - 11 = 0 \tag{11.14}$$

Observe that (11.14) gives an equation of line ℓ in standard form. Incidentally, we get the same answer if we use the point (−3,5), instead of (1,2), in the above computation.

ample 11.3 Find an equation of the line ℓ with slope −4 and which passes through (−5,−1).

Solution By the point-slope form, we know that an equation of line ℓ is $y - (-1) = -4[x - (-5)]$, and thus $y + 1 = -4x - 20$, which is equivalent to $4x + y + 21 = 0$.

Example 11.4 Find an equation of the line ℓ with slope -5 and with y intercept 7.

Solution By the slope-intercept form, we know that an equation of line ℓ is $y = -5x + 7$, or, equivalently, $5x + y - 7 = 0$.

Example 11.5 Find an equation of the line ℓ with x intercept -4 and with y intercept 3.

Solution By the two-intercept form, we know that an equation of line ℓ is

$$\frac{x}{-4} + \frac{y}{3} = 1$$

or, equivalently,

$$3x - 4y + 12 = 0$$

Example 11.6 Find the slopes (if they exist) of the following lines: (*a*) $x = -1$, (*b*) $y = -5$, and (*c*) $2x - 3y + 1 = 0$.

Solution

(*a*) The line $x = -1$ is a vertical line and hence has no slope.

(*b*) The line $y = -5$ is a horizontal line and hence has slope zero.

(*c*) The equation $2x - 3y + 1 = 0$ is equivalent to $y = \frac{2}{3}x + \frac{1}{3}$ and hence the slope of this line is $\frac{2}{3}$.

Example 11.7 Prove that the points $P_1: (1,2)$, $P_2: (3,-4)$, $P_3: (2,-1)$ are *collinear* (i.e., lie on the same line).

Solution By the slope formula (11.4), we know that

$$\text{Slope of } P_1P_2 = \frac{-4 - 2}{3 - 1} = -3$$

$$\text{Slope of } P_2P_3 = \frac{-1 - (-4)}{2 - 3} = -3$$

Since P_1P_2 and P_2P_3 have the same slope, it follows by high school geometry that P_1, P_2, P_3 must all lie on the same line; that is, they are collinear.

It is instructive to inquire when two nonvertical lines are parallel. Thus suppose that ℓ_1 and ℓ_2 are two nonvertical lines with slopes m_1 and m_2, respectively. Suppose, further, that the inclination of line ℓ_1 is θ_1 and the inclination of line ℓ_2 is θ_2. Then, by definitions of slope and inclination, we have

$$m_1 = \tan \theta_1 \qquad m_2 = \tan \theta_2 \qquad 0 \le \theta_1 < \pi, 0 \le \theta_2 < \pi \qquad (11.1$$

Now, by high school geometry, we know that ℓ_1 is parallel to ℓ_2 if and only if $\theta_1 = \theta_2$ (see Fig. 11.5). Since both θ_1 and θ_2 lie in the first or second quadrant, it is easily seen that $\tan \theta_1 = \tan \theta_2$ if and only if $\theta_1 = \theta_2$. The net result, then, is

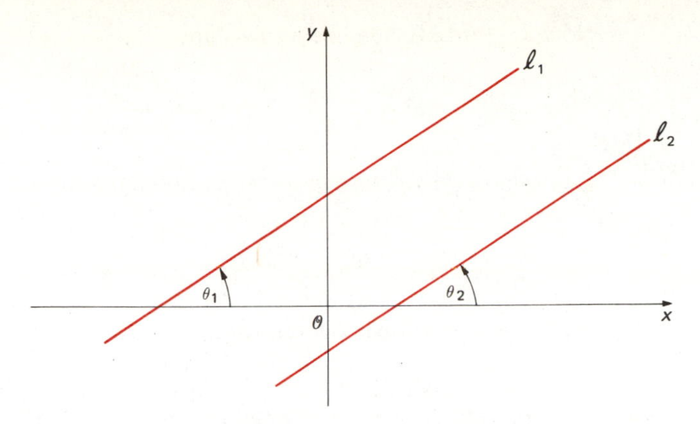

Fig. 11.5 Parallel lines have equal slopes.

that ℓ_1 is parallel to ℓ_2 if and only if $\tan\theta_1 = \tan\theta_2$, that is, if and only if $m_1 = m_2$ [see (11.15)]. We have thus shown the following:

Two nonvertical lines are parallel if and only if they have equal slopes. (11.16)

Next we ask: when are two lines ℓ_1, ℓ_2 perpendicular? To begin with, if the slope m_1 of line ℓ_1 is zero, then ℓ_1 is a horizontal line (i.e., parallel to the x axis) and hence, in this case, ℓ_2 is perpendicular to ℓ_1 if and only if ℓ_2 is vertical (i.e., parallel to the y axis). A similar argument applies if the slope m_2 of ℓ_2 is zero. Thus, we shall now assume that $m_1 \neq 0$ and $m_2 \neq 0$. Now, suppose that θ_1 is the inclination of line ℓ_1, and θ_2 is the inclination of line ℓ_2. Suppose that the lines are labeled such that $\theta_1 > \theta_2$. Then, by high school geometry, ℓ_1 is perpendicular to ℓ_2 if and only if (see Fig. 11.6) the inclination θ_1 of line ℓ_1 is equal to the inclination θ_2 of line ℓ_2 plus $\pi/2$; that is (see Fig. 11.6)

$$\theta_1 = \theta_2 + \frac{\pi}{2}$$

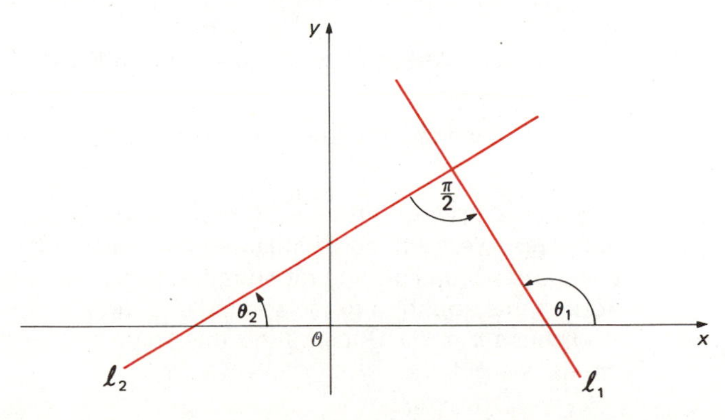

Fig. 11.6 ℓ_1 and ℓ_2 are perpendicular if and only if the product of their slopes is equal to -1.

Now, by Prob. 9.2.6*a*, we know that

$$\tan \theta_1 = \tan\left(\theta_2 + \frac{\pi}{2}\right) = -\cot \theta_2 = -\frac{1}{\tan \theta_2} \tag{11.17}$$

Combining (11.17) with $m_1 = \tan \theta_1$ and $m_2 = \tan \theta_2$, we obtain

$$m_1 = -\frac{1}{m_2} \qquad \text{or} \qquad m_1 m_2 = -1 \tag{11.18}$$

We have thus shown the following:

> Two lines ℓ_1 and ℓ_2, with nonzero slopes m_1 and m_2, are perpendicular if and only if $m_1 m_2 = -1$. If either $m_1 = 0$ or $m_2 = 0$, then ℓ_1 is perpendicular to ℓ_2 if and only if the other slope is undefined. (11.19)

Example 11.8 Which of the following lines are parallel, and which are perpendicular: (*a*) $2x - 3y + 6 = 0$, (*b*) $9x + 6y - 18 = 0$, (*c*) $4x - 6y - 12 = 0$?

Solution By (11.13), we know that (*a*) slope of line $2x - 3y + 6 = 0$ is equal to $2/3$; (*b*) slope of line $9x + 6y - 18 = 0$ is equal to $-3/2$; (*c*) slope of line $4x - 6y - 12 = 0$ is equal to $2/3$. Hence, by (11.16) and (11.19), we conclude that first, the lines in parts *a* and *c* are parallel, then the lines in parts *a* and *b* are perpendicular [since $(2/3)(-3/2) = -1$], and finally the lines in parts *b* and *c* are perpendicular.

Example 11.9 Sketch the graphs of the lines in Example 11.8.

Solution Since two distinct points suffice to determine a straight line, we can draw the graph of each line ℓ by finding any two points on ℓ. Now, observe that

(0,2) and (−3,0) lie on line of Example 11.8*a*

(0,3) and (2,0) lie on line of Example 11.8*b*

(0,−2) and (3,0) lie on line of Example 11.8*c*

The graphs of these lines are sketched in Fig. 11.7.

Example 11.10 Find the point of intersection of the lines (*a*) $2x - 3y + 6 = 0$, (*b*) $9x + 6y - 18 = 0$

Solution In general, to find the point of intersection (if any) of two lines whose equations are given, we try to solve these equations simultaneously (Why?). Now, multiplying both sides of the first equation in our example by 2, we get $4x - 6y + 12 = 0$. Adding this equation to the second equation in our example, we obtain $13x - 6 = 0$, and hence $x = 6/13$. Substituting this value of x in the first equation of the example, we get

$$2(6/13) - 3y + 6 = 0$$

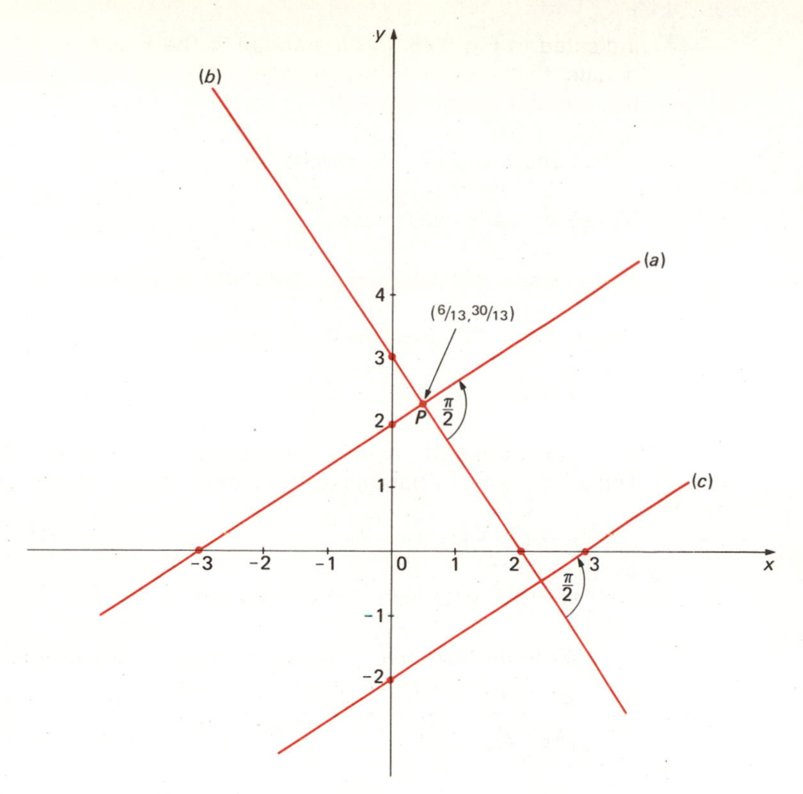

Fig. 11.7 Graphs of the lines (*a*) $2x - 3y + 6 = 0$, (*b*) $9x + 6y - 18 = 0$, (*c*) $4x - 6y - 12 = 0$.

and thus $y = {}^{30}/_{13}$. Hence, the desired point of intersection is $({}^{6}/_{13}, {}^{30}/_{13})$. These lines were sketched in Fig. 11.7. The student should compare $({}^{6}/_{13}, {}^{30}/_{13})$ with the approximate coordinates of the point P in Fig. 11.7.

It should be pointed out that the graphs of $2x - 3y + 6 = 0$ and $4x - 6y - 12 = 0$, being *parallel* lines as we have seen in Example 11.8, have *no* point of intersection. This can also be seen algebraically by observing that *equations in Example 11.8a and c have no common solution.* Indeed, by multiplying the first equation by 2, we obtain $4x - 6y + 12 = 0$, which has *no* solution in common with $4x - 6y - 12 = 0$ (Why?). Thus, we may say that *two lines are parallel if and only if their equations have no common solution.*

We recall that in Eq. (9.49) we established the so-called *distance formula.* For convenience, we restate this formula:

> ***Distance formula:*** *The positive distance (or length) of the line segment joining* $P_1: (x_1, y_1)$ *and* $P_2: (x_2, y_2)$ *is* $\sqrt{(x_2 - x_1)^2 + (y_2 - y_1)^2}$.

There is another formula, known as the *midpoint formula*, which is often useful and which we now would like to establish. Thus suppose that $P_1: (x_1, y_1)$, $P_2: (x_2, y_2)$ are any two given points. What are the coordinates of the midpoint P of the line segment P_1P_2? To answer this question, suppose that $P: (x, y)$, and suppose, as indicated in Fig. 11.8, P_1R is parallel to the x axis, while PQ and P_2R are both parallel to the y axis. Now, since P_1P and PP_2 have equal lengths, it follows, by high school geometry, that

$$\text{Length of } P_1Q = \text{length of } QR \tag{11.20}$$

Moreover, as is easily seen,

$$\text{Length of } P_1Q = x - x_1 \qquad \text{length of } QR = x_2 - x \tag{11.21}$$

Hence, by (11.20) and (11.21), we obtain $x - x_1 = x_2 - x$, or $2x = x_1 + x_2$, and hence

$$x = (x_1 + x_2)/2$$

A similar argument (in which we now draw lines from P, P_2 parallel to the x axis and a line from P_1 parallel to the y axis, etc.) shows that

$$y = (y_1 + y_2)/2$$

We have thus established the following:

Midpoint formula: *The coordinates of the midpoint of the line segment joining* $P_1:(x_1,y_1)$ *and* $P_2:(x_2,y_2)$ *are*

$$\left(\frac{x_1 + x_2}{2}, \frac{y_1 + y_2}{2}\right)$$

(Thus the coordinates of the midpoint are obtained by taking the "average" of the coordinates of the points.)

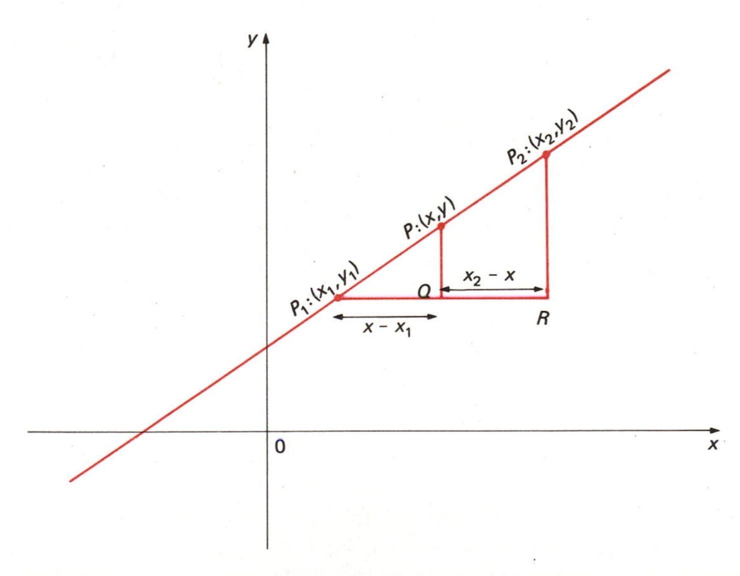

Fig. 11.8 *Midpoint formula*: The midpoint of line segment joining $P_1:(x_1,y_1)$ and $P_2:(x_2,y_2)$ is $\left(\frac{x_1 + x_2}{2}, \frac{y_1 + y_2}{2}\right)$.

These formulas are extremely useful in establishing algebraic proofs of certain well-known theorems from high school geometry as well as proofs of new theorems. The method of proof we are about to give is often called an *analytic proof*.

Example 11.11 Prove analytically that the line segment joining the midpoints of the sides AB and BC in a triangle ABC is parallel to AC and that its length is half the length of the side AC.

Solution In order to be able to apply the formulas we have learned, we must first make up our mind where the coordinate axes will be. Now, although we can choose the coordinate axes arbitrarily, it is certainly to our advantage to choose them such that the computation involved is as simple as possible. So, a reasonable choice for the coordinate axes appears to be as follows: Let the x axis lie along one of the sides, say AB, of the triangle ABC, and let the point A (say) be the origin, as indicated in Fig. 11.9. Let D be the midpoint of the line segment AB, and let E be the midpoint of the line segment BC. Now, since B lies on the x axis, the coordinates of B are $(c,0)$ for some real number c. Suppose that the coordinates of C are (a,b), as indicated in Fig. 11.9. Then, by the midpoint formula, we have

$$\text{Coordinates of } D \text{ are } \left(\frac{c+0}{2}, \frac{0+0}{2}\right) = \left(\frac{c}{2}, 0\right)$$

$$\text{Coordinates of } E \text{ are } \left(\frac{a+c}{2}, \frac{b+0}{2}\right) = \left(\frac{a+c}{2}, \frac{b}{2}\right)$$

Now, if $a = 0$, then it is easy to see that both DE and AC are vertical, and hence parallel. Thus, suppose $a \neq 0$. Then, by the slope formula (11.4), we get

$$\text{Slope of } DE = \frac{b/2 - 0}{(a+c)/2 - c/2} = \frac{b}{a}$$

that is,

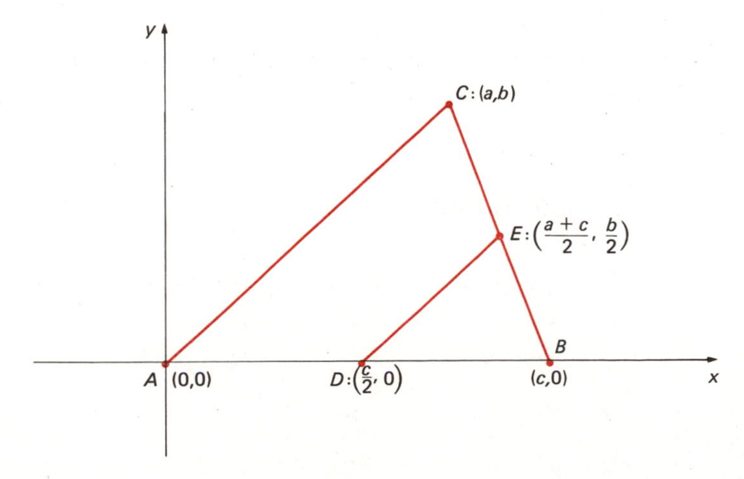

Fig. 11.9 An illustration of an analytic proof

$$\text{Slope of } DE = \frac{b}{a} \tag{11.22}$$

Moreover, by the slope formula again, we have

$$\text{Slope of } AC = \frac{b-0}{a-0} = \frac{b}{a} \tag{11.23}$$

Hence, by (11.22) and (11.23), DE and AC have equal slopes, and therefore by (11.16) DE and AC are parallel. Thus, *in any case*, DE and AC are parallel.

Now, by the distance formula, we have (regardless of the value of a)

$$\begin{aligned}\text{Length of } DE &= \sqrt{\left(\frac{b}{2} - 0\right)^2 + \left(\frac{a+c}{2} - \frac{c}{2}\right)^2} \\ &= \sqrt{\frac{b^2}{4} + \frac{a^2}{4}} \\ &= \tfrac{1}{2}\sqrt{a^2 + b^2}\end{aligned}$$

that is,

$$\text{Length of } DE = \tfrac{1}{2}\sqrt{a^2 + b^2} \tag{11.24}$$

Moreover, by the distance formula again, we have

$$\text{Length of } AC = \sqrt{(a-0)^2 + (b-0)^2} = \sqrt{a^2 + b^2} \tag{11.25}$$

Comparing (11.24) and (11.25), we conclude that the length of the line segment DE is half the length of the line segment AC. This completes the proof.

Remark The result in the above example is well known from high school geometry. It is interesting to remark that analytic proofs often yield simpler proofs than the geometric ones, provided we make a judicious choice of the location of the coordinate axes.

In the remainder of this chapter, we shall be concerned with the study of *conic sections.*

Problem Set 11.1

11.1.1 Find the slopes (if they exist) of the following lines:

(*a*) $y = 0$

(*b*) $y = -5$

(*c*) $x = 0$

(*d*) $y = 2x - 1$

(*e*) $x = 2y - 3$

(*f*) $3x - 4y + 1 = 0$

(*g*) $\frac{x}{2} - \frac{y}{3} = 1$

(*h*) $2x - 3 = 0$

11.1.2 Find the slopes (if they exist) of the lines passing through:

(*a*) $(1,-2)$ and $(-1,4)$
(*b*) $(-3,-2)$ and $(0,0)$
(*c*) $(1,2)$ and parallel to x axis
(*d*) $(-1,-5)$ and parallel to y axis
(*e*) $(1,2)$ and $(1,-3)$
(*f*) $(-1,3)$ and $(-5,3)$

11.1.3 Find the equations for the lines (*a*) passing through $(-5,3)$ with slope -1; (*b*) passing through the origin with slope -5; (*c*) passing through $(1,-4)$ with slope zero; (*d*) passing through $(1,5)$ parallel to the x axis; and (*e*) passing through $(-1,3)$ parallel to the y axis.

11.1.4 Find equations for the lines (*a*) with slope -4 and y intercept 1; (*b*) with slope -4 and x intercept 1; (*c*) with y intercept 3 and parallel to the x axis; and (*d*) with x intercept -2 and parallel to the y axis.

11.1.5 Find equations for the lines passing through these points:

(*a*) $(1,-1)$ and $(-2,4)$
(*b*) $(2,3)$ and $(0,0)$
(*c*) $(5,5)$ and $(5,-8)$
(*d*) $(1,-2)$ and $(4,-2)$

11.1.6 Find equations for the lines

(*a*) with x intercept 1 and y intercept 2
(*b*) with x intercept -1 and y intercept 3
(*c*) with x intercept 1 and y intercept -5
(*d*) with x intercept -4 and y intercept -3

11.1.7 Find an equation of a line passing through $(1,2)$ and with equal intercepts.

11.1.8 Find an equation of a line parallel to $2x-3y+1=0$ and passing through $(-2,-3)$.

11.1.9 Find an equation of a line passing through $(1,-4)$ and perpendicular to $-3x+4y-5=0$.

11.1.10 Sketch the graphs of the lines $3x-2y=6$ and $x+4y=-8$. Find the point of intersection of these lines.

11.1.11 Find an equation of a line passing through the intersection of the lines in Prob. 11.1.10 and with slope -10. Sketch this line.

11.1.12 Find an equation of a line passing through the intersection of $x-y=1$ and $x+y=3$, and parallel to the line $3x+4y+5=0$. Sketch all four lines.

11.1.13 Find an equation of a line passing through the intersection of $2x-y=1$ and $x-2y=-1$, and perpendicular to the line $-2x+5y+3=0$.

11.1.14 Find the slopes, x intercepts, and y intercepts, if they exist, of the following lines. Explain.

(*a*) $x=0$
(*b*) $y=0$
(*c*) $x=-1$
(*d*) $y=4$
(*e*) $y=2x-5$
(*f*) $x=-3y+7$
(*g*) $2x-5=0$
(*h*) $-3y+8=0$
(*i*) $-2x+5y-8=0$
(*j*) $3x-2y=1$

11.1.15 Sketch all the lines in Prob. 11.1.14.

11.1.16 Do the points (3,4), (5,0), (−3,−4) form a right triangle? Explain.

11.1.17 Find an equation of the perpendicular bisector of the line segment joining (1,2) and (−4,7). Also, find the x intercept and y intercept of the bisector, and sketch both lines.

11.1.18 Show that (3,4), (−3,4), (3,−4), (−3,−4) are vertices of a rectangle.

11.1.19 Show that (1,0), (0,1), (−1,0), (0,−1) are vertices of a square.

11.1.20 Show that (0,0), $(1,\sqrt{3})$, (2,0) are vertices of an equilateral triangle.

11.1.21 Show that (1,1), (−1,−1), (2,0), (−2,0) are vertices of a parallelogram.

11.1.22 Find the point of intersection of the diagonals of the parallelogram in Prob. 11.1.21.

11.1.23 Suppose that the line passing through $(0,b)$ and $(a,0)$ is the same as the line passing through (2,−3) and (−5,7). Find a and b.

11.1.24 Find equations for the lines determined by the three sides of a triangle with vertices (1,−1), (2,5), (−3,4). How do you determine whether a given point (x_1,y_1) is outside, on the boundary of, or inside the given triangle?

11.1.25 Find equations for the median lines of the triangle in Prob. 11.1.24.

11.1.26 Find equations for the altitude lines of the triangle in Prob. 11.1.24.

11.1.27 Find the vertices of the triangle whose sides are parts of lines with equations $2x - y = 1$, $2x + y = 3$, and $x + y = 5$. Sketch these lines.

11.1.28 The line $y = mx + 3$ has x intercept −4. Find m.

11.1.29 The point (2,−3) lies on the line $y = mx - 7$. Find m.

11.1.30 The line $x/a - y/5 = 1$ has slope −10. Find a.

11.1.31 (*a*) Find the distance between (1,−7) and (−3,−4).

(*b*) Find the midpoint of the line segment joining (1,−7) and (−3,−4).

11.1.32 The midpoint of (1,3) and (x,y) is (5,−8). Find x and y.

11.1.33 Prove that the following points are collinear, using (1) slope formula and (2) distance formula.

(*a*) (2,−1), (5,1), (8,3)

(*b*) (0,6), (4,0), (−2,9)

11.1.34 Are the following lines concurrent (i.e., pass through the same point)? Explain.

(*a*) $-2x + 5y = 8$, $3x - 2y = 7$, $-x + 8y = 23$

(*b*) $2x - 3y = 4$, $6y = 4x + 9$, $x + y = 1$

11.1.35 Sketch all lines in Prob. 11.1.34. Also, find the x intercept, y intercept, and slope of each of these lines.

11.1.36 Are the following pairs of lines parallel? Are they perpendicular? Explain. Also, sketch all these lines.

(*a*) $3x - 4y = 1$, $-6x + 8y = 3$

(*b*) $2x - 5y = 7$, $-y = \frac{5}{2}x + 1$

(*c*) $2x + 3y = 4$, $3x + 2y = 10$

(*d*) $2x - 7y = 10$, $2x + 7y = 5$

11.1.37 Prove analytically that the diagonals of a parallelogram bisect each other.

11.1.38 Prove analytically that if the diagonals of a parallelogram are equal in length, then the parallelogram is a rectangle.

11.1.39 Prove analytically that the midpoints of the four sides of any quadrilateral form a parallelogram.

11.1.40 Prove analytically that the three medians of any triangle are concurrent.

11.1.41 Prove analytically that the three altitudes of any triangle are concurrent.

11.1.42 Prove analytically that the diagonals of a rhombus intersect at right angles.

11.1.43 Prove that the linear function f given by $f(x) = mx + b$ (m, b real, $m \neq 0$) is strictly increasing if and only if $m > 0$, and is strictly decreasing if and only if $m < 0$ (for the definition of strictly increasing and strictly decreasing, see Sec. 8.2).

11.1.44 Find the distance between the point $P:(4,5)$ and the line ℓ with equation $2x + 3y - 6 = 0$. [*Hint*: Let PQ be the line segment perpendicular to ℓ and meeting ℓ at Q. Then, the slope of PQ is the negative reciprocal of the slope of ℓ. Now, find the equation of PQ, and then solve simultaneously the equations of lines ℓ and PQ. This gives the coordinates of Q. Finally, find the length of PQ.]

11.1.45 Show that the distance between the point $P:(h,k)$ and the line ℓ with equation $Ax + By + C = 0$ is $|Ah + Bk + C|/\sqrt{A^2 + B^2}$. [For a neat proof of this result, see the *American Mathematical Monthly*, Vol. 59 (1952), pp. 242, 248.]

11.2 Conic Sections: Circles

Parabolas, ellipses, and hyperboles are called *conic sections* because they can be obtained by intersecting a plane with a cone. Which of the conic sections we obtain this way depends, of course, on the angle which the plane makes with the axis of the cone. Since a circle may be viewed as a special case of an ellipse, the study of circles will be the starting point of our investigation of conic sections.

Suppose that $P_0:(h,k)$ is any fixed point in the plane and r is any fixed positive real number. We recall from high school geometry that a *circle with center* $P_0:(h,k)$ *and radius* r is simply the set S of all points $P:(x,y)$ in the plane such that the distance between $P_0:(h,k)$ and $P:(x,y)$ is equal to r. Now, recalling the distance formula, we know that if $P:(x,y)$ is any point on S, then

$$\sqrt{(x-h)^2 + (y-k)^2} = r$$

which is equivalent to

$$(x-h)^2 + (y-k)^2 = r^2 \qquad \text{provided } r \geq 0 \tag{11.26}$$

Thus, a point $P:(x,y)$ lies on circle S if and only if (11.26) holds. For this reason, we say that (11.26) is an *equation of the circle* S. We thus have the following:

An equation of a circle S with center (h,k) and radius r, $r > 0$, is

$$(x-h)^2 + (y-k)^2 = r^2 \tag{11.27}$$

This is known as the ***standard form of the equation of a circle.***

Now, if we expand the equation in (11.27), we get

$$x^2 + y^2 - 2hx - 2ky + (h^2 + k^2 - r^2) = 0 \tag{11.28}$$

By the "graph" of an equation $f(x,y) = 0$ we mean the set of all points (a,b) whose coordinates "satisfy" the given equation in the sense that $f(a,b) = 0$. Equation (11.28) suggests that the graph of the equation

$$x^2 + y^2 + Dx + Ey + F = 0 \tag{11.29}$$

where D, E, F are real numbers, is always a circle. Is this true? It turns out that the answer is yes *provided we impose a condition on D, E, F*. What is this condition? To answer this question, let us apply the technique of "completing the square" to the left side of (11.29). Indeed, we have

$$\begin{aligned} x^2 + y^2 + Dx + Ey + F &= (x^2 + Dx \quad) + (y^2 + Ey \quad) + F \\ &= \left(x^2 + Dx + \frac{D^2}{4}\right) + \left(y^2 + Ey + \frac{E^2}{4}\right) + \left(F - \frac{D^2}{4} - \frac{E^2}{4}\right) \\ &= \left(x + \frac{D}{2}\right)^2 + \left(y + \frac{E}{2}\right)^2 + \left(F - \frac{D^2}{4} - \frac{E^2}{4}\right) \end{aligned}$$

and hence (11.29) now becomes

$$\left(x + \frac{D}{2}\right)^2 + \left(y + \frac{E}{2}\right)^2 + \left(F - \frac{D^2}{4} - \frac{E^2}{4}\right) = 0$$

or equivalently,

$$\left(x + \frac{D}{2}\right)^2 + \left(y + \frac{E}{2}\right)^2 = \frac{D^2}{4} + \frac{E^2}{4} - F \tag{11.30}$$

Suppose, for convenience in notation, we let

$$k = \frac{D^2}{4} + \frac{E^2}{4} - F$$

Now, for all real numbers x and y, the left side of (11.30), being the sum of squares, is nonnegative (recall that D and E are also real), and hence

$$k = \frac{D^2}{4} + \frac{E^2}{4} - F \geq 0$$

We have thus shown the following [keep in mind (11.27) and (11.30)]:

The graph of the equation

$$x^2 + y^2 + Dx + Ey + F = 0 \qquad D, E, F \text{ all real}$$

is a circle with center $(-D/2, -E/2)$ and radius $\sqrt{k}$, assuming that (11.31)

$$k = \frac{D^2}{4} + \frac{E^2}{4} - F > 0$$

This circle "degenerates" to the single point $(-D/2, -E/2)$ if $k = 0$, while there is no graph (i.e., the graph is the empty set) if $k < 0$.

Example 11.12 Find an equation of the circle S which passes through $(-1,1)$, $(1,-1)$, $(0,2)$.

Solution Let the desired equation be

$$x^2 + y^2 + Dx + Ey + F = 0 \tag{11.32}$$

Now, since $(-1,1)$, $(1,-1)$, and $(0,2)$ are all on S, the coordinates of each of these three points must satisfy (11.32). Substituting these coordinates in (11.32), we obtain

$$(-1)^2 + 1^2 + D(-1) + E(1) + F = 0$$

$$1^2 + (-1)^2 + D(1) + E(-1) + F = 0$$

$$0^2 + 2^2 + D(0) + E(2) + F = 0$$

These equations, in turn, are equivalent to

$$-D + E + F = -2 \tag{11.33}$$

$$D - E + F = -2 \tag{11.34}$$

$$2E + F = -4 \tag{11.35}$$

Subtracting (11.34) from (11.33), we obtain

$$-2D + 2E = 0 \tag{11.36}$$

Subtracting (11.35) from (11.34), we get

$$D - 3E = 2 \tag{11.37}$$

Now, (11.36) shows at once that $D = E$. Substituting this in (11.37), we obtain $-2E = 2$, and hence

$$E = -1$$

Hence, we also have

$$D = -1$$

Finally, substituting these values in (11.33), say, we obtain

$$F = -2$$

Finally, substituting the above values in (11.32), we conclude that

$$x^2 + y^2 - x - y - 2 = 0 \tag{11.38}$$

is an equation of our circle. Observe that, by (11.31), the circle described in (11.38) has center

$$\left(-\frac{D}{2}, -\frac{E}{2}\right) = \left(\frac{1}{2}, \frac{1}{2}\right)$$

and radius

$$\sqrt{\frac{D^2}{4} + \frac{E^2}{4} - F} = \sqrt{\frac{5}{2}} = \frac{\sqrt{10}}{2}$$

Problem Set 11.2

11.2.1 Find an equation of a circle such that the

(*a*) center is (0,0) and radius is 4

(*b*) center is (1,3) and radius is 2

(*c*) center is (−1,−5) and radius is 3

(*d*) center is (1,−2) and radius is 7

11.2.2 Find an equation of the circle which passes through (0,0), (1,−2), and (7,5).

11.2.3 Show that there are two circles passing through (1,0) and (0,2), each of radius 5. Find equations for these circles and sketch both circles.

11.2.4 Find an equation of the circle with center (−1,2) and passing through (1,4).

11.2.5 Find an equation of the circle with center on the x axis and passing through (3,2) and (6,0).

11.2.6 Sketch the graph of each of the following circles. In each case, find the radius and the center.

(*a*) $x^2 + y^2 = 49$

(*b*) $x^2 + y^2 - 2x = 3$

(*c*) $x^2 + y^2 + 4y = 12$

(*d*) $x^2 + y^2 - 6x + 2y = 26$

(*e*) $3x^2 + 3y^2 - 6x + 8y + 3 = 0$

11.3 Conic Sections: Parabolas

In this section we consider another type of a conic section, called a *parabola*. Thus, suppose that d is any fixed line and F is any fixed point not lying on d. This fixed line d is called the *directrix*, while the fixed point F is called the *focus* By a *parabola with focus F and directrix d,* we mean the set of all points P in the plane containing F and d such that the distance from P to F is equal to the positive length of the perpendicular line segment from P to d. In order to obtain an *equation of a parabola*, that is, an equation satisfied by the coordinates of all points on the parabola but by no others, we must first agree on a choice for the coordinate axes. Although such a choice can be made in an arbitrary fashion, we naturally would like to choose the coordinate axes in such a way as to make the desired equation assume a form which is as simple as possible. With this in mind, we choose the x axis to be the line through the focus F which is perpendicular to the directrix d and which meets d at G. We also agree to let the origin, $\mathcal{O}$, be the midpoint of the line segment FG, as indicated in Fig. 11.10. In view of this choice of coordinate axes, the coordinates of F are now of the form $(p,0)$, and the equation of the directrix d is $x = -p$ (thus, $2p$ denotes the distance from the focus F to the directrix d; see Fig. 11.10).

Now, suppose $P:(x,y)$ is any point on our parabola, and suppose PQ is perpendicular to the directrix d and meets d at Q, as indicated in Fig. 11.10. Then, by *definition* of a parabola, we have

Length of FP = length of QP

Applying the distance formula, this becomes

$$\sqrt{(x-p)^2 + (y-0)^2} = |x+p| \tag{11.39}$$

which is equivalent to

$$(x-p)^2 + y^2 = (x+p)^2 \tag{11.40}$$

Simplifying, we see that (11.39) is equivalent to

$$y^2 = 4px$$

and hence

An equation of the parabola with focus $(p,0)$ and directrix $x = -p$ *is* $y^2 = 4px$, $(p \neq 0)$. (11.41)

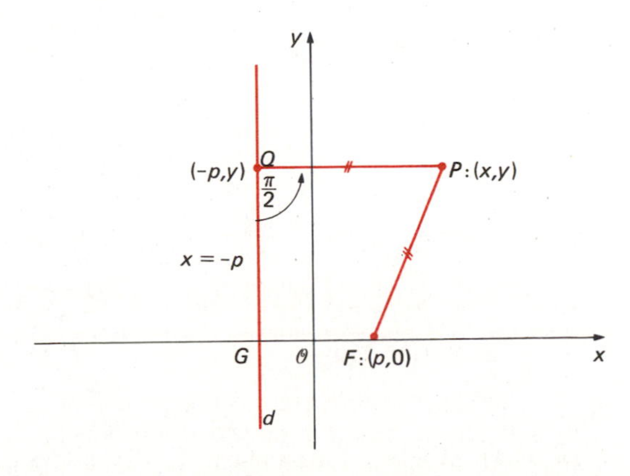

Fig. 11.10 A choice for the coordinate axes in the case of a parabola ($p > 0$)

We also have shown the following:

> The graph of the equation
>
> $$y^2 = 4px \qquad (p \neq 0) \tag{11.42}$$
>
> is a parabola with focus $F:(p,0)$ and with directrix d, where d is the vertical line whose equation is $x = -p$. By definition, the origin, $\mathcal{O}$, is the vertex of this parabola.

It is easy to check that the graph of the parabola described in (11.42) is as indicated in Figs. 11.11 and 11.12, depending upon whether $p > 0$ or $p < 0$, respectively. Observe also that the graph of the equation $y^2 = 4px$ is *symmetric with respect to the x axis* in the sense that if (x,y) lies on the graph, so does $(x,-y)$. For this reason, we say that the x axis is an *axis of symmetry* for the graph of the parabola whose equation is $y^2 = 4px$.

Now, if in our choice of coordinate axes which led to (11.42) we interchange the roles of the x axis and y axis, we obtain the following result:

> The graph of the equation
>
> $$x^2 = 4py \qquad (p \neq 0) \tag{11.43}$$
>
> is a parabola with focus $F:(0,p)$ and with directrix d, where d is the horizontal line whose equation is $y = -p$. By definition, the origin, $\mathcal{O}$, is the vertex of this parabola.

It is readily verified that the graph of the parabola described in (11.43) is as indicated in Figs. 11.13 and 11.14, depending upon whether $p > 0$ or $p < 0$, respectively. Observe that the graph of the equation $x^2 = 4py$ is *symmetric with respect to the y axis* in the sense that if (x,y) lies on the graph, so does $(-x,y)$. For this reason, we say that the y axis is an *axis of symmetry* for the graph of the parabola whose equation is $x^2 = 4py$.

How are the equations of the above parabolas affected if the vertex of the parabola is the point $P_0:(h,k)$ instead of the origin? The answer is given below, and is easily justified, using arguments similar to those above. Thus, analogous to (11.42), we have

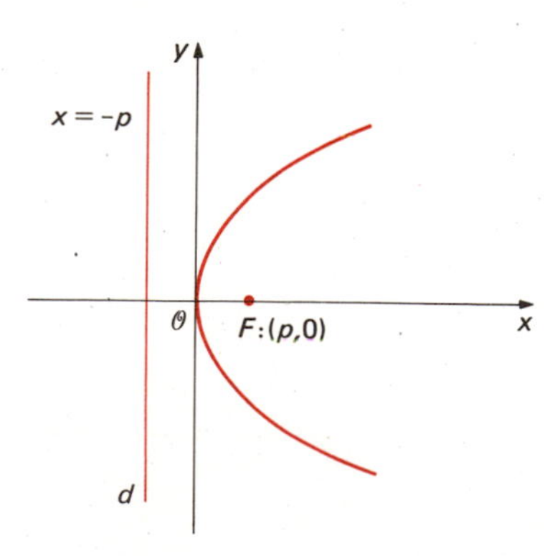

Fig. 11.11 Graph of the parabola $y^2 = 4px$, if $p > 0$

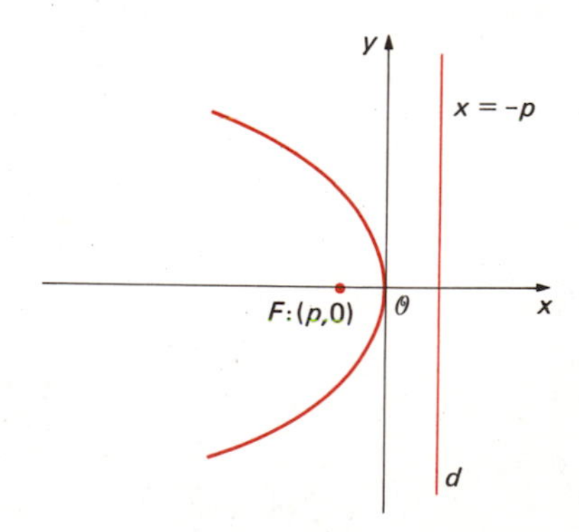

Fig. 11.12 Graph of the parabola $y^2 = 4px$, if $p < 0$

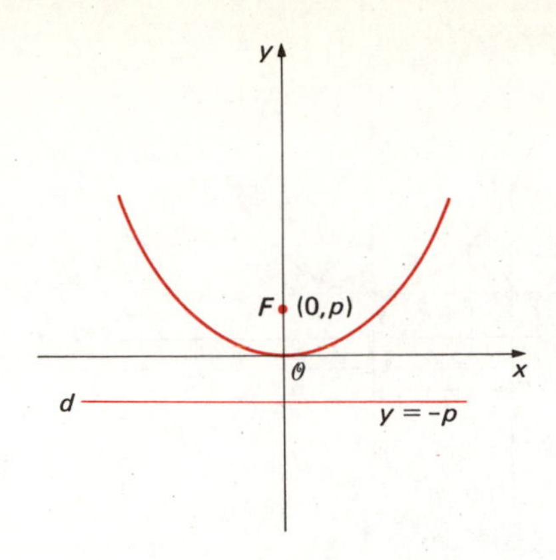

Fig. 11.13 Graph of the parabola $x^2 = 4py$, if $p > 0$

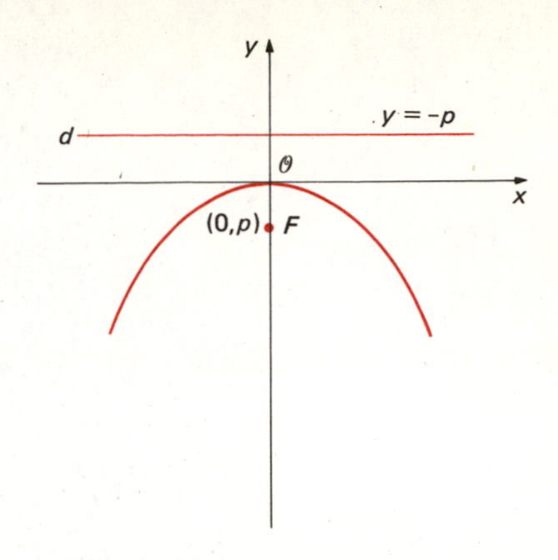

Fig. 11.14 Graph of the parabola $x^2 = 4py$, if $p < 0$

The graph of the equation

$$(y - k)^2 = 4p(x - h) \qquad (p \neq 0) \tag{11.44}$$

is a parabola with focus $F:(h + p,k)$ and with directrix d, where d is the vertical line whose equation is $x = h - p$. The vertex of this parabola is $P_0:(h,k)$. (See Fig. 11.15.)

Analogous to (11.43), we have

The graph of the equation

$$(x - h)^2 = 4p(y - k) \qquad (p \neq 0) \tag{11.45}$$

is a parabola with focus $F:(h,k + p)$ and with directrix d, where d is the horizontal line whose equation is $y = k - p$. The vertex of this parabola is $P_0:(h,k)$. (See Fig. 11.16.)

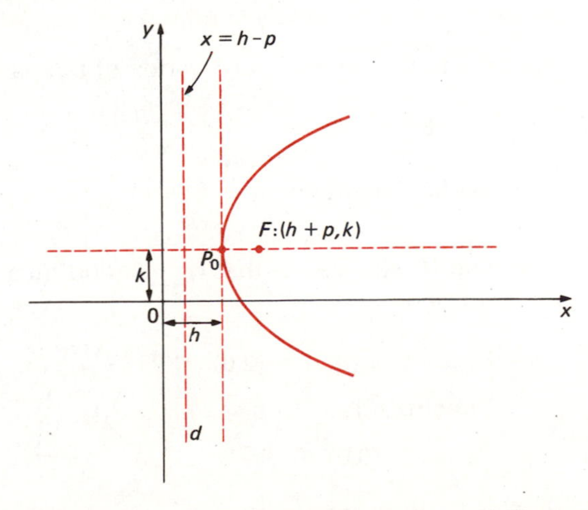

Fig. 11.15 Graph of the parabola $(y - k)^2 = 4p(x - h)$, if $p > 0$. The parabola opens to the left if $p < 0$.

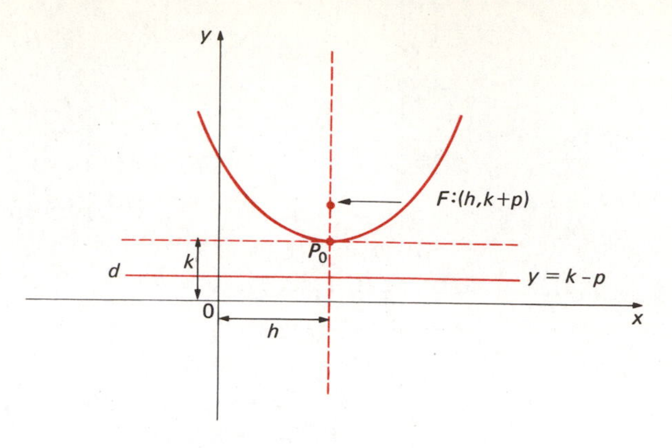

Fig. 11.16 Graph of the parabola $(x-h)^2 = 4p(y-k)$, if $p > 0$. The parabola opens downward if $p < 0$.

Expanding and reordering the terms in Eqs. (11.41) to (11.45), which are know as the standard forms of the equations of a parabola, suggest the following:

> The graph of every equation of the form
>
> $Ax^2 + Bx + Cy + D = 0 \qquad A \neq 0;\ A,\ B,\ C,\ D$ real
>
> or of the form
>
> $Ay^2 + Bx + Cy + D = 0 \qquad A \neq 0;\ A,\ B,\ C,\ D$ real (11.4
>
> is always a parabola. This parabola may "degenerate" to two parallel or coincident lines (consider $x^2 - x = 0$ or $y^2 = 0$) or to an empty set (consider $x^2 + 1 = 0$).

It turns out that (11.46) is, in fact, true, and can easily be shown, using the "completing-the-square" technique, as illustrated by the following examples.

Example 11.13 Find the focus, directrix, and vertex of the parabola given by

$$y^2 = 8x \qquad (11.4$$

Also, sketch the graph.

Solution Recalling (11.42), we see that $4p = 8$, and thus $p = 2$. Hence, by (11.42), we have the following:

Focus F is $(p,0) = (2,0)$

Directrix d is $x = -p = -2$

Vertex is $(0,0)$

Keeping in mind Fig. 11.11, we see that the graph of the parabola (11.47) is as sketched in Fig. 11.17.

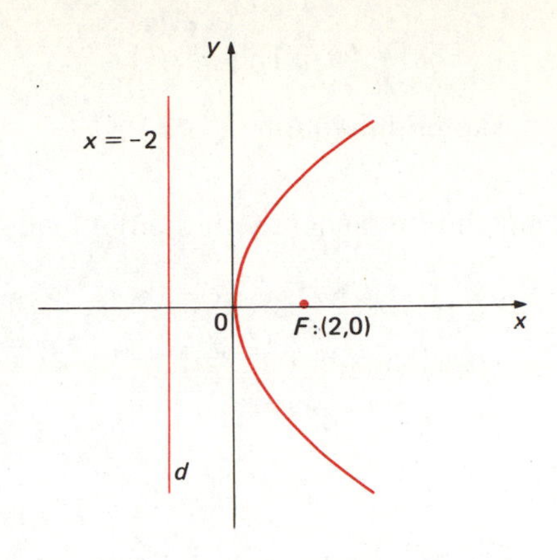

Fig. 11.17 Graph of $y^2 = 8x$

Example 11.14 Find the focus, directrix, and vertex of the parabola given by

$$x^2 = -12y \tag{11.48}$$

Also, sketch the graph.

Solution Recalling (11.43), we see that $4p = -12$ and thus $p = -3$. Hence, by (11.43), we have the following:

Focus F is $(0,p) = (0,-3)$

Directrix d is $y = -p = 3$

Vertex is $(0,0)$

Keeping in mind Fig. 11.14, we easily see that the graph of the parabola (11.48) is as sketched in Fig. 11.18.

Example 11.15 Find the focus, directrix, and vertex of the parabola given by

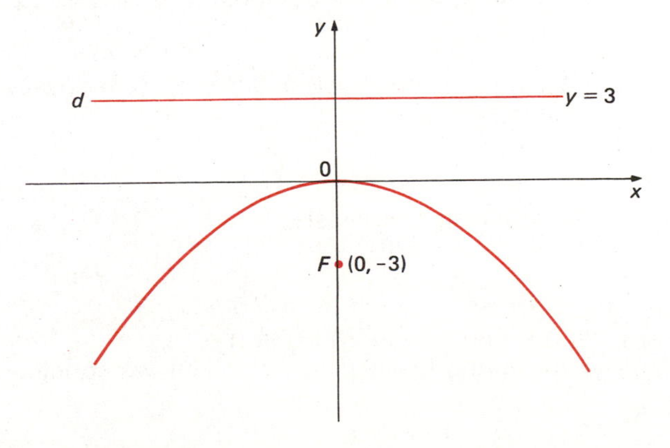

Fig. 11.18 Graph of $x^2 = -12y$

$$y = 2x^2 - 6x + 1 \tag{11.49}$$

Also, sketch the graph.

Solution Completing the square on the right side of (11.49), we obtain

$$\begin{aligned} y &= 2(x^2 - 3x \qquad\qquad) + 1 \\ &= 2[x^2 - 3x + (-3/2)^2] + 1 - 2(-3/2)^2 \\ &= 2(x - 3/2)^2 - 7/2 \end{aligned}$$

that is,

$$y = 2(x - 3/2)^2 - 7/2 \tag{11.50}$$

This, of course, reminds us of the form in (11.45). Keeping (11.45) in mind, we rewrite (11.50) in the following equivalent form:

$$(x - 3/2)^2 = 1/2(y + 7/2) \tag{11.51}$$

Now, identifying (11.51) with the equation given in (11.45), we get

$$h = 3/2 \qquad k = -7/2 \qquad 4p = 1/2 \qquad \text{(and hence } p = 1/8\text{)} \tag{11.52}$$

Therefore, by (11.45), we have the following:

Focus F is $(h,k + p) = (3/2, -7/2 + 1/8) = (3/2, -27/8)$

Directrix d is $y = k - p = -7/2 - 1/8 = -29/8$

Vertex is $P_0 : (h,k) = (3/2, -7/2)$

The graph of the parabola given in (11.50), or, equivalently (11.49), is sketched in Fig. 11.19.

Example 11.16 Find an equation of a parabola with vertex (2,1) and focus (5,1).

Solution Keeping in mind (11.44) and Fig. 11.15, we easily see that our desired equation is of the form

$$(y - k)^2 = 4p(x - h) \tag{11.53}$$

Moreover, by hypothesis,

$$h = 2 \qquad k = 1 \qquad p = 5 - 2 = 3$$

Substituting these values in (11.53), we conclude that our desired equation is

$$(y - 1)^2 = 12(x - 2)$$

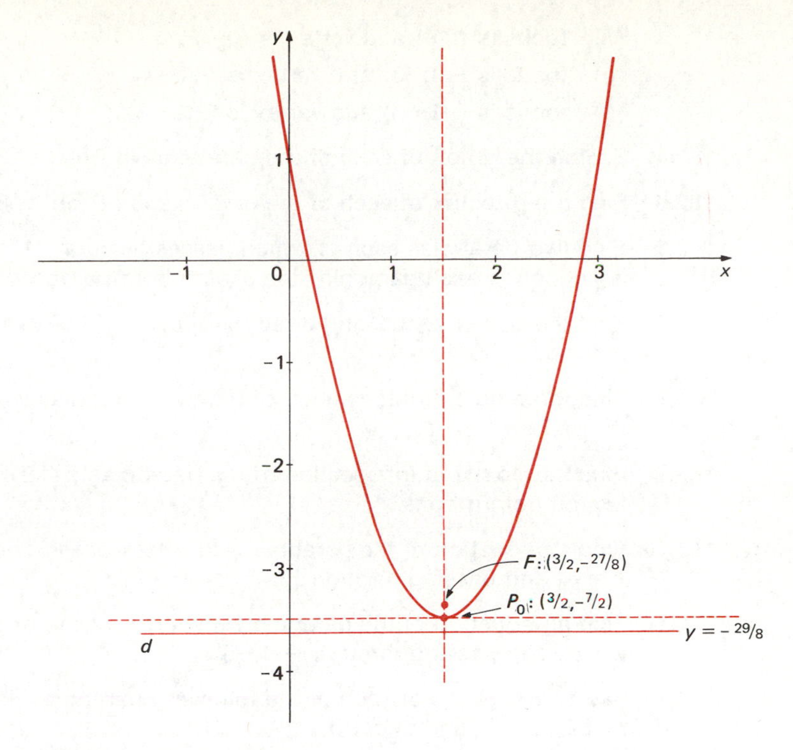

Fig. 11.19 Graph of $y = 2x^2 - 6x + 1$

Problem Set 11.3

11.3.1 Find an equation of a parabola such that the

(*a*) focus is (1,0) and directrix is $x = -1$

(*b*) focus is (0,1) and directrix is $y = -1$

(*c*) focus is (1,2) and directrix is $x = -1$

(*d*) focus is (2,1) and directrix is $y = -1$

(*e*) focus is (3,−5) and directrix is $x = 1$

(*f*) focus is (3,−5) and directrix is $y = -1$

(*g*) focus is (3,−5) and directrix is $x = 5$

(*h*) focus is (3,−5) and directrix is $y = -7$

11.3.2 Sketch the graph of each of the parabolas in Prob. 11.3.1.

11.3.3 Find an equation of a parabola such that the

(*a*) focus is (2,0) and vertex is (1,0)

(*b*) focus is (0,2) and vertex is (0,1)

(*c*) focus is (0,0) and vertex is (2,0)

(*d*) focus is (0,0) and vertex is (−2,0)

(*e*) focus is (0,0) and vertex is (0,2)

(f) focus is (0,0) and vertex is (0,−2)

(g) focus is (−1,−3) and vertex is (−1,−4)

(h) focus is (−1,−3) and vertex is (−2,−3)

11.3.4 Sketch the graph of each of the parabolas in Prob. 11.3.3.

11.3.5 Find the directrix of each of the parabolas in Prob. 11.3.3.

11.3.6 Find two parabolas each of which passes through (1,2), (−1,−1), (0,1). Find an equation for each parabola, and sketch both parabolas.

11.3.7 Find the vertex, focus, and directrix of each of the two parabolas you obtained in Prob. 11.3.6.

11.3.8 Find the points of intersection of the two parabolas $y = x^2$ and $x = y^2$. Sketch the graphs.

11.3.9 Find the points of intersection of the parabolas $y = 8 - x^2$ and $y = -10 + x^2$. Sketch both graphs.

11.3.10 Find an equation of the parabola with vertex at the origin, with focus on the x axis, and passing through (5,4).

11.3.11 Find an equation of the parabola with vertex at the origin, with focus on the y axis, and passing through (−4,−5).

11.3.12 Sketch the graph of each of the following parabolas. In each case, find the focus, vertex, and directrix.

(a) $x^2 = 8y$

(b) $y^2 = -16x$

(c) $x^2 + 4y = 0$

(d) $y^2 + 8x = 0$

(e) $x^2 - 2x + y = 0$

(f) $y^2 - 2y + x = 0$

(g) $y = 2x^2 - 4x + 10$

(h) $x = 2y^2 - 4y + 10$

(i) $4x^2 - 8x + 6y + 1 = 0$

(j) $4y^2 - 8y + 6x + 1 = 0$

11.3.13 Show that the equations $x = t^2$ and $y = 2t$, where t is any real number, represent a parabola. Find the focus, vertex, and directrix of this parabola. Also, sketch the graph. [*Hint*: Eliminate t.]

11.3.14 Show that the equations $x = 2t$ and $y = t^2$, where t is any real number, represent a parabola. Find the focus, vertex, and directrix of this parabola. Also, sketch the graph. [*Hint*: Eliminate t.]

11.4 Conic Sections: Ellipses

In this section, we study another type of a conic section, known as the ellipse. Suppose F_1 and F_2 are two fixed points, and suppose that the distance between F_1 and F_2 is $2c$. Suppose, further, that a is any positive real number such that $a > c > 0$. We define an *ellipse with foci F_1 and F_2* to be the set of all points P in the plane such that

$$\text{Length of } PF_1 + \text{length of } PF_2 = 2a \tag{11.54}$$

In order to obtain an *equation of an ellipse*, that is, an equation which is satisfied by the coordinates of all points on the ellipse but by no others, we must first decide on a choice for the coordinate axes. Naturally, we seek a choice of these axes in such a way as to obtain an equation which is as simple as possible. For this reason, we choose the x axis to be the line joining the two foci F_1 and F_2, and we choose the origin, $\mathcal{O}$, to be the midpoint of the line segment F_1F_2, as indicated in Fig. 11.20. Now, suppose that $P:(x,y)$ is any point on our ellipse. Then, by the distance formula, we have (see Fig. 11.20)

$$\text{Length of } PF_1 = \sqrt{[x-(-c)]^2+(y-0)^2} = \sqrt{(x+c)^2+y^2} \tag{11.55}$$

$$\text{Length of } PF_2 = \sqrt{(x-c)^2+(y-0)^2} = \sqrt{(x-c)^2+y^2} \tag{11.56}$$

Hence, substituting (11.55) and (11.56) in (11.54), we get

$$\sqrt{(x+c)^2+y^2} + \sqrt{(x-c)^2+y^2} = 2a \tag{11.57}$$

and we now proceed to simplify (11.57). Subtracting the second radical on the left side of (11.57) from both sides, we obtain

$$\sqrt{(x+c)^2+y^2} = 2a - \sqrt{(x-c)^2+y^2} \tag{11.58}$$

Now, squaring both sides of (11.58), we get

$$(x+c)^2+y^2 = 4a^2 + [(x-c)^2+y^2] - 4a\sqrt{(x-c)^2+y^2} \tag{11.59}$$

Solving for the radical on the right side of (11.59), we obtain

$$(x+c)^2+y^2-4a^2-(x-c)^2-y^2 = -4a\sqrt{(x-c)^2+y^2}$$

which, after simplifying, becomes

$$cx - a^2 = -a\sqrt{(x-c)^2+y^2} \tag{11.60}$$

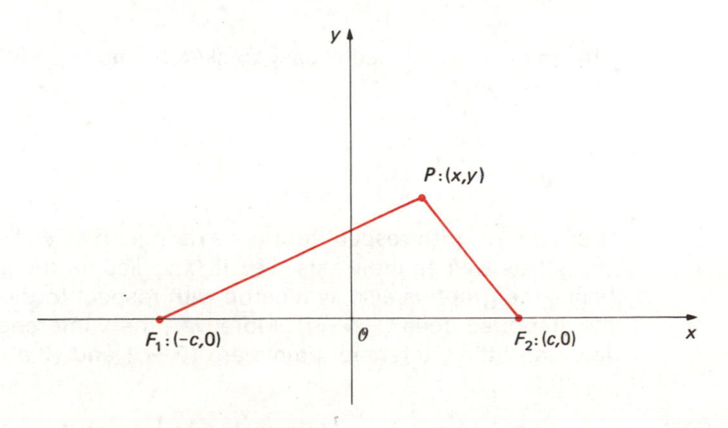

Fig. 11.20 A choice of axes for an ellipse

Squaring both sides of (11.60), we get

$$c^2x^2 + a^4 - 2a^2cx = a^2(x-c)^2 + a^2y^2$$
$$= a^2x^2 + a^2c^2 - 2a^2cx + a^2y^2$$

and hence, simplifying again, we obtain

$$c^2x^2 + a^4 = a^2x^2 + a^2c^2 + a^2y^2$$

or

$$(a^2 - c^2)x^2 + a^2y^2 = a^4 - a^2c^2 = a^2(a^2 - c^2) \tag{11.61}$$

Thus, dividing both sides of (11.61) by $a^2(a^2 - c^2)$, we get

$$\frac{x^2}{a^2} + \frac{y^2}{a^2 - c^2} = 1 \tag{11.62}$$

Now, recall that by hypothesis $a > c > 0$, and hence $a^2 - c^2$ is *positive*. In view of this, we may write

$$a^2 - c^2 = b^2 \qquad \text{(where } b \text{ is some real number)}$$

If we do so, then (11.62) becomes

$$\frac{x^2}{a^2} + \frac{y^2}{b^2} = 1 \qquad (b^2 = a^2 - c^2 < a^2;\ a,\ b,\ c \text{ nonzero real numbers}) \tag{11.63}$$

It can also be verified that the above steps are reversible, and hence we have the following:

> An equation of the ellipse with foci $F_1:(-c,0)$ and $F_2:(c,0)$ and with $PF_1 + PF_2 = 2a$ for all points P on the graph, is (11.64)
>
> $$\frac{x^2}{a^2} + \frac{y^2}{b^2} = 1 \qquad (b^2 = a^2 - c^2 < a^2;\ a,\ b,\ c \text{ nonzero real numbers})$$

The graph of an ellipse is easy to sketch (see Fig. 11.21). Indeed, the graph of the equation

$$\frac{x^2}{a^2} + \frac{y^2}{b^2} = 1 \tag{11.65}$$

is symmetric with respect to the x axis [i.e., if (x,y) lies on the graph, so does $(x,-y)$] as well as the y axis [i.e., if (x,y) lies on the graph, so does $(-x,y)$], and hence the graph is also symmetric with respect to the origin [i.e., if (x,y) lies on the graph, so does $(-x,-y)$]. Moreover, the x intercept points are $(-a,0)$ and $(a,0)$, and the y intercept points are $(0,-b)$ and $(0,b)$. Also, by (11.65),

$$\frac{x^2}{a^2} = 1 - \frac{y^2}{b^2} \le 1 \qquad \text{and} \qquad \frac{y^2}{b^2} = 1 - \frac{x^2}{a^2} \le 1$$

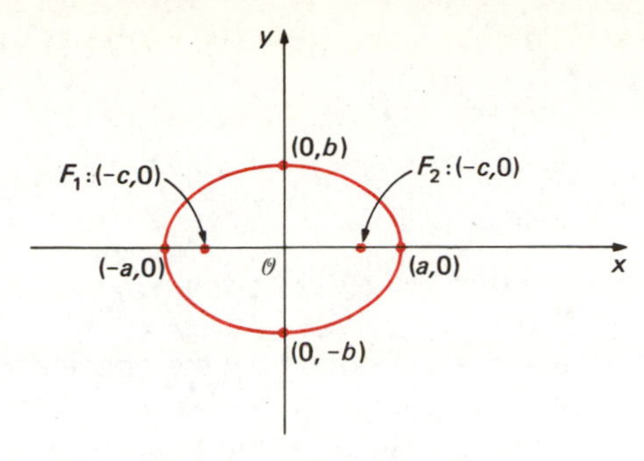

Fig. 11.21 Graph of the ellipse $x^2/a^2 + y^2/b^2 = 1$ $(a^2 > b^2)$

and hence, for all points (x,y) on the ellipse (11.65), we have

$$-a \leq x \leq a \qquad \text{and} \qquad -b \leq y \leq b \tag{11.66}$$

Keeping the above facts in mind, we easily see that the graph of the ellipse (11.65) is as indicated in Fig. 11.21. We call the points $(-a,0)$, $(a,0)$ the *vertices*, and the point $(0,0)$ the *center*, of the ellipse. Moreover, the line on which the vertices, foci, and center lie is called the *major axis*. The line perpendicular to the major axis and passing through the center is called the *minor axis*. Thus, for the ellipse given in (11.65), the major axis is the x axis and the minor axis is the y axis (see Fig. 11.21). We summarize our results in the following:

The graph of the equation

$$\frac{x^2}{a^2} + \frac{y^2}{b^2} = 1 \qquad (a^2 > b^2;\ a,\ b \text{ nonzero real numbers}) \tag{11.67}$$

is an ellipse with the following properties:

Vertices: $(-a,0)$, $(a,0)$

Foci: $(-\sqrt{a^2 - b^2},0)$, $(\sqrt{a^2 - b^2},0)$

Center: $(0,0)$

Major axis: x axis (or, $y = 0$)

Minor axis: y axis (or, $x = 0$)

Now, if in our choice of coordinate axes which led to (11.67) we interchange the roles of the x axis and y axis, we obtain the following result [compare with (11.67)]:

The graph of the equation

$$\frac{y^2}{a^2} + \frac{x^2}{b^2} = 1 \qquad (a^2 > b^2;\ a,\ b \text{ nonzero real numbers}) \tag{11.68}$$

is an ellipse with the following properties:

Vertices: $(0,-a)$, $(0,a)$

Foci:	$(0, -\sqrt{a^2 - b^2})$, $(0, \sqrt{a^2 - b^2})$
Center:	$(0,0)$
Major axis:	y axis (or, $x = 0$)
Minor axis:	x axis (or $y = 0$)

The graph of the ellipse given by $y^2/a^2 + x^2/b^2 = 1$, where $a^2 > b^2$, appears in Fig. 11.22.

How are the equations of the above ellipses affected if the center of the ellipse is the point $P_0:(h,k)$ instead of the origin? The answer is given below, and is easily justified, using arguments similar to the ones above. Thus, analogous to (11.67), we have

The graph of the equation

$$\frac{(x-h)^2}{a^2} + \frac{(y-k)^2}{b^2} = 1 \qquad (a^2 > b^2;\ a,\ b \text{ nonzero};\ a,\ b,\ h,\ k \text{ real}) \tag{11.69}$$

is an ellipse with the following properties:

Vertices:	$(h - a,k)$, $(h + a,k)$
Foci:	$(h - \sqrt{a^2 - b^2},k)$, $(h + \sqrt{a^2 - b^2},k)$
Center:	(h,k)
Major axis:	the line $y = k$
Minor axis:	the line $x = h$

Moreover, analogous to (11.68), we have:

The graph of the equation

$$\frac{(y-k)^2}{a^2} + \frac{(x-h)^2}{b^2} = 1 \qquad (a^2 > b^2;\ a,\ b \text{ nonzero};\ a,\ b,\ h,\ k \text{ real}) \tag{11.70}$$

is an ellipse with the following properties:

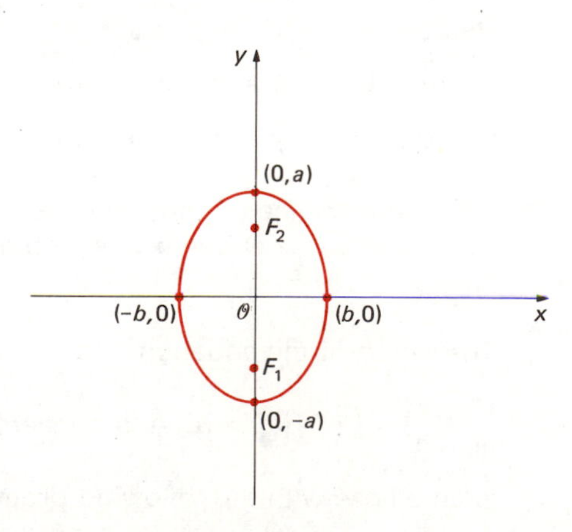

Fig. 11.22 Graph of the ellipse $y^2/a^2 + x^2/b^2 = 1$ $(a^2 > b^2)$

Vertices:	$(h,k-a)$, $(h,k+a)$
Foci:	$(h,k-\sqrt{a^2-b^2})$, $(h,k+\sqrt{a^2-b^2})$
Center:	(h,k)
Major axis:	the line $x=h$
Minor axis:	the line $y=k$

The student should sketch graphs illustrating the results given in (11.69) and (11.70).

Expanding and reordering the terms in each of Eqs. (11.67) to (11.70), which are known as the standard forms of the equations of an ellipse, suggest the following:

> The graph of every equation of the form
>
> $$Ax^2+By^2+Cx+Dy+E=0 \tag{11.71}$$
>
> where A, B, C, D, E are any real numbers such that A and B are both nonzero and have the same signs, is always an ellipse. This ellipse may "degenerate" to a circle (consider $x^2+y^2-1=0$), or to a single point (consider $x^2+2y^2=0$), or to an empty set (consider $x^2+2y^2+1=0$).

It turns out that (11.71) is, in fact, true, and can easily be shown using the completing-the-square technique, as illustrated in the following examples.

xample 11.17 Find the vertices, foci, center, major axis, and minor axis of the ellipse given by

$$\frac{x^2}{25}+\frac{y^2}{16}=1 \tag{11.72}$$

Also, sketch the graph.

olution Recalling (11.67), we see that $a=5$, $b=4$, and hence $\sqrt{a^2-b^2}=3$. Therefore, by (11.67), we have

Vertices:	$(-5,0)$, $(5,0)$	Major axis:	x axis
Foci:	$(-3,0)$, $(3,0)$	Minor axis:	y axis
Center:	$(0,0)$		

The graph of the ellipse (11.72) is sketched in Fig. 11.23.

xample 11.18 Find the vertices, foci, center, major axis, and minor axis of the ellipse given by

$$9x^2+4y^2+18x-16y-11=0 \tag{11.73}$$

Also, sketch the graph.

olution In order to put (11.73) in the form (11.69) or (11.70), whichever happens to be appropriate, we first complete the squares on the left side of (11.73), as indicated below.

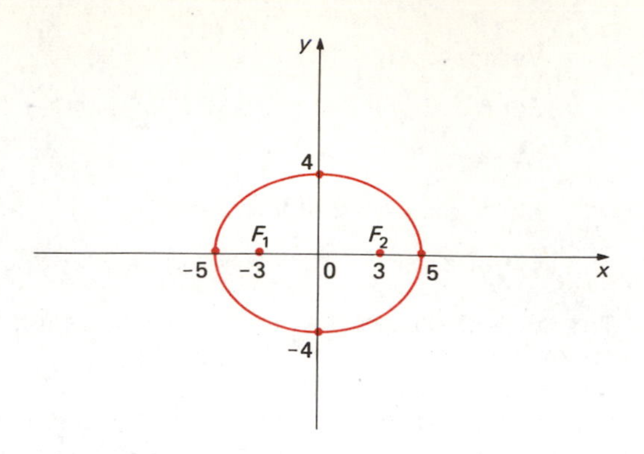

Fig. 11.23 Graph of $x^2/25 + y^2/16 = 1$

$$
\begin{aligned}
9x^2 + 4y^2 + 18x - 16y - 11 &= 9(x^2 + 2x) + 4(y^2 - 4y) - 11 \\
&= 9(x^2 + 2x + 1) + 4(y^2 - 4y + 4) - 11 - 9 - 16 \\
&= 9(x + 1)^2 + 4(y - 2)^2 - 36
\end{aligned}
$$

We have thus shown that (11.73) is equivalent to $9(x+1)^2 + 4(y-2)^2 - 36 = 0$, which, when divided by 36, is seen to be equivalent to

$$\frac{(y-2)^2}{3^2} + \frac{(x+1)^2}{2^2} = 1 \tag{11.74}$$

Now, it is easy to see that (11.74) is of the form given in (11.70), where $a = 3$, $b = 2$, $h = -1$, $k = 2$, and hence $\sqrt{a^2 - b^2} = \sqrt{5}$. Hence, by (11.70), we have:

Vertices:	$V_1: (-1,-1)$, $V_2: (-1,5)$
Foci:	$F_1: (-1,2 - \sqrt{5})$, $F_2: (-1,2 + \sqrt{5})$
Center:	$C: (-1,2)$
Major axis:	the line $x = -1$
Minor axis:	the line $y = 2$

The graph of the ellipse (11.73) is sketched in Fig. 11.24.

Example 11.19 Find an equation of the ellipse with vertices $V_1: (4,1)$, $V_2: (4,9)$, and with foci $F_1: (4,2)$, $F_2: (4,8)$. Also, sketch the graph.

Solution Since the vertices and foci lie on the vertical line $x = 4$, we know that the major axis of our ellipse is the line $x = 4$ itself, and hence the desired equation has the form given in (11.70). It also follows from (11.70) that $h = 4$ and, moreover, the desired equation is

$$\frac{(y-k)^2}{a^2} + \frac{(x-4)^2}{b^2} = 1 \qquad (a^2 > b^2) \tag{11.75}$$

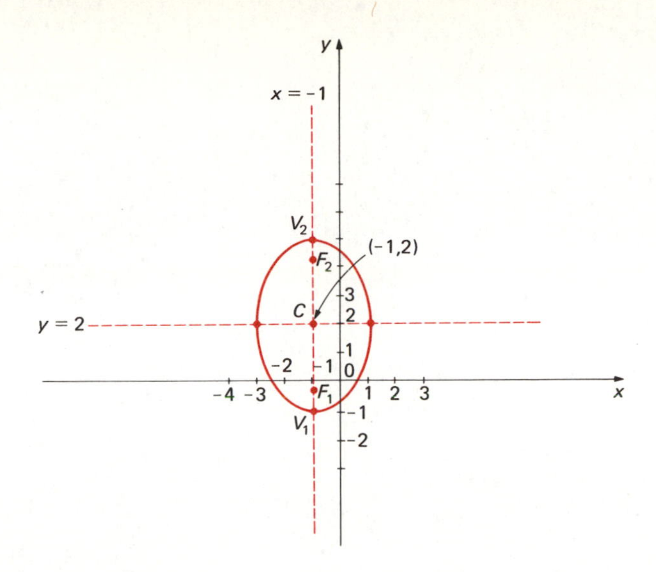

Fig. 11.24 Graph of the ellipse $9x^2 + 4y^2 + 18x - 16y - 11 = 0$

Now, by (11.70), we have:

$$\begin{aligned} &\text{Vertices:} \quad (4,k-a),\ (4,k+a) \\ &\text{Foci:} \quad (4,k-\sqrt{a^2-b^2}),\ (4,k+\sqrt{a^2-b^2}) \end{aligned} \tag{11.76}$$

But, by hypothesis, we also have:

$$\text{Vertices:} \quad (4,1),\ (4,9) \qquad \text{Foci:} \quad (4,2),\ (4,8) \tag{11.77}$$

Identifying (11.76) and (11.77), we get

$$k - a = 1 \qquad k + a = 9 \tag{11.78}$$

$$k - \sqrt{a^2 - b^2} = 2 \qquad k + \sqrt{a^2 - b^2} = 8 \tag{11.79}$$

Now, adding the two equations in (11.78) gives $2k = 10$, and hence $k = 5$. Therefore, by (11.78), $a = 4$. Moreover, subtracting the equations in (11.79) gives $\sqrt{a^2 - b^2} = 3$, and hence $a^2 - b^2 = 9$. Thus, $b^2 = a^2 - 9 = 7$. The net result, then, is $h = 4$, $k = 5$, $a = 4$, and $b = \sqrt{7}$ (since $b > 0$, by definition). Substituting these values in (11.75), we see that our desired equation is

$$\frac{(y-5)^2}{4^2} + \frac{(x-4)^2}{(\sqrt{7})^2} = 1 \tag{11.80}$$

The graph of (11.80) appears in Fig. 11.25.

We conclude this section with the following remark.

Remark A circle may be viewed as an ellipse in which $a^2 = b^2$ [see (11.69)]. In other words, we may view a circle as the "limiting position" of an ellipse in which the two foci become identical to each other. For, in this case (11.69) and (11.27) are equivalent.

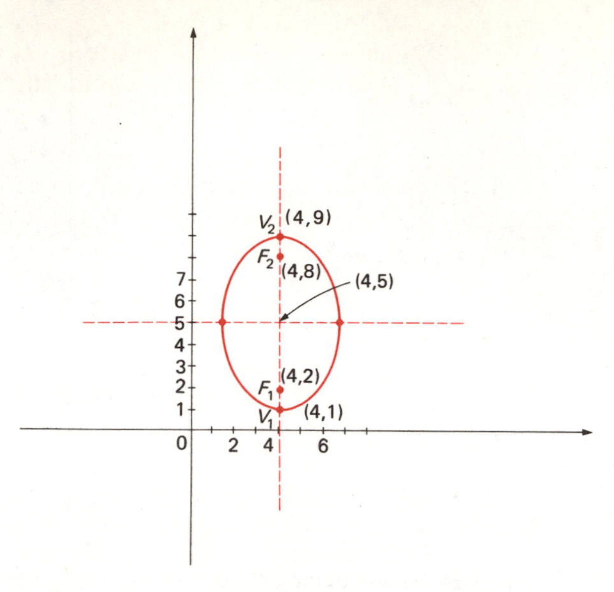

Fig. 11.25 Graph of an ellipse with vertices V_1: (4,1), V_2: (4,9) and foci F_1: (4,2), F_2: (4,8)

Problem Set 11.4

11.4.1 Find an equation of an ellipse such that the

(*a*) foci are (2,0), (−2,0) and vertices are (3,0), (−3,0)

(*b*) foci are (0,2), (0,−2) and vertices are (0,3), (0,−3)

(*c*) foci are (4,2), (4,10) and vertices are (4,0), (4,12)

(*d*) foci are (6,−2), (10,−2) and vertices are (5,−2), (11,−2)

11.4.2 Sketch the graph of each ellipse in Prob. 11.4.1.

11.4.3 Find the center, major axis, and minor axis, of each ellipse in Prob. 11.4.1.

11.4.4 Find an equation of the ellipse with center at the origin, with foci on a coordinate axis, and passing through (−5,0), (3,−3). Sketch the graph.

11.4.5 Find an equation of the ellipse with center at the origin, with foci on a coordinate axis, and passing through (0,5), (−3,−3). Sketch the graph.

11.4.6 Find the vertices, foci, major axis, and minor axis of the ellipse in Prob. 11.4.4.

11.4.7 Find the vertices, foci, major axis, and minor axis of the ellipse in Prob. 11.4.5.

11.4.8 Sketch the graph of each of the following ellipses. In each case, find the vertices, foci, center, major axis, and minor axis.

(*a*) $4x^2 + 9y^2 = 36$

(*b*) $9x^2 + 4y^2 = 36$

(*c*) $x^2 + 4y^2 - 24y + 20 = 0$

(*d*) $4x^2 + y^2 - 24x + 20 = 0$

(*e*) $4x^2 + 9y^2 - 8x - 36y - 24 = 0$

(*f*) $16x^2 + 4y^2 + 64x - 8y + 19 = 0$

11.4.9 Show that the equations $x = 3 \sin t$ and $y = 2 \cos t$, where t is any real number, represent an ellipse. Find the foci, vertices, center, major axis, minor axis, and sketch the graph of this ellipse. [*Hint*: Square both equations and then eliminate t.]

11.4.10 Show that the equations $x = 4 \sin t$ and $y = 7 \cos t$, where t is any real number, represent an ellipse. Find the foci, vertices, center, major axis, minor axis, and sketch the graph of this ellipse. [*Hint*: See Prob. 11.4.9.]

11.5 Conic Sections: Hyperbolas

The last type of a conic section which we shall study is called a *hyperbola.* Suppose F_1 and F_2 are two fixed points, and suppose that the distance between F_1 and F_2 is $2c$. Suppose, further, that a is any positive real number such that $a < c$. We define a *hyperbola with foci F_1 and F_2* to be the set of all points P in the plane such that

Absolute value of the difference between the lengths of PF_1 and PF_2 is equal to $2a$. (11.81)

In order to obtain an *equation of a hyperbola*, that is, an equation which is satisfied by the coordinates of all points on the hyperbola but by no others, we must first decide on a choice for the coordinate axes. Naturally, we seek to choose these axes in such a way as to obtain an equation which is as simple as possible. For this reason, we choose the x axis to be the line joining the two foci F_1, F_2, and we choose the origin, $\mathcal{O}$, to be the midpoint of the line segment F_1F_2, as indicated in Fig. 11.26.

Now, suppose that $P: (x,y)$ is any point on our hyperbola. Then, by the distance formula, we have (see Fig. 11.26)

$$\text{Length of } PF_1 = \sqrt{(x+c)^2 + y^2} \qquad \text{length of } PF_2 = \sqrt{(x-c)^2 + y^2}$$

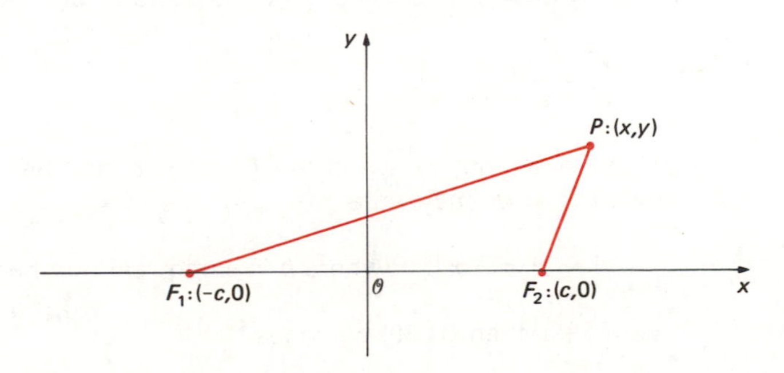

Fig. 11.26 A choice of axes for a hyperbola

and hence (11.81) becomes

$$|\sqrt{(x+c)^2+y^2}-\sqrt{(x-c)^2+y^2}|=2a \tag{11.82}$$

Equation (11.82) is clearly equivalent to

$$\sqrt{(x+c)^2+y^2}-\sqrt{(x-c)^2+y^2}=\pm 2a \tag{11.83}$$

and we now proceed to simplify (11.83). Adding the second radical on the left side of (11.83) to both sides of (11.83), we obtain

$$\sqrt{(x+c)^2+y^2}=\pm 2a+\sqrt{(x-c)^2+y^2} \tag{11.84}$$

Now, squaring both sides of (11.84), we get

$$(x+c)^2+y^2=4a^2+[(x-c)^2+y^2]\pm 4a\sqrt{(x-c)^2+y^2} \tag{11.85}$$

Solving for the radical on the right side of (11.85), we obtain

$$(x+c)^2+y^2-4a^2-(x-c)^2-y^2=\pm 4a\sqrt{(x-c)^2+y^2}$$

which, after simplifying, becomes

$$cx-a^2=\pm a\sqrt{(x-c)^2+y^2} \tag{11.86}$$

Squaring both sides of (11.86), we get

$$\begin{aligned} c^2x^2+a^4-2a^2cx &= a^2(x-c)^2+a^2y^2 \\ &= a^2x^2+a^2c^2-2a^2cx+a^2y^2 \end{aligned}$$

and hence, simplifying again, we obtain

$$c^2x^2+a^4=a^2x^2+a^2c^2+a^2y^2$$

and, equivalently,

$$(a^2-c^2)x^2+a^2y^2=a^4-a^2c^2=a^2(a^2-c^2) \tag{11.87}$$

Thus, dividing both sides of (11.87) by $a^2(a^2-c^2)$, we get

$$\frac{x^2}{a^2}+\frac{y^2}{a^2-c^2}=1 \tag{11.88}$$

Now, recall that by hypothesis $0<a<c$, and hence a^2-c^2 is *negative.* In view of this, we may write

$$a^2-c^2=-b^2 \qquad \text{(where } b \text{ is some real number)} \tag{11.89}$$

If we do so, then (11.88) becomes

$$\frac{x^2}{a^2}+\frac{y^2}{-b^2}=1 \qquad (b^2=c^2-a^2)$$

and equivalently,

$$\frac{x^2}{a^2} - \frac{y^2}{b^2} = 1 \qquad (b^2 = c^2 - a^2;\ a,\ b,\ c \text{ nonzero real numbers})$$

It can also be shown that the above steps are reversible, and hence we have the following:

An equation of a hyperbola with foci $F_1:(-c,0)$, $F_2:(c,0)$, and with $|PF_1 - PF_2| = 2a$ for all points P on the graph, is

$$\frac{x^2}{a^2} - \frac{y^2}{b^2} = 1 \qquad (b^2 = c^2 - a^2;\ a,\ b,\ c, \text{ nonzero real numbers}) \tag{11.90}$$

(Here we do not require that $a^2 > b^2$.)

The graph of a hyperbola is easy to sketch (see Fig. 11.27). Indeed, the graph of the equation

$$\frac{x^2}{a^2} - \frac{y^2}{b^2} = 1 \tag{11.91}$$

is symmetric with respect to both x and y axes as well as the origin. Moreover, the x intercept points are $V_1:(-a,0)$, $V_2:(a,0)$. Also, by (11.91)

$$\frac{x^2}{a^2} = 1 + \frac{y^2}{b^2} \geq 1 \qquad \text{and hence} \qquad x^2 \geq a^2$$

Thus, for all points (x,y) on the hyperbola (11.91), we have (recall that $a > 0$)

$$x \geq a \qquad \text{or} \qquad x \leq -a \tag{11.92}$$

In view of (11.92), *no points of the graph of the hyperbola (11.91) lie inside the "vertical strip" bounded by the vertical lines* $x = -a$ *and* $x = a$. Keeping the above facts in mind, we see that the graph of the hyperbola (11.91) is as indicated in Fig. 11.27. Just as in the case of an ellipse, we call the points $(-a,0)$, $(a,0)$ the *vertices*, and the point $(0,0)$ the *center*, of the hyperbola. Moreover, the line

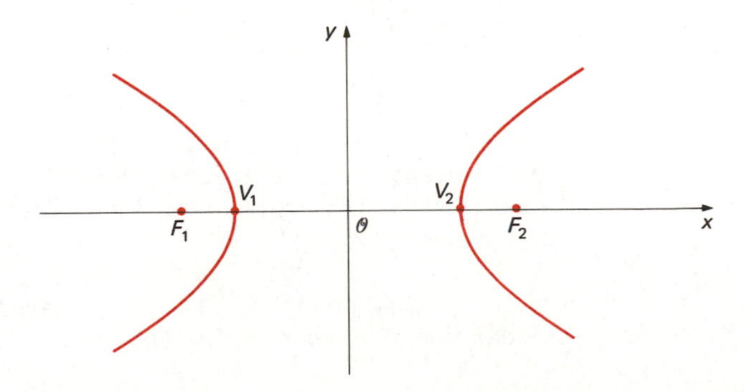

Fig. 11.27 Graph of the hyperbola $x^2/a^2 - y^2/b^2 = 1$. The vertices are $V_1: (-a,0)$, $V_2: (a,0)$, and the foci are $F_1: (-c,0)$, $F_2: (c,0)$, where $c^2 = a^2 + b^2$.

on which the vertices, foci, and center lie is called the *transverse axis*, while the line perpendicular to the transverse axis and passing through the center is called the *conjugate axis*. Thus, for the hyperbola given in (11.91), the transverse axis is the x axis while the conjugate axis is the y axis (see Fig. 11.27). We summarize our results in the following:

> The graph of the equation
>
> $$\frac{x^2}{a^2} - \frac{y^2}{b^2} = 1 \qquad (a, b \text{ nonzero real numbers, } a^2 \text{ not necessarily greater than } b^2) \tag{11.93}$$
>
> is a hyperbola with the following properties:
>
> | | |
> |---|---|
> | Vertices: | $(-a,0)$, $(a,0)$ |
> | Foci: | $(-\sqrt{a^2+b^2},0)$, $(\sqrt{a^2+b^2},0)$ |
> | Center: | $(0,0)$ |
> | Transverse axis: | x axis |
> | Conjugate axis: | y axis |

The equation in (11.93) and its analogs [see (11.94) below] are known as the standard forms of the equations of a hyperbola.

Now, comparing (11.93) and (11.67), we see that the only difference in the equations and information given in these two situations is that b^2 and $-b^2$ are interchanged throughout. For example, the foci in (11.67) (found in Sec. 11.4) are $(-\sqrt{a^2-b^2},0)$, $(\sqrt{a^2-b^2},0)$, while the foci in (11.93) are $(-\sqrt{a^2+b^2},0)$, $(\sqrt{a^2+b^2},0)$, etc. In view of this, we have the following general facts about hyperbolas:

> All the information given in ellipse equations (11.67) to (11.70) about vertices, foci, and center remains valid for hyperbolas provided that we replace $-b^2$ by b^2 throughout. Furthermore, "major axis" is now replaced by "transverse axis" and "minor axis" is replaced by "conjugate axis." (We do not require that $a^2 > b^2$ for a hyperbola.) (11.94)

Moreover, analogous to (11.71), we now have

> The graph of every equation of the form
>
> $$Ax^2 + By^2 + Cx + Dy + E = 0 \tag{11.95}$$
>
> where A, B, C, D, E are any real numbers such that A and B are both nonzero and have opposite signs, is always a hyperbola. This hyperbola may degenerate to two intersecting lines (consider $x^2 - y^2 = 0$).

The proof of (11.95), like that of (11.71), is achieved by using the completing-the-square technique, illustrated in the following examples.

Example 11.20 Find the vertices, foci, center, transverse axis, and conjugate axis of the hyperbola given by

$$\frac{x^2}{2^2} - \frac{y^2}{3^2} = 1 \tag{11.96}$$

Solution Identifying (11.96) with (11.93), we observe that $a = 2$, $b = 3$, and $\sqrt{a^2 + b^2} = \sqrt{13}$. Hence, by (11.93), we have the following:

Vertices:	$(-2,0)$, $(2,0)$	Transverse axis:	x axis
Foci:	$(-\sqrt{13},0)$, $(\sqrt{13},0)$	Conjugate axis:	y axis
Center:	$(0,0)$		

Example 11.21 Find the vertices, foci, center, transverse axis, and conjugate axis of the hyperbola

$$\frac{y^2}{3^2} - \frac{x^2}{2^2} = 1 \tag{11.97}$$

Solution In view of (11.94), we only need to look up the information given in (11.68) and simply change b^2 to $-b^2$ throughout. Keeping this fact, together with (11.68), in mind, we observe that $a = 3$, $b = 2$, and $\sqrt{a^2 + b^2} = \sqrt{13}$. Hence we have the following:

Vertices:	$(0,-3)$, $(0,3)$	Transverse axis:	y axis
Foci:	$(0,-\sqrt{13})$, $(0,\sqrt{13})$	Conjugate axis:	x axis
Center:	$(0,0)$		

Example 11.22 Use the same coordinate axes to sketch the graphs of the hyperbolas given in (11.96) and (11.97).

Solution Recalling the discussion which led to Fig. 11.27, we can easily sketch the graph of the hyperbola in (11.96). A similar discussion leads to the graph of the hyperbola in (11.97). Both graphs are sketched in Fig. 11.28.

Remark The hyperbolas $x^2/a^2 - y^2/b^2 = 1$ and $y^2/b^2 - x^2/a^2 = 1$ are called *conjugate hyperbolas* (see Fig. 11.28). Observe that the conjugate axis of the hyperbola $x^2/a^2 - y^2/b^2 = 1$ (namely, the y axis) is the same as the transverse axis of the conjugate hyperbola $y^2/b^2 - x^2/a^2 = 1$, and this explains, at least in part, the reason for naming the pair *conjugate hyperbolas.*

Example 11.23 Find the vertices, foci, center, transverse axis, and conjugate axis of the hyperbola

$$9x^2 - 4y^2 + 18x + 16y - 43 = 0 \tag{11.98}$$

Also, sketch the graph.

Solution First, we complete the squares on the left side of (11.98). Thus,

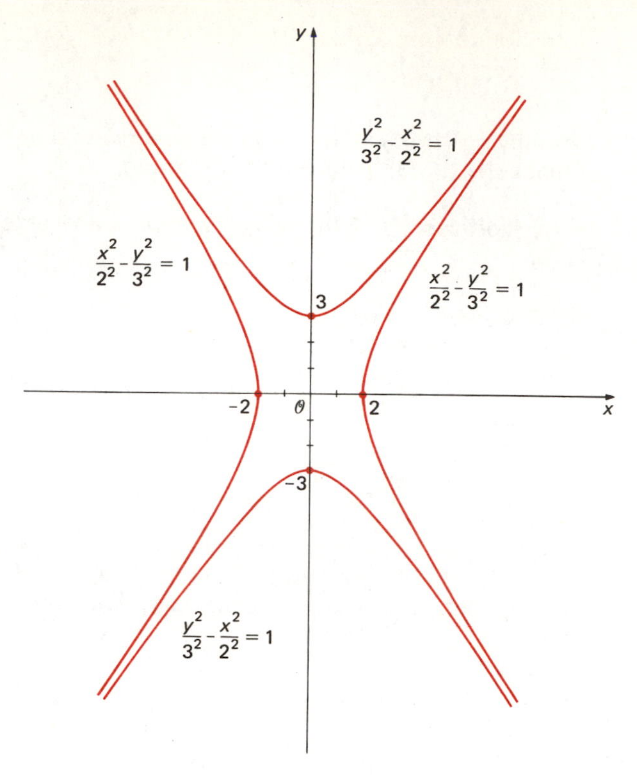

Fig. 11.28 Graphs of the hyperbolas $x^2/2^2 - y^2/3^2 = 1$ and $y^2/3^2 - x^2/2^2 = 1$

$$\begin{aligned} 9x^2 - 4y^2 + 18x + 16y - 43 &= 9(x^2 + 2x) - 4(y^2 - 4y) - 43 \\ &= 9(x^2 + 2x + 1) - 4(y^2 - 4y + 4) - 9 + 16 - 43 \\ &= 9(x + 1)^2 - 4(y - 2)^2 - 36 \end{aligned}$$

In view of this, (11.98) is equivalent to $9(x + 1)^2 - 4(y - 2)^2 = 36$, which, when both sides are divided by 36, gives

$$\frac{(x + 1)^2}{2^2} - \frac{(y - 2)^2}{3^2} = 1 \tag{11.99}$$

Now, in view of (11.94), the information we are after is essentially contained in (11.69) *provided that we replace $-b^2$ by b^2 throughout*. Moreover, keeping (11.69) in mind, we have $a = 2$, $b = 3$, $h = -1$, $k = 2$, and $\sqrt{a^2 + b^2} = \sqrt{13}$. And hence by (11.69) (found in Sec. 11.4) we have the following:

$$\begin{aligned} &\text{Vertices:} && V_1: (h - a,k) = (-3,2),\ V_2: (h + a,k) = (1,2) \\ &\text{Foci:} && F_1: (h - \sqrt{a^2 + b^2},k) = (-1 - \sqrt{13},2) \\ & && F_2: (h + \sqrt{a^2 + b^2},k) = (-1 + \sqrt{13},2) \\ &\text{Center:} && C: (h,k) = (-1,2) \end{aligned} \tag{11.100}$$

$$\begin{aligned} &\text{Transverse axis:} \quad \text{the line } y = k = 2 \\ &\text{Conjugate axis:} \quad \text{the line } x = h = -1 \end{aligned} \tag{11.100}$$

The graph of the hyperbola in (11.99) [and hence also (11.98)] is sketched in Fig. 11.29.

Example 11.24 Find an equation of a hyperbola with vertices (2,3), (2,7) and with foci (2,1), (2,9).

Solution Since the vertices and foci lie on the vertical line $x = 2$, we know that the transverse axis of our hyperbola is the line $x = 2$ itself, and hence, in view of (11.94), the desired equation has the form in (11.70) *except that b^2 is now to be replaced by $-b^2$*. In other words, the desired equation is

$$\frac{(y-k)^2}{a^2} - \frac{(x-h)^2}{b^2} = 1 \tag{11.101}$$

By (11.70) we know that $h = 2$, and we also have the following facts:

$$\begin{aligned} &\text{Vertices:} \quad (2, k-a),\ (2, k+a) \\ &\text{Foci:} \quad (2, k-\sqrt{a^2+b^2}),\ (2, k+\sqrt{a^2+b^2}) \end{aligned} \tag{11.102}$$

But, by hypothesis, we also have:

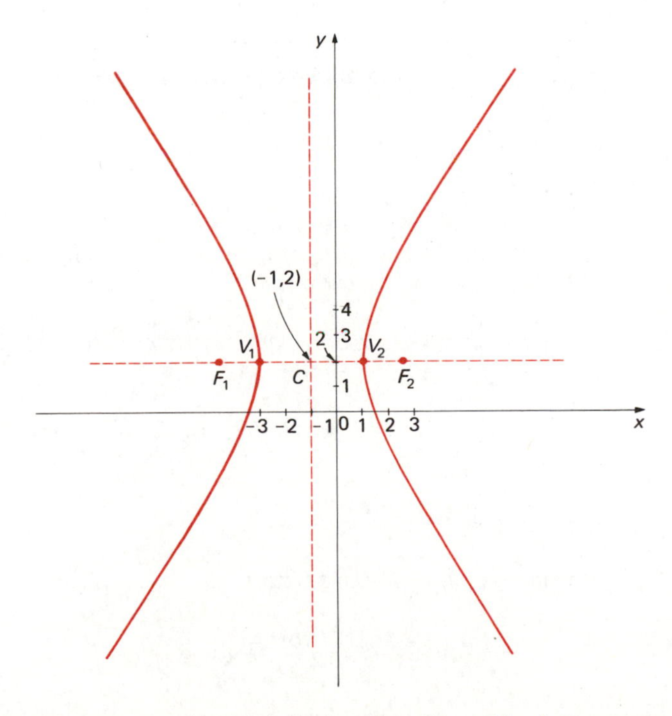

Fig. 11.29 Graph of the hyperbola $9x^2 - 4y^2 + 18x + 16y - 43 = 0$. The vertices V_1, V_2, the foci F_1, F_2 and the center C are given in (11.100).

$$\text{Vertices:} \quad (2,3), (2,7) \qquad \text{Foci:} \quad (2,1), (2,9) \tag{11.103}$$

Identifying (11.102) and (11.103), we get

$$k - a = 3 \qquad k + a = 7$$

$$k - \sqrt{a^2 + b^2} = 1 \qquad k + \sqrt{a^2 + b^2} = 9$$

Now, solving the above equations, we get $k = 5$, $a = 2$, $b^2 = 12$. Substituting these values in (11.101), we conclude that our desired equation is (recall that $h = 2$)

$$\frac{(y-5)^2}{4} - \frac{(x-2)^2}{12} = 1$$

Unlike other conic sections, a hyperbola has two oblique asymptotes. In fact, the hyperbola

$$\frac{x^2}{a^2} - \frac{y^2}{b^2} = 1 \qquad (a, b \text{ nonzero real numbers}) \tag{11.104}$$

has the lines $y = \pm (b/a)x$ as asymptotes. To see this, let us solve Eq. (11.104) for y, we get

$$y = \pm \frac{b}{a}\sqrt{x^2 - a^2} = \pm \frac{b}{a}\sqrt{x^2}\sqrt{1 - \frac{a^2}{x^2}}$$

Since $\sqrt{x^2} = |x|$, the above equation is equivalent to

$$y = \pm \frac{b}{a} x \sqrt{1 - \frac{a^2}{x^2}} \tag{11.105}$$

Now, as x becomes very large numerically, we see intuitively that $\sqrt{1 - a^2/x^2}$ becomes very close to 1, and hence by (11.105), y becomes very close to $\pm (b/a)x$. We thus have the following:

> The lines $y = \pm (b/a)x$ are asymptotes for the graph of the hyperbola $x^2/a^2 - y^2/b^2 = 1$. (11.106)

Next, consider the *conjugate hyperbola* of the one given in (11.104), namely,

$$\frac{y^2}{b^2} - \frac{x^2}{a^2} = 1 \tag{11.107}$$

Solving (11.107) for y, we obtain

$$y = \pm \frac{b}{a}\sqrt{x^2 + a^2} = \pm \frac{b}{a}\sqrt{x^2}\sqrt{1 + \frac{a^2}{x^2}}$$

and thus

$$y = \pm\frac{b}{a}x\sqrt{1+\frac{a^2}{x^2}} \qquad (11.108)$$

Again, as x becomes very large numerically, we see intuitively that $\sqrt{1 + a^2/x^2}$ becomes very close to 1, and hence by (11.108), y becomes very close to $\pm(b/a)x$. We thus have the following:

> The lines $y = \pm(b/a)x$ are also asymptotes for the graph of the hyperbola $y^2/b^2 - x^2/a^2 = 1$; hence a hyperbola and its conjugate hyperbola have the same asymptotes. (11.109)

For example, the lines $y = \pm 3/2\, x$ are asymptotes for both of the hyperbolas sketched in Fig. 11.28, namely,

$$\frac{x^2}{2^2} - \frac{y^2}{3^2} = 1 \qquad \text{and} \qquad \frac{y^2}{3^2} - \frac{x^2}{2^2} = 1$$

Now, a careful examination of the general formulas for parabolas, ellipses, and hyperbolas which we obtained in the last three sections [see (11.46) in Sec. 11.3, (11.71) in Sec. 11.4, and (11.95)] shows that we have classified those conic sections for which the quadratic equation in x and y under consideration does *not* involve the term xy. If an arbitrary quadratic polynomial equation in two variables x *and* y does involve a term in xy (with a nonzero coefficient), its graph is still a conic section. For a proof of this fact, the student may consult more advanced texts.

Problem Set 11.5

11.5.1 Find an equation of a hyperbola such that the

(*a*) foci are (3,0), (−3,0) and vertices are (2,0), (−2,0)

(*b*) foci are (0,3), (0,−3) and vertices are (0,2), (0,−2)

11.5.2 Sketch the graph of each of the hyperbolas in Prob. 11.5.1.

11.5.3 Find the center, transverse axis, conjugate axis, and asymptotes of each of the hyperbolas in Prob. 11.5.1.

11.5.4 Find an equation of the hyperbola with center at the origin, with foci on a coordinate axis, and passing through (3,0), (−6,−9). Sketch the graph.

11.5.5 Find an equation of the hyperbola with center at the origin, with foci on a coordinate axis, and passing through (0,−4), (−9,8). Sketch the graph.

11.5.6 Find the vertices, foci, transverse axis, and conjugate axis of the hyperbola in Prob. 11.5.4.

11.5.7 Find the vertices, foci, transverse axis, and conjugate axis of the hyperbola in Prob. 11.5.5.

11.5.8 Sketch the graph of each of the following hyperbolas. In each case, find the vertices, foci, center, transverse axis, and conjugate axis.

(*a*) $4x^2 - 9y^2 = 36$ (*d*) $x^2 - 4y^2 - 8y = 0$

(*b*) $16y^2 - 9x^2 = 144$ (*e*) $x^2 - 2y^2 - 2x + 4y = 5$

(*c*) $4x^2 - y^2 + 24x = 0$ (*f*) $-4x^2 + 2y^2 - 8x - 8y = 12$

11.5.9 Find the points of intersection (if any) of the line $y = 2x + 1$ and the circle whose equation is $x^2 + y^2 - 10x - 2y + 6 = 0$. Sketch both the line and the circle.

11.5.10 Find the points of intersection (if any) of the parabola whose equation is $y = x^2 +$ and (*a*) the line $y = 2x$; (*b*) the line $y = -6x - 8$. Sketch the parabola and both lines.

11.5.11 Find the points of intersection (if any) of the line $x + y = 1$ and the circle whose equation is $x^2 + y^2 - 2x - 4y = 4$.

11.5.12 Suppose that $f: R \to R$ (R is the set of real numbers) is given by $f(x) = x^2$. Suppose, moreover, that $A = (1, f(1))$, and $P = (1 + h, f(1 + h))$, where $h \neq 0$. Show:

(*a*) slope of $AP = \dfrac{f(1+h) - f(1)}{h}$ $(h \neq 0)$

(*b*) slope of $AP = 2 + h$ $(h \neq 0)$

11.5.13 Sketch the graph of the parabola whose equation is $y = x^2$. Use the results of Prob. 11.5.12 to draw to the graph of this parabola the secant lines which join $A = (1,1)$ to P, where P is the point on the parabola given by

(*a*) $P = (1.1, (1.1)^2)$ (*b*) $P = (0.9, (0.9)^2)$

In each case, find the slope of AP. What are the values of these slopes close to?

11.5.14 The expression $[f(1 + h) - f(1)]/h$ we encountered in Prob. 11.5.12 is an example of what is known as a *quotient difference*. Geometrically, it can be interpreted as a slope (see Prob. 11.5.12). Find $[f(1 + h) - f(1)]/h$, where $h \neq 0$, for each of the following functions (a, b, c are constants).

(*a*) $f(x) = c$ (*f*) $f(x) = x^3$

(*b*) $f(x) = x$ (*g*) $f(x) = cx^3$

(*c*) $f(x) = cx$ (*h*) $f(x) = ax + b$

(*d*) $f(x) = x^2$ (*i*) $f(x) = ax^2 + bx + c$

(*e*) $f(x) = cx^2$

11.5.15 What are the values of the quotient differences $[f(1 + h) - f(1)]/h$ in Prob. 11.5.1 close to when h is very close to zero, but $h \neq 0$? Recalling that $[f(1 + h) - f(1)]/h$ is the slope of the line joining the points $(1, f(1))$, $(1 + h, f(1 + h))$ on the graph of $y = f(x)$, what do the results of Prob. 11.5.14 say geometrically when h is very close to zero, but $h \neq 0$ (also see Prob. 11.5.13).

11.5.16 Suppose that f is a quadratic function given by

$$f(x) = ax^2 + bx + c \qquad a, b, c \text{ real}, a \neq 0$$

Prove that the graph of f [i.e., the graph of the equation $y = f(x)$] has a maximum point at

$$\left(\frac{-b}{2a}, f\left(\frac{-b}{2a}\right)\right) \qquad \text{if } a < 0$$

and has a minimum point at

$$\left(\frac{-b}{2a}, f\left(\frac{-b}{2a}\right)\right) \qquad \text{if } a > 0$$

[*Hint*: Complete the square.]

11.5.17 Suppose that f is a quadratic function given by

$$f(x) = ax^2 + bx + c \qquad a, b, c \text{ real}, a \neq 0$$

and suppose that x_1 and x_2 are the roots of the equation $f(x) = 0$. Prove that the maximum point (when $a < 0$) or minimum point (when $a > 0$) of the graph of f is the point

$$\left(\frac{x_1 + x_2}{2}, f\left(\frac{x_1 + x_2}{2}\right)\right)$$

[*Hint*: Use the result of Prob. 11.5.16.]

11.5.18 Suppose that f is a quadratic function given by

$$f(x) = ax^2 + bx + c \qquad a, b, c \text{ real}, a \neq 0$$

and suppose that the graph of f passes through the origin, and has a maximum point at (1,2). Find $f(x)$. Also, sketch the graph of $y = f(x)$. [*Hint*: Use the result of Prob. 11.5.16.]

11.5.19 Suppose that f is a quadratic function given by

$$f(x) = ax^2 + 4x + 2a \qquad a \text{ real}, a \neq 0$$

Find the value of a so that (a) the graph of f has a maximum point with equal coordinates and (b) the graph of f has a minimum point with equal coordinates. [*Hint*: Use the result of Prob. 11.5.16.]

11.6 Summary

In this chapter, we saw that every straight line has an equation of the form $Ax + By + C = 0$, where A, B, C are fixed real numbers such that not both A and B are zero. This line has slope $-A/B$ (if $B \neq 0$) but has no slope if $B = 0$. Various other forms of a line were also established. For example, the equation of a line with slope m and passing through the point (x_1,y_1) is $y - y_1 = m(x - x_1)$. Moreover, the equation of a line with slope m and y intercept b is $y = mx + b$. Also, we saw that two lines are parallel if and only if their slopes are equal, while two nonvertical lines with slopes m_1 and m_2 are perpendicular if and only if $m_1 m_2 = -1$. The midpoint formula, which states that the midpoint of the line segment joining (x_1,y_1) and (x_2,y_2) is

$$\left(\frac{x_1 + x_2}{2}, \frac{y_1 + y_2}{2}\right)$$

was also established.

We then studied conic sections. In particular, we showed the following:

1. The graph of the equation

   $$(x - h)^2 + (y - k)^2 = r^2$$

   is a circle with radius r and center (h,k).

2. The graph of the equation

   $$y^2 = 4px \qquad (p \neq 0)$$

   is a parabola with focus $(p,0)$, vertex $(0,0)$, and directrix $x = -p$.

3. The graph of the equation

   $$\frac{x^2}{a^2} + \frac{y^2}{b^2} = 1 \qquad (a^2 > b^2,\ a,\ b \text{ nonzero real numbers})$$

   is an ellipse with vertices $(-a,0)$ and $(a,0)$, foci $(-\sqrt{a^2 - b^2},0)$ and $(\sqrt{a^2 - b^2},0)$, and center $(0,0)$. Moreover, the major axis is the x axis and the minor axis is the y axis.

4. The graph of the equation

   $$\frac{x^2}{a^2} - \frac{y^2}{b^2} = 1 \qquad a,\ b \text{ nonzero real numbers, } a^2 \textit{ not necessarily greater than } b$$

   is a hyperbola with vertices $(-a,0)$ and $(a,0)$, foci $(-\sqrt{a^2 + b^2},0)$ and $(\sqrt{a^2 + b^2},0)$, and center $(0,0)$. Moreover, the transverse axis is the x axis, and the conjugate axis is the y axis. Also, the lines $y = \pm(b/a)x$ are asymptotes for this hyperbola.

By interchanging the roles of x and y in (2), (3) and (4) above, we obtained additional standard forms for the equations of conic sections. Further standard forms were also obtained by assuming that the center of the conic is at the point (h,k) [instead of $(0,0)$]. Finally, we saw that the method of completing the square was very effective in putting the equation of a conic section in standard form.

12 Mathematical Induction, Progressions, and the Binomial Theorem

12.1 Mathematical Induction

Consider for a moment a statement $P(n)$ about a positive integer. For example, $P(n)$ might be "$2^n > n$," or "$n^2 + n$ is even," etc. Suppose we are told that $P(n)$ possesses the following two properties:

1. $P(1)$ is true; that is, the statement $P(n)$ is true when $n = 1$.
2. If k is any positive integer and $P(k)$ is true, then $P(k+1)$ is also true. That is, whenever $P(n)$ is true for $n = k$, then $P(n)$ is also true for the next bigger integra value of n (namely, $n = k + 1$).

We now ask: Is $P(n)$ true for *all* positive integers n? In attempting to answer this question intuitively, we proceed as follows. To begin with, $P(1)$ is true, by hypothesis 1. Now, reading hypothesis (2) with $k = 1$, we obtain

$P(2)$ is true whenever $P(1)$ is true (12.1

But $P(1)$ is true [by (1)], and hence (12.1) now implies that

$P(2)$ is true (12.2

How about $P(3)$? Nothing to it. Just put $k = 2$ in hypothesis 2, to get

$P(3)$ is true whenever $P(2)$ is true (12.3

But $P(2)$ is true [by (12.2)], and hence (12.3) now implies that

$P(3)$ is true (12.4

Now, that we have seen that $P(1)$, $P(2)$, $P(3)$ are all true, we naturally wonder if $P(4)$ is also true. Imitating the above argument, put $k = 3$ in hypothesis (2), to get

$P(4)$ is true whenever $P(3)$ is true (12.5

But $P(3)$ is true [by (12.4)], and hence (12.5) now implies that

$P(4)$ is true (12.6

It is now clear that we can continue this process without stopping. Thus, if we put $k = 4$ in hypothesis 2, we obtain

$P(5)$ is true whenever $P(4)$ is true (12.7

Since $P(4)$ is true [by (12.6)], (12.7) implies at once that $P(5)$ is true. In this same way (i.e., putting $k = 5$, then $k = 6$, then $k = 7$, etc., in hypothesis 2), we readily see that $P(6)$ is true, $P(7)$ is true, $P(8)$ is true, and so on. We thus see, intuitively, that $P(n)$ is true for $n = 1, 2, 3, 4, 5, 6, 7, 8$, and so on. It appears, then, that $P(n)$ is true for *all* positive integers n.

Keeping the above plausibility argument in mind, we now state the following, known as *principle of mathematical induction* (*PMI*).

rinciple of Mathematical nduction PMI)

Suppose that $P(n)$ is a statement involving a positive integer n and suppose that

(1) $P(1)$ is true.

(2) If $P(k)$ is true, then $P(k+1)$ must also be true, where k is any positive integer.

Then, $P(n)$ is true for all positive integers n.

Let us illustrate this principle by some examples.

xample 12.1

Prove, by mathematical induction, that for all positive integers n,

$$1+2+3+\cdots+n=\frac{n(n+1)}{2} \tag{12.8}$$

[The dots in (12.8) denote all the omitted integers, namely, those which are *strictly* between 3 and n.]

olution

Suppose that

$$P(n) \text{ is the statement "}1+2+3+\cdots+n=\frac{n(n+1)}{2}\text{"} \tag{12.9}$$

Now, if we set $n=1$ in (12.8), we obtain

$$1=\frac{1(1+1)}{2}$$

which is certainly true. Hence $P(n)$ is true when $n=1$ [see (12.9)]; that is,

$$P(1) \text{ is true} \tag{12.10}$$

Next, suppose that $P(k)$ is true, where k is a positive integer; that is, suppose that [see (12.9)]

$$1+2+3+\cdots+k=\frac{k(k+1)}{2} \tag{12.11}$$

Then, by adding $k+1$ to both sides of (12.11), we get

$$\begin{aligned}
1+2+3+\cdots+k+(k+1) &= \frac{k(k+1)}{2}+(k+1) \\
&= \frac{k(k+1)+2(k+1)}{2} \\
&= \frac{k^2+k+2k+2}{2} \\
&= \frac{k^2+3k+2}{2} \\
&= \frac{(k+1)(k+2)}{2} \\
&= \frac{(k+1)[(k+1)+1]}{2}
\end{aligned}$$

We have thus shown that

$$1 + 2 + 3 + \cdots + (k + 1) = \frac{(k + 1)[(k + 1) + 1]}{2} \tag{12.12}$$

But (12.12) is precisely the same as (12.8) with $n = k + 1$, and hence (12.12) is equivalent to saying [see (12.9)] that

$$P(k + 1) \text{ is true} \tag{12.13}$$

We have thus shown that

$$\text{If } P(k) \text{ is true, then } P(k + 1) \text{ is also true} \tag{12.14}$$

where k is *any* positive integer. In view of (12.10) and (12.14), we see that both (1) and (2) of PMI hold for (12.8), and hence, by the principle of mathematical induction, we conclude that $P(n)$ is true for all positive integers n; that is [see (12.9)],

$$1 + 2 + 3 + \cdots + n = \frac{n(n + 1)}{2}$$

for all positive integers n. This proves the above assertion.

Example 12.2 Prove, by PMI, that for all positive integers n

$$1^2 + 2^2 + 3^2 + \cdots + n^2 = \frac{n(n + 1)(2n + 1)}{6} \tag{12.15}$$

Solution Suppose that

$$P(n) \text{ is the statement } ``1^2 + 2^2 + 3^2 + \cdots + n^2 = \frac{n(n + 1)(2n + 1)}{6}" \tag{12.16}$$

Now, if we set $n = 1$ in (12.15), we obtain

$$1^2 = \frac{1(1 + 1)(2 \cdot 1 + 1)}{6}$$

which is certainly true. Hence $P(n)$ is true when $n = 1$ [see (12.16)]; that is,

$$P(1) \text{ is true} \tag{12.17}$$

Next, suppose that $P(k)$ is true, where k is a positive integer; that is, suppose that [see (12.16)]

$$1^2 + 2^2 + 3^2 + \cdots + k^2 = \frac{k(k + 1)(2k + 1)}{6} \tag{12.18}$$

Then, by adding $(k + 1)^2$ to both sides of (12.18), we get

$$1^2 + 2^2 + 3^2 + \cdots + k^2 + (k+1)^2 = \frac{k(k+1)(2k+1)}{6} + (k+1)^2$$

$$= \frac{k(k+1)(2k+1) + 6(k+1)^2}{6}$$

$$= \frac{(k+1)[k(2k+1) + 6(k+1)]}{6}$$

$$= \frac{(k+1)(2k^2 + 7k + 6)}{6}$$

$$= \frac{(k+1)(k+2)(2k+3)}{6}$$

We therefore have

$$1^2 + 2^2 + 3^2 + \cdots + (k+1)^2 = \frac{(k+1)[(k+1)+1][2(k+1)+1]}{6} \tag{12.19}$$

But (12.19) is precisely the same as (12.15) with $n = k + 1$, and hence (12.19) is equivalent to [see (12.16)]

$$P(k+1) \text{ is true} \tag{12.20}$$

We have thus shown that

$$P(k+1) \text{ is true whenever } P(k) \text{ is true} \tag{12.21}$$

where k is *any* positive integer. In view of (12.17) and (12.21), we see that both (1) and (2) of PMI hold for (12.15), and hence, by PMI, we conclude that $P(n)$ is true for all positive integers n; that is [see (12.16)],

$$1^2 + 2^2 + 3^2 + \cdots + n^2 = \frac{n(n+1)(2n+1)}{6}$$

for all positive integers n. This proves the above assertion.

Example 12.3 Prove, by PMI, that for all positive integers n

$$2^n > n \tag{12.22}$$

Solution Suppose that

$$P(n) \text{ is the statement "} 2^n > n \text{"} \tag{12.23}$$

Now, if we set $n = 1$ in (12.22), we obtain

$$2^1 > 1$$

which is certainly true. Hence $P(n)$ is true when $n = 1$ [see (12.23)]; that is,

$$P(1) \text{ is true} \tag{12.24}$$

Next, suppose that $P(k)$ is true, where k is a positive integer; that is, suppose that [see (12.23)]

$$2^k > k \tag{12.2}$$

Then, by multiplying both sides of (12.25) by 2, we get

$$2^{k+1} > 2k \tag{12.2}$$

Now, since $k \geq 1$, $k + k \geq k + 1$, and hence $2k \geq k + 1$. Combining this with (12.26), we obtain

$$2^{k+1} > k + 1 \tag{12.2}$$

But (12.27) is precisely the same as (12.22) with $n = k + 1$, and hence (12.27) is equivalent to [see (12.23)]

$P(k+1)$ is true (12.2

We have thus shown that

If $P(k)$ is true, then $P(k+1)$ is also true (12.2

where k is any positive integer. In view of (12.24) and (12.29), we see that both (1) and (2) of PMI are true in the above example, and hence we conclude that $P(n)$ is true for all positive integers n; that is [see (12.23)]

$$2^n > n$$

for all positive integers n. This proves the above assertion.

In all the above examples, we observe that there are two basic steps which must be verified:

Step 1 Show that $P(1)$ is true.

Step 2 *Assume* that $P(k)$ is true (this is called the *induction assumption*, or *induction hypothesis*), and then *prove* that $P(k+1)$ is true, where k is any positive intege

Observe that step 1 is usually very easy to verify, while the verification of step 2 is, generally speaking, more involved. However, *both steps are required and needed before any conclusion can be drawn.* Let us illustrate this point by an example.

Suppose that

$P(n)$ is the statement "$n > n + 1$" (12.3

Then, $P(k)$ is the statement "$k > k + 1$," and $P(k+1)$ is the statement "$k + 1 >$ $k + 2$." Now, *if* $P(k)$ is true, *then* $k > k + 1$, and hence, by adding 1 to both sides, we get $k + 1 > k + 2$; that is, $P(k+1)$ is true. In other words, we have shown that

$$\text{If } P(k) \text{ is true, then } P(k+1) \text{ is also true} \tag{12.31}$$

where k is *any* positive integer. In our present situation, then, we see that *hypothesis* (2) *in the principle of mathematical induction is indeed true* [see (12.31)]. However, hypothesis (1) in this principle is *not* true [since, by (12.30), $P(1)$ means $1 > 1 + 1$, which is certainly false]. Observe that in our present example hypothesis (2) of PMI is true, but hypothesis (1) is false, and, moreover, the inequality in (12.30) is, in fact, false. Thus, we do need to show that *both* (1) and (2) in the principle of mathematical induction are true before we can draw a positive conclusion regarding $P(n)$. The student should also be warned against drawing any conclusions regarding a statement $P(n)$, say, just because $P(n)$ happens to be true for some (even many) particular values of n. For example, the statement "$n^2 - n + 41$ is prime" can be shown to be true for the first 40 values of n (namely, $n = 1, 2, 3, 4, \ldots, 40$). However, it is certainly false to say that "$n^2 - n + 41$ is prime for all positive integers n" (take $n = 41$), and we cannot therefore expect to be able to prove such a false statement by mathematical induction (or, for that matter, by any other means).

The above two examples show that there are *two* essential ingredients in an induction proof (these were stated in Step 1 and Step 2 above), and both these ingredients must be present before any conclusion regarding $P(n)$ can be made.

The student might have wondered how formulas, such as (12.8) and (12.15), are obtained in the first place. Whatever the answer is, *these formulas are certainly not discovered by mathematical induction!* Indeed, the method of mathematical induction is essentially concerned with *proving formulas which we have already obtained* somehow—perhaps by guessing after making a lot of observations. However, once we have discovered (or guessed) a formula, mathematical induction often furnishes an effective tool for verifying such a formula.

Returning to formula (12.8), we can easily establish this formula as follows: Let

$$x = 1 + 2 + 3 + \cdots + n \tag{12.32}$$

Now, rewriting the sum at the right side of (12.32) in the reverse order, we get

$$x = n + (n-1) + (n-2) + \cdots + 1 \tag{12.33}$$

Next, let us add Eqs. (12.32) and (12.33) *by columns*; we obtain

$$\begin{aligned} 2x &= (1+n) + [2 + (n-1)] + [3 + (n-2)] + \cdots + (n+1) \\ &= (n+1) + (n+1) + (n+1) + \cdots + (n+1) \qquad (n \text{ times}) \\ &= n(n+1) \end{aligned}$$

Thus, $2x = n(n+1)$, and hence $x = n(n+1)/2$. Therefore, recalling (12.32), we conclude that

$$1 + 2 + 3 + \cdots + n = \frac{n(n+1)}{2} = \frac{n^2 + n}{2} \tag{12.34}$$

as asserted in Example 12.1.

It is interesting to note that the great German mathematician Gauss (1777–1855) discovered Eq. (12.34) when he was in the fifth grade. The formula was in print long before that time, but Gauss discovered it without reading about it.

How does one go about getting Eq. (12.15)? To answer this question, recall that Eq. (12.34) states that the sum $1 + 2 + 3 + \cdots + n$ is equal to a *quadratic* polynomial in n. This suggests that, perhaps, the sum of the squares of the first n positive integers, namely, $1^2 + 2^2 + 3^2 + \cdots + n^2$, is equal to a *cubic* polynomial in n. Accordingly, we try to find, if possible, a cubic polynomial

$$an^3 + bn^2 + cn + d$$

where a, b, c, d are some unknown real numbers, such that

$$1^2 + 2^2 + 3^2 + \cdots + n^2 = an^3 + bn^2 + cn + d \tag{12.35}$$

for *all* positive integers n. Now, substituting $n = 1$, $n = 2$, $n = 3$, and $n = 4$, successively in (12.35), we get

$$\begin{aligned} 1^2 &= a \cdot 1^3 + b \cdot 1^2 + c \cdot 1 + d \\ 1^2 + 2^2 &= a \cdot 2^3 + b \cdot 2^2 + c \cdot 2 + d \\ 1^2 + 2^2 + 3^2 &= a \cdot 3^3 + b \cdot 3^2 + c \cdot 3 + d \\ 1^2 + 2^2 + 3^2 + 4^2 &= a \cdot 4^3 + b \cdot 4^2 + c \cdot 4 + d \end{aligned}$$

The above equations can be solved for the unknowns a, b, c, d. In fact, the solution turns out to be

$$a = 1/3 \qquad b = 1/2 \qquad c = 1/6 \qquad d = 0 \tag{12.36}$$

Substituting (12.36) into (12.35), we get

$$\begin{aligned} 1^2 + 2^2 + 3^2 + \cdots + n^2 &= \frac{1}{3}n^3 + \frac{1}{2}n^2 + \frac{1}{6}n \\ &= \frac{n(n+1)(2n+1)}{6} \end{aligned}$$

In view of this, it is reasonable to *conjecture* that

$$1^2 + 2^2 + 3^2 + \cdots + n^2 = \frac{n(n+1)(2n+1)}{6} \tag{12.37}$$

One of the most effective ways to verify this conjecture is to appeal to mathematical induction. In fact, we have done so in Example 12.2, and found out that the conjecture we arrived at in (12.37) is indeed true.

In the next section, we shall use mathematical induction effectively to provide proofs for the formulas for the sum of arithmetic and geometric progressions.

Problem Set 12.1

Prove the following by mathematical induction (here n is any positive integer).

12.1.1 $1 + 3 + 5 + 7 + \cdots + (2n - 1) = n^2$

12.1.2 $2 + 4 + 6 + 8 + \cdots + 2n = n^2 + n$

12.1.3 $1^3 + 2^3 + 3^3 + 4^3 + \cdots + n^3 = [n(n+1)/2]^2$

12.1.4 $1 + 2 + 2^2 + 2^3 + \cdots + 2^{n-1} = 2^n - 1$

12.1.5 $\frac{1}{1 \cdot 2} + \frac{1}{2 \cdot 3} + \frac{1}{3 \cdot 4} + \frac{1}{4 \cdot 5} + \cdots + \frac{1}{n(n+1)} = \frac{n}{n+1}$

12.1.6 $1 \cdot 2 + 2 \cdot 3 + 3 \cdot 4 + 4 \cdot 5 + \cdots + n(n+1) = n(n+1)(n+2)/3$

12.1.7 $1^2 + 3^2 + 5^2 + 7^2 + \cdots + (2n-1)^2 = (4n^3 - n)/3$

12.1.8 $1 + 2 \cdot 2 + 3 \cdot 2^2 + 4 \cdot 2^3 + \cdots + n \cdot 2^{n-1} = 1 + (n-1)2^n$

12.1.9 $2n \geq n + 1$

12.1.10 $n^2 + 1 > n$

12.1.11 $(n+1)^2 > n^2 + 1$

12.1.12 $(x + y)^n \geq x^n + y^n$ $\quad$ x, y any positive real numbers

12.1.13 $(1 + x)^n \geq 1 + nx$ $\quad$ x any real number such that $x > -1$

12.1.14 $2n \leq 2^n$

12.1.15 $n^3 - n + 3$ is divisible by 3

12.1.16 Verify that the values of a, b, c, d in (12.36) give a solution for the four equations under consideration.

12.1.17 Suppose that a and b are fixed real numbers with $a > 0$. Suppose that A is the statement "$an + b > 0$ for all positive integers n." Is A true? If so, prove A by mathematical induction. If A is false, explain how the induction procedure breaks down.

12.1.18 Consider Probs. 12.1.1 and 12.1.2. If we add the two left sides together and the two right sides together, do we have an example of Eq. (12.8) in this section? Give reasons.

12.2 Progressions

In this section, we give further applications of the principle of mathematical induction. Indeed, we use PMI to establish formulas for the sum of the first n terms in an arithmetic progression and for the sum of the first n terms in a geometric progression.

Suppose that N is the set of positive integers and R is the set of real numbers. A function $F: N \to R$ (with domain N and codomain R) is called a *sequence*. For example, the function $f: N \to R$ given by

$$f(n) = \frac{1}{n} \qquad \text{for all positive integers } n$$

is a sequence. We also say in this case that

$$\frac{1}{1}, \frac{1}{2}, \frac{1}{3}, \ldots, \frac{1}{n}, \ldots \tag{12.38}$$

is a sequence. In general, if a sequence f is given by

$$f(n) = a_n \qquad \text{for all positive integers } n$$

we say that

$$a_1, a_2, a_3, \ldots, a_n, \ldots$$

is a sequence.

Now, suppose that a and d are any real numbers, and suppose n is any positiv integer. The numbers in the sequence

$$a, a + d, a + 2d, \ldots, a + (n - 1)d, \ldots \tag{12.39}$$

are said to form an *arithmetic progression*. Each number in the sequence is called a *term*. In particular, a is the first term, while $a + (n - 1)d$ is the nth term. The number d is called the *common difference*. This terminology is well justified since every term in the sequence (after the first term) is obtained by adding d to the term immediately preceding it. In fact, it is precisely this property that distinguishes a sequence of numbers forming an arithmetic progression from other sequences.

Example 12.4 Do the following sequences form arithmetic progressions: (*a*) 1, 4, 7, 10, . . . , $3n - 2$, . . . ; (*b*) $-5, -3, -1, 1, 3, 5, \ldots, 2n - 7, \ldots$; (*c*) $2, 0, -2, -4, -6, \ldots, -2n + 4, \ldots$? If so, find the common difference.

Solution (*a*) It is easily seen that this sequence forms an arithmetic progression with common difference 3, since the difference between any term and the one that immediately precedes it is always equal to 3.

(*b*) This sequence forms an arithmetic progression of common difference 2, as is readily verified.

(*c*) This sequence is easily seen to form an arithmetic progression of commo difference -2.

Now, suppose that x denotes the sum of the first n terms in the arithmetic progression in (12.39). Suppose, further, we form this sum in two ways. First, add the terms in the same order in which they appear in (12.39). Next, add these terms in the reverse order. This results in the following two equations:

$$x = a + (a + d) + (a + 2d) + \cdots + [a + (n - 3)d] + [a + (n - 2)d] + [a + (n - 1)d]$$

$$x = [a + (n - 1)d] + [a + (n - 2)d] + [a + (n - 3)d] + \cdots + (a + 2d) + (a + d) + a$$

Adding these two equations *by columns*, we obtain

$$2x = [2a + (n - 1)d] + [2a + (n - 1)d] + [2a + (n - 1)d] + \cdots + [2a + (n - 1)d] \qquad (n \text{ times})$$

and hence

$$2x = n[2a + (n - 1)d]$$

Therefore,

$$x = \frac{n}{2}[2a + (n - 1)d]$$

We have thus shown that the sum of the first n terms of the arithmetic progression in (12.39) is given by

$$a + (a + d) + (a + 2d) + \cdots + [a + (n - 1)d] = \frac{n}{2}[2a + (n - 1)d] \qquad (12.40)$$

Let us give another proof of (12.40), using mathematical induction. Thus, suppose that

$$P(n) \text{ denotes the formula in (12.40)} \qquad (12.41)$$

Then, setting $n = 1$ in (12.40), we obtain $a = \frac{1}{2}(2a)$, which is certainly true. Thus,

$$P(1) \text{ is true} \qquad (12.42)$$

Now, suppose (induction hypothesis) that $P(k)$ is true; that is, suppose that

$$a + (a + d) + (a + 2d) + \cdots + [a + (k - 1)d] = \frac{k}{2}[2a + (k - 1)d] \qquad (12.43)$$

Then, by adding $a + kd$ to both sides of (12.43), we get

$$a+(a+d)+(a+2d)+\cdots+[a+(k-1)d]+(a+kd)=\frac{k}{2}[2a+(k-1)d]+(a+kd)$$

$$=\frac{k[2a+kd-d]+2(a+kd)}{2}$$

$$=\frac{2ka+k^2d-kd+2a+2kd}{2}$$

$$=\frac{(2ka+2a)+(k^2d+kd)}{2}$$

$$=\frac{(k+1)2a+(k+1)kd}{2}$$

$$=\frac{(k+1)}{2}(2a+kd)$$

$$=\frac{(k+1)}{2}\{2a+[(k+1)-1]d\}$$

We have thus shown that

$$a+(a+d)+(a+2d)+\cdots+(a+kd)=\frac{(k+1)}{2}\{2a+[(k+1)-1]d\}$$

which is precisely formula (12.40) with $n=k+1$. In other words, we have proved that

$$\text{If } P(k) \text{ is true, then } P(k+1) \text{ must also be true} \tag{12.4}$$

where k is *any* positive integer. In view of (12.42) and (12.44), we now know, by PMI, that $P(n)$ is true for *all* positive integers n; that is, (12.40) is always true [see (12.41)].

Example 12.5 Find the sum of the first n terms in each of the arithmetic progressions in Example 12.4.

Solution (*a*) Here $a=1$, $d=3$, the nth term is $3n-2$, and hence by (12.40) we have

$$1+4+7+10+\cdots+(3n-2)=\frac{n}{2}[2(1)+(n-1)3] \tag{12.4}$$

$$=\frac{n(3n-1)}{2}$$

Thus, the desired sum is equal to $n(3n-1)/2$.

(*b*) Here $a=-5$, $d=2$, nth term is $2n-7$, and hence by (12.40) we have

$$(-5)+(-3)+(-1)+1+3+5+\cdots+(2n-7)=\frac{n}{2}[2(-5)+(n-1)2]$$

$$=\frac{n(2n-12)}{2} \tag{12.}$$

$$=n(n-6)$$

Thus, the desired sum is equal to $n(n-6)$.

(*c*) Here $a=2$, $d=-2$, nth term is $-2n+4$ and hence by (12.40) we have

$$2+0+(-2)+(-4)+(-6)+\cdots+(-2n+4)=\frac{n}{2}[2(2)+(n-1)(-2)]$$

$$=\frac{n(-2n+6)}{2} \qquad (12.47)$$

$$=n(-n+3)$$

Thus, the desired sum is equal to $n(-n+3)$.

Example 12.6 Find the sum of the first ten terms in each of the arithmetic progressions in Example 12.4.

Solution (*a*) Setting $n=10$ in (12.45), we get $1+4+7+10+\cdots+28=10(29)/2=145$.

(*b*) Setting $n=10$ in (12.46), we get $(-5)+(-3)+(-1)+1+3+5+\cdots+13$ $=10(4)=40$.

(*c*) Setting $n=10$ in (12.47), we get $2+0+(-2)+(-4)+(-6)+\cdots+(-16)$ $=10(-7)=-70$.

We define the *average* of two numbers a and b as follows:

$$\text{Average of } a \text{ and } b=\frac{a+b}{2} \qquad (12.48)$$

Now, a careful examination of formula (12.40) shows that

$$\text{Sum of the first } n \text{ terms in an arithmetic progression} = n\cdot(\text{average of first and } n\text{th terms}). \qquad (12.49)$$

[To see this, just apply the definition of average given in (12.48). Recall that, by (12.39), the nth term is $a+(n-1)d$.]

Example 12.7 In an arithmetic progression, the first term is 2 and the thirtieth term is 118. Find (*a*) the sum of the first 30 terms of this arithmetic progression; (*b*) the common difference of this arithmetic progression; (*c*) the nineteenth term of this arithmetic progression.

Solution (*a*) By (12.48) and (12.49), we have the desired sum $=30\cdot(2+118)/2=1800$.

(*b*) Since the thirtieth term is 118, we have [see (12.39)]

$$a+(n-1)d=118$$

where, by hypothesis, $a=2$ and $n=30$. Hence the above equation gives $2+(30-1)d=118$. Therefore, $d=4$, and thus the common difference is 4.

(*c*) By (12.39), we have (recall that from part *b*, $d=4$) that the nineteenth term $=a+(19-1)d=2+18(4)=74$.

We now direct our attention to geometric progressions. Thus, suppose that a and r are real numbers. A *geometric progression* is a sequence of numbers of the form

$$a, ar, ar^2, \ldots, ar^{n-1}, \ldots \qquad (12.50)$$

Each number in the sequence is called a *term*. In particular, the number a is the first term, while the number ar^{n-1} is the nth term of the progression. The number r is called the *common ratio*. This terminology is well justified, since every term in the sequence (after the first term) is obtained by multiplying the term immediately preceding it by r. In fact, it is precisely this property that distinguishes a sequence of numbers forming a geometric progression from other sequences.

Example 12.8 Do the following sequences form geometric progressions: (a) $1, 2, 4, 8, \ldots, 2^{n-1}, \ldots$; ($b$) $3, 1, 1/3, 1/9, \ldots, 3(1/3)^{n-1}, \ldots$; ($c$) $1, -1/2, 1/4, -1/8, \ldots, (-1/2)^{n-1}, \ldots$? If so, find the common ratio.

Solution (a) Since the quotient of any term by the term immediately preceding it is always equal to 2, it follows from the definition that this sequence forms a geometric progression with common ratio 2.

(b) Again, since the quotient of any term by the term immediately preceding it is always equal to $1/3$, it follows from the definition that this sequence forms a geometric progression with common ratio $1/3$.

(c) It is readily verified that the quotient of any term by the term immediately preceding it is always equal to $-1/2$, and hence, by definition, this sequence forms a geometric progression with common ratio $-1/2$.

Now, suppose that x denotes the sum of the first n terms in the geometric progression in (12.50); that is,

$$x = a + ar + ar^2 + \cdots + ar^{n-1} \qquad (12.51)$$

Then, by multiplying both sides of (12.51) by r, we obtain

$$rx = ar + ar^2 + ar^3 + \cdots + ar^{n-1} + ar^n \qquad (12.52)$$

Now, subtracting (12.51) from (12.52), we get (upon canceling identical terms)

$$rx - x = ar^n - a$$

and hence

$$(r-1)x = a(r^n - 1)$$

Therefore,

$$x = \frac{a(r^n - 1)}{r - 1} \qquad \text{if } r \neq 1 \qquad (12.53)$$

Combining (12.51), (12.53), we obtain

$$a + ar + ar^2 + \cdots + ar^{n-1} = \frac{a(r^n - 1)}{r - 1} \quad \text{if } r \neq 1 \tag{12.54}$$

Let us prove this formula, using PMI. Thus, suppose that

$$P(n) \text{ denotes the formula in Eq. (12.54)} \tag{12.55}$$

Then, setting $n = 1$ in (12.54), we obtain

$$a = \frac{a(r^1 - 1)}{r - 1} \qquad (r \neq 1)$$

which is certainly true. Thus,

$$P(1) \text{ is true} \tag{12.56}$$

Now, suppose (induction hypothesis) that $P(k)$ is true; that is, suppose that

$$a + ar + ar^2 + \cdots + ar^{k-1} = \frac{a(r^k - 1)}{r - 1} \qquad (r \neq 1) \tag{12.57}$$

Then, by adding ar^k to both sides of (12.54), we get

$$\begin{aligned} a + ar + ar^2 + \cdots + ar^{k-1} + ar^k &= \frac{a(r^k - 1)}{r - 1} + ar^k \\ &= \frac{a(r^k - 1) + ar^k(r - 1)}{r - 1} \\ &= \frac{ar^k - a + ar^{k+1} - ar^k}{r - 1} \\ &= \frac{a(r^{k+1} - 1)}{r - 1} \end{aligned}$$

We have thus shown that

$$a + ar + ar^2 + \cdots + ar^k = \frac{a(r^{k+1} - 1)}{r - 1} \qquad (r \neq 1)$$

which is precisely (12.54) with $n = k + 1$. In other words, we have proved that

$$\text{If } P(k) \text{ is true, then } P(k + 1) \text{ must also be true} \tag{12.58}$$

where k is *any* positive integer. In view of (12.56) and (12.58), we now know, by mathematical induction, that $P(n)$ is true for *all* positive integers n; that is, (12.54) is always true [see (12.55)].

emark

In (12.54) we assumed that the common ratio r satisfied the condition $r \neq 1$. What happens if $r = 1$? In this case, the right side of (12.54) does *not* make any sense, even though the left side of (12.54) is perfectly meaningful. In this situation, instead of (12.54), we have the following formula:

$$a + ar + ar^2 + \cdots + ar^{n-1} = na \qquad \text{if } r = 1 \tag{12.59}$$

Equation (12.59) is obvious, since every term at the left side of (12.59) is now equal to a (recall that $r = 1$). Notice that a geometric progression with $r = 1$ is also an arithmetic progression with $d = 0$. Is there any other sequence which is both an arithmetic progression and a geometric progression? Give reasons.

Example 12.9 Find the sum of the first eight terms in each of the geometric progressions given in Example 12.8.

Solution (*a*) Here $a = 1$, $r = 2$, $n = 8$, and hence, by (12.54), we have

$$1 + 2 + 4 + 8 + \cdots + 1(2)^{8-1} = \frac{1(2^8 - 1)}{2 - 1} = 255$$

(*b*) Here $a = 3$, $r = 1/3$, $n = 8$, and hence, by (12.54), we have

$$3 + 1 + \frac{1}{3} + \frac{1}{9} + \cdots + 3\left(\frac{1}{3}\right)^{8-1} = \frac{3[(1/3)^8 - 1]}{1/3 - 1}$$

$$= \frac{3280}{729}$$

(*c*) Here $a = 1$, $r = -1/2$, $n = 8$, and hence, by (12.54), we have

$$1 - \frac{1}{2} + \frac{1}{4} - \frac{1}{8} + \cdots + 1\left(\frac{-1}{2}\right)^{8-1} = \frac{1[(-1/2)^8 - 1]}{-1/2 - 1} = \frac{85}{128}$$

Once again, consider the geometric progression $a, ar, ar^2, \ldots, ar^{n-1}, \ldots$. Let us denote the nth term by ℓ; that is,

$$\ell = ar^{n-1} \qquad \text{(the } n\text{th term)} \tag{12.60}$$

Then $\ell r = ar^n$, and hence by substitution in (12.54), we have

$$a + ar + ar^2 + \cdots + ar^{n-1} = \frac{\ell r - a}{r - 1} \qquad \text{if } r \neq 1,\ \ell \text{ the } n\text{th term} \tag{12.61}$$

Example 12.10 In a geometric progression, the first term is 2 and the tenth term is 1024. Find (*a*) the common ratio of this geometric progression; (*b*) the sum of the first 10 terms of this geometric progression; (*c*) the seventh term of this geometric progression.

Solution (*a*) Here we have $a = 2$, $\ell = 1024$, $n = 10$, and hence (12.60) gives

$$1024 = 2r^{10-1} \qquad \text{or} \qquad r^9 = 512 = 2^9$$

Therefore $r = 2$; that is, the common ratio of our geometric progression is 2.

(*b*) By (12.61), we know that

$$\text{Sum of first 10 terms} = \frac{\ell r - a}{r - 1}$$

$$= \frac{(1024)2 - 2}{2 - 1}$$

$$= 2046$$

Therefore the sum of the first 10 terms in our geometric progression is 2046.

(*c*) By (12.60), we have (recall from part *a* that $r = 2$), the seventh term $= ar^{n-1} = 2(2)^{7-1} = 128$.

A very interesting case of Eq. (12.54) arises when the common ratio r is strictly less than 1 in absolute value. As an illustration of what we have in mind, suppose that $r = 1/2$. Then $r^{20} = 1/1{,}048{,}576$, and thus $r^{20} < 0.000001$. Intuitively, it is apparent that r^n can be made as close to zero as we please by choosing the positive integer n sufficiently large. This is typical of what happens when $-1 < r < 1$. Indeed, we have the following fact:

If r is any fixed real number such that $-1 < r < 1$, then r^n is very close to zero when n is a very large positive integer. (12.62)

A glance at (12.54) shows that we may rewrite it as follows:

$$a + ar + ar^2 + \cdots + ar^{n-1} = \frac{a(1 - r^n)}{1 - r} = \frac{a}{1 - r} - \frac{ar^n}{1 - r} \qquad \text{if } r \neq 1 \tag{12.63}$$

Now, suppose that $-1 < r < 1$, *r fixed*. Then, in view of (12.62), r^n is very close to zero when n is a very large positive integer, and hence $ar^n/(1 - r)$, too, is very close to zero for such large positive n. Combining this fact with (12.63), we see intuitively that

If $-1 < r < 1$, r fixed, and if n is a very large positive integer, then

$$a + ar + ar^2 + \cdots + ar^{n-1} \text{ is approximately equal to } \frac{a}{1 - r} \tag{12.64}$$

Indeed, the above sum can be made as close as we please to $a/(1 - r)$ by choosing n large enough.

We define an *infinite geometric series* to be the *sum* of the terms in a geometric progression (the number of such terms is infinite). Also, when we write, for example,

$$1, \frac{1}{2}, \frac{1}{4}, \frac{1}{8}, \frac{1}{16}, \ldots, \frac{1}{2^{n-1}}, \ldots \tag{12.65}$$

we shall understand that (12.65) contains an *infinite* number of terms [this is what is intended by the dots at the *end* of (12.65)]. In fact, (12.65) is an example of an *infinite geometric sequence* of common ratio $1/2$. In view of (12.64), we have the following:

If r is any fixed real number such that $-1 < r < 1$, then the sum of the infinite geometric sequence (12.50) is the geometric series

$$a + ar + ar^2 + \cdots + ar^{n-1} + \cdots \tag{12.66}$$

which is equal to $a/(1-r)$. We also say in this case that the infinite geometric series above converges to $a/(1-r)$.

Example 12.11 Find the following infinite geometric series:

(a) $1 + \frac{1}{2} + \frac{1}{4} + \frac{1}{8} + \cdots + \frac{1}{2^{n-1}} + \cdots$

(b) $1 - \frac{1}{3} + \frac{1}{9} - \frac{1}{27} + \cdots + (-1)^{n-1}\left(\frac{1}{3}\right)^{n-1} + \cdots$

Solution (a) Here $a = 1$, $r = 1/2$, and hence, by (12.66), we have

$$1 + \frac{1}{2} + \frac{1}{4} + \frac{1}{8} + \cdots + \frac{1}{2^{n-1}} + \cdots = \frac{a}{1-r} = \frac{1}{1 - 1/2} = 2$$

(b) Here $a = 1$, $r = -1/3$, and hence, by (12.66), we have

$$1 - \frac{1}{3} + \frac{1}{9} - \frac{1}{27} + \cdots + (-1)^{n-1}\left(\frac{1}{3}\right)^{n-1} + \cdots = \frac{1}{1 - (-1/3)} = \frac{3}{4}$$

Example 12.12 Evaluate the following infinite repeating decimals: (a) 0.1111 . . . ; (b) 0.14141414 . . . ; (c) 0.32222 . . . .

Solution (a) Observe that

$$0.1111\ldots = \frac{1}{10} + \frac{1}{10^2} + \frac{1}{10^3} + \frac{1}{10^4} + \cdots$$

which is an infinite geometric series in which $a = 1/10$ and $r = 1/10$. Hence, by (12.66), we have

$$0.1111\ldots = \frac{a}{1-r} = \frac{1/10}{1 - 1/10} = \frac{1}{9}$$

(b) Observe that

$$0.14141414\ldots = \frac{14}{10^2} + \frac{14}{10^4} + \frac{14}{10^6} + \frac{14}{10^8} + \cdots$$

which is an infinite geometric series in which $a = 14/10^2$ and $r = 1/10^2$. Hence, by (12.66), we have

$$0.14141414\ldots = \frac{14/10^2}{1 - 1/10^2} = \frac{14}{99}$$

(c) Observe that

$$0.32222\ldots = 0.3 + \frac{2}{10^2} + \frac{2}{10^3} + \frac{2}{10^4} + \frac{2}{10^5} + \cdots$$

Now, if we ignore the term 0.3, the remaining sum is an infinite geometric series with first term $2/10^2$ and with common ratio 1/10. Hence, by (12.66), we get,

$$\begin{aligned} 0.32222\ldots &= 0.3 + \frac{2/10^2}{1 - 1/10} \\ &= 0.3 + \frac{2}{90} = \frac{27}{90} + \frac{2}{90} \\ &= \frac{29}{90} \end{aligned}$$

Thus,

$$0.32222\ldots = \frac{29}{90}$$

Using (12.66), as indicated in the above example, we can show the following interesting fact, which is probably familiar to the student:

Every infinite repeating decimal is a rational number (i.e., a number of the form a/b, where a, b are integers).

Next we study harmonic progressions. We say that the sequence

$$a_1, a_2, a_3, \ldots, a_n, \ldots \tag{12.67}$$

of nonzero real numbers forms a *harmonic progression* if the sequence of reciprocals

$$\frac{1}{a_1}, \frac{1}{a_2}, \frac{1}{a_3}, \ldots, \frac{1}{a_n}, \ldots \tag{12.68}$$

forms an arithmetic progression. For example, the sequence

$$1, \frac{1}{2}, \frac{1}{3}, \ldots, \frac{1}{n}, \ldots$$

forms a harmonic progression, since the sequence of reciprocals 1, 2, 3, . . . , n, . . . forms an arithmetic progression. To deal with a problem involving a harmonic progression (12.67), we often consider instead the sequence of reciprocals given in (12.68) which, by definition, forms an arithmetic progression. We then apply to (12.68) the general facts we learned about arithmetic progressions.

xample 12.13 The numbers $1/x$, $(x+1)/x$, $(3x+1)/x$ form a harmonic progression. Find x.

Solution Since $1/x$, $(x+1)/x$, $(3x+1)/x$ form a harmonic progression, it follows from the definition that $x/1$, $x/(x+1)$, $x/(3x+1)$ form an arithmetic progression. Hence,

$$\frac{x}{x+1}-\frac{x}{1}=\frac{x}{3x+1}-\frac{x}{x+1}$$

and thus

$$\frac{x}{1}+\frac{x}{3x+1}=\frac{2x}{x+1}$$

Now, multiplying both sides of this last equation by $(3x+1)(x+1)$, we obtain

$$x(3x+1)(x+1)+x(x+1)=2x(3x+1)$$

Simplifying the last equation, we get $3x^3-x^2=0$. Hence $x=0$ or $x=1/3$. But $x=0$ is impossible here (Why?). Thus, $x=1/3$ is the only possible value of x.

We conclude this section with a discussion of arithmetic, geometric, and harmonic means. Thus, suppose that a and b are any real numbers.

*The **arithmetic mean** of a and b is simply a real number x such that a, x, b form an arithmetic progression.*

Now, since the numbers a, x, b form an arithmetic progression, we obtain at once

$$x-a=b-x$$

and hence $x=(a+b)/2$. Thus, for any real numbers a and b, the

$$\text{Arithmetic mean of } a \text{ and } b=\frac{a+b}{2} \tag{12.69}$$

Next, we define geometric mean. Thus, suppose that a and b are any *positive* real numbers.

*The **geometric mean** of a and b is simply a real number x such that a, x, b form a geometric progression.*

Now, since the numbers a, x, b form a geometric progression, and since a and b are positive, we conclude that $x\neq 0$ and, moreover,

$$\frac{x}{a}=\frac{b}{x}$$

Hence $x^2=ab$, and thus $x=\pm\sqrt{ab}$. The net result, then, is the following:

There are two geometric means of the positive real numbers a and b, namely, $\sqrt{ab}$ and $-\sqrt{ab}$. (Observe that $\pm\sqrt{ab}$ are real since a and b are positive.) (12.70)

Incidentally, a, $\sqrt{ab}$, b form a geometric progression in which the common ratio is $\sqrt{ab}/a$ $[= b/\sqrt{ab}]$, while $a, -\sqrt{ab}, b$ form a geometric progression in which the common ratio is $-\sqrt{ab}/a$ $[= b/(-\sqrt{ab})]$. Observe that the two common ratios we just obtained are negatives of each other. Observe also that we could have defined the geometric mean of a and b if a and b are *both* negative, but not if one is negative and the other is positive (Why?). Finally, we define:

> *The* ***harmonic mean*** *of any two nonzero real numbers a and b is a real number x such that a, x, b form a harmonic progression.* (12.71)

Now, since a, x, b form a harmonic progression, it follows from the definition that

$$\frac{1}{a}, \frac{1}{x}, \frac{1}{b} \text{ form an arithmetic progression} \tag{12.72}$$

Hence,

$$\frac{1}{x} - \frac{1}{a} = \frac{1}{b} - \frac{1}{x}$$

and thus,

$$\frac{1}{x} = \frac{1/a + 1/b}{2} = \frac{a+b}{2ab}$$

Therefore,

$$x = \frac{2ab}{a+b} \qquad \text{provided } a + b \neq 0$$

We have thus shown that

> If a and b are any two nonzero real numbers such that $a + b \neq 0$, then the harmonic mean of a and b is $2ab/(a + b)$. (12.73)

Let us illustrate these concepts by an example.

Example 12.14 Find the arithmetic, geometric, and harmonic means of the numbers 2 and 8.

Solution By Eqs. (12.69), (12.70), and (12.73) we have

$$\text{Arithmetic mean of 2 and 8} = \frac{2+8}{2} = 5$$

$$\text{Geometric means of 2 and 8} = \pm\sqrt{(2)(8)} = \pm 4$$

$$\text{Harmonic mean of 2 and 8} = \frac{2(2)(8)}{2+8} = 3.2$$

In the above example, we have seen that

Arithmetic mean > positive geometric mean > harmonic mean

Is this a coincidence? The answer is no. In fact, the following is always true:

Let a and b be any positive real numbers, and let A, G, H be the arithmetic mean, positive geometric mean, and harmonic mean of a and b, respectively. Then

$$A \geq G \geq H \tag{12.74}$$

Moreover, equality holds in the above inequalities if and only if $a = b$.

To prove (12.74), observe that, by (12.69) and (12.70), we have

$$\begin{aligned} A - G &= \frac{a+b}{2} - \sqrt{ab} \\ &= \frac{a + b - 2\sqrt{ab}}{2} \\ &= \frac{(\sqrt{a} - \sqrt{b})^2}{2} \\ &\geq 0 \qquad \text{(since the square of any real number } \geq 0) \end{aligned}$$

Hence $A - G \geq 0$, and thus $A \geq G$. Moreover, the above calculation shows that $A - G = 0$ if and only if $(\sqrt{a} - \sqrt{b})^2 = 0$, that is, if and only if $\sqrt{a} - \sqrt{b} = 0$, or $a = b$. On the other hand, by (12.70) and (12.73) we have

$$\begin{aligned} G - H &= \sqrt{ab} - \frac{2ab}{a+b} \\ &= \frac{(a+b)\sqrt{ab} - 2ab}{a+b} \\ &= \frac{\sqrt{ab}(a + b - 2\sqrt{ab})}{a+b} \\ &= \frac{\sqrt{ab}(\sqrt{a} - \sqrt{b})^2}{a+b} \\ &\geq 0 \qquad [\text{since } \sqrt{ab} > 0,\ a + b > 0, \text{ and } (\sqrt{a} - \sqrt{b})^2 \geq 0] \end{aligned}$$

Hence $G - H \geq 0$, and thus $G \geq H$. Moreover, the above calculation shows that $G - H = 0$ if and only if $(\sqrt{a} - \sqrt{b})^2 = 0$, that is, if and only if $\sqrt{a} - \sqrt{b} = 0$, or $a = b$. This completes the proof of (12.74).

In the next section, we shall use mathematical induction again—this time to prove the so-called binomial theorem.

Problem Set 12.2

Find the sum of the following arithmetic progressions.

12.2.1 $2 + 4 + 6 + \cdots + 2n$

12.2.2 $3 + 5 + 7 + \cdots + (2n + 1)$

12.2.3 $4 + 7 + 10 + \cdots + (3n + 1)$

12.2.4 $2 + 5 + 8 + \cdots + (3n - 1)$

12.2.5 $1 + 5 + 9 + \cdots + (4n - 3)$

12.2.6 $3 + 6 + 9 + \cdots + 3n$

12.2.7 $-5 - 1 + 3 + \cdots + (4n - 9)$

Find the sum of the following geometric progressions.

12.2.8 $1 - 1/3 + 1/9 - \cdots + (-1)^{n-1}(1/3)^{n-1}$

12.2.9 $1 + 3 + 3^2 + \cdots + 3^{n-1}$

12.2.10 $1 + 1/2 + 1/4 + \cdots + (1/2)^{n-1}$

12.2.11 $5 + 5^2 + 5^3 + \cdots + 5^n$

12.2.12 $-5 + 5^2 - 5^3 + \cdots + (-1)^n 5^n$

12.2.13 Find the sum of an arithmetic progression consisting of 20 terms in which the first term is 1 and the common difference is 2.

12.2.14 Find the sum of a geometric progression consisting of 20 terms in which the first term is 1 and the common ratio is 2. Compare your answer with the answer you obtained for Prob. 12.2.13.

12.2.15 Find the sum of the first 1000 positive integers.

12.2.16 In an arithmetic progression consisting of 30 terms, the first term is 2 and the last term is 89. Find the sum of this progression.

12.2.17 In a geometric progression consisting of 12 terms, the first term is 1 and the last term is 2048. Find the common ratio and the sum of this progression.

12.2.18 Find the arithmetic mean, the geometric mean, and the harmonic mean of 8 and 32.

12.2.19 Let A, G, H be as in (12.74). Show that $G^2 = AH$.

12.2.20 Suppose that x, y, z are distinct real numbers such that $1/(y-x)$, $1/(2y)$, $1/(y-z)$ form an arithmetic progression. Prove that x, y, z form a geometric progression. Is the converse true? Explain.

12.2.21 For what values of x do the numbers x, $2x + 3$, $3x + 22$ form a geometric progression? Find the common ratio in each case. Will these numbers ever form an arithmetic progression? Why or why not?

12.2.22 For what values of x do the numbers x^2, $x^2 + 11$, $4x^2 + 3x + 16$ form an arithmetic progression? Find the common difference in each case.

12.2.23 Suppose that x is any real number such that $x \neq 1$ and $x \neq -1$. Prove that the numbers $1 + x$, $1 - x^2$, $1 - x$ always form a harmonic progression.

12.2.24 The numbers x, $x+1$, $x+3$ form a harmonic progression. Find x.

12.2.25 The arithmetic mean of two numbers is 15, and their positive geometric mean is 9. Find the numbers.

12.2.26 The arithmetic mean of two numbers is 4, and their harmonic mean is 3. Find the numbers.

12.2.27 The harmonic mean of two numbers is 8, and their positive geometric mean is 10. Find the numbers.

12.2.28 Express the following infinite repeating decimals as rational numbers.

(*a*) 0.131313 . . . (*c*) 0.15555 . . .

(*b*) 0.2222 . . . (*d*) 0.123333 . . .

12.2.29 Find the following infinite geometric series.

(*a*) $1+\frac{1}{3}+\frac{1}{3^2}+\cdots+\frac{1}{3^{n-1}}+\cdots$

(*b*) $1-\frac{1}{2}+\frac{1}{2^2}-\cdots+(-1)^{n-1}\left(\frac{1}{2}\right)^{n-1}+\cdots$

(*c*) $5+1+\frac{1}{5}+\cdots+5\left(\frac{1}{5}\right)^{n-1}+\cdots$

(*d*) $1-\frac{1}{7}+\frac{1}{7^2}-\cdots+(-1)^{n-1}\left(\frac{1}{7}\right)^{n-1}+\cdots$

12.2.30 An infinite geometric series is 3, and the first term is 2. Find the common ratio. Also, find the first five terms.

12.2.31 An infinite geometric series is 8, and the common ratio is $1/4$. Find the first six terms.

12.2.32 To show that $x = 0.23545454\ldots$ is a rational number, we may proceed as follows. Since $x = 0.23545454\ldots$, we have

$$(10^4)x = 2354.5454\ldots \qquad (10^2)x = 23.5454\ldots$$

Hence, upon subtracting these two equations, we get $(10^4-10^2)x = 2331$, and thus

$$x=\frac{2331}{10^4-10^2}=\frac{2331}{9900}$$

Work out Prob. 12.2.28, using this method.

12.3 Binomial Theorem

In this final section, we prove a very interesting and important theorem, known as the *Binomial theorem*. This theorem will be proved by mathematical induction.

Suppose a and b are any real numbers, and suppose n is any positive integer. We recall from high school algebra that $a+b$ is called a *binomial*. We now ask:

what is $(a+b)^n$ equal to? Let us try a few cases, say, $n = 2, 3, 4, 5, 6$. It is readily verified that

$$(a+b)^2 = a^2 + 2ab + b^2$$

$$(a+b)^3 = a^3 + 3a^2b + 3ab^2 + b^3$$

$$(a+b)^4 = a^4 + 4a^3b + 6a^2b^2 + 4ab^3 + b^4$$

$$(a+b)^5 = a^5 + 5a^4b + 10a^3b^2 + 10a^2b^3 + 5ab^4 + b^5$$

$$(a+b)^6 = a^6 + 6a^5b + 15a^4b^2 + 20a^3b^3 + 15a^2b^4 + 6ab^5 + b^6$$

Several facts now emerge:

1. The *number of terms* on the right side of each of the above equalities is exactly *1 more than the exponent* appearing on the left side of the same equality.
2. The degree in a and b (i.e., the sum of the exponents of a and b) of each term on the right side of each equality is exactly the same as the exponent appearing on the left side of the same equality. For example, in the first equality, each of the terms a^2, $2ab$, b^2 is of degree 2 in a and b, and this integer is the same as the exponent on the left side of the first equality.
3. The coefficients on the right side of each equality, known as the *binomial coefficients*, read exactly the same sequence forward or backward.

How do these binomial coefficients behave? As an illustration, let us take the coefficients on the right side of the last equality above. These coefficients are 1, 6, 15, 20, 15, 6, 1. We observe the following facts:

$$6 = \frac{6}{1}$$

$$15 = \frac{6 \cdot 5}{1 \cdot 2}$$

$$20 = \frac{6 \cdot 5 \cdot 4}{1 \cdot 2 \cdot 3}$$

$$15 = \frac{6 \cdot 5 \cdot 4 \cdot 3}{1 \cdot 2 \cdot 3 \cdot 4}$$

$$6 = \frac{6 \cdot 5 \cdot 4 \cdot 3 \cdot 2}{1 \cdot 2 \cdot 3 \cdot 4 \cdot 5}$$

$$1 = \frac{6 \cdot 5 \cdot 4 \cdot 3 \cdot 2 \cdot 1}{1 \cdot 2 \cdot 3 \cdot 4 \cdot 5 \cdot 6}$$

This phenomenon is actually true no matter what the positive integer exponent n is in $(a+b)^n$. In fact, it turns out that for every integer r such that $1 \le r \le n$

$$\textit{Coefficient} \text{ of } a^{n-r}b^r \text{ in } (a+b)^n = \frac{n(n-1)(n-2)\cdots(n-r+1)}{1 \cdot 2 \cdot 3 \cdots r} \tag{12.75}$$

The student should verify this fact in each of the formulas for $(a+b)^2$, $(a+b)^3$, $(a+b)^4$, $(a+b)^5$, $(a+b)^6$, given above. The above remarks now make the following theorem plausible. This theorem, known as the *Binomial theorem*, is of considerable interest and importance in mathematics, and its proof gives a further illustration of the technique of proof by mathematical induction.

Binomial Theorem

Suppose n is any positive integer, and suppose a, b are any real numbers. Then

$$(a+b)^n = a^n + \frac{n}{1}a^{n-1}b + \frac{n(n-1)}{1\cdot 2}a^{n-2}b^2 + \frac{n(n-1)(n-2)}{1\cdot 2\cdot 3}a^{n-3}b^3 + \cdots + \frac{n(n-1)(n-2)\cdots(n-r+2)}{1\cdot 2\cdot 3\cdots(r-1)}a^{n-r+1}b^{r-1} + \frac{n(n-1)(n-2)\cdots(n-r+1)}{1\cdot 2\cdot 3\cdots r}a^{n-r}b^r + \cdots + b^n \tag{12.76}$$

This formula for $(a+b)^n$ is known as the *binomial formula*.

We shall now prove that Binomial theorem by mathematical induction. Thus, suppose that

$P(n)$ denotes the above binomial formula (12.76) for $(a+b)^n$ (12.77)

Then, $P(1)$ is simply the statement $(a+b)^1 = a^1 + b^1$, which is certainly true [observe that all the terms in the middle in Eq. (12.76) drop when $n = 1$]. Thus, $P(n)$ is true when $n = 1$; that is,

$P(1)$ is true (12.78)

Now, suppose that the binomial formula is true for $n = k$, where k is a positive integer. In other words, suppose (induction hypothesis) that

$$(a+b)^k = a^k + \frac{k}{1}a^{k-1}b + \frac{k(k-1)}{1\cdot 2}a^{k-2}b^2 + \frac{k(k-1)(k-2)}{1\cdot 2\cdot 3}a^{k-3}b^3 + \cdots + \frac{k(k-1)(k-2)\cdots(k-r+2)}{1\cdot 2\cdot 3\cdots(r-1)}a^{k-r+1}b^{r-1} + \frac{k(k-1)(k-2)\cdots(k-r+1)}{1\cdot 2\cdot 3\cdots r}a^{k-r}b^r + \cdots + b^k \tag{12.79}$$

Multiplying (12.79) by $(a+b)$ and collecting similar terms, we obtain

$$(a+b)^{k+1} = a^{k+1} + \left(\frac{k}{1}+1\right)a^k b + \left[\frac{k(k-1)}{1\cdot 2} + \frac{k}{1}\right]a^{k-1}b^2 + \left[\frac{k(k-1)(k-2)}{1\cdot 2\cdot 3} + \frac{k(k-1)}{1\cdot 2}\right]a^{k-2}b^3 + \cdots + \left[\frac{k(k-1)(k-2)\cdots(k-r+1)}{1\cdot 2\cdot 3\cdots r} + \frac{k(k-1)(k-2)\cdots(k-r+2)}{1\cdot 2\cdot 3\cdots(r-1)}\right]a^{k-r+1}b^r + \cdots + b^{k+1} \tag{12.80}$$

In verifying (12.80), observe that when the right side of (12.79) is multiplied by $(a+b)$, the first four terms in the resulting product are

$$aa^k = a^{k+1}$$

$$a\left(\frac{k}{1}a^{k-1}b\right) + ba^k = \left(\frac{k}{1}+1\right)a^kb$$

$$a\left[\frac{k(k-1)}{1\cdot 2}a^{k-2}b^2\right] + b\left(\frac{k}{1}a^{k-1}b\right) = \left[\frac{k(k-1)}{1\cdot 2}+\frac{k}{1}\right]a^{k-1}b^2$$

$$a\left[\frac{k(k-1)(k-2)}{1\cdot 2\cdot 3}a^{k-3}b^3\right] + b\left[\frac{k(k-1)}{1\cdot 2}a^{k-2}b^2\right] = \left[\frac{k(k-1)(k-2)}{1\cdot 2\cdot 3}+\frac{k(k-1)}{1\cdot 2}\right]a^{k-2}b^3$$

and these are precisely the first four terms which appear on the right side of (12.80). In general, the typical term we obtain when the right side of (12.79) is multiplied by $a+b$ is

$$a\left[\frac{k(k-1)(k-2)\cdots(k-r+1)}{1\cdot 2\cdot 3\cdots r}a^{k-r}b^r\right]$$
$$+\,b\left[\frac{k(k-1)(k-2)\cdots(k-r+2)}{1\cdot 2\cdot 3\cdots(r-1)}a^{k-r+1}b^{r-1}\right]$$
$$=\left[\frac{k(k-1)(k-2)\cdots(k-r+1)}{1\cdot 2\cdot 3\cdots r}+\frac{k(k-1)(k-2)\cdots(k-r+2)}{1\cdot 2\cdot 3\cdots(r-1)}\right]a^{k-r+1}b^r$$

and this is precisely the typical term which appears on the right side of (12.80). Now that we have verified (12.80), let us simplify the coefficients on the right side of (12.80); we get

$$\frac{k}{1}+1=\frac{k+1}{1}$$

$$\frac{k(k-1)}{1\cdot 2}+\frac{k}{1}=\frac{k(k-1)+2k}{1\cdot 2}=\frac{(k+1)k}{1\cdot 2}$$

$$\frac{k(k-1)(k-2)}{1\cdot 2\cdot 3}+\frac{k(k-1)}{1\cdot 2}=\frac{k(k-1)(k-2)+3k(k-1)}{1\cdot 2\cdot 3}=\frac{(k+1)k(k-1)}{1\cdot 2\cdot 3}$$

In general, the typical expression in brackets in (12.80), when simplified, gives

$$\frac{k(k-1)(k-2)\cdots(k-r+1)}{1\cdot 2\cdot 3\cdots r}+\frac{k(k-1)(k-2)\cdots(k-r+2)}{1\cdot 2\cdot 3\cdots(r-1)}$$
$$=\frac{k(k-1)(k-2)\cdots(k-r+2)(k-r+1)+k(k-1)(k-2)\cdots(k-r+2)r}{1\cdot 2\cdot 3\cdots r}$$
$$=\frac{k(k-1)(k-2)\cdots(k-r+2)[(k-r+1)+r]}{1\cdot 2\cdot 3\cdots r}$$
$$=\frac{(k+1)k(k-1)(k-2)\cdots(k-r+2)}{1\cdot 2\cdot 3\cdots r}$$

Substituting the above results into (12.80), we get

$$(a+b)^{k+1} = a^{k+1} + \frac{(k+1)}{1}a^k b + \frac{(k+1)k}{1\cdot 2}a^{k-1}b^2 + \frac{(k+1)k(k-1)}{1\cdot 2\cdot 3}a^{k-2}b^3$$
$$+\cdots+\frac{(k+1)k(k-1)(k-2)\cdots(k-r+2)}{1\cdot 2\cdot 3\cdots r}a^{k-r+1}b^r+\cdots+b^{k+1}$$

which is precisely what the binomial formula for $(a+b)^n$ reads when $n=k+1$. [We did not display the term involving $a^{k-r+2}b^{r-1}$ in the above expansion of $(a+b)^{k+1}$.] In other words, we have shown that

$$P(k+1) \text{ is true whenever } P(k) \text{ is true} \tag{12.81}$$

for *all* positive integers k. In view of (12.78) and (12.81), it follows, by mathematical induction, that $P(n)$ is true for *all* positive integers n, which [see (12.77)] proves the Binomial theorem.

Example 12.15 Expand $(a+b)^7$.

Solution Applying the Binomial theorem, we get

$$(a+b)^7 = a^7 + \frac{7}{1}a^6b + \frac{7\cdot 6}{1\cdot 2}a^5b^2 + \frac{7\cdot 6\cdot 5}{1\cdot 2\cdot 3}a^4b^3 + \frac{7\cdot 6\cdot 5\cdot 4}{1\cdot 2\cdot 3\cdot 4}a^3b^4$$
$$+\frac{7\cdot 6\cdot 5\cdot 4\cdot 3}{1\cdot 2\cdot 3\cdot 4\cdot 5}a^2b^5 + \frac{7\cdot 6\cdot 5\cdot 4\cdot 3\cdot 2}{1\cdot 2\cdot 3\cdot 4\cdot 5\cdot 6}ab^6 + b^7$$
$$= a^7 + 7a^6b + 21a^5b^2 + 35a^4b^3 + 35a^3b^4 + 21a^2b^5 + 7ab^6 + b^7$$

Hence,

$$(a+b)^7 = a^7 + 7a^6b + 21a^5b^2 + 35a^4b^3 + 35a^3b^4 + 21a^2b^5 + 7ab^6 + b^7 \tag{12.82}$$

Example 12.16 Find, correct to three decimals, $(1.01)^7$.

Solution By setting $a=1$ and $b=0.01$ in (12.82), we get

$$\begin{aligned}(1.01)^7 &= (1+0.01)^7 = 1 + 7(0.01) + 21(0.01)^2 + 35(0.01)^3 + \cdots \\ &\approx 1.072\end{aligned} \tag{12.83}$$

Thus, $(1.01)^7$ is approximately equal to 1.072. Observe that all the terms after the third term on the right side of (12.83) are negligible if we are interested in only the first three decimals. Thus, we need not compute these terms in this case.

Example 12.17 Find the coefficient of a^5b^7 in the binomial formula expansion of $(a+b)^{12}$.

Solution By (12.75), we have

$$\text{Coefficient of } a^{12-7}b^7 = \frac{12 \cdot 11 \cdot 10 \cdots (12-7+1)}{1 \cdot 2 \cdot 3 \cdots 7}$$

$$= \frac{12 \cdot 11 \cdot 10 \cdot 9 \cdot 8 \cdot 7 \cdot 6}{1 \cdot 2 \cdot 3 \cdot 4 \cdot 5 \cdot 6 \cdot 7}$$

$$= 792$$

Thus, the coefficient of a^5b^7 in the binomial formula expansion of $(a+b)^{12}$ is 792.

Example 12.18 Suppose that the function $f: R \to R$, where R is the set of real numbers, is given by $f(x) = x^n$, n is a positive integer. Find an approximate value of the quotient difference

$$\frac{f(x+h) - f(x)}{h}$$

when h is very close to zero but $h \neq 0$.

Solution By definition of the function f, we have

$$f(x+h) = (x+h)^n, \; f(x) = x^n$$

and hence

$$\frac{f(x+h) - f(x)}{h} = \frac{(x+h)^n - x^n}{h} \qquad (h \neq 0) \tag{12.84}$$

Now, by the Binomial theorem, we have

$$(x+h)^n = x^n + nx^{n-1}h + h^2g(x) \tag{12.85}$$

where $g(x)$ is a polynomial in x with coefficients involving h. Now, substituting (12.85) into (12.84), we obtain

$$\frac{f(x+h) - f(x)}{h} = \frac{[x^n + nx^{n-1}h + h^2g(x)] - x^n}{h}$$

$$= \frac{[nx^{n-1} + hg(x)]h}{h}$$

$$= nx^{n-1} + hg(x)$$

if $h \neq 0$. We have thus shown that

$$\frac{f(x+h) - f(x)}{h} = nx^{n-1} + hg(x) \qquad \text{if } h \neq 0 \tag{12.86}$$

Now, suppose that h is very close to zero, but $h \neq 0$. Then the right side of (12.86) becomes very close to nx^{n-1}, since the polynomial $hg(x)$ is very close to zero when h is close to zero. We have thus shown the following interesting result:

Suppose that $f: R \to R$ is the function given by $f(x) = x^n$. Then the quotient difference

$$\frac{f(x+h) - f(x)}{h}$$

is approximately equal to nx^{n-1} when h is very close to zero but $h \neq 0$.

The binomial coefficients which appear on the right side of the binomial formula in the Binomial theorem are often denoted by $\binom{n}{r}$, (read: n above r). Thus, by definition, if n and r are positive integers and if $r \leq n$, then [see (12.75)]

$$\binom{n}{r} = \frac{n(n-1)(n-2)\cdots(n-r+1)}{1 \cdot 2 \cdot 3 \cdots r} = \begin{matrix}\text{coefficient of} \\ a^{n-r}b^r \text{ in } (a+b)^n\end{matrix} \tag{12.87}$$

Another convenient notation is the so-called *factorial notation*, which is defined as follows:

> Suppose that k is any positive integer. Then $k!$ (read: k factorial) is defined by
>
> $$k! = 1 \cdot 2 \cdot 3 \cdots k \tag{12.88}$$
>
> For convenience, we define $0! = 1$.

Thus, $1! = 1$; $2! = 1 \cdot 2 = 2$; $3! = 1 \cdot 2 \cdot 3 = 6$, and so on. With the aid of this factorial notation, we are now able to cast the binomial coefficient $\binom{n}{r}$ given in (12.87) in another useful form. Indeed, we have

$$\binom{n}{r} = \frac{n!}{r!(n-r)!} \qquad n, r \text{ positive integers, } r \leq n \tag{12.89}$$

The proof of (12.89) is as follows: By definition of factorial given in (12.88), the right side of (12.89) becomes

$$\begin{aligned}
\frac{n!}{r!(n-r)!} &= \frac{1 \cdot 2 \cdot 3 \cdots (n-r)(n-r+1)(n-r+2)\cdots n}{(1 \cdot 2 \cdot 3 \cdots r)[1 \cdot 2 \cdot 3 \cdots (n-r)]} \\
&= \frac{(n-r+1)(n-r+2)\cdots n}{1 \cdot 2 \cdot 3 \cdots r} \qquad \text{(after canceling)} \\
&= \frac{n(n-1)(n-2)\cdots(n-r+1)}{1 \cdot 2 \cdot 3 \cdots r} \qquad \begin{matrix}\text{(by reversing the order of} \\ \text{the factors in numerator)}\end{matrix} \\
&= \binom{n}{r} \qquad \text{by (12.87)}
\end{aligned}$$

This proves (12.89).

A very interesting corollary of (12.89) is

$$\binom{n}{r} = \binom{n}{n-r} \qquad n \text{ a positive integer, } 1 \leq r \leq n-1 \tag{12.90}$$

The proof of (12.90) is quite easy. In fact, by replacing r by $n-r$ in (12.89), we obtain

$$\binom{n}{n-r} = \frac{n!}{(n-r)![n-(n-r)]!} = \frac{n!}{(n-r)!r!} = \frac{n!}{r!(n-r)!}$$

$$= \binom{n}{r} \qquad \text{by (12.89)}$$

which proves (12.90). Incidentally, if we further agree, as is customary, to *define*

$$\binom{n}{0} = 1 \quad n \text{ any positive integer} \tag{12.91}$$

then (12.90) is easily seen to be valid for all r such that $0 \leq r \leq n$. Now, recalling (12.87), we have

$$\binom{n}{r} = \begin{array}{l}\text{coefficient of } a^{n-r}b^r \text{ in the} \\ \text{binomial formula expansion of } (a+b)^n\end{array} \tag{12.92}$$

$$\binom{n}{n-r} = \begin{array}{l}\text{coefficient of } a^{n-(n-r)}b^{n-r} \ (= a^r b^{n-r}) \text{ in} \\ \text{the binomial formula expansion of } (a+b)^n\end{array} \tag{12.93}$$

In view of (12.90) to (12.93), we now see that the binomial coefficients are "symmetric" in the sense that they read in the same sequence in either order (i.e., forward or backward). Moreover, taking advantage of the above notation and, again, using (12.90) to (12.93), we may cast the binomial formula in the Binomial theorem, as follows:

$$(a+b)^n = \binom{n}{0}a^n + \binom{n}{1}a^{n-1}b + \binom{n}{2}a^{n-2}b^2 + \cdots + \binom{n}{r}a^{n-r}b^r + \cdots + \binom{n}{n}b^n \tag{12.94}$$

where n is any positive integer.

Example 12.19 Prove that, for every positive integer n,

(*a*) $\binom{n}{0} + \binom{n}{1} + \binom{n}{2} + \cdots + \binom{n}{n} = 2^n$

(*b*) $\binom{n}{0} - \binom{n}{1} + \binom{n}{2} - \cdots + (-1)^n\binom{n}{n} = 0$

Solution (*a*) Setting $a = 1$ and $b = 1$ in (12.94), we obtain at once the equation in (*a*).

(*b*) Setting $a = 1$, $b = -1$ in (12.94), we readily see that the equation in (*b*) is true.

Example 12.20 Evaluate $\binom{15}{12}$.

Solution It would be unnecessarily complicated to apply (12.87). Instead, we first use (12.90), to obtain

$$\binom{15}{12} = \binom{15}{15-12} = \binom{15}{3} \tag{12.95}$$

Now, by (12.87),

$$\binom{15}{3} = \frac{15 \cdot 14 \cdot 13}{1 \cdot 2 \cdot 3} = \frac{15 \cdot 14 \cdot 13}{6} = 455 \tag{12.96}$$

and hence, by (12.95) and (12.96) we get $\binom{15}{12} = 455$.

Example 12.21 Find the coefficient of x^4y^{16} in the binomial formula expansion of $(2x - y)^{20}$.

Solution Setting $a = 2x$, $b = -y$, $n = 20$, $r = 16$, in (12.94), and keeping in mind (12.87), we conclude that

$$\text{Coefficient of } (2x)^{20-16}(-y)^{16} = \binom{20}{16} = \binom{20}{20-16} = \binom{20}{4} \tag{12.97}$$

by (12.90). In view of (12.97), we know that

$$\binom{20}{4}(2x)^{20-16}(-y)^{16} = \binom{20}{4}2^4(-1)^{16}x^4y^{16}$$

is the term involving x^4y^{16} in the binomial formula expansion of $(2x - y)^{20}$. Since

$$\binom{20}{4}2^4(-1)^{16} = 16\binom{20}{4} = 16 \cdot \frac{(20)(19)(18)(17)}{4!}$$
$$= (16)(4845) = 77{,}520$$

we conclude that the desired coefficient is 77,520.

We conclude this section with the following remark.

Remark In Prob. 12.3.9 you will be asked to show that, for all positive integers n and r, where $r \leq n$,

$$\binom{n}{r-1} + \binom{n}{r} = \binom{n+1}{r} \tag{12.98}$$

Equation (12.98) can be used effectively to obtain the binomial coefficients $\binom{n}{r}$, for the various values of n, as indicated below in what is known as *Pascal's triangle*:

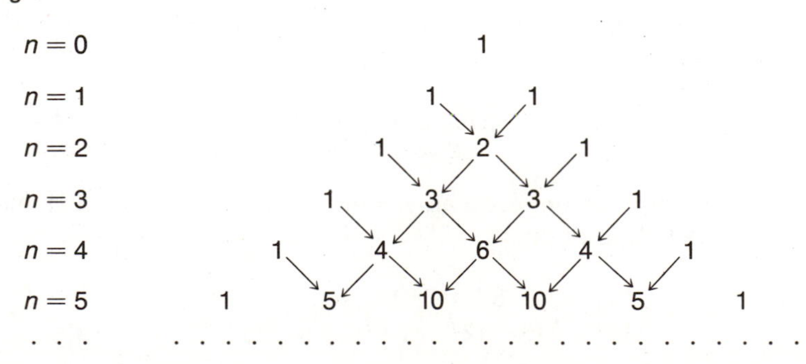

Observe that *the integers in any row n are precisely the binomial coefficients in* $(a+b)^n$. For example, the integers in the row headed by $n=4$ are 1, 4, 6, 4, 1, and these are precisely the binomial coefficients in the binomial expansion of $(a+b)^4$. The above assertion [regarding the binomial coefficients in $(a+b)^n$] can be justified by appealing to Eq. (12.98). For, observe that the sum of each two consecutive integers in the same row has been written in the following row in the position indicated by the arrows—precisely as prescribed in Eq. (12.98).

Problem Set 12.3

12.3.1 Find the binomial formula expansion of each of the following:

(*a*) $(x-y)^8$

(*b*) $(3x+y)^7$

(*c*) $(2x-3y)^5$

(*d*) $(1-x)^{10}$

(*e*) $(x-1)^9$

(*f*) $(3-y)^6$

12.3.2 Find, correct to three decimals, the following: (*a*) $(1.02)^8$; (*b*) $(0.98)^8$; $(1.1)^{10}$; (*d*) $(0.9)^{10}$.

12.3.3 Find the coefficient of x^3y^7 in the binomial formula expansion of $(3x+y)^{10}$.

12.3.4 Find the coefficient of x^7y^3 in the binomial formula expansion of $(x-3y)^{10}$.

12.3.5 Show that, for all positive integers k, $k! = k \cdot (k-1)!$.

12.3.6 Evaluate:

(*a*) $6!$

(*b*) $\dfrac{10!}{7!}$

(*c*) $\dfrac{12!}{8!4!}$

(*d*) $\dfrac{9!-8!}{7!}$

12.3.7 Evaluate:

(*a*) $\binom{5}{3}+\binom{5}{2}-\binom{6}{3}$

(*b*) $\binom{10}{10}-\binom{5}{0}$

(*c*) $\binom{5}{0}+\binom{5}{1}+\binom{5}{2}+\binom{5}{3}+\binom{5}{4}+\binom{5}{5}$

(*d*) $\binom{5}{0}-\binom{5}{1}+\binom{5}{2}-\binom{5}{3}+\binom{5}{4}-\binom{5}{5}$

(e) $\binom{4}{0}+\binom{4}{1}+\binom{4}{2}+\binom{4}{3}$

(f) $-\binom{4}{0}+\binom{4}{1}-\binom{4}{2}+\binom{4}{3}-\binom{4}{4}$

12.3.8 Evaluate:

(a) $\binom{100}{97}$

(b) $\binom{50}{47}$

12.3.9 Show that, for all positive integers n, r where $r \leq n$,

$$\binom{n}{r}+\binom{n}{r-1}=\binom{n+1}{r}$$

12.3.10 Suppose that $f:R \to R$, where R is the set of real numbers, is given by $f(x) = cx^n$, c is a constant, n is a positive integer. Find an approximate value of the quotient difference $[f(x+h) - f(x)]/h$ when h is very close to zero but h is not zero.

12.3.11 Suppose that $f:R \to R$, where R is the set of real numbers, is given by $f(x) = 3x^2 - 4x + 7$. Find an approximate value of the quotient difference $[f(x+h) - f(x)]/h$ when h is very close to zero but h is not zero.

12.3.12 Suppose that $f:R \to R$, where R is the set of real numbers, is given by $f(x) = 2x^3$. Find the value of each of the quotient differences $[f(1+h) - f(1)]/h$ when $h = 0.1, 0.01, -0.1, -0.01$. What are these values close to? How do your answers compare with the result in Prob. 12.3.10?

12.3.13 Sketch the graph of $y = f(x) = 2x^3$, and draw each of the two secant lines joining the points on the graph of f with coordinates $(1, f(1))$ and $(1+h, f(1+h))$ when $h = 0.1, -0.1$. Find the slope of each of these two secant lines. Interpret the results geometrically. [*Hint*: Use the results of Prob. 12.3.12.]

12.3.14 Prove that, for all positive integers n and all nonnegative real numbers x, $(1+x)^n \geq 1 + nx$.

12.3.15 Let S be a set consisting of exactly n elements. Show that the number of distinct subsets T of S of k elements each is $\binom{n}{k}$, where $0 \leq k \leq n$.

Hint: Let $T = \{a_1, a_2, \ldots, a_k\}$. Show that there are exactly n choices for a_1. Having chosen a_1, show that there are exactly $n-1$ choices for a_2, and so on. Thus, there are exactly $n(n-1) \cdots (n-k+1)$ choices for $a_1, a_2, \ldots, a_k$. Because the *order* of the elements $a_1, a_2, \ldots, a_k$ is *not* important in forming a subset, each rearrangement of $a_1, a_2, \ldots, a_k$ still gives the *same* subset T. But there are $k(k-1) \cdots 1$ ways to rearrange the elements $a_1, a_2, \ldots, a_k$ (Why?), and all these rearrangements give rise to the *same* subset T. Hence the number of *distinct* subsets T of k elements each is

$$\frac{n(n-1) \cdots (n-k+1)}{k \cdot (k-1) \cdots 1} = \binom{n}{k}$$

12.3.16 Use the result in Prob. 12.3.15 to give an interpretation of the result in Example 12.19*a* in terms of subsets. [Compare with the result in Prob. 1.1.6.]

12.4 Summary

An important technique for proving theorems, known as the principle of mathematical induction, states the following:

> Suppose $P(n)$ is a statement involving a positive integer n. Suppose that (1) $P(1)$ is true, and (2) if $P(k)$ is true, then $P(k+1)$ is also true, where k is any positive integer. Then $P(n)$ is true for all positive integers n.

We also saw that *both* of the above conditions (1) and (2) are needed in any proof which uses mathematical induction.

An *arithmetic progression* is a sequence of numbers of the form

$$a, a+d, a+2d, \ldots, a+(n-1)d, \ldots \tag{12.99}$$

The sum S_n of the first n terms of an arithmetic progression (12.99) is given by

$$S_n = \frac{n}{2}[2a + (n-1)d] \tag{12.100}$$

A *geometric progression* is a sequence of numbers of the form

$$a, ar, ar^2, \ldots, ar^{n-1}, \ldots \tag{12.101}$$

The sum T_n of the first n terms of a geometric progression (12.101) is given by

$$T_n = \frac{a(r^n - 1)}{r-1} \qquad \text{if } r \neq 1 \tag{12.102}$$

(If $r = 1$, the sum T_n is now given by $T_n = na$.)

We saw that both Eqs. (12.100) and (12.102) can be proved by mathematical induction (as well as by other means).

Suppose that the value of r in the infinite geometric sequence in (12.101) is such that $-1 < r < 1$. Then the *sum* of this infinite geometric sequence, called an *infinite geometric series*, is equal to $a/(1-r)$.

A sequence $a_1, a_2, a_3, \ldots, a_n, \ldots$ of nonzero real numbers is said to form a *harmonic progression* if the sequence of reciprocals $1/a_1, 1/a_2, 1/a_3, \ldots, 1/a_n, \ldots$ forms an arithmetic progression. To deal with a problem involving a harmonic progression, we often consider instead the sequence of reciprocals which, by definition, forms an arithmetic progression, and then apply the general facts we learned about arithmetic progressions.

Suppose that a and b are *positive* real numbers. The *arithmetic mean* A, the *positive geometric mean* G, and the *harmonic mean* H, of a and b, are given by $A = (a+b)/2$, $G = \sqrt{ab}$, $H = 2ab/(a+b)$. Moreover, we always have $A \geq G \geq H$, with equality holding if and only if $a = b$.

Finally, suppose that a and b are any real numbers and n is any positive integer. The *Binomial theorem*, proved by mathematical induction, states that

$$(a+b)^n = a^n + \binom{n}{1}a^{n-1}b + \binom{n}{2}a^{n-2}b^2 + \cdots + \binom{n}{n-1}ab^{n-1} + b^n \qquad (12.103)$$

where $\binom{n}{r} = [n(n-1)(n-2)\cdots(n-r+1)]/(1\cdot 2\cdot 3\cdots r)$. These numbers $\binom{n}{r}$ are called the *binomial coefficients*, and are symmetrically located in (12.103) in the sense that $\binom{n}{r} = \binom{n}{n-r}$. Another interesting property of these binomial coefficients is the following: $\binom{n}{r-1} + \binom{n}{r} = \binom{n+1}{r}$. This property can be used effectively to obtain the binomial coefficients for the various values of n. The binomial coefficients may also be obtained systematically by means of *Pascal's triangle*.

Tables

Table A Squares and Square Roots

x	x^2	$\sqrt{x}$	x	x^2	$\sqrt{x}$
1	1	1.000	51	2,601	7.141
2	4	1.414	52	2,704	7.211
3	9	1.732	53	2,809	7.280
4	16	2.000	54	2,916	7.348
5	25	2.236	55	3,025	7.416
6	36	2.449	56	3,136	7.483
7	49	2.646	57	3,249	7.550
8	64	2.828	58	3,364	7.616
9	81	3.000	59	3,481	7.681
10	100	3.162	60	3,600	7.746
11	121	3.317	61	3,721	7.810
12	144	3.464	62	3,844	7.874
13	169	3.606	63	3,969	7.937
14	196	3.742	64	4,096	8.000
15	225	3.873	65	4,225	8.062
16	256	4.000	66	4,356	8.124
17	289	4.123	67	4,489	8.185
18	324	4.243	68	4,624	8.246
19	361	4.359	69	4,761	8.307
20	400	4.472	70	4,900	8.367
21	441	4.583	71	5,041	8.426
22	484	4.690	72	5,184	8.485
23	529	4.796	73	5,329	8.544
24	576	4.899	74	5,476	8.602
25	625	5.000	75	5,625	8.660
26	676	5.099	76	5,776	8.718
27	729	5.196	77	5,929	8.775
28	784	5.292	78	6,084	8.832
29	841	5.385	79	6,241	8.888
30	900	5.477	80	6,400	8.944
31	961	5.568	81	6,561	9.000
32	1,024	5.657	82	6,724	9.055
33	1,089	5.745	83	6,889	9.110
34	1,156	5.831	84	7,056	9.165
35	1,225	5.916	85	7,225	9.220
36	1,296	6.000	86	7,396	9.274
37	1,369	6.083	87	7,569	9.327
38	1,444	6.164	88	7,744	9.381
39	1,521	6.245	89	7,921	9.434
40	1,600	6.325	90	8,100	9.487
41	1,681	6.403	91	8,281	9.539
42	1,764	6.481	92	8,464	9.592
43	1,849	6.557	93	8,649	9.644
44	1,936	6.633	94	8,836	9.695
45	2,025	6.708	95	9,025	9.747
46	2,116	6.782	96	9,216	9.798
47	2,209	6.856	97	9,409	9.849
48	2,304	6.928	98	9,604	9.899
49	2,401	7.000	99	9,801	9.950
50	2,500	7.071	100	10,000	10.000

Table B Exponential Functions

x	e^x	e^{-x}	x	e^x	e^{-x}
0.00	1.0000	1.0000	1.5	4.4817	0.2231
0.01	1.0101	0.9901	1.6	4.9530	0.2019
0.02	1.0202	0.9802	1.7	5.4739	0.1827
0.03	1.0305	0.9705	1.8	6.0496	0.1653
0.04	1.0408	0.9608	1.9	6.6859	0.1496
0.05	1.0513	0.9512	2.0	7.3891	0.1353
0.06	1.0618	0.9418	2.1	8.1662	0.1225
0.07	1.0725	0.9324	2.2	9.0250	0.1108
0.08	1.0833	0.9331	2.3	9.9742	0.1003
0.09	1.0942	0.9139	2.4	11.0230	0.0907
0.10	1.1052	0.9048	2.5	12.182	0.0821
0.11	1.1163	0.8958	2.6	13.464	0.0743
0.12	1.1275	0.8869	2.7	14.880	0.0672
0.13	1.1388	0.8781	2.8	16.445	0.0608
0.14	1.1503	0.8694	2.9	18.174	0.0550
0.15	1.1618	0.8607	3.0	20.086	0.0498
0.16	1.1735	0.8521	3.1	22.198	0.0450
0.17	1.1853	0.8437	3.2	24.533	0.0408
0.18	1.1972	0.8353	3.3	27.113	0.0369
0.19	1.2092	0.8270	3.4	29.964	0.0334
0.20	1.2214	0.8187	3.5	33.115	0.0302
0.21	1.2337	0.8106	3.6	36.598	0.0273
0.22	1.2461	0.8025	3.7	40.447	0.0247
0.23	1.2586	0.7945	3.8	44.701	0.0224
0.24	1.2712	0.7866	3.9	49.402	0.0202
0.25	1.2840	0.7788	4.0	54.598	0.0183
0.30	1.3499	0.7408	4.1	60.340	0.0166
0.35	1.4191	0.7047	4.2	66.686	0.0150
0.40	1.4918	0.6703	4.3	73.700	0.0136
0.45	1.5683	0.6376	4.4	81.451	0.0123
0.50	1.6487	0.6065	4.5	90.017	0.0111
0.55	1.7333	0.5769	4.6	99.484	0.0101
0.60	1.8221	0.5488	4.7	109.950	0.0091
0.65	1.9155	0.5220	4.8	121.510	0.0082
0.70	2.0138	0.4966	4.9	134.290	0.0074
0.75	2.1170	0.4724	5.0	148.41	0.0067
0.80	2.2255	0.4493	5.5	244.69	0.0041
0.85	2.3396	0.4274	6.0	403.43	0.0025
0.90	2.4596	0.4066	6.5	665.14	0.0015
0.95	2.5857	0.3867	7.0	1096.60	0.0009
1.0	2.7183	0.3679	7.5	1808.0	0.0006
1.1	3.0042	0.3329	8.0	2981.0	0.0003
1.2	3.3201	0.3012	8.5	4914.8	0.0002
1.3	3.6693	0.2725	9.0	8103.1	0.0001
1.4	4.0552	0.2466	10.0	22026.0	0.00005

Table C
Common Logarithms

x	0	1	2	3	4	5	6	7	8	9
1.0	.0000	.0043	.0086	.0128	.0170	.0212	.0253	.0294	.0334	.0374
1.1	.0414	.0453	.0492	.0531	.0569	.0607	.0645	.0682	.0719	.0755
1.2	.0792	.0828	.0864	.0899	.0934	.0969	.1004	.1038	.1072	.1106
1.3	.1139	.1173	.1206	.1239	.1271	.1303	.1335	.1367	.1399	.1430
1.4	.1461	.1492	.1523	.1533	.1584	.1614	.1644	.1673	.1703	.1732
1.5	.1761	.1790	.1818	.1847	.1875	.1903	.1931	.1959	.1987	.2014
1.6	.2041	.2068	.2095	.2122	.2148	.2175	.2201	.2227	.2253	.2279
1.7	.2304	.2330	.2355	.2380	.2405	.2430	.2455	.2480	.2504	.2529
1.8	.2553	.2577	.2601	.2625	.2648	.2672	.2695	.2718	.2742	.2765
1.9	.2788	.2810	.2833	.2856	.2878	.2900	.2923	.2945	.2967	.2989
2.0	.3010	.3032	.3054	.3075	.3096	.3118	.3139	.3160	.3181	.3201
2.1	.3222	.3243	.3263	.3284	.3304	.3324	.3345	.3365	.3385	.3404
2.2	.3424	.3444	.3464	.3483	.3502	.3522	.3541	.3560	.3579	.3598
2.3	.3617	.3636	.3655	.3674	.3692	.3711	.3729	.3747	.3766	.3784
2.4	.3802	.3820	.3838	.3856	.3874	.3892	.3909	.3927	.3945	.3962
2.5	.3979	.3997	.4014	.4031	.4048	.4065	.4082	.4099	.4116	.4133
2.6	.4150	.4166	.4183	.4200	.4216	.4232	.4249	.4265	.4281	.4298
2.7	.4314	.4330	.4346	.4362	.4378	.4393	.4409	.4425	.4440	.4456
2.8	.4472	.4487	.4502	.4518	.4533	.4548	.4564	.4579	.4594	.4609
2.9	.4624	.4639	.4654	.4669	.4683	.4698	.4713	.4728	.4742	.4757
3.0	.4771	.4786	.4800	.4814	.4829	.4843	.4857	.4871	.4886	.4900
3.1	.4914	.4928	.4942	.4955	.4969	.4983	.4997	.5011	.5024	.5038
3.2	.5051	.5065	.5079	.5092	.5105	.5119	.5132	.5145	.5159	.5172
3.3	.5185	.5198	.5211	.5224	.5237	.5250	.5263	.5276	.5289	.5302
3.4	.5315	.5328	.5340	.5353	.5366	.5378	.5391	.5403	.5416	.5428
3.5	.5441	.5453	.5465	.5478	.5490	.5502	.5514	.5527	.5539	.5551
3.6	.5563	.5575	.5587	.5599	.5611	.5623	.5635	.5647	.5658	.5670
3.7	.5682	.5694	.5705	.5717	.5729	.5740	.5752	.5763	.5775	.5786
3.8	.5798	.5809	.5821	.5832	.5843	.5855	.5866	.5877	.5888	.5899
3.9	.5911	.5922	.5933	.5944	.5955	.5966	.5977	.5988	.5999	.6010
4.0	.6021	.6031	.6042	.6053	.6064	.6075	.6085	.6096	.6107	.6117
4.1	.6128	.6138	.6149	.6160	.6170	.6180	.6191	.6201	.6212	.6222
4.2	.6232	.6243	.6253	.6263	.6274	.6284	.6294	.6304	.6314	.6325
4.3	.6335	.6345	.6355	.6365	.6375	.6385	.6395	.6405	.6415	.6425
4.4	.6435	.6444	.6454	.6464	.6474	.6484	.6493	.6503	.6513	.6522
4.5	.6532	.6542	.6551	.6561	.6571	.6580	.6590	.6599	.6609	.6618
4.6	.6628	.6637	.6646	.6656	.6665	.6675	.6684	.6693	.6702	.6712
4.7	.6721	.6730	.6739	.6749	.6758	.6767	.6776	.6785	.6794	.6803
4.8	.6812	.6821	.6830	.6839	.6848	.6857	.6866	.6875	.6884	.6893
4.9	.6902	.6911	.6920	.6928	.6937	.6946	.6955	.6964	.6972	.6981
5.0	.6990	.6998	.7007	.7016	.7024	.7033	.7042	.7050	.7059	.7067
5.1	.7076	.7084	.7093	.7101	.7110	.7118	.7126	.7135	.7143	.7152
5.2	.7160	.7168	.7177	.7185	.7193	.7202	.7210	.7218	.7226	.7235
5.3	.7243	.7251	.7259	.7267	.7275	.7284	.7292	.7300	.7308	.7316
5.4	.7324	.7332	.7340	.7348	.7356	.7364	.7372	.7380	.7388	.7396
x	0	1	2	3	4	5	6	7	8	9

Table C (continued)

x	0	1	2	3	4	5	6	7	8	9
5.5	.7404	.7412	.7419	.7427	.7435	.7443	.7451	.7459	.7466	.7474
5.6	.7482	.7490	.7497	.7505	.7513	.7520	.7528	.7536	.7543	.7551
5.7	.7559	.7566	.7574	.7582	.7589	.7597	.7604	.7612	.7619	.7627
5.8	.7634	.7642	.7649	.7657	.7664	.7672	.7679	.7686	.7694	.7701
5.9	.7709	.7716	.7723	.7731	.7738	.7745	.7752	.7760	.7767	.7774
6.0	.7782	.7789	.7796	.7803	.7810	.7818	.7825	.7832	.7839	.7846
6.1	.7853	.7860	.7868	.7875	.7882	.7889	.7896	.7903	.7910	.7917
6.2	.7924	.7931	.7938	.7945	.7952	.7959	.7966	.7973	.7980	.7987
6.3	.7993	.8000	.8007	.8014	.8021	.8028	.8035	.8041	.8048	.8055
6.4	.8062	.8069	.8075	.8082	.8089	.8096	.8102	.8109	.8116	.8122
6.5	.8129	.8136	.8142	.8149	.8156	.8162	.8169	.8176	.8182	.8189
6.6	.8195	.8202	.8209	.8215	.8222	.8228	.8235	.8241	.8248	.8254
6.7	.8261	.8267	.8274	.8280	.8287	.8293	.8299	.8306	.8312	.8319
6.8	.8325	.8331	.8338	.8344	.8351	.8357	.8363	.8370	.8376	.8382
6.9	.8388	.8395	.8401	.8407	.8414	.8420	.8426	.8432	.8439	.8445
7.0	.8451	.8457	.8463	.8470	.8476	.8482	.8488	.8494	.8500	.8506
7.1	.8513	.8519	.8525	.8531	.8537	.8543	.8549	.8555	.8561	.8567
7.2	.8573	.8579	.8585	.8591	.8597	.8603	.8609	.8615	.8621	.8627
7.3	.8633	.8639	.8645	.8651	.8657	.8663	.8669	.8675	.8681	.8686
7.4	.8692	.8698	.8704	.8710	.8716	.8722	.8727	.8733	.8739	.8745
7.5	.8751	.8756	.8762	.8768	.8774	.8779	.8785	.8791	.8797	.8802
7.6	.8808	.8814	.8820	.8825	.8831	.8837	.8842	.8848	.8854	.8859
7.7	.8865	.8871	.8876	.8882	.8887	.8893	.8899	.8904	.8910	.8915
7.8	.8921	.8927	.8932	.8938	.8943	.8949	.8954	.8960	.8965	.8971
7.9	.8976	.8982	.8987	.8993	.8998	.9004	.9009	.9015	.9020	.9025
8.0	.9031	.9036	.9042	.9047	.9053	.9058	.9063	.9069	.9074	.9079
8.1	.9085	.9090	.9096	.9101	.9106	.9112	.9117	.9122	.9128	.9133
8.2	.9138	.9143	.9149	.9154	.9159	.9165	.9170	.9175	.9180	.9186
8.3	.9191	.9196	.9201	.9206	.9212	.9217	.9222	.9227	.9232	.9238
8.4	.9243	.9248	.9253	.9258	.9263	.9269	.9274	.9279	.9284	.9289
8.5	.9294	.9299	.9304	.9309	.9315	.9320	.9325	.9330	.9335	.9340
8.6	.9345	.9350	.9355	.9360	.9365	.9370	.9375	.9380	.9385	.9390
8.7	.9395	.9400	.9405	.9410	.9415	.9420	.9425	.9430	.9435	.9440
8.8	.9445	.9450	.9455	.9460	.9465	.9469	.9474	.9479	.9484	.9489
8.9	.9494	.9499	.9504	.9509	.9513	.9518	.9523	.9528	.9533	.9538
9.0	.9542	.9547	.9552	.9557	.9562	.9566	.9571	.9576	.9581	.9586
9.1	.9590	.9595	.9600	.9605	.9609	.9614	.9619	.9624	.9628	.9633
9.2	.9638	.9643	.9647	.9652	.9657	.9661	.9666	.9671	.9675	.9680
9.3	.9685	.9689	.9694	.9699	.9703	.9708	.9713	.9717	.9722	.9727
9.4	.9731	.9736	.9741	.9745	.9750	.9754	.9759	.9763	.9768	.9773
9.5	.9777	.9782	.9786	.9791	.9795	.9800	.9805	.9809	.9814	.9818
9.6	.9823	.9827	.9832	.9836	.9841	.9845	.9850	.9854	.9859	.9863
9.7	.9868	.9872	.9877	.9881	.9886	.9890	.9894	.9899	.9903	.9908
9.8	.9912	.9917	.9921	.9926	.9930	.9934	.9939	.9943	.9948	.9952
9.9	.9956	.9961	.9965	.9969	.9974	.9978	.9983	.9987	.9991	.9996
x	0	1	2	3	4	5	6	7	8	9

Table D Natural Logarithms

x	ln x	x	ln x	x	ln x
	*	4.5	1.5041	9.0	2.1972
0.1	7.6974	4.6	1.5261	9.1	2.2083
0.2	8.3906	4.7	1.5476	9.2	2.2192
0.3	8.7960	4.8	1.5686	9.3	2.2300
0.4	9.0837	4.9	1.5892	9.4	2.2407
0.5	9.3069	5.0	1.6094	9.5	2.2513
0.6	9.4892	5.1	1.6292	9.6	2.2618
0.7	9.6433	5.2	1.6487	9.7	2.2721
0.8	9.7769	5.3	1.6677	9.8	2.2824
0.9	9.8946	5.4	1.6864	9.9	2.2925
1.0	0.0000	5.5	1.7047	10	2.3026
1.1	0.0953	5.6	1.7228	11	2.3979
1.2	0.1823	5.7	1.7405	12	2.4849
1.3	0.2624	5.8	1.7579	13	2.5649
1.4	0.3365	5.9	1.7750	14	2.6391
1.5	0.4055	6.0	1.7918	15	2.7081
1.6	0.4700	6.1	1.8083	16	2.7726
1.7	0.5306	6.2	1.8245	17	2.8332
1.8	0.5878	6.3	1.8405	18	2.8904
1.9	0.6419	6.4	1.8563	19	2.9444
2.0	0.6931	6.5	1.8718	20	2.9957
2.1	0.7419	6.6	1.8871	25	3.2189
2.2	0.7885	6.7	1.9021	30	3.4012
2.3	0.8329	6.8	1.9169	35	3.5553
2.4	0.8755	6.9	1.9315	40	3.6889
2.5	0.9163	7.0	1.9459	45	3.8067
2.6	0.9555	7.1	1.9601	50	3.9120
2.7	0.9933	7.2	1.9741	55	4.0073
2.8	1.0296	7.3	1.9879	60	4.0943
2.9	1.0647	7.4	2.0015	65	4.1744
3.0	1.0986	7.5	2.0149	70	4.2485
3.1	1.1314	7.6	2.0281	75	4.3175
3.2	1.1632	7.7	2.0412	80	4.3820
3.3	1.1939	7.8	2.0541	85	4.4427
3.4	1.2238	7.9	2.0669	90	4.4998
3.5	1.2528	8.0	2.0794	100	4.6052
3.6	1.2809	8.1	2.0919	110	4.7005
3.7	1.3083	8.2	2.1041	120	4.7875
3.8	1.3350	8.3	2.1163	130	4.8676
3.9	1.3610	8.4	2.1282	140	4.9416
4.0	1.3863	8.5	2.1401	150	5.0106
4.1	1.4110	8.6	2.1518	160	5.0752
4.2	1.4351	8.7	2.1633	170	5.1358
4.3	1.4586	8.8	2.1748	180	5.1930
4.4	1.4816	8.9	2.1861	190	5.2470

*Subtract 10 for $x < 1$. Thus $\ln 0.3 = 8.7960 - 10 = -1.2040$.

Table E Values of Trigonometric Functions

Angle θ Degrees	Radians	sin θ	csc θ	tan θ	cot θ	sec θ	cos θ		
0° 00′	.0000	.0000	No value	.0000	No value	1.000	1.0000	1.5708	90° 00′
10	029	029	343.8	029	343.8	000	000	679	50
20	058	058	171.9	058	171.9	000	000	650	40
30	087	087	114.6	087	114.6	000	1.0000	621	30
40	116	116	85.94	116	85.94	000	.9999	592	20
50	145	145	68.76	145	68.75	000	999	563	10
1° 00′	.0175	.0175	57.30	.0175	57.29	1.000	.9998	1.5533	89° 00′
10	204	204	49.11	204	49.10	000	998	504	50
20	233	233	42.98	233	42.96	000	997	475	40
30	262	262	38.20	262	38.19	000	997	446	30
40	291	291	34.38	291	34.37	000	996	417	20
50	320	320	31.26	320	31.24	001	995	388	10
2° 00′	.0349	.0349	28.65	.0349	28.64	1.001	.9994	1.5359	88° 00′
10	378	378	26.45	378	26.43	001	993	330	50
20	407	407	24.56	407	24.54	001	992	301	40
30	436	436	22.93	437	22.90	001	990	272	30
40	465	465	21.49	466	21.47	001	989	243	20
50	495	494	20.23	495	20.21	001	988	213	10
3° 00′	.0524	.0523	19.11	.0524	19.08	1.001	.9986	1.5184	87° 00′
10	553	552	18.10	553	18.07	002	985	155	50
20	582	581	17.20	582	17.17	002	983	126	40
30	611	610	16.38	612	16.35	002	981	097	30
40	640	640	15.64	641	15.60	002	980	068	20
50	669	669	14.96	670	14.92	002	978	039	10
4° 00′	.0698	.0698	14.34	.0699	14.30	1.002	.9976	1.5010	86° 00′
10	727	727	13.76	729	13.73	003	974	981	50
20	756	765	13.23	758	13.20	003	971	952	40
30	785	785	12.75	787	12.71	003	969	923	30
40	814	814	12.29	816	12.25	003	967	893	20
50	844	843	11.87	846	11.83	004	964	864	10
5° 00′	.0873	.0872	11.47	.0875	11.43	1.004	.9962	1.4835	85° 00′
10	902	901	11.10	904	11.06	004	959	806	50
20	931	929	10.76	934	10.71	004	957	777	40
30	960	958	10.43	963	10.39	005	954	748	30
40	.0989	.0987	10.13	.0992	10.08	005	951	719	20
50	.1018	.1016	9.839	.1022	9.788	005	948	690	10
6° 00′	.1047	.1045	9.567	.1051	9.514	1.006	.9945	1.4661	84° 00′
10	076	074	9.309	080	9.255	006	942	632	50
20	105	103	9.065	110	9.010	006	939	603	40
30	134	132	8.834	139	8.777	006	936	573	30
40	164	161	8.614	169	8.556	007	932	544	20
50	193	190	8.405	198	8.345	007	929	515	10
7° 00′	.1222	.1219	8.206	.1228	8.144	1.008	.9925	1.4486	83° 00′
10	251	248	8.016	257	7.953	008	922	457	50
20	280	276	7.834	287	7.770	008	918	428	40
30	309	305	7.661	317	7.596	009	914	399	30
40	338	334	7.496	346	7.429	009	911	370	20
50	367	363	7.337	376	7.269	009	907	341	10
8° 00′	.1396	.1392	7.185	.1405	7.115	1.010	.9903	1.4312	82° 00′
		cos θ	sec θ	cot θ	tan θ	csc θ	sin θ	Radians	Degrees Angle θ

Table E
(continued)

Angle θ									
Degrees	Radians	sin θ	csc θ	tan θ	cot θ	sec θ	cos θ		
8° 00′	.1396	.1392	7.185	.1405	7.115	1.010	.9930	1.4312	82° 00′
10	425	421	7.040	435	6.968	010	899	283	50
20	454	449	6.900	465	827	011	894	254	40
30	484	478	765	495	691	011	890	224	30
40	513	507	636	524	561	012	886	195	20
50	452	536	512	554	435	012	881	166	10
9° 00′	.1571	.1564	6.392	.1584	6.314	1.012	.9877	1.4137	81° 00′
10	600	593	277	614	197	013	872	108	50
20	629	622	166	644	6.084	013	868	079	40
30	658	650	6.059	673	5.976	014	863	050	30
40	687	679	5.955	703	871	014	858	1.4021	20
50	716	708	855	733	769	015	853	1.3992	10
10° 00′	.1745	.1736	5.759	.1763	5.671	1.015	.9848	1.3963	80° 00′
10	774	765	665	793	576	016	843	934	50
20	804	794	575	823	485	016	838	904	40
30	833	822	487	853	396	017	833	875	30
40	862	851	403	883	309	018	827	846	20
50	891	880	320	914	226	018	822	817	10
11° 00′	.1920	.1908	5.241	.1944	5.145	1.019	.9816	1.3788	79° 00′
10	949	937	164	.1974	5.066	019	811	759	50
20	.1978	965	089	.2004	4.989	020	805	730	40
30	.2007	.1994	5.016	035	915	020	799	701	30
40	036	.2022	4.945	065	843	021	793	672	20
50	065	051	876	095	773	022	787	643	10
12° 00′	.2094	.2079	4.810	.2126	4.705	1.022	.9781	1.3614	78° 00′
10	123	108	745	156	638	023	775	584	50
20	153	136	682	186	574	024	769	555	40
30	182	164	620	217	511	024	763	526	30
40	211	193	560	247	449	025	757	497	20
50	240	221	502	278	390	026	750	468	10
13° 00′	.2269	.2250	4.445	.2309	4.331	1.026	.9744	1.3439	77° 00′
10	298	278	390	339	275	027	737	410	50
20	327	306	336	370	219	028	730	381	40
30	356	334	284	401	165	028	724	352	30
40	385	363	232	432	113	029	717	323	20
50	414	391	182	462	061	030	710	294	10
14° 00′	.2443	.2419	4.134	.2493	4.011	1.031	.9703	1.3265	76° 00′
10	473	447	086	524	3.962	031	696	235	50
20	502	476	4.039	555	914	032	689	206	40
30	531	504	3.994	586	867	033	681	177	30
40	560	532	950	617	821	034	674	148	20
50	589	560	906	648	776	034	667	119	10
15° 00′	.2618	.2588	3.864	.2679	3.732	1.035	.9659	1.3090	75° 00′
10	647	616	822	711	689	036	652	061	50
20	676	644	782	742	647	037	644	032	40
30	705	672	742	773	606	038	636	1.3003	30
40	734	700	703	805	566	039	628	1.2974	20
50	763	728	665	836	526	039	621	945	10
16° 00′	.2793	.2756	3.628	.2867	3.487	1.040	.9613	1.2915	74° 00′
		cos θ	sec θ	cot θ	tan θ	csc θ	sin θ	Radians	Degrees
								Angle θ	

Table E
(continued)

Angle θ Degrees	Radians	sin θ	csc θ	tan θ	cot θ	sec θ	cos θ		
16° 00′	.2793	.2756	3.628	.2867	3.487	1.040	.9613	1.2915	74° 00′
10	822	784	592	899	450	041	605	886	50
20	851	812	556	931	412	042	596	857	40
30	880	840	521	962	376	043	588	828	30
40	909	868	487	.2944	340	044	580	799	20
50	938	896	453	.3026	305	045	572	770	10
17° 00′	.2967	.2924	3.420	.3057	3.271	1.046	.9563	1.2741	73° 00′
10	.2996	952	388	089	237	047	555	712	50
20	.3025	.2979	357	121	204	048	546	683	40
30	054	.3007	326	153	172	048	537	654	30
40	083	035	295	185	140	049	528	625	20
50	113	062	265	217	108	050	520	595	10
18° 00′	.3142	.3090	3.236	.3249	3.078	1.051	.9511	1.2566	72° 00′
10	171	118	207	281	047	052	502	537	50
20	200	145	179	314	3.018	053	492	508	40
30	229	173	152	346	2.989	054	483	479	30
40	258	201	124	378	960	056	474	450	20
50	287	228	098	411	932	057	465	421	10
19° 00′	.3316	.3256	3.072	.3443	2.904	1.058	.9455	1.2392	71° 00′
10	345	283	046	476	877	059	446	363	50
20	374	311	3.021	508	850	060	436	334	40
30	403	338	2.996	541	824	061	426	305	30
40	432	365	971	574	798	062	417	275	20
50	462	393	947	607	773	063	407	246	10
20° 00′	.3491	.3420	2.924	.3640	2.747	1.064	.9397	1.2217	70° 00′
10	520	448	901	673	723	065	387	188	50
20	549	475	878	706	699	066	377	159	40
30	578	502	855	739	675	068	367	130	30
40	607	529	833	772	651	069	356	101	20
50	636	557	812	805	628	070	346	072	10
21° 00′	.3665	.3584	2.790	.3839	2.605	1.071	.9336	1.2043	69° 00′
10	694	611	769	872	583	072	325	1.2014	50
20	723	638	749	906	560	074	315	1.1985	40
30	752	665	729	939	539	075	304	956	30
40	782	692	709	.3973	517	076	293	926	20
50	811	719	689	.4006	496	077	283	897	10
22° 00′	.3840	.3746	2.669	.4040	2.475	1.079	.9272	1.1868	68° 00′
10	869	773	650	074	455	080	261	839	50
20	898	800	632	108	434	081	250	810	40
30	927	827	613	142	414	082	239	781	30
40	956	854	595	176	394	084	228	752	20
50	985	881	577	210	375	085	216	723	10
23° 00′	.4014	.3907	2.559	.4245	2.356	1.086	.9205	1.1694	67° 00′
10	043	934	542	279	337	088	194	665	50
20	072	961	525	314	318	089	182	636	40
30	102	.3987	508	348	300	090	171	606	30
40	131	.4014	491	383	282	092	159	577	20
50	160	041	475	417	264	093	147	548	10
24° 00′	.4189	.4067	2.459	.4452	2.246	1.095	.9135	1.1519	66° 00′
		cos θ	sec θ	cot θ	tan θ	csc θ	sin θ	Radians	Angle θ Degrees

Table E (*continued*)

Angle θ Degrees	Angle θ Radians	sin θ	csc θ	tan θ	cot θ	sec θ	cos θ		
24° 00′	.4189	.4067	2.459	.4452	2.246	1.095	.9135	1.1519	66° 00′
10	218	094	443	487	229	096	124	490	50
20	247	120	427	522	211	097	112	461	40
30	276	147	411	557	194	099	100	432	30
40	305	173	396	592	177	100	088	403	20
50	334	200	381	628	161	102	075	374	10
25° 00′	.4363	.4226	2.366	.4663	2.145	1.103	.9063	1.1345	65° 00′
10	392	253	352	699	128	105	051	316	50
20	422	279	337	734	112	106	038	286	40
30	451	305	323	770	097	108	026	257	30
40	480	331	309	806	081	109	013	228	20
50	509	358	295	841	066	111	.9001	199	10
26° 00′	.4538	.4384	2.281	.4877	2.050	1.113	.8988	1.1170	64° 00′
10	567	410	268	913	035	114	975	141	50
20	596	436	254	950	020	116	962	112	40
30	625	462	241	.4986	2.006	117	949	083	30
40	654	488	228	.5022	1.991	119	936	054	20
50	683	514	215	059	977	121	923	1.1025	10
27° 00′	.4712	.4540	2.203	.5095	1.963	1.122	.8910	1.0996	63° 00′
10	741	566	190	132	949	124	897	966	50
20	771	592	178	169	935	126	884	937	40
30	800	617	166	206	921	127	870	908	30
40	829	643	154	243	907	129	857	879	20
50	858	669	142	280	894	131	843	850	10
28° 00′	.4887	.4695	2.130	.5317	1.881	1.133	.8829	1.0821	62° 00′
10	916	720	118	354	868	134	816	792	50
20	945	746	107	392	855	136	802	763	40
30	.4974	772	096	430	842	138	788	734	30
40	.5003	797	085	467	829	140	774	705	20
50	032	823	074	505	816	142	760	676	10
29° 00′	.5061	.4848	2.063	.5543	1.804	1.143	.8746	1.0647	61° 00′
10	091	874	052	581	792	145	732	617	50
20	120	899	041	619	780	147	718	588	40
30	149	924	031	658	767	149	704	559	30
40	178	950	020	696	756	151	689	530	20
50	207	.4975	010	735	744	153	675	501	10
30° 00′	.5236	.5000	2.000	.5774	1.732	1.155	.8660	1.0472	60° 00′
10	265	025	1.990	812	720	157	646	443	50
20	294	050	980	851	709	159	631	414	40
30	323	075	970	890	698	161	616	385	30
40	352	100	961	930	686	163	601	356	20
50	381	125	951	.5969	675	165	587	327	10
31° 00′	.5411	.5150	1.942	.6009	1.664	1.167	.8572	1.0297	59° 00′
10	440	175	932	048	653	169	557	268	50
20	469	200	923	088	643	171	542	239	40
30	498	225	914	128	632	173	526	210	30
40	527	250	905	168	621	175	511	181	20
50	556	275	896	208	611	177	496	152	10
32° 00′	.5585	.5299	1.887	.6249	1.600	1.179	.8480	1.0123	58° 00′
		cos θ	sec θ	cot θ	tan θ	csc θ	sin θ	Angle θ Radians	Angle θ Degrees

Table E (continued)

Angle θ Degrees	Radians	sin θ	csc θ	tan θ	cot θ	sec θ	cos θ		
32° 00′	.5585	.5299	1.887	.6249	1.600	1.179	.8480	1.0123	58° 00′
10	614	324	878	289	590	181	465	094	50
20	643	348	870	330	580	184	450	065	40
30	672	373	861	371	570	186	434	036	30
40	701	398	853	412	560	188	418	1.0007	20
50	730	422	844	452	550	190	403	.9977	10
33° 00′	.5760	.5446	1.836	.6494	1.540	1.192	.8387	.9948	57° 00′
10	789	471	828	536	530	195	371	919	50
20	818	495	820	577	520	197	355	890	40
30	847	519	812	619	511	199	339	861	30
40	876	544	804	661	501	202	323	832	20
50	905	568	796	703	492	204	307	803	10
34° 00′	.5934	.5592	1.788	.6745	1.483	1.206	.8290	.9774	56° 00′
10	963	616	781	787	473	209	274	745	50
20	.5992	640	773	830	464	211	258	716	40
30	.6021	664	766	873	455	213	241	687	30
40	050	688	758	916	446	216	225	657	20
50	080	712	751	.6959	437	218	208	628	10
35° 00′	.6109	.5736	1.743	.7002	1.428	1.221	.8192	.9599	55° 00′
10	138	760	736	046	419	223	175	570	50
20	167	783	729	089	411	226	158	541	40
30	196	807	722	133	402	228	141	512	30
40	225	831	715	177	393	231	124	483	20
50	254	854	708	221	385	233	107	454	10
36° 00′	.6283	.5878	1.701	.7265	1.376	1.236	.8090	.9425	54° 00′
10	312	901	695	310	368	239	073	396	50
20	341	925	688	355	360	241	056	367	40
30	370	948	681	400	351	244	039	338	30
40	400	972	675	445	343	247	021	308	20
50	429	.5995	668	490	335	249	.8004	279	10
37° 00′	.6458	.6018	1.662	.7536	1.327	1.252	.7986	.9250	53° 00′
10	487	041	655	581	319	255	696	221	50
20	516	065	649	627	311	258	951	192	40
30	545	088	643	673	303	260	934	163	30
40	574	111	636	720	295	263	916	134	20
50	603	134	630	766	288	266	898	105	10
38° 00′	.6632	.6157	1.624	.7813	1.280	1.269	.7880	.9076	52° 00′
10	661	180	618	860	272	272	862	047	50
20	690	202	612	907	265	275	844	.9018	40
30	720	225	606	.7954	257	278	826	.8988	30
40	749	248	601	.8002	250	281	808	959	20
50	778	271	595	050	242	284	790	930	10
39° 00′	.6807	.6293	1.589	.8098	1.235	1.287	.7771	.8901	51° 00′
10	836	316	583	146	228	290	753	872	50
20	865	338	578	195	220	293	735	843	40
30	894	361	572	243	213	296	716	814	30
40	923	383	567	292	206	299	698	785	20
50	952	406	561	342	199	302	679	756	10
40° 00′	.6981	.6428	1.556	.8391	1.192	1.305	.7660	.8727	50° 00′
		cos θ	sec θ	cot θ	tan θ	csc θ	sin θ	Radians	Angle θ Degrees

Table E (*continued*)

Angle θ									
Degrees	Radians	sin θ	csc θ	tan θ	cot θ	sec θ	cos θ		
40° 00′	.6981	.6428	1.556	.8391	1.192	1.305	.7660	.8727	50° 00′
10	.7010	450	550	441	185	309	642	698	50
20	039	472	545	491	178	312	623	668	40
30	069	494	540	541	171	315	604	639	30
40	098	517	535	591	164	318	585	610	20
50	127	539	529	642	157	322	566	581	10
41° 00′	.7156	.6561	1.524	.8693	1.150	1.325	.7547	.8552	49° 00′
10	185	583	519	744	144	328	528	523	50
20	214	604	514	796	137	332	509	494	40
30	243	626	509	847	130	335	490	465	30
40	272	648	504	899	124	339	470	436	20
50	301	670	499	.8952	117	342	451	407	10
42° 00′	.7330	.6691	1.494	.9004	1.111	1.346	.7431	.8378	48° 00′
10	359	713	490	057	104	349	412	348	50
20	389	734	485	110	098	353	392	319	40
30	418	756	480	163	091	356	373	290	30
40	447	777	476	217	085	360	353	261	20
50	476	799	471	271	079	364	333	232	10
43° 00′	.7505	.6820	1.466	.9325	1.072	1.367	.7314	.8203	47° 00′
10	534	841	462	380	066	371	294	174	50
20	563	862	457	435	060	375	274	145	40
30	592	884	453	490	054	379	254	116	30
40	621	905	448	545	048	382	234	087	20
50	650	926	444	601	042	386	214	058	10
44° 00′	.7679	.6947	1.440	.9657	1.036	1.390	.7193	.8029	46° 00′
10	709	967	435	713	030	394	173	.7999	50
20	738	.6988	431	770	024	398	153	970	40
30	767	.7009	427	827	018	402	133	941	30
40	796	030	423	884	012	406	112	912	20
50	825	050	418	.9942	006	410	092	883	10
45° 00′	.7854	.7071	1.414	1.000	1.000	1.414	.7071	.7854	45° 00′
		cos θ	sec θ	cot θ	tan θ	csc θ	sin θ	Radians	Degrees
								Angle θ	

Answers to Odd-numbered Problems

Chapter 1

1.1.1 $S \cup T = \{1, 2, 4\}$; $S \cap T = \{1\}$; $S \times T = \{(1,1), (1,4), (2,1), (2,4)\}$;
$T \times S = \{(1,1), (1,2), (4,1), (4,2)\}$; $S \times S = \{(1,1), (1,2), (2,1), (2,2)\}$;
$T \times T = \{(1,1), (1,4), (4,1), (4,4)\}$; $S' = \{3, 4, 5, 6, 7, 8, 9\}$; $T' = \{2, 3, 5, 6, 7, 8, 9\}$.

1.1.3 Yes. No, since 0 is neither positive nor negative.

1.1.5 The subsets of S_1 are $\varnothing$, S_1. The subsets of S_2 are $\varnothing$, S_2, $\{b\}$, $\{c\}$. The subsets of S_3 are $\varnothing$, S_3, $\{d\}$, $\{e\}$, $\{f\}$, $\{d, e\}$, $\{d, f\}$, $\{e, f\}$. There are exactly 2 subsets of S_1, 2^2 subsets of S_2, and 2^3 subsets of S_3.

1.1.11 $S = T$

1.2.1 Yes. Yes. Yes. Yes.

1.2.3 f is not onto, not one-to-one, and not a bijection.

1.2.5 f is onto, one-to-one, and a bijection.

1.2.7 (a) For *each* of the elements $y_1, y_2, \ldots, y_n$ in T, there corresponds *at least one* of the elements $x_1, \ldots, x_m$ in S.

(b) *Distinct* elements x_i, x_j in S have *distinct* images $f(x_i)$, $f(x_j)$ in T. Hence, $f(x_1)$, $f(x_2)$, $\ldots$, $f(x_m)$ are all *distinct* elements in T.

(c) Since f is a bijection, f is both one-to-one and onto. Hence, by parts a and b, $n \le m$ and $n \ge m$, and thus $n = m$.

1.2.9 Let $y = f(x) = 5x - 8$. Then $x = (y + 8)/5$, and hence $f((y + 8)/5) = y$. Thus, f is onto. Moreover, if $f(x_1) = f(x_2)$, then $5x_1 - 8 = 5x_2 - 8$, and thus $x_1 = x_2$. Hence f is also one-to-one. Therefore, $f: R \to R$ is a bijection.

1.2.11 f is a function, since $1 - 0^2 = 1 + 0^2$, and thus $f(0)$ is well defined.

1.2.13 $n(n-1)(n-2)\cdots 3 \cdot 2 \cdot 1$

1.2.15 $p = 4s$, where p denotes the perimeter and s denotes the length of one of the sides.

1.2.17 $v = s^3$, where v denotes the volume and s denotes the length of one of the sides.

1.3.1

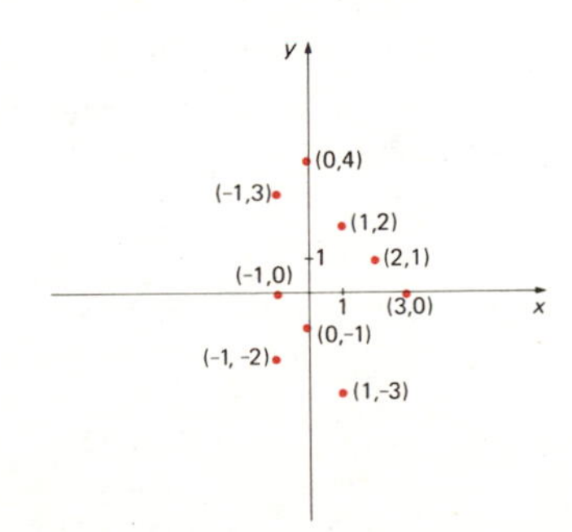

1.3.3 The coordinates of such points are all of the form $(1,a)$, where a is *any* real number.

1.3.5 In parts a to e and g, the domain of the function is the set of all real numbers. The domain of the function in part f is the set of all real numbers *except zero*. The domain of the function in part h is the set of all real numbers x such that $x \ge 1$.

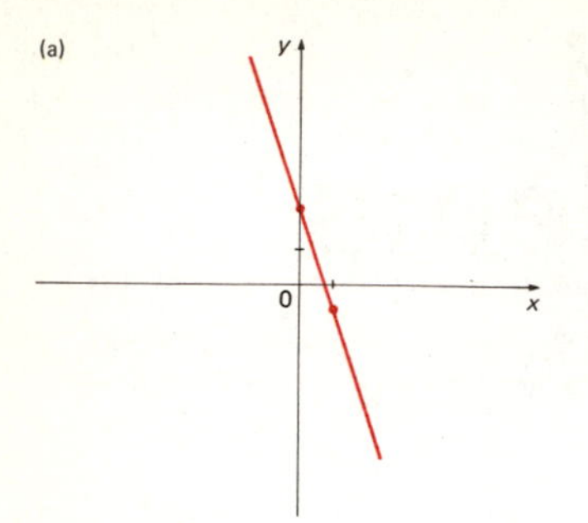

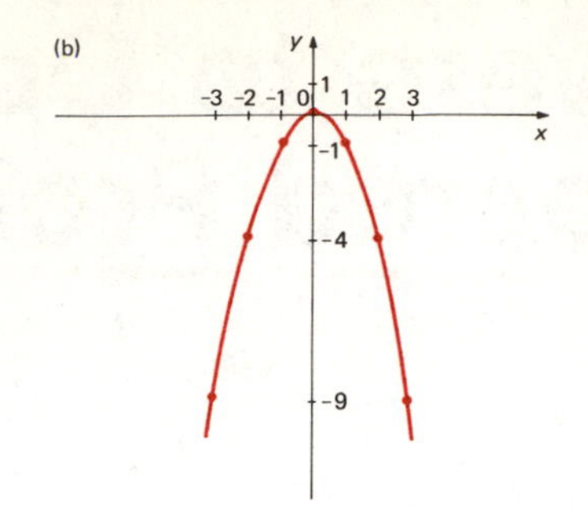

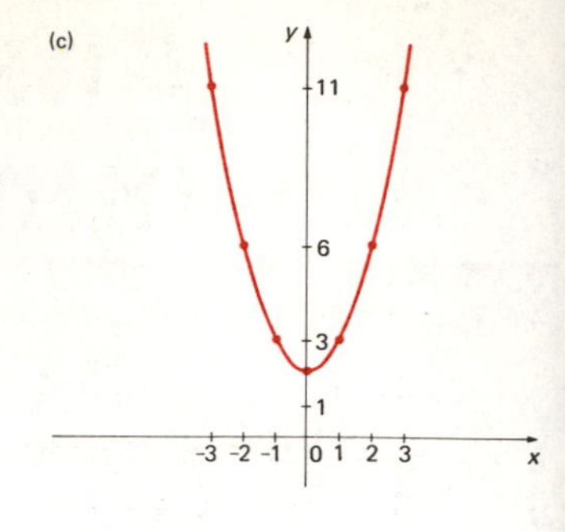

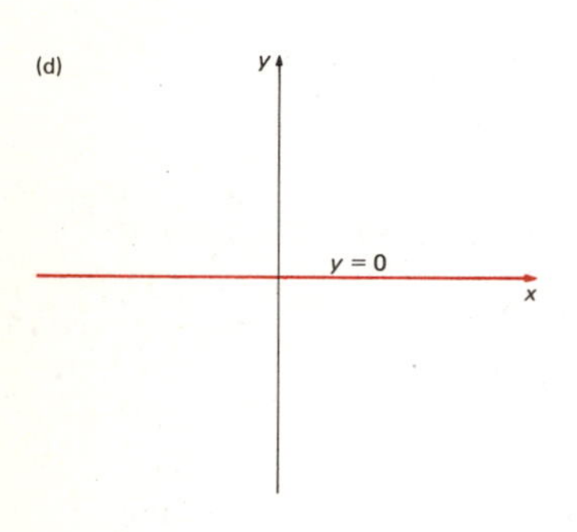

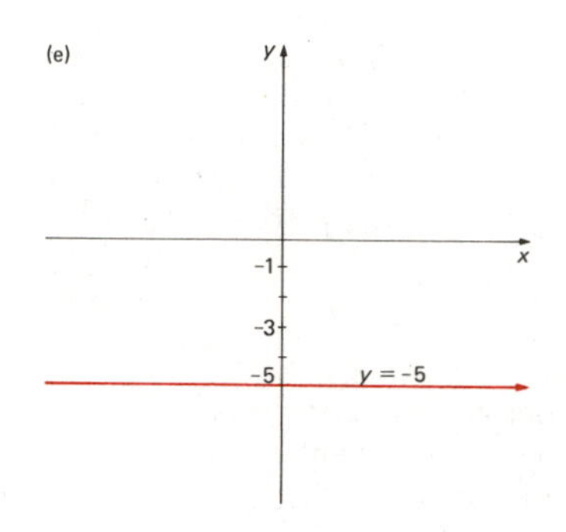

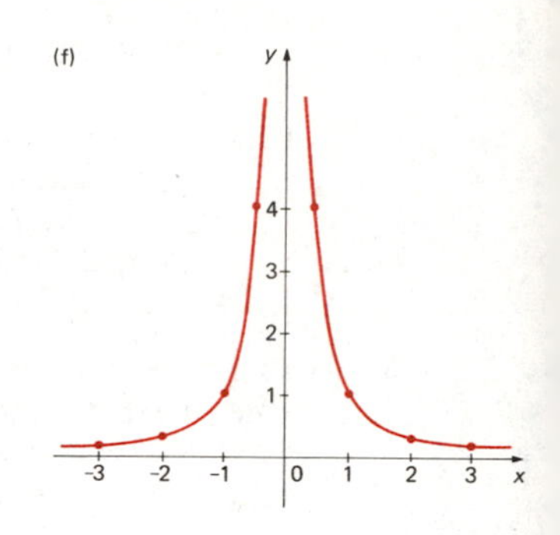

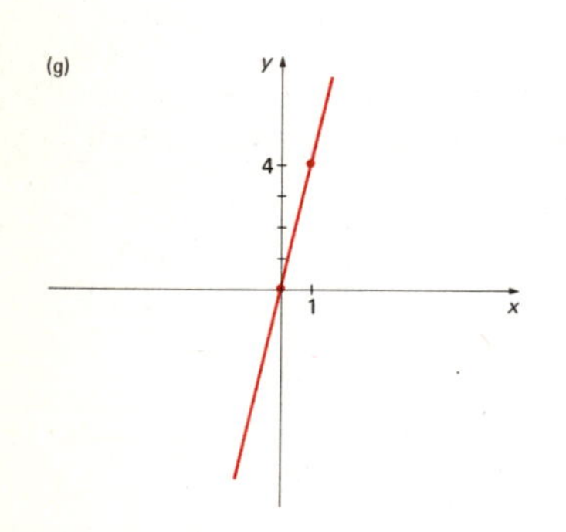

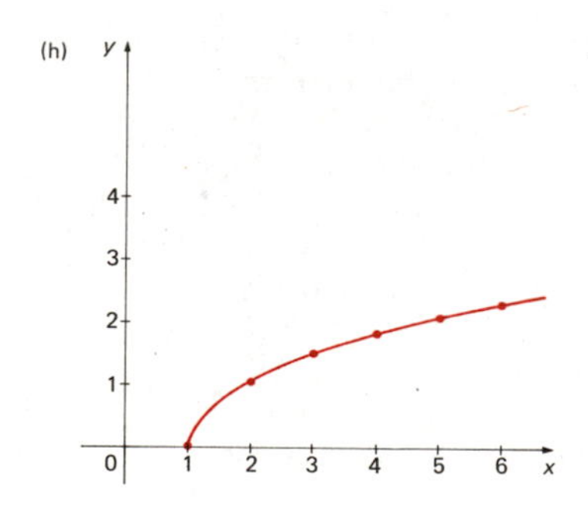

1.3.7 The domain of f is the set of all *nonzero* real numbers. There are no x intercept points and no y intercept points.

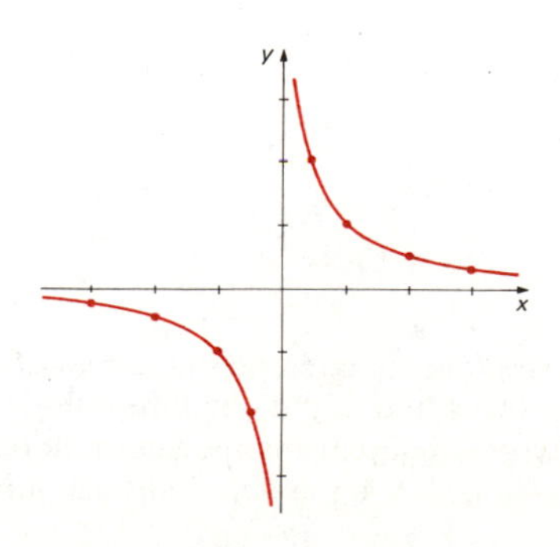

1.3.9 (*a*) No; (*b*) yes; (*c*) no; (*d*) yes; (*e*) yes; (*f*) no. The graphs in parts *b*, *d*, and *e* represent functions, since for each x, $f(x)$ has a *unique* value. In parts *a*, *c*, and *f*, the graphs do *not* represent functions, since $f(x)$ does *not* have a unique value.

1.4.1 $(f+g)(x) = x^2 + 2x + 4$ $\quad (fg)(x) = (x^2+1)(2x+3) = 2x^3 + 3x^2 + 2x + 3$
$(f-g)(x) = x^2 - 2x - 2$ $\quad (f/g)(x) = (x^2+1)/(2x+3)$
$(g-f)(x) = -x^2 + 2x + 2$ $\quad (g/f)(x) = (2x+3)/(x^2+1)$
The domain of each of the functions $f+g$, $f-g$, $g-f$, fg is the set of all real numbers. The domain of f/g is the set of all real numbers *with the exception of* $-3/2$. The domain of g/f is the set of all real numbers (with no exception).

1.4.3 No. No. No.

1.4.5 $f(0) = 1$ $\quad (f \circ g)(4) = f(g(4)) = 122$
$g(0) = 3$ $\quad (g \circ f)(4) = g(f(4)) = 37$
$f(-x) = x^2 + 1$ $\quad (f \circ f)(3) = f(f(3)) = 101$
$g(x^2) = 2x^2 + 3$ $\quad (g \circ g)(5) = g(g(5)) = 29$

1.4.7 $(f \circ g)(x) = f(g(x)) = [1/(x+1)]^2$
$(g \circ f)(x) = g(f(x)) = 1/(x^2+1)$
$(f \circ f)(x) = f(f(x)) = x^4$
$(g \circ g)(x) = g(g(x)) = (x+1)/(x+2)$
The domain of $f \circ g$ is the set of all real numbers except -1. The domain of $g \circ f$ is the set of *all* real numbers. The domain of $f \circ f$ is the set of all real numbers. The domain of $g \circ g$ is the set of all real numbers except -1 and -2. Observe that $f \circ g \neq g \circ f$.

1.4.9 f is not one-to-one, but g is.

1.4.11 $f(0) = 0$ $\quad f(1) = 1$ $\quad f(-1) = 1$ $\quad f(-x) = x^2$ $\quad f(g(2)) = 1/9$ $\quad g(f(2)) = 1/5$
$f(f(5)) = 625$ $\quad g(g(0)) = 1/2$ $\quad g(1/x) = x/(x+1)$ $\quad g(5x-1) = 1/(5x)$

1.4.13 (*a*) $f^{-1}(x) = x$ $\quad$ (*d*) $f^{-1}(x) = -1/7(x+2)$
(*b*) $f^{-1}(x) = -x$ $\quad$ (*e*) $f^{-1}(x) = \sqrt[3]{x-5}$
(*c*) $f^{-1}(x) = 1/3(x-1)$ $\quad$ (*f*) $f^{-1}(x) = \sqrt[3]{4-x}$
The domain of each function f^{-1} is the set of *all* real numbers.

1.4.15 $g(x) = f^{-1}(x) = 9x/5 + 32$

Chapter 2

2.1.1

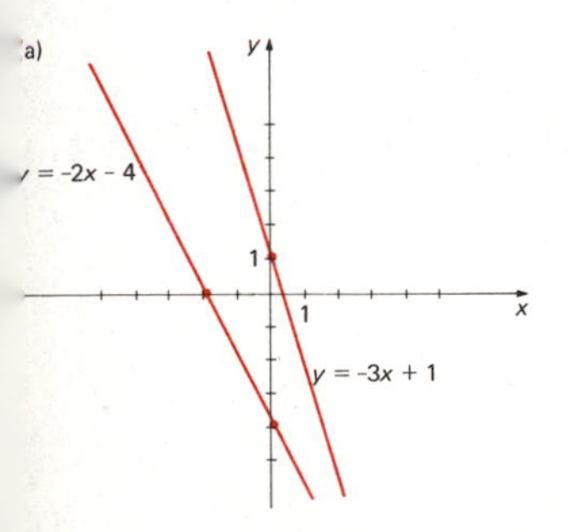

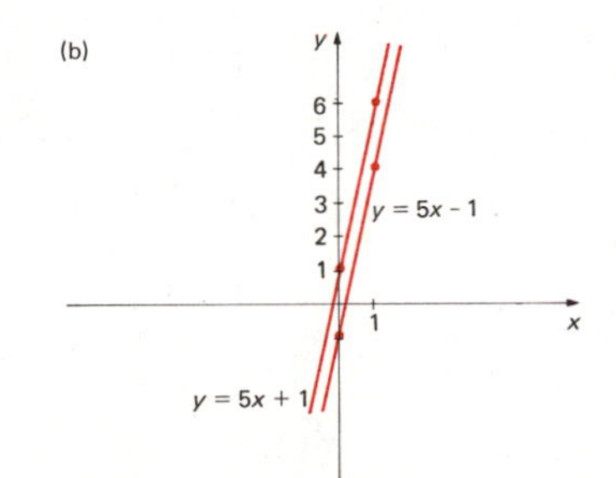

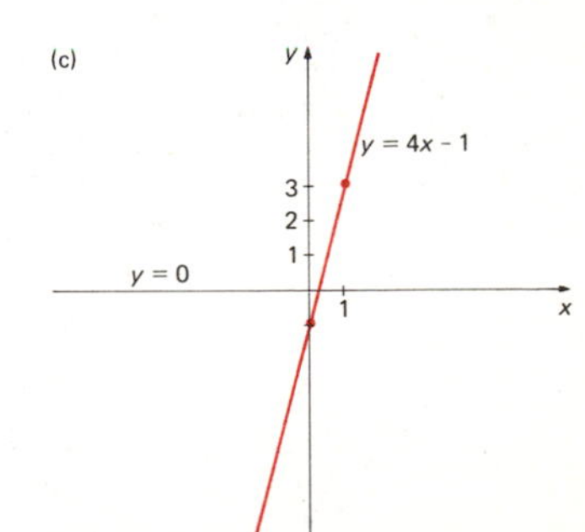

2.1.3 $(1/2, 1/2)$, $(0,1)$, $(-3/4, -2)$

2.1.5 $(0,5)$ and $(2,11)$ are not on the line $y = 3x - 5$, since the coordinates of these points do not satisfy the equation of the line. The other points are on the given line, since their coordinates satisfy the equation of this line.

2.1.7 $a = 2$

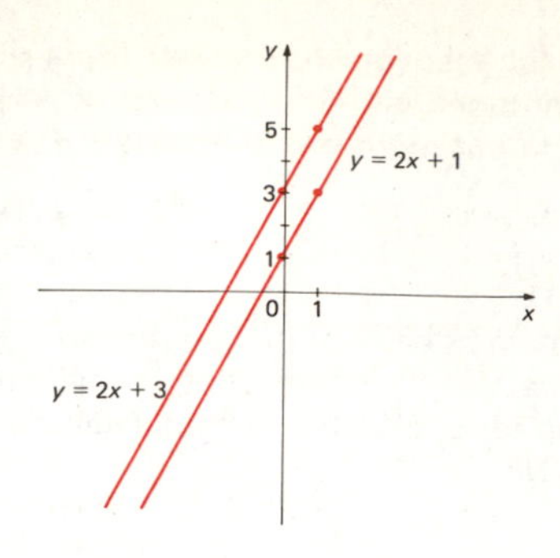

2.1.9 $f(x) = (-b/a)x + b$

2.1.11 (*a*) $a = 0$, $b = 0$, c is any real number. (*b*) $a = 0$, $b \neq 0$, c is any real number. These results follow from the definitions of a constant function and a linear function.

2.1.13 $f^{-1}(x) = (1/a)(x - b)$, for all real numbers x

2.2.1 (*a*) $-1/2, 1$ (*b*) $1/3, 1/3$ (*c*) $2 - \sqrt{5}, 2 + \sqrt{5}$ (*d*) $1 - \sqrt{-3}, 1 + \sqrt{-3}$

2.2.3 $(-3,-10)$, $(2,5)$

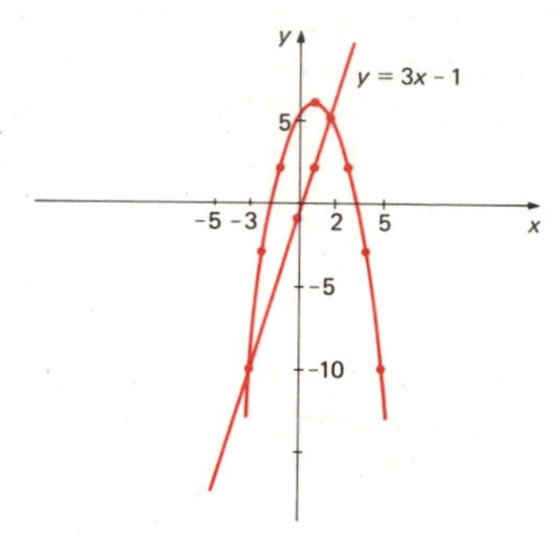

2.2.5 No, since $1^2 + 1 - 1 \neq -1$, and thus the coordinates of the point $(1,-1)$ do not satisfy the given equation.

2.2.7 (*a*) Graph has a maximum point at $(-1/4, 9/8)$.
(*b*) Graph has a minimum point at $(2,6)$.
(*c*) Graph has a maximum point at $(-1,0)$.

2.2.11 Roots are equal when $b^2 - 4ac = 0$. Roots are real when $b^2 - 4ac \geq 0$. Roots are rational when a, b, c are rational numbers and $b^2 - 4ac$ is a (perfect) square of a rational number.

2.2.13 -2

2.2.15 $c \leq 9/4$

2.2.17 $-4\sqrt{2}, 4\sqrt{2}$

2.2.19 The x intercept points are $[(3 - \sqrt{7})/2, 0]$ and $[(3 + \sqrt{7})/2, 0]$, and the y intercept point is $(0,1)$.

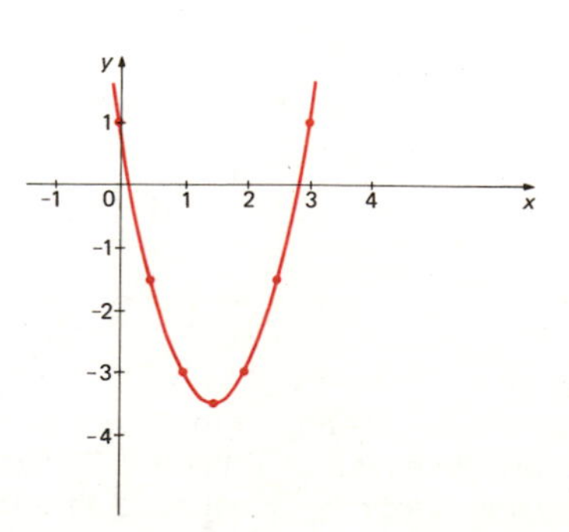

2.2.21 By high school geometry, we get $(8-y)/8 = x/6$, and hence $y = -\frac{4}{3}(x-6)$. Thus the area of the rectangle is given by $A = xy = -4x(x-6)/3 = -\frac{4}{3}x^2 + 8x$. Therefore the rectangle with the largest area has dimensions given by $x = 3$, $y = 4$.

Chapter 3

3.1.1 $f(0) = 0$ $f(1) = 1$ $f(-1) = 1$ $g(0) = 3$ $g(1) = 1$ $g(-1) = 5$ $f(g(-1)) = 5$ $g(f(-1)) = 1$ $f(f(-4)) = 4$ $g(g(-4)) = 19$ $g(\frac{3}{2}) = 0$

3.1.3 $f(x) = f(-x)$, since $|x| = |-x|$, for all real numbers x. $g(x) \neq g(-x)$ for all real numbers x, since $|2x-3| \neq |2(-x)-3|$. For example, $g(1) = 1$, but $g(-1) = 5$.

3.1.5 Since $|\pm 5| = 5$, both $f(5)$ and $f(-5)$ are well defined, and hence f is a function. The domain of f is the set of *all* real numbers. The range of f is the set of all real numbers between 0 and 5, *inclusive* of both 0 and 5.

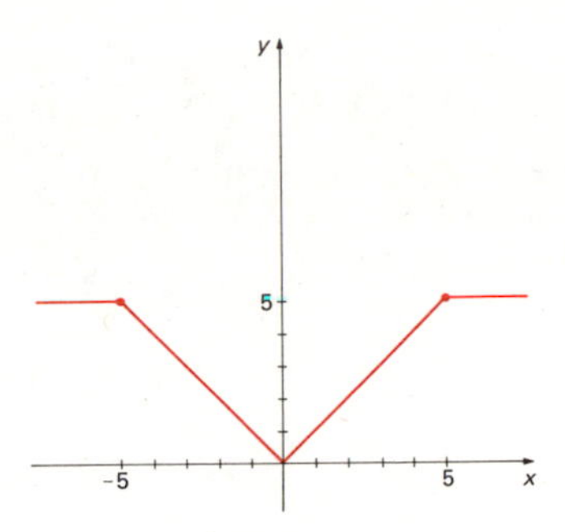

3.1.7 $f(x) = \sqrt{x^2} = x$, if $x \geq 0$. Also, $f(x) = \sqrt{x^2} = -x$, if $x < 0$. Hence, $f(x) = |x|$.

3.1.9 (*a*) $x \geq 0$; (*b*) $x \leq 0$; (*c*) x is *any* real number; (*d*) $x \geq 0$; (*e*) $x \leq 0$; (*f*) $x = 0$; (*g*) $x = 0$; (*h*) x is *any* real number; (*i*) $x = 0$; (*j*) there is *no* such real number x; (*k*) $x = 1$ or $x = -1$; (*l*) $x = 4$ or $x = 6$; (*m*) $x = -5$; (*n*) there is *no* such real number x; (*o*) $x = \frac{1}{2}$ or $x = -\frac{1}{2}$.

3.2.1 (*a*) $\{x | x \leq 3\}$ (*c*) $\{x | x > -\frac{3}{5}\}$ (*e*) $\{x | x > -3\}$
(*b*) $\{x | x \geq \frac{2}{3}\}$ (*d*) $\{x | x > -\frac{9}{2}\}$ (*f*) $\{x | x \leq -\frac{5}{2}\}$

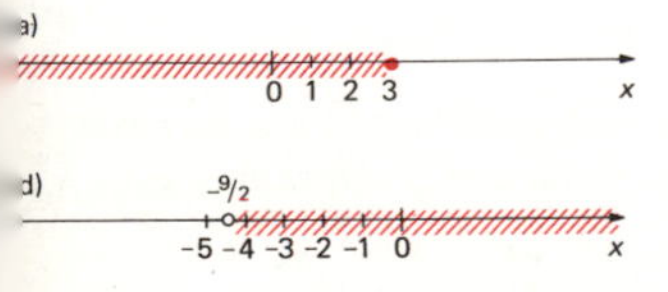

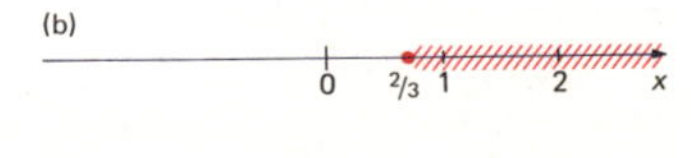

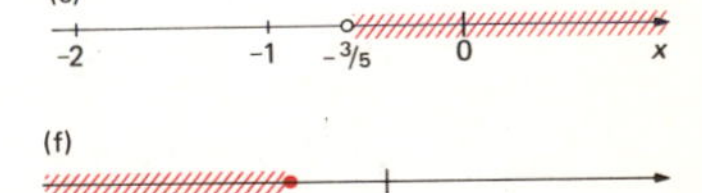

3.2.3 (*a*) $\{x | x \geq \frac{1}{2}\} \cup \{x | x < 0\}$
(*b*) $\{x | 0 < x < \frac{1}{5}\}$
(*c*) $\{x | x < -1\}$
(*d*) $\{x | -1 \leq x < 1\} \cup \{x | x > 3\}$
(*e*) $\{x | x \leq 1\} \cup \{x | 2 \leq x \leq 3\}$
(*f*) $\{x | x \leq -3\} \cup \{x | -2 \leq x < 0\} \cup \{x | x > 1\}$
(*g*) $\{x | -\frac{5}{2} < x < -2\} \cup \{x | 2 < x < \frac{5}{2}\}$
(*h*) $\{x | (1-\sqrt{5})/2 \leq x \leq (1+\sqrt{5})/2\}$
(*i*) $\{x | x < 1\} \cup \{x | x > \frac{5}{2}\}$

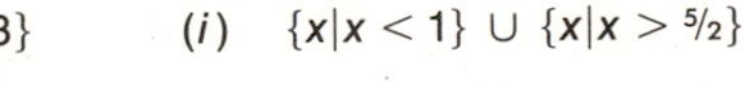

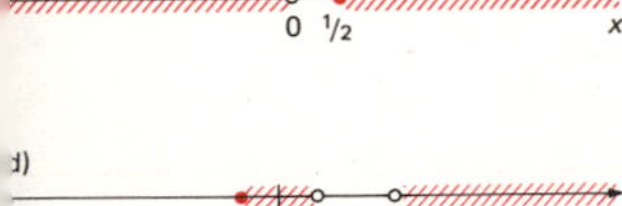

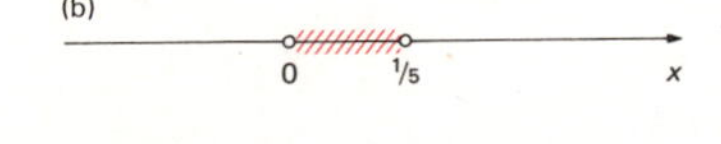

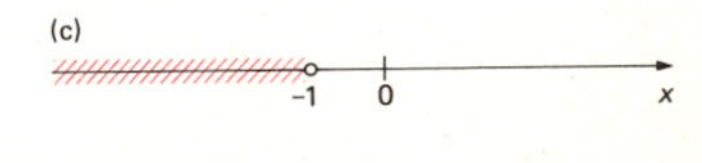

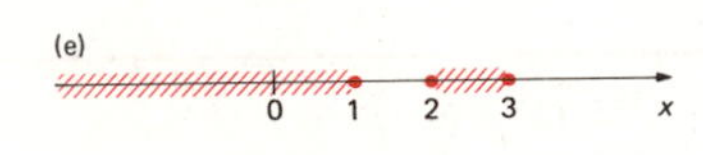

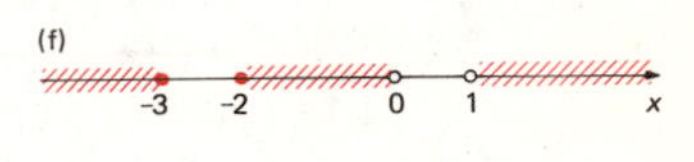

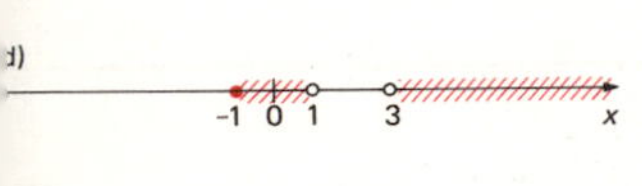

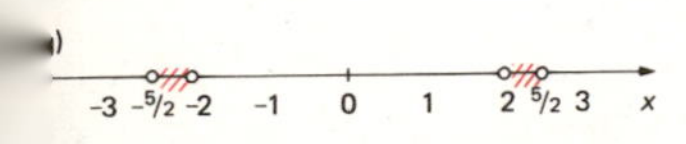

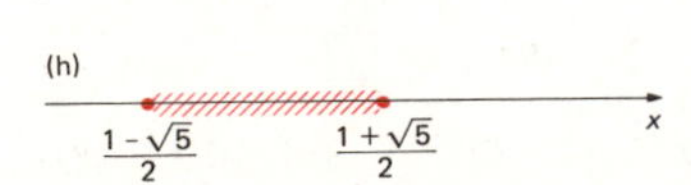

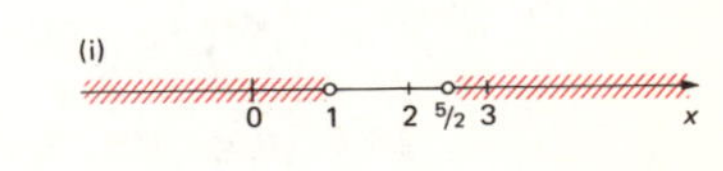

3.2.5 (*a*) Equality holds when, and only when, $x = 1$.

3.2.7 Since $a > 1$, a is positive. Hence, multiplying both sides of the inequality $1 < a$ by the *positive* number a, we obtain $a \cdot 1 < a \cdot a$, that is, $a < a^2$.

3.2.9 $-3 < 3x - 6 < 3$; $2 > 4 - 2x > -2$.

3.3.1 (*a*) $\{x | -1 \le x \le 1\}$. Thus, the locus of all points $(x,0)$ whose distances from the origin are at most 1 is the set indicated above.
(*b*) $\{x | x < -2\} \cup \{x | x > 2\}$. Thus, the locus of all points $(x,0)$ whose distances from the origin are greater than 2 is the set indicated above.
(*c*) $\{x | 0 < x < 2 \text{ but } x \ne 1\}$. Thus, the locus of all points $(x,0)$ whose distances from the point $(1,0)$ are strictly positive and less than 1 is the set indicated above.
(*d*) $\{x | x \le -3\} \cup \{x | x \ge -1\}$. Thus, the locus of all points $(x,0)$ whose distances from the point $(-2,0)$ are at least 1 is the set indicated above.

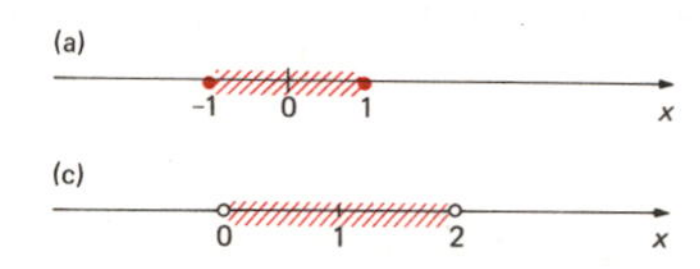

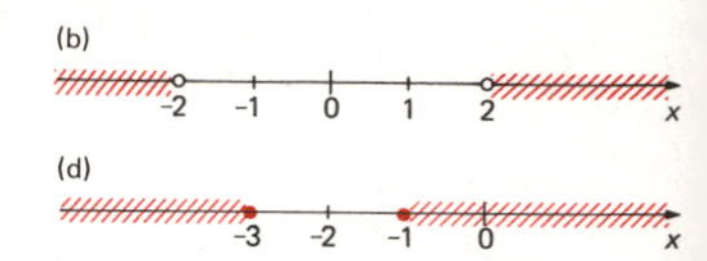

3.3.3 (*a*) $\{x | x \le -2\} \cup \{x | x \ge 0\}$
(*b*) $\{x | 1 \le x \le 5 \text{ but } x \ne 3\}$
(*c*) $\{x | -1/7 < x < 3\}$
(*d*) $\{x | x < -2\} \cup \{x | x > 0 \text{ but } x \ne 1/2\}$
(*e*) $\{x | -3 \le x \le 2\}$
(*f*) $\{x | x \le -2\} \cup \{x | -1 \le x \le 0\} \cup \{x | x \ge 1\}$

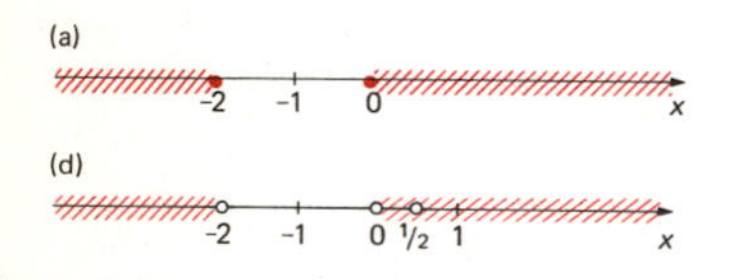

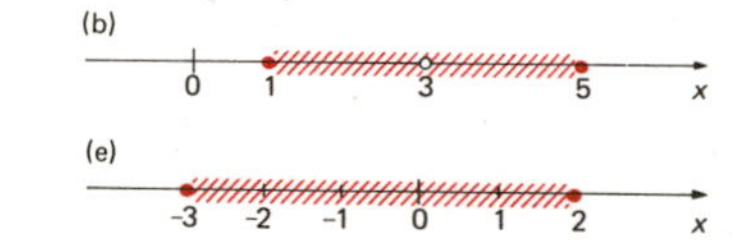

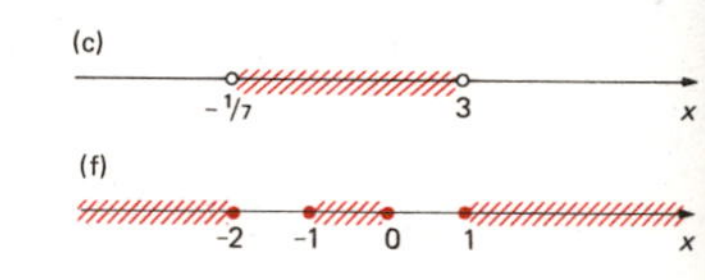

3.3.5 (*a*) The domain of f is the set of all real numbers, and the range of f is $\{x | -2 \le x \le 2\}$.
(*b*) The domain of f is the set of all real numbers, and the range of f is $\{x | x \ge 5\}$.

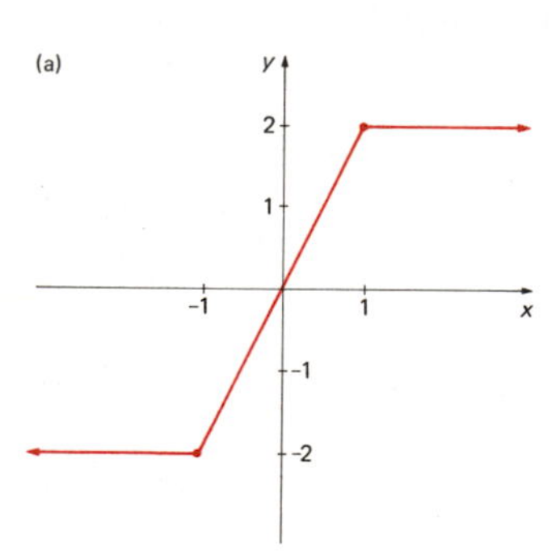

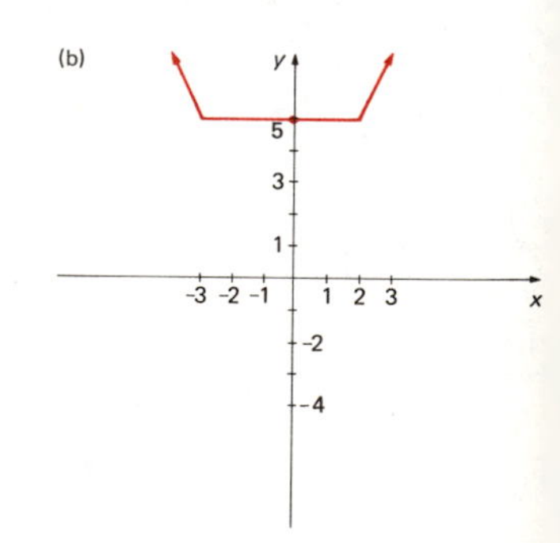

3.3.9 The domain of f is the set of all real numbers, and the range of f is the set of all real numbers x such that $x \ge 2$.

Chapter 4

4.1.1 (*a*) $1 + 0i$
(*b*) $-3 + 2i$
(*c*) $3 - 2i$
(*d*) $(-1/2) + (-1/2)i$
(*e*) $(-3/2) + (-1/2)i$
(*f*) $(-3/5) + (1/5)i$
(*g*) $-9 + 7i$
(*h*) $-9 + 7i$
(*i*) $-6 + 4i$
(*j*) $4 + 13i$

4.1.3 (*a*) $0 + 2i$ (*b*) $-4 + 0i$ (*c*) $0 + (-8)i$ (*d*) $16 + 0i$

4.1.5 (*a*) $0 + (-2)i$ (*b*) $-4 + 0i$ (*c*) $0 + 8i$ (*d*) $16 + 0i$

4.1.11 (*a*) $(1/2) + (-1/2)i$ (*f*) $0 + 32i$
(*b*) $(1/2) + (1/2)i$ (*g*) $0 + (-32)i$
(*c*) $(-1/2) + (1/2)i$ (*h*) $4096 + 0i$
(*d*) $0 + (-1)i$ (*i*) $4096 + 0i$
(*e*) $0 + 1i$

4.2.1 (*a*) $-2; -3; -2 + 3i; \sqrt{13}; \sqrt{13}; -2; 3$

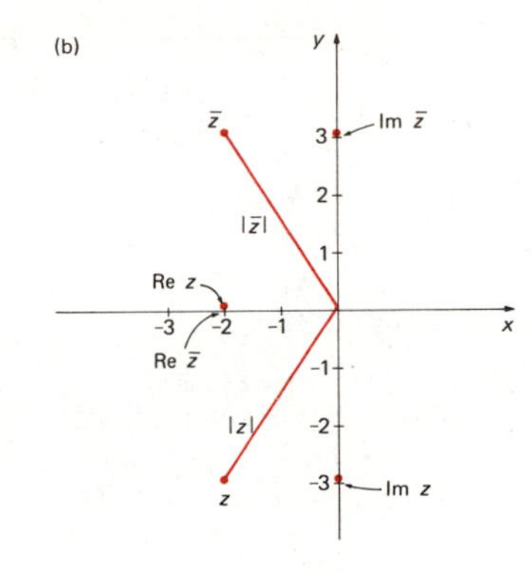

4.2.3 $\overline{z_1 + z_2} = \overline{0 - 6i} = 0 + 6i$ $\quad \bar{z}_1 + \bar{z}_2 = (-1 + 2i) + (1 + 4i) = 0 + 6i$
Thus, $\overline{z_1 + z_2} = \bar{z}_1 + \bar{z}_2$. Also, $\overline{z_1 - z_2} = \overline{-2 + 2i} = -2 - 2i$; $\bar{z}_1 - \bar{z}_2 = (-1 + 2i) - (1 + 4i) = -2 - 2i$. Thus, $\overline{z_1 - z_2} = \bar{z}_1 - \bar{z}_2$

4.2.5 Re $z = -2$ $\quad |z| = \sqrt{13}$ $\quad$ Im $z = -3$ $\quad |\bar{z}| = \sqrt{13}$ $\quad z + \bar{z} = -4$
Observe that Re $z < |z|$, Im $z < |z|$, $|\bar{z}| = |z|$, and $z + \bar{z} = 2$ Re z. Thus, Eq. (4.30) is verified. Moreover, $z\bar{z} = (-2 - 3i)(-2 + 3i) = 13 = |z|^2$; that is, $z\bar{z} = |z|^2$, and Eq. (4.31) is verified.

4.2.7 $|z_1 + z_2| = |0 - 6i| = \sqrt{36} = 6$ $\quad |z_1| + |z_2| = \sqrt{5} + \sqrt{17}$
Observe that $|z_1 + z_2| < |z_1| + |z_2|$, and thus Eq. (4.34) is verified. Moreover, $|z_1 - z_2| = |-2 + 2i| = \sqrt{8}$; $||z_1| - |z_2|| = |\sqrt{5} - \sqrt{17}| = \sqrt{17} - \sqrt{5}$. Observe that $|z_1 - z_2| > ||z_1| - |z_2||$. (Consult a table of square roots.)

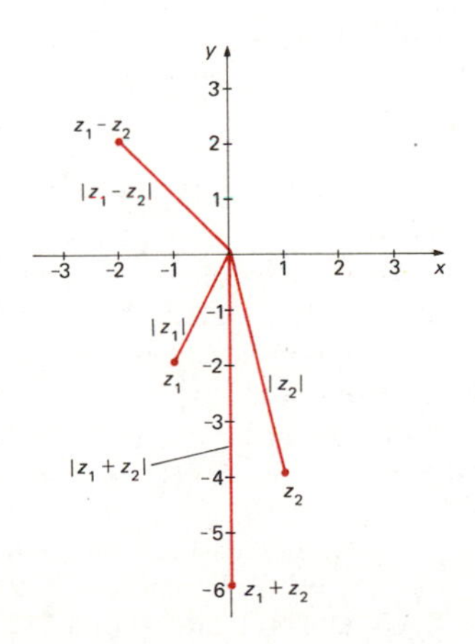

4.2.9 (*a*) $0 + \sqrt{2}i$ and $0 - \sqrt{2}i$; (*b*) $-1 + i$ and $-1 - i$; (*c*) $1/4 + (\sqrt{7}/4)i$ and $1/4 - (\sqrt{7}/4)i$.

4.2.11 If z is a positive real number, or zero, then, clearly, $\bar{z} = z$ and $|z| = z$. Hence, $z + \bar{z} = z + z = 2z = 2|z|$; that is, $z + \bar{z} = 2|z|$. Conversely, suppose $z + \bar{z} = 2|z|$, and let $z = x + iy$ (x, y real). Then $\bar{z} = x - iy$, and $|z| = \sqrt{x^2 + y^2}$. Thus, $z + \bar{z} = 2|z|$ is equivalent to

$x + iy + x - iy = 2\sqrt{x^2 + y^2}$, which, in turn, is equivalent to $x = \sqrt{x^2 + y^2}$. Hence $y = 0$ and thus $x = \sqrt{x^2} = |x|$. Therefore $x \geq 0$, and hence $z = x + iy = x \geq 0$. In other words, z is a positive real number, or zero.

4.2.13 (a) $z = a + 0i$ (d) $z = a + 0i$, where $a \leq 0$
(b) $z = 0 + bi$ (e) $0 + 0i$
(c) $z = a + 0i$, where $a \geq 0$ (f) $0 + 0i$

Chapter 5

5.1.1 (a) 5, 2, 4 (b) 4, −1, −2 (c) 3, 1, 0 (d) 0, 1, 1

5.1.3 (a) Let $f(x) = 2x^3 - x + 1$ and $g(x) = -2x^3 + 4x^2 + 3$. Then the degree of $f + g$ is 2, which is less than the degree of f. Also, the degree of $f - g$ is 3, which is equal to the degree of f (as well as the degree of g).
(b) Let $f(x)$ and $g(x)$ be as in part *a* above.

5.1.5 (a) $x^4 + 2x^3 + 4x^2 + 8x + 16$ (c) $x^n - 1$
(b) $x^4 - 2x^3 + 4x^2 - 8x + 16$

5.2.1 $f(x) + g(x) = 2x^4 + x^2 + 6x - 3$
$f(x) - g(x) = 2x^4 - 2x^3 + 5x^2 + 4x - 5$
$g(x) - f(x) = -2x^4 + 2x^3 - 5x^2 - 4x + 5$
$f(x)g(x) = 2x^7 - 5x^6 + 7x^5 - 12x^3 + 16x^2 + x - 4$
$\frac{f(x)}{g(x)} = (2x + 3) + \frac{7x^2 - 7}{x^3 - 2x^2 + x + 1}$
All these are polynomials, except $f(x)/g(x)$.

5.2.3 $f(0) = -4$ $g(-1) = -3$
$g(0) = 1$ $f(g(2)) = f(3) = 173$
$f(1) = 5$ $g(f(2)) = g(42) = 70{,}603$

5.2.5 $f(x) = (x - 1)(x - 3)(x + 2)(x + 4)$

5.2.7 $f(1) = 0$ $f(5) = 0$
$f(2) = -45$ $f(-1) = 0$
$f(3) = -96$ $f(-2) = -21$
$f(4) = -105$ $f(-3) = 0$

5.2.9 The zeros of $f(x)$ are 1, 5, −1, −3. All these zeros are real. The polynomial $f(x)$ has exactly four zeros.

5.2.11 The zeros of $f(x)$ are 2, −3, −4.

5.2.17 $a = -1$, $b = 1$

5.2.19 $k = 3$

Chapter 6

6.1.1 $f(-2) = -260$, $f(-1) = -27$, $f(0) = 10$, $f(1) = -5$, $f(2) = 72$

6.1.3 The real zeros of $f(x)$ lie between −1 and 0; 0 and 1; and 1 and 2.

6.1.5 (a) The quotient is $x^2 - x + 3$ and the remainder is 13.
(b) The quotient is $\frac{2}{3}x^2 - \frac{11}{9}x + \frac{74}{27}$, and the remainder is $\frac{196}{27}$.

6.1.7 $a = -4$, $b = 2$

6.1.9 (a) 1 and −1 (b) 1

6.1.11 (a) $-\frac{1}{2}, \frac{4}{5}, -\frac{4}{3}$; all zeros are rational.
(b) $2, \frac{1}{3}, -\frac{1}{3}, i, -i$; 2, $\frac{1}{3}$, and $-\frac{1}{3}$ are the rational zeros.
(c) 2, −1, −3; all zeros are rational.
(d) $5, -4, \sqrt{3}i, -\sqrt{3}i$; 5, −4 are the rational zeros.
(e) $1, \frac{3}{2}, -3, -\frac{3}{2}$; all zeros are rational.
(f) $-1, 4, \sqrt{2}, -\sqrt{2}$; −1 and 4 are the rational zeros.
(g) $3, -4, (-1 + \sqrt{3}i)/2, (-1 - \sqrt{3}i)/2$; 3 and −4 are the rational zeros.

(*h*) $-2, -2, (1+\sqrt{5})/2, (1-\sqrt{5})/2$; -2 and -2 are the rational zeros.
(*i*) $-1, (1+\sqrt{3}i)/2, (1-\sqrt{3}i)/2$; -1 is the only rational zero.
(*j*) $2, -1+\sqrt{3}i, -1-\sqrt{3}i$; 2 is the only rational zero.

6.1.13 (*a*) $-1, 3, (-1+\sqrt{7}i)/2, (-1-\sqrt{7}i)/2$; -1 and 3 are the integral zeros.
(*b*) $1, -5, (1+\sqrt{3}i)/2, (1-\sqrt{3}i)/2$; 1 and -5 are the integral zeros.
(*c*) $-3, (1+\sqrt{5})/2, (1-\sqrt{5})/2$; -3 is the only integral zero.
(*d*) $5, (-1+\sqrt{3}i)/2, (-1-\sqrt{3}i)/2$; 5 is the only integral zero.

6.1.15 $f(x) = (x-1)(x-2)(x+1)(5x+1)$

6.1.17 Let $y = 3x-1$. Then $x = (y+1)/3$, and hence

$$f(x) = f\left(\frac{y+1}{3}\right) = \left(\frac{y+1}{3}\right)^3 + 2\frac{y+1}{3} + 5 = \frac{y^3+3y^2+21y+154}{27}$$

Hence the zeros of the polynomial $g(y) = y^3+3y^2+21y+154$ are $3a-1$, $3b-1$, and $3c-1$.

6.2.1 One zero is between -3 and -2, another zero is between -2 and -1, while the third zero is between 0 and 1.

6.2.3 $-2.5, -1.3, 0.9$

6.2.5

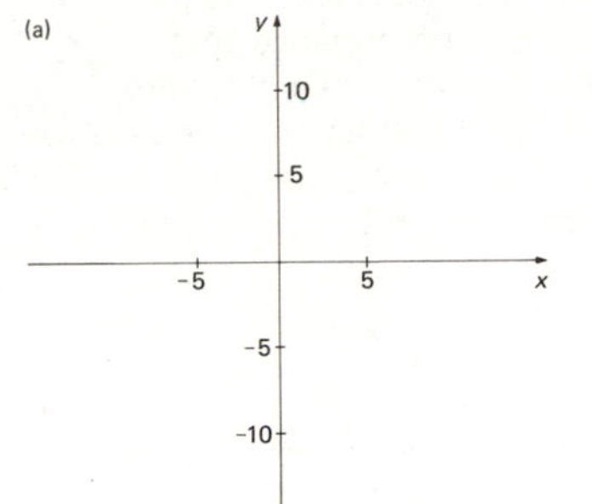

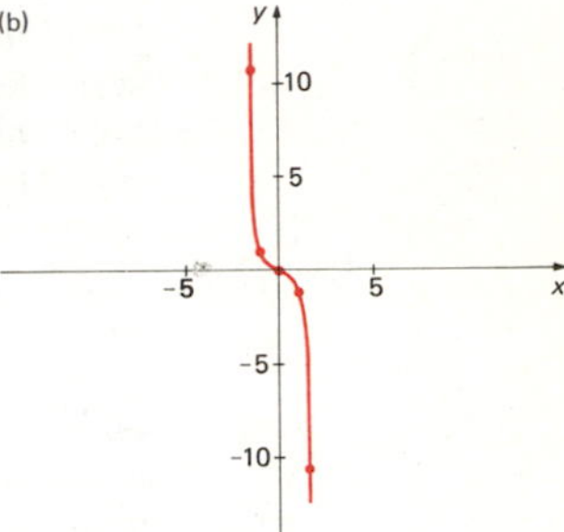

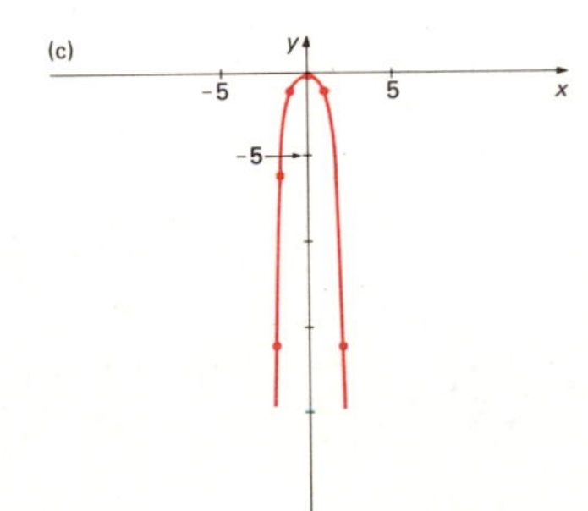

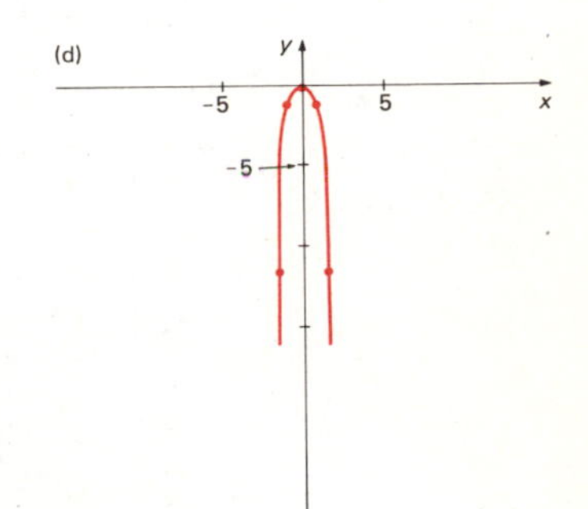

6.2.9 1.8, 2.2

6.2.13 $[(x-1)^2-2][(x+1)^2-18]$

6.2.15 Suppose that $f(x)$ is a polynomial *with rational coefficients*, and suppose p is a *positive prime*. If $a+b\sqrt{p}$ (a, b rational) is a zero of $f(x)$, then $a-b\sqrt{p}$ is also a zero of $f(x)$. For a proof, see hint given in Prob. 6.2.12.

6.3.1 $f(i) = 0$ $f(2i) = -4-8i$ $f(-i) = 2-2i$ $f(-2i) = 0$ $f(1) = 6+2i$ $f(-1) = 0$

6.3.3 $f(x) = (x-i)(x+2i)(x+1)$

6.3.5 No, for in the Complex Conjugates theorem it is assumed that all the coefficients of the polynomial are *real*, which is *not* the case here.

6.3.7 $z = [(2 \pm \sqrt{2}) \pm \sqrt{2}i]/4$

6.3.9 $f(x) = [x-(1+i)][x-(5-2i)]$

6.3.13 $2, -6$

6.3.15 -1 is a zero of $f(x)$ of multiplicity 3, and 2 is a zero of $f(x)$ of multiplicity 2.

6.3.17 $f(x) = [x - (1+i)]^2[x - (3-2i)]^3$

6.3.19 (*a*) $(x+2)(x-1)$

(*b*) $\left(x - \frac{-1+\sqrt{21}}{2}\right)\left(x - \frac{-1-\sqrt{21}}{2}\right)$

(*c*) $\left(x - \frac{1+\sqrt{11}i}{2}\right)\left(x - \frac{1-\sqrt{11}i}{2}\right)$

(*d*) $\left(x - \frac{-3+\sqrt{23}i}{-4}\right)\left(x - \frac{-3-\sqrt{23}i}{-4}\right)$

6.3.21 (*a*) $(x+1)(x-2)\left(x - \frac{-1+\sqrt{3}i}{2}\right)\left(x - \frac{-1-\sqrt{3}i}{2}\right)$

(*b*) $(x+1)(x-3)\left(x - \frac{1+\sqrt{5}}{2}\right)\left(x - \frac{1-\sqrt{5}}{2}\right)$

(*c*) $(x+1)(x-1)(x+\sqrt{5}i)(x-\sqrt{5}i)$

6.3.23 Suppose that $a + bi$ is *any* complex number. By the Fundamental Theorem of Algebra, the polynomial equation $f(z) = a + bi$ has a solution z_0; that is, $f(z_0) = a + bi$ (where z_0 is a suitably chosen complex number). Thus, $f: C \to C$ is *onto*.

6.3.25 $[x - (1+\sqrt{2})][x - (1-\sqrt{2})][x - (2-3i)][x - (2+3i)] = [(x-1)^2 - 2][(x-2)^2 + 9]$

6.3.27 (*a*) There is at most one positive zero and no negative zero.

(*b*) There are at most two positive zeros and at most one negative zero.

(*c*) There is at most one positive zero and no negative zero.

(*d*) There are no positive zeros and at most one negative zero.

(*e*) There is at most one positive zero and at most one negative zero.

(*f*) There are no positive zeros and no negative zeros.

(*g*) There are no positive zeros and at most one negative zero.

Chapter 7

7.1.1 $(f \circ g)(x) = f(g(x)) = f(x-2) = (x-2)^3 - 2(x-2) + 1$

$(g \circ f)(x) = g(f(x)) = g(x^3 - 2x + 1) = (x^3 - 2x + 1) - 2$

The domain of each of $f \circ g$ and $g \circ f$ is the set R of real numbers. Moreover, $f \circ g \neq g \circ f$.

7.1.3 $\left(\frac{f}{g}\right)(x) = \frac{f(x)}{g(x)} = \frac{x^3 - 2x + 1}{x - 2} = (x^2 + 2x + 2) + \frac{5}{x-2}$

7.1.5 (*a*) $\frac{-2x+9}{x^2-1}$

(*b*) $\frac{x^2 + 18x - 28}{(x-1)(x+1)(x-2)}$

(*c*) $\frac{7x^4 + 32x^3 + 73x^2 + 81x + 22}{(x-2)(x^2+2x+4)(x+2)}$

(*d*) $\frac{-5x^4 - 6x^3 - 39x - 58}{(x+2)^2(x^2-2x+4)}$

7.1.7 (*a*) $\frac{(4x^2+6x+9)(3x+1)}{3x-1}$ (*b*) $\frac{(x-2)(4-x)}{(x-1)(x^2-x+1)}$

7.1.9 $\frac{x^n + a^n}{x + a} = (x^{n-1} - x^{n-2}a + x^{n-3}a^2 - \cdots - a^{n-1}) + \frac{2a^n}{x+a}$

7.1.11 $\frac{x^n + a^n}{x - a} = (x^{n-1} + x^{n-2}a + x^{n-3}a^2 + \cdots + a^{n-1}) + \frac{2a^n}{x-a}$

7.1.13 (*a*) No solution. (*b*) No solution. (*c*) $x = -2$ or $x = 2$

7.1.15 (*a*) No; $\frac{x+2}{x+1}$, which is not a polynomial expression

(*b*) No; $x^2 + x - 1$, which is a polynomial expression

(*c*) No; $x^9 + x^8a + x^7a^2 + x^6a^3 + x^5a^4 + x^4a^5 + x^3a^6 + x^2a^7 + xa^8 + a^9$, which is a polynomial expression.

(*d*) Yes

(*e*) No; $\frac{x^2+3x-4}{x+3}$, which is not a polynomial expression

7.2.1 (*a*) Yes; −1. (*b*) Yes; −3. (*c*) Yes; $\sqrt{2}$. (*d*) No.

7.2.3 The domain of $f \circ g$ is the set of all real numbers *except* $-D/C$ (if $C \neq 0$) and $-d'/c'$ (if $c' \neq 0$).

7.2.5 Suppose that $f(x) = c$ for *all* real numbers x, where c is a constant, and suppose f has an inverse function g. Then $(f \circ g)(x) = f(g(x)) = c$, for *all* real numbers x. Moreover, $(f \circ g)(x) = x$. Hence $x = c$ for *all* real numbers x, which, of course, is impossible.

7.2.7

	Domain of f	*Range of f*	*Inverse of f*
(*a*)	$\{x \mid x \neq 0\}$	$\{x \mid x \neq 0\}$	$f^{-1}(x) = \frac{5}{x}$
(*b*)	$\{x \mid x \neq 0\}$	$\{x \mid x \neq 0\}$	$f^{-1}(x) = \frac{-4}{x}$
(*c*)	$\{x \mid x \neq 1\}$	$\{x \mid x \neq 0\}$	$f^{-1}(x) = \frac{x+3}{x}$
(*d*)	$\{x \mid x \neq 5/3\}$	$\{x \mid x \neq 0\}$	$f^{-1}(x) = \frac{5x+2}{3x}$
(*e*)	$\{x \mid x \neq -7/3\}$	$\{x \mid x \neq 2/3\}$	$f^{-1}(x) = \frac{-7x}{3x-2}$
(*f*)	$\{x \mid x \neq 0\}$	$\{x \mid x \neq 5/3\}$	$f^{-1}(x) = \frac{-2}{3x-5}$
(*g*)	$\{x \mid x \neq -1/7\}$	$\{x \mid x \neq 4/7\}$	$f^{-1}(x) = \frac{-x-3}{7x-4}$
(*h*)	$\{x \mid x \neq 2/3\}$	$\{x \mid x \neq -2/3\}$	$f^{-1}(x) = \frac{2x+1}{3x+2}$
(*i*)	$\{x \mid x \neq 2\}$	$\{x \mid x \neq 3\}$	$f^{-1}(x) = \frac{2x-5}{x-3}$

7.2.9 Let $y = f(x) = (ax+b)/(cx+d)$. Solving this equation for x in terms of y, we obtain $x = (-dy+b)/(cy-a) = g(y)$ (say). Observe that since $ad - bc \neq 0$ and $c \neq 0$, $cy - a \neq 0$ (Why?). Thus, f^{-1} exists and, in fact, $f^{-1} = g$. Now suppose $f = f^{-1}$. Then $f = g$, and hence $f(x) = g(x)$ for all real numbers for which these functions are defined. But $f(x) = g(x)$ is equivalent to $(ax+b)/(cx+d) = (-dx+b)/(cx-a)$, which, in turn, is equivalent to $(ac+cd)x^2 + (d^2 - a^2)x - (ab+bd) = 0$. Now, in order for this last equation to hold for *all* x, we must have $ac + cd = 0$, $d^2 - a^2 = 0$, $ab + bd = 0$. Since $c \neq 0$, the equation $ac + cd = 0$ implies that $(a+d)c = 0$ and hence $a + d = 0$; that is, $d = -a$. We have thus shown that if $f = f^{-1}$, then $d = -a$. Conversely, if $d = -a$, then $a = -d$, and hence $f(x) = (ax+b)/(cx+d) = (-dx+b)/(cx-a) = g(x) = f^{-1}(x)$. Thus, $f = f^{-1}$.

7.2.11 Since $b = 0$ and $c = 0$, $f(x) = ax/d$ ($d \neq 0$ and $a \neq 0$, since $ad - bc \neq 0$). Let $y = f(x) = ax/d$. Then $x = dy/a = g(y)$ (say), and thus $f^{-1} = g$ exists. Now, $f = f^{-1}$ if and only if $f(x) = f^{-1}(x) = g(x)$ for *all* real numbers x, that is, $ax/d = dx/a$ for *all* real numbers x. Hence $f = f^{-1}$ if and only if $a/d = d/a$, that is, $a^2 = d^2$, or, equivalently, $d = \pm a$.

7.2.13 That f^{-1} exists was essentially shown in Prob. 7.2.9. Moreover, by Prob. 7.2.9, $f = f^{-1}$ if and only if $d = -a$. Now, by Prob. 7.2.9,

$$f(x) = \frac{ax+b}{cx+d}$$

$$f^{-1}(x) = \frac{-dx+b}{cx-a}$$

Hence [see (7.55)]

Asymptotes of $y = f(x)$ are $y = \frac{a}{c}$ and $x = \frac{-d}{c}$

Asymptotes of $y = f^{-1}(x)$ are $y = \frac{-d}{c}$ and $x = \frac{a}{c}$

Thus, the graphs of $y = f(x)$ and $y = f^{-1}(x)$ have the same asymptotes if and only if $d = -a$, and this holds (by what we have shown above) if and only if $f = f^{-1}$.

7.2.15 The linear rational function f is given by $f(x) = (4x - 4)/(x - 4)$. The domain of f is the set of all real numbers *except* 4. The range of f is the set of all real numbers *except* 4. The asymptotes of the graph of $y = f(x)$ are $y = 4$ and $x = 4$.

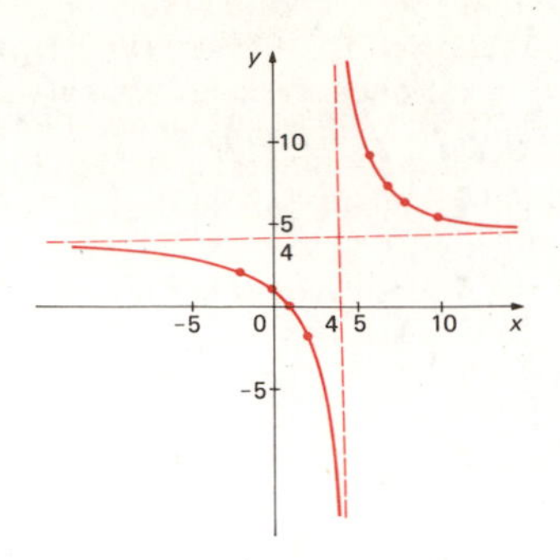

7.2.17 $f(x) = (2x - 10)/(x - 1)$

7.3.1 (*a*) Odd; (*b*) even; (*c*) even; (*d*) neither; (*e*) neither; (*f*) odd; (*g*) neither; (*h*) neither; (*i*) odd; (*j*) even; (*k*) even *and* odd; (*l*) odd.

7.3.3 (*a*) $f(x) = \dfrac{1 - x^2}{x}$; asymptotes are $x = 0$ and $y = -x$

(*b*) $f(x) = \dfrac{1 + x^4}{x^2}$; asymptotes are $x = 0$ and $y = x^2$

(*g*) $f(x) = \dfrac{1 - x}{1 + x}$; asymptotes are $x = -1$ and $y = -1$

(*h*) $f(x) = \dfrac{1 - x^2}{1 + x} = 1 - x$; there are no asymptotes

(*i*) $f(x) = \dfrac{x^2 + 1}{x^3}$; asymptotes are $x = 0$ and $y = 0$

(*j*) $f(x) = 1$; there are no asymptotes

(*k*) $f(x) = 0$; there are no asymptotes

The functions given in parts *c* to *f* and *l* cannot be reduced to rational functions. (Observe that the rational functions given in parts *h*, *j*, and *k* are also *polynomial* functions.)

7.3.5 (*a*) Verify that $g(-x) = g(x)$ and $h(-x) = -h(x)$.

(*b*) Observe that $f(x) = g(x) + h(x)$ for all x, and hence $f = g + h$.

7.3.7 Suppose that f and g are even functions; that is, $f(-x) = f(x)$ and $g(-x) = g(x)$, for all real numbers x for which these functions are defined. Then

$$(f + g)(-x) = f(-x) + g(-x) = f(x) + g(x) = (f + g)(x)$$

and hence $(f + g)(-x) = (f + g)(x)$. Thus, $f + g$ is an even function. The arguments for the difference, product, and quotient are similar.

7.3.9 Suppose that f is an even function and g is an odd function. Then $f(-x) = f(x)$ and $g(-x) = -g(x)$, for all real numbers x for which these functions are defined. Hence

$$(fg)(-x) = f(-x)g(-x) = f(x)[-g(x)] = -(fg)(x)$$

Thus, $(fg)(-x) = -(fg)(x)$, and hence fg is an odd function. The argument for the quotient is similar.

7.3.11 (*a*) The x intercept point is $(-3,0)$; the y intercept point is $(0, -3/2)$. The asymptotes are $x = -1 + \sqrt{3}$, $x = -1 - \sqrt{3}$ (these are the zeros of $x^2 + 2x - 2$), and $y = 0$. The turning points of the graph are $(-2,-1/2)$ and $(-4,-1/6)$. The domain of f is the set of all real numbers *except* $-1 + \sqrt{3}$ and $-1 - \sqrt{3}$. The range of f is the set of all real numbers y *except* those for which $-1/6 < y < -1/2$.

(*b*) The x intercept points are $(-1 + \sqrt{3},0)$ and $(-1 - \sqrt{3},0)$; the y intercept point is $(0,-2/3)$. The asymptotes are $x = -3$ and $y = x - 1$. The turning points of the graph are $(-2,-2)$ and $(-4,-6)$. The domain of f is the set of all real numbers *except* -3.

The range of f is the set of all real numbers y *except* those for which $-6 < y < -2$.
(*c*) The x intercept point is (1,0); the y intercept point is $(0,-1/6)$. The asymptotes are $x = 2$, $x = 3$, and $y = 0$. The turning points of the graph are

$$\left(\frac{-7+5\sqrt{2}}{-3+2\sqrt{2}}, -3+2\sqrt{2}\right) \quad \text{and} \quad \left(\frac{7+5\sqrt{2}}{3+2\sqrt{2}}, -3-2\sqrt{2}\right)$$

The domain of f is the set of all real numbers *except* 2 and 3. The range of f is the set of all real numbers y *except* those for which $-3-2\sqrt{2} < y < -3+2\sqrt{2}$. It can be shown (see Chap. 8) that $(-7+5\sqrt{2})/(-3+2\sqrt{2}) = 1-\sqrt{2}$ and $(7+5\sqrt{2})/(3+2\sqrt{2}) = 1+\sqrt{2}$. Thus, the turning points of the graph are $(1-\sqrt{2}, -3+2\sqrt{2})$ and $(1+\sqrt{2}, -3-2\sqrt{2})$.

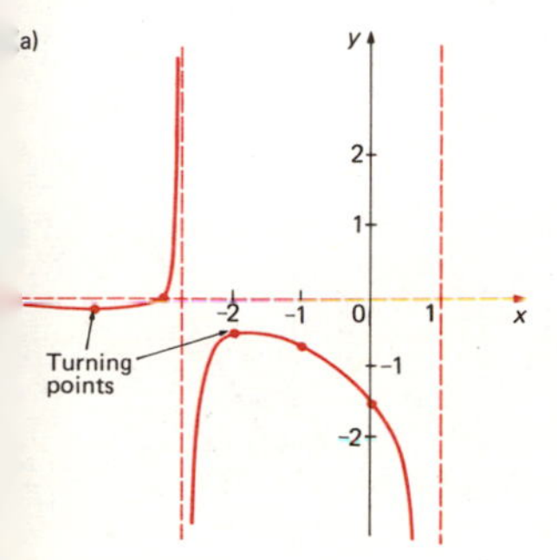

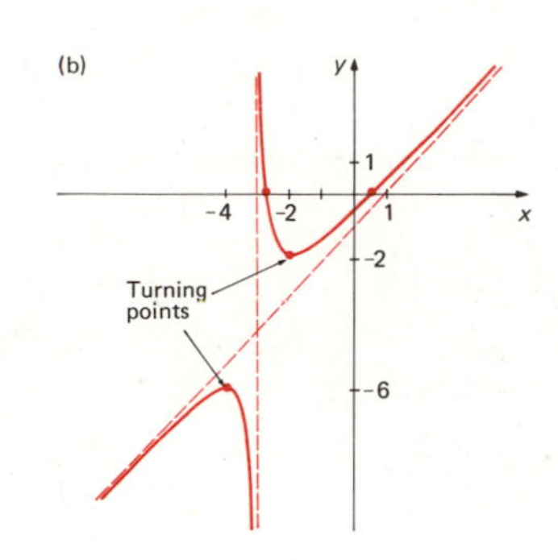

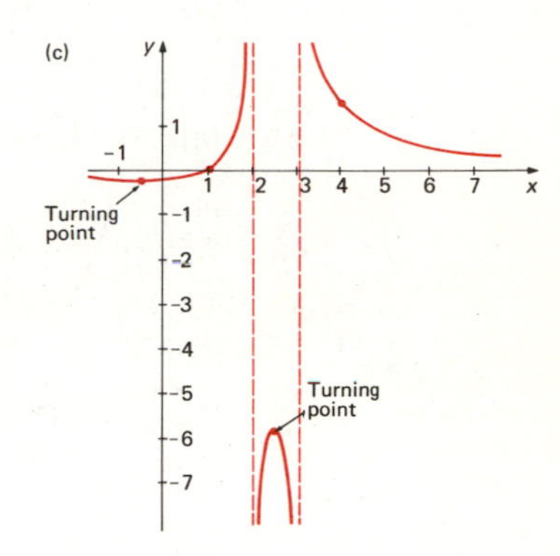

7.3.13 $A = 1$, $B = -2$
7.3.15 $A = 3$, $B = 2$, $C = -5$

Chapter 8

8.1.1 (*a*) 243 (*b*) $1/5$ (*c*) 4 (*d*) 512
8.1.3 (*a*) $1/6$ (*b*) $5\sqrt{5}$ (*c*) $7\sqrt[3]{2}$ (*d*) $205/4$
8.1.5 (*a*) -2 (*b*) $-3/4$ (*c*) $2/5$ (*d*) $5/6$
8.1.7 (*a*) $-3-2\sqrt{2}$ (*c*) $(60\sqrt{2}+45\sqrt{7}-8\sqrt{70}-42\sqrt{5})/-155$
(*b*) $5-2\sqrt{6}$ (*d*) $(80\sqrt{3}+60\sqrt{5}+12\sqrt{30}+45\sqrt{2})/5$

8.2.1

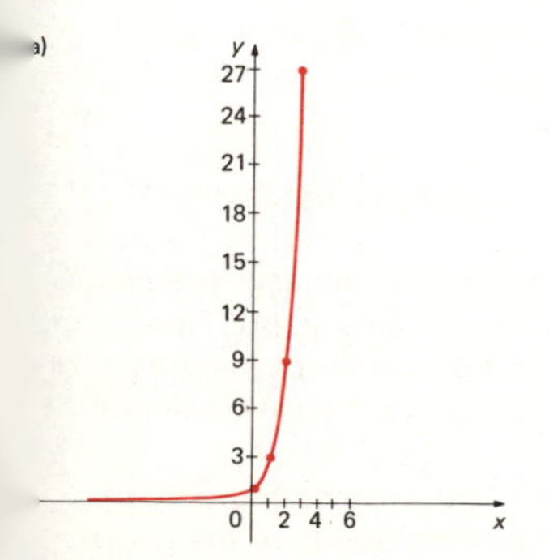

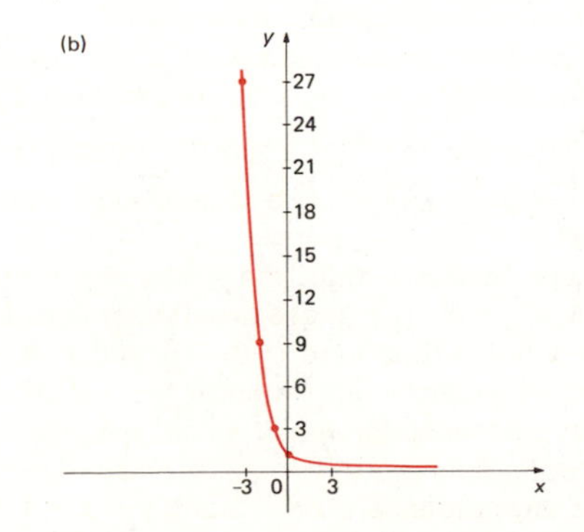

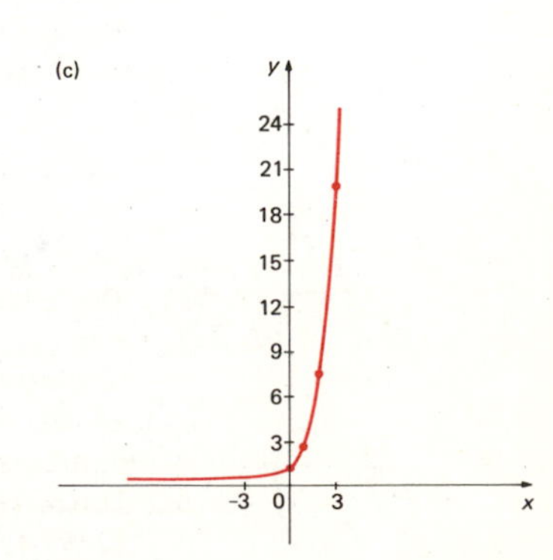

8.2.1

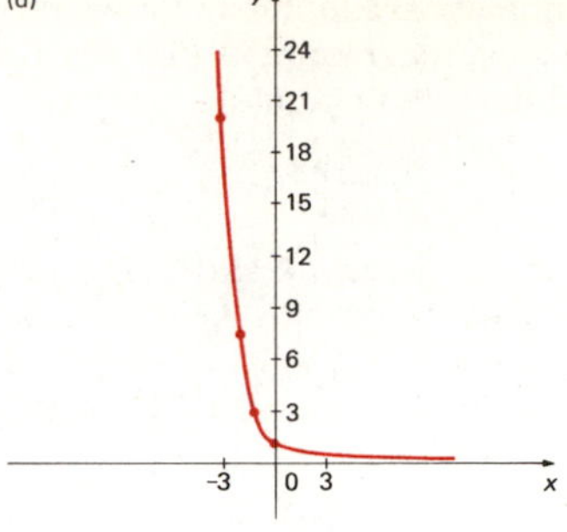

(*e*) See the graph for part *b*.
(*f*) See the graph for part *a*.

8.2.3 (*a*) The function *f* is even; (*b*) the function *f* is odd.

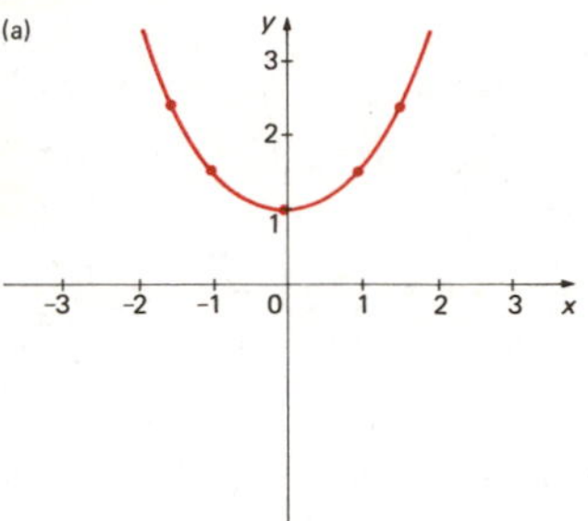

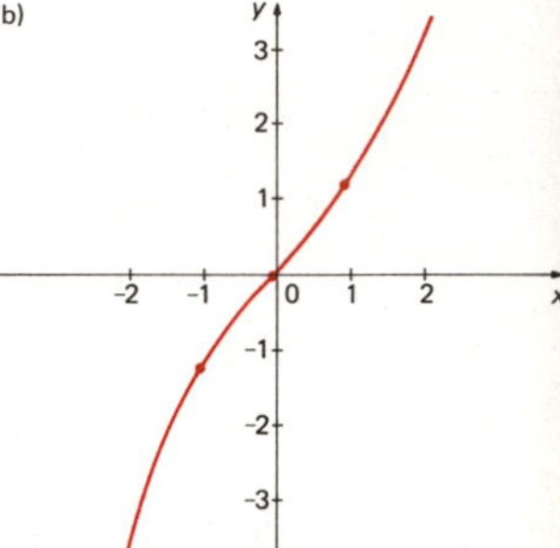

8.2.5

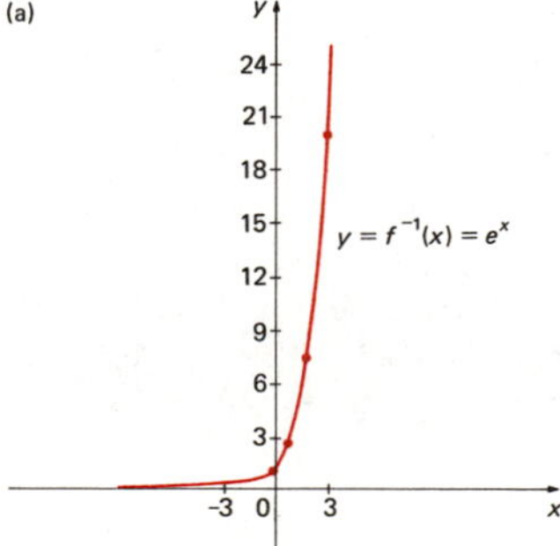

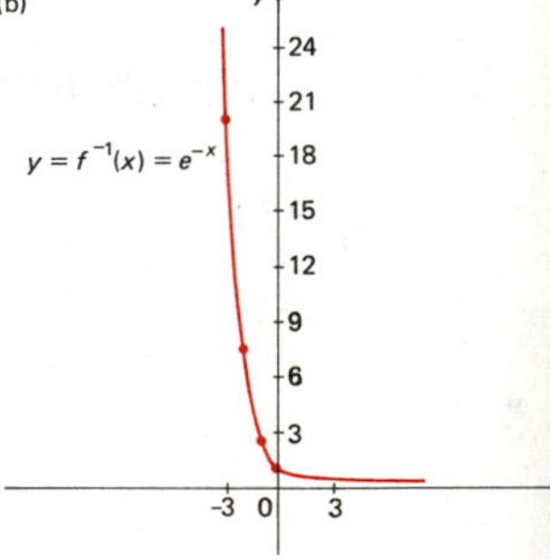

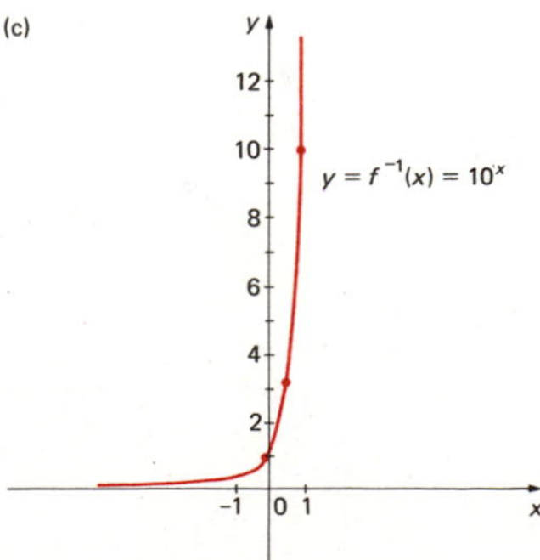

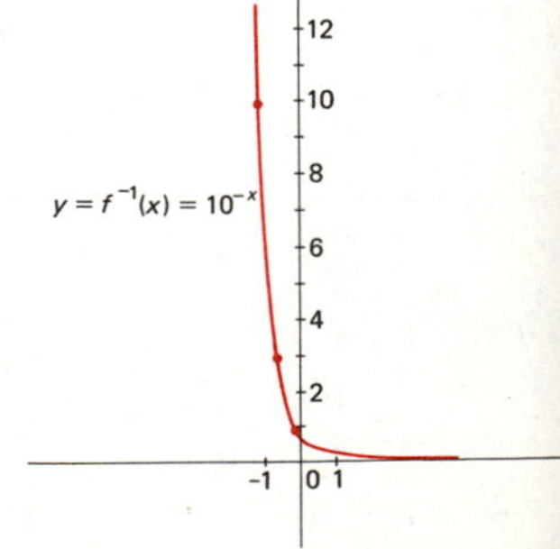

8.2.7 $\ln 10 = 1/\log e = 1/0.4343 = 2.3026$ approximately

8.2.9 (*a*) $\log 14 = \log 2 + \log 7$
(*b*) $\log 35 = \log {}^{70}\!/_2 = \log 7 - \log 2 + 1$
(*c*) $\log {}^{1}\!/_{15} = \log {}^{2}\!/_{30} = \log 2 - \log 3 - 1$
(*d*) $\log 56 = \log(2^3 \cdot 7) = 3 \log 2 + \log 7$
(*e*) $\log \sqrt[3]{7/5} = {}^{1}\!/_3 \log 7 - {}^{1}\!/_3 \log {}^{10}\!/_2$
$= {}^{1}\!/_3 \log 7 - {}^{1}\!/_3 + {}^{1}\!/_3 \log 2$
(*f*) $\log({}^{4}\!/_{63})^{1/5} = {}^{1}\!/_5[\log 2^2 - \log(3^2 \cdot 7)]$
$= {}^{2}\!/_5 \log 2 - {}^{2}\!/_5 \log 3 - {}^{1}\!/_5 \log 7$

8.3.1 $x \approx 0.73$

8.3.3 If $b = 1$, then *any* real number x is a solution of $b^x = 1$. If $b \neq 1$ and $b \neq 0$, then $x = 0$ is the only solution of $b^x = 1$.

8.3.5 (*a*) $x \approx 9.45$ (*b*) $x \approx 1.55$ (*c*) $x \approx 0.7$

8.4.1 2.56

8.4.3 The number of bacteria at the beginning is approximately equal to 98. The number of bacteria at the end of 6 hr is approximately equal to 1643. It takes this bacteria 1.48 hr, approximately, to double its number.

8.4.5 (*a*) $500(1.01)^{40} \approx 744.43$ (*b*) $500\, e^{0.4} \approx 745.90$

8.4.7 (*a*) 7.17 percent approximately (*b*) 6.93 percent approximately

8.4.9 $m \approx 34.7$, $P = 250$

Chapter 9

9.1.1 36°, 24°, 112°, −15°, −42°

9.1.3

θ	$\sin\theta$	$\cos\theta$	$\tan\theta$	$\cot\theta$	$\sec\theta$	$\csc\theta$
−30°	$-\frac{1}{2}$	$\frac{\sqrt{3}}{2}$	$-\frac{1}{\sqrt{3}}$	$-\sqrt{3}$	$\frac{2}{\sqrt{3}}$	-2
−45°	$-\frac{1}{\sqrt{2}}$	$\frac{1}{\sqrt{2}}$	-1	-1	$\sqrt{2}$	$-\sqrt{2}$
−60°	$-\frac{\sqrt{3}}{2}$	$\frac{1}{2}$	$-\sqrt{3}$	$-\frac{1}{\sqrt{3}}$	2	$-\frac{2}{\sqrt{3}}$
−120°	$-\frac{\sqrt{3}}{2}$	$-\frac{1}{2}$	$\sqrt{3}$	$\frac{1}{\sqrt{3}}$	-2	$-\frac{2}{\sqrt{3}}$
−135°	$-\frac{1}{\sqrt{2}}$	$-\frac{1}{\sqrt{2}}$	1	1	$-\sqrt{2}$	$-\sqrt{2}$
−150°	$-\frac{1}{2}$	$-\frac{\sqrt{3}}{2}$	$\frac{1}{\sqrt{3}}$	$\sqrt{3}$	$-\frac{2}{\sqrt{3}}$	-2

9.1.5 $\sin 150° = \cos 60°$, $\cos 150° = -\sin 60°$, $\tan 150° = -\cot 60°$, and so on. These results follow from the properties of similar triangles (draw a sketch).

9.1.7 $\sin 225° = -\sin 45°$, $\cos 225° = -\cos 45°$, $\tan 225° = \tan 45°$, and so on. These results follow from the properties of similar triangles (draw a sketch).

9.1.9 $\sin 210° = -\sin 30°$, $\cos 210° = -\cos 30°$, $\tan 210° = \tan 30°$, and so on. These results, as a sketch indicates, follow from the properties of similar triangles.

9.1.11 These angles have identical trigonometric functions, since $30° = 360° - 330°$.

9.1.13 (*a*) $\cos\theta = \sqrt{3}/2$, $\tan\theta = 1/\sqrt{3}$, $\cot\theta = \sqrt{3}$, $\sec\theta = 2/\sqrt{3}$, $\csc\theta = 2$
(*b*) $\sin\theta = \sqrt{15}/4$, $\tan\theta = \sqrt{15}$, $\cot\theta = 1/\sqrt{15}$, $\sec\theta = 4$, $\csc\theta = 4/\sqrt{15}$
(*c*) $\sin\theta = {}^{3}\!/_5$, $\cos\theta = {}^{4}\!/_5$, $\cot\theta = {}^{4}\!/_3$, $\sec\theta = {}^{5}\!/_4$, $\csc\theta = {}^{5}\!/_3$
(*d*) $\sin\theta = 1/\sqrt{10}$, $\cos\theta = 3/\sqrt{10}$, $\tan\theta = {}^{1}\!/_3$, $\sec\theta = \sqrt{10}/3$, $\csc\theta = \sqrt{10}$
(*e*) $\sin\theta = \sqrt{2}/\sqrt{3}$, $\cos\theta = 1/\sqrt{3}$, $\tan\theta = \sqrt{2}$, $\cot\theta = 1/\sqrt{2}$, $\csc\theta = \sqrt{3}/\sqrt{2}$
(*f*) $\sin\theta = 1/\sqrt{5}$, $\cos\theta = 2/\sqrt{5}$, $\tan\theta = {}^{1}\!/_2$, $\cot\theta = 2$, $\sec\theta = \sqrt{5}/2$

9.2.1 $\sin\frac{5\pi}{12} = \sqrt{\frac{1 - \cos(5\pi/6)}{2}} = \sqrt{\frac{1 - (-\sqrt{3}/2)}{2}} = \frac{\sqrt{2 + \sqrt{3}}}{2}$

$$\cos\frac{5\pi}{12}=\sqrt{\frac{1+\cos(5\pi/6)}{2}}=\frac{\sqrt{2-\sqrt{3}}}{2}$$

$$\tan\frac{5\pi}{12}=\frac{\sin(5\pi/12)}{\cos(5\pi/12)}=\sqrt{\frac{2+\sqrt{3}}{2-\sqrt{3}}}$$

$$\cot\frac{5\pi}{12}=\frac{1}{\tan(5\pi/12)}=\sqrt{\frac{2-\sqrt{3}}{2+\sqrt{3}}}$$

$$\sec\frac{5\pi}{12}=\frac{1}{\cos(5\pi/12)}=\frac{2}{\sqrt{2-\sqrt{3}}}$$

$$\csc\frac{5\pi}{12}=\frac{1}{\sin(5\pi/12)}=\frac{2}{\sqrt{2+\sqrt{3}}}$$

9.2.3 First, we find $\sin(\pi/8)$ and $\cos(\pi/8)$. Observe that

$$\sin\frac{\pi}{8}=\sqrt{\frac{1-\cos(\pi/4)}{2}}=\sqrt{\frac{1-\sqrt{2}/2}{2}}=\frac{\sqrt{2-\sqrt{2}}}{2}$$

$$\cos\frac{\pi}{8}=\sqrt{\frac{1+\cos(\pi/4)}{2}}=\frac{\sqrt{2+\sqrt{2}}}{2}$$

Now,

$$\begin{aligned}\sin(\pi/24)&=\sin(\pi/6-\pi/8)\\&=\sin(\pi/6)\cos(\pi/8)-\cos(\pi/6)\sin(\pi/8)\\&=\frac{1}{2}\cdot\frac{\sqrt{2+\sqrt{2}}}{2}-\frac{\sqrt{3}}{2}\cdot\frac{\sqrt{2-\sqrt{2}}}{2}\\&=\frac{\sqrt{2+\sqrt{2}}-\sqrt{6-3\sqrt{2}}}{4}\end{aligned}$$

Thus, $\sin(\pi/24)=(\sqrt{2+\sqrt{2}}-\sqrt{6-3\sqrt{2}})/4$. Similarly, we readily verify that

$$\begin{aligned}\cos(\pi/24)&=\cos(\pi/6)\cos(\pi/8)+\sin(\pi/6)\sin(\pi/8)\\&=\frac{\sqrt{3}}{2}\cdot\frac{\sqrt{2+\sqrt{2}}}{2}+\frac{1}{2}\cdot\frac{\sqrt{2-\sqrt{2}}}{2}\end{aligned}$$

Thus, $\cos(\pi/24)=(\sqrt{6+3\sqrt{2}}+\sqrt{2-\sqrt{2}})/4$. Therefore,

$$\tan\frac{\pi}{24}=\frac{\sqrt{2+\sqrt{2}}-\sqrt{6-3\sqrt{2}}}{\sqrt{6+3\sqrt{2}}+\sqrt{2-\sqrt{2}}}\qquad\sec\frac{\pi}{24}=\frac{4}{\sqrt{6+3\sqrt{2}}+\sqrt{2-\sqrt{2}}}$$

$$\cot\frac{\pi}{24}=\frac{\sqrt{6+3\sqrt{2}}+\sqrt{2-\sqrt{2}}}{\sqrt{2+\sqrt{2}}-\sqrt{6-3\sqrt{2}}}\qquad\csc\frac{\pi}{24}=\frac{4}{\sqrt{2+\sqrt{2}}-\sqrt{6-3\sqrt{2}}}$$

It can be shown that the answers for $\sin(\pi/24)$ and $\cos(\pi/24)$ given here are equivalent to those given in Example 9.8.

9.2.9 (*a*)

$$\begin{aligned}\frac{2\tan(x/2)}{1+\tan^2(x/2)}&=\frac{2\tan(x/2)}{\sec^2(x/2)}=2\tan(x/2)\cos^2(x/2)\\&=2\frac{\sin(x/2)}{\cos(x/2)}\cos^2(x/2)=2\sin(x/2)\cos(x/2)\\&=\sin x\end{aligned}$$

(*b*)

$$\begin{aligned}\frac{1-\tan^2(x/2)}{1+\tan^2(x/2)}&=\frac{1-\tan^2(x/2)}{\sec^2(x/2)}=\cos^2(x/2)\,[1-\tan^2(x/2)]\\&=\cos^2(x/2)-\cos^2(x/2)\tan^2(x/2)=\cos^2(x/2)-\sin^2(x/2)\\&=\cos x\end{aligned}$$

(*c*) $\tan x=\dfrac{\sin x}{\cos x}=\dfrac{2\tan(x/2)}{1-\tan^2(x/2)}$ by parts *a* and *b* above

9.2.11 (*a*) $\sin(x+y)=\sin x\cos y+\cos x\sin y$ $\quad\sin(x-y)=\sin x\cos y-\cos x\sin y$

$\therefore\ \sin(x+y)+\sin(x-y)=2\sin x\cos y$

(*b*) Subtracting the first two identities in part *a*, we obtain the result at once.
(*c*) Argue as in part *a*, starting with the identities for $\cos(x + y)$ and $\cos(x - y)$.
(*d*) Argue as in part *b*, starting with the identities for $\cos(x + y)$ and $\cos(x - y)$.

9.2.13 $\sin 15° = \sqrt{\dfrac{1 - \cos 30°}{2}} = \sqrt{\dfrac{1 - \sqrt{3}/2}{2}} = \dfrac{\sqrt{2 - \sqrt{3}}}{2}$

$\cos 15° = \sqrt{\dfrac{1 + \cos 30°}{2}} = \dfrac{\sqrt{2 + \sqrt{3}}}{2}$

9.3.3 Suppose that $f(x) = c$ for *all* real numbers x (c is a constant). Then $f(x + a) = f(x)$ for *all* real numbers a. Hence f is a periodic function, and every nonzero real number a is a period of f. Clearly, f has *no* fundamental period, since a can be chosen to be arbitrarily small.

9.3.5

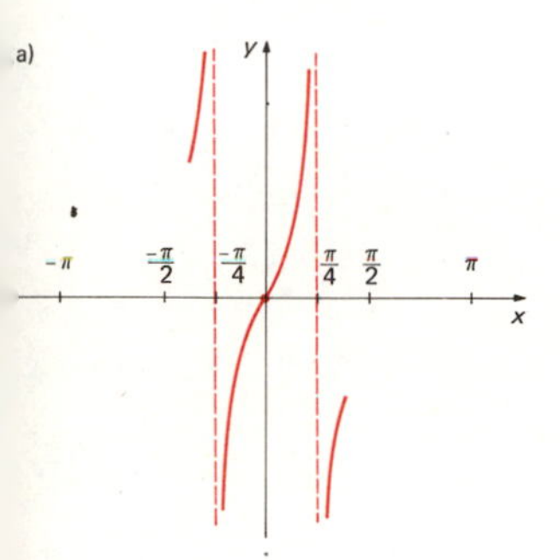

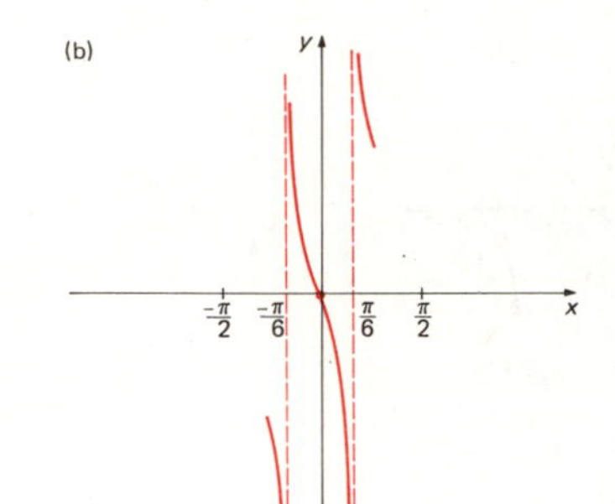

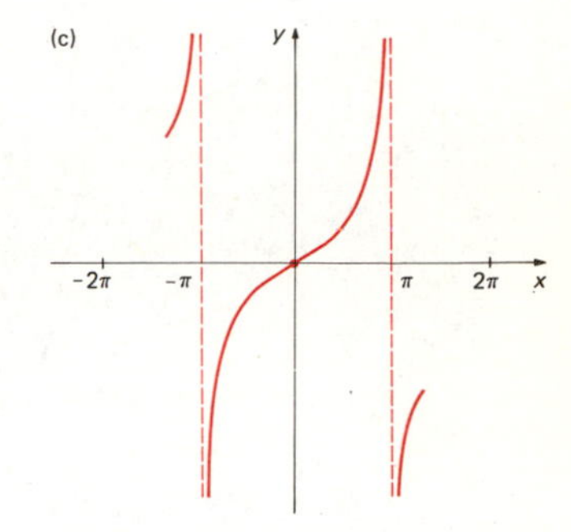

9.3.7

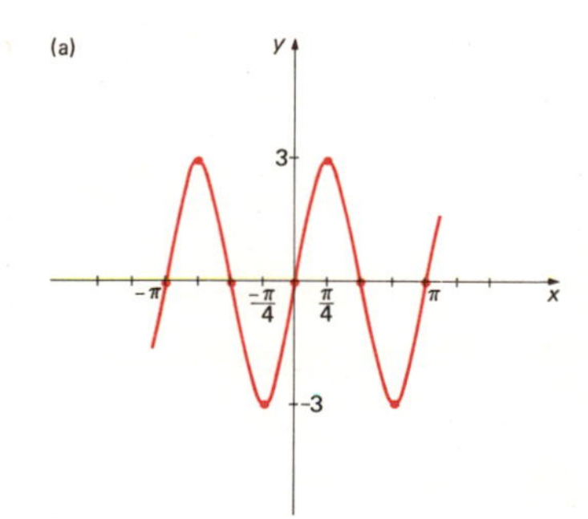

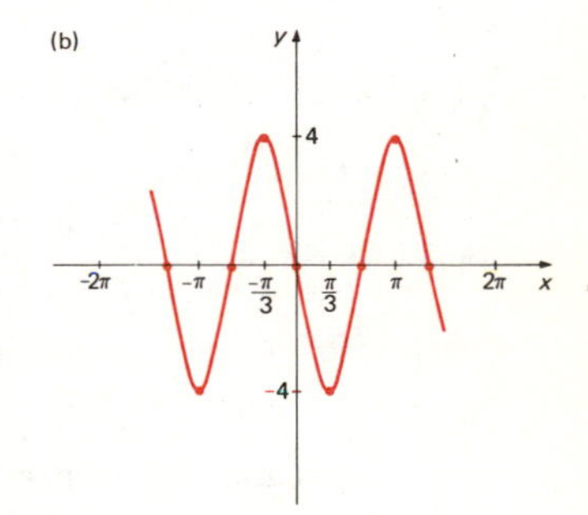

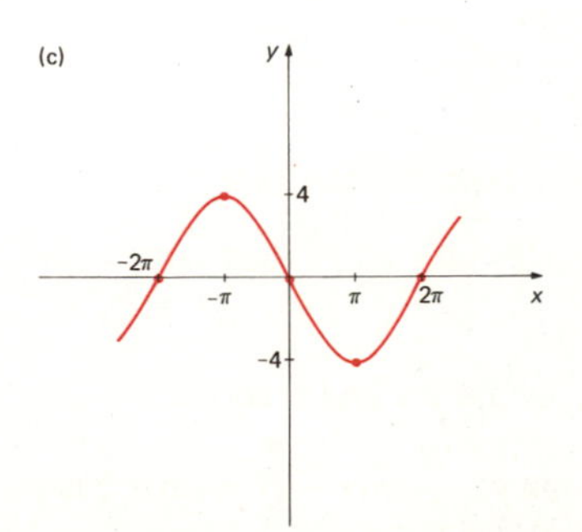

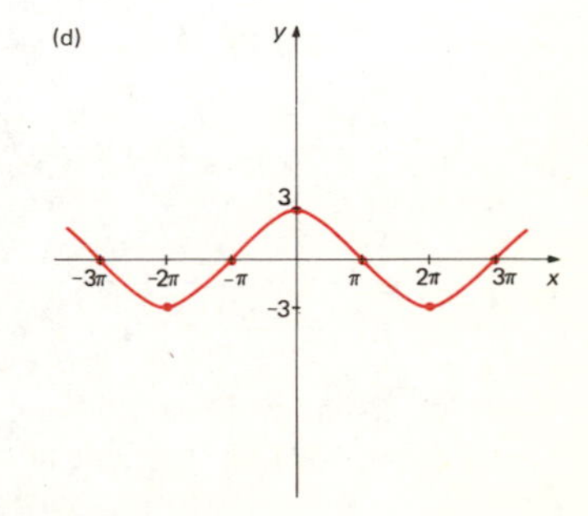

9.3.9

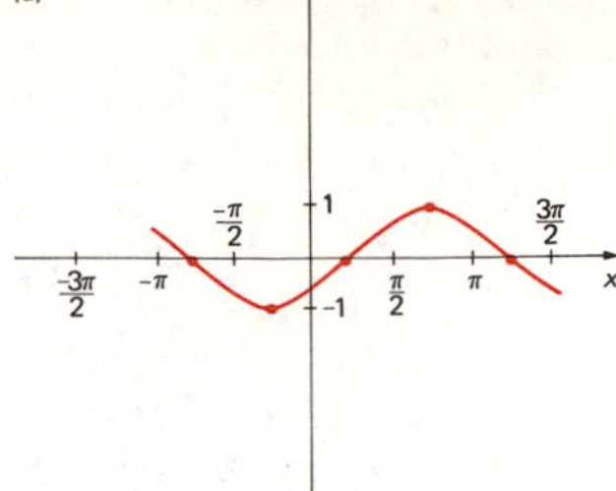

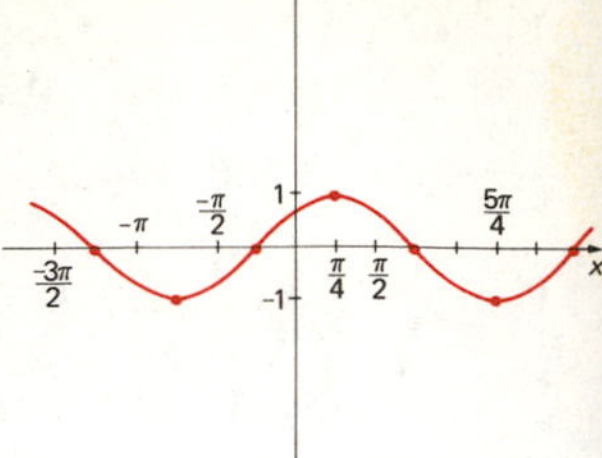

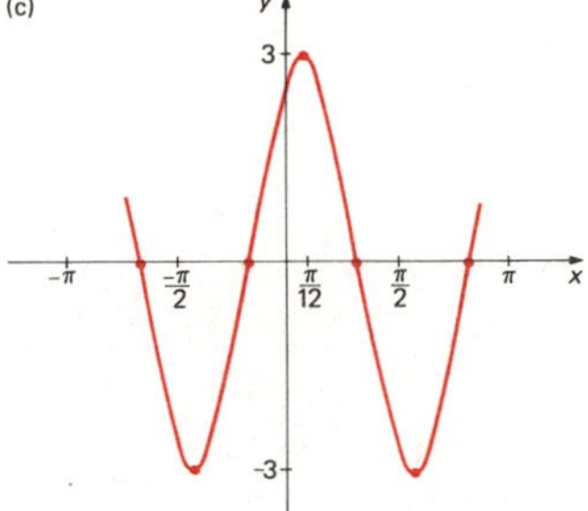

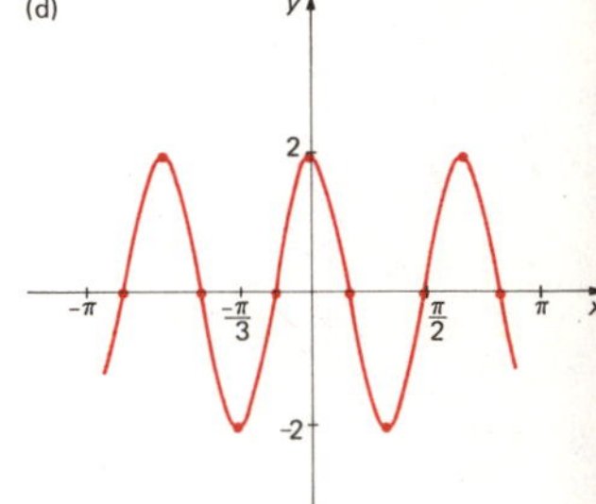

9.3.11

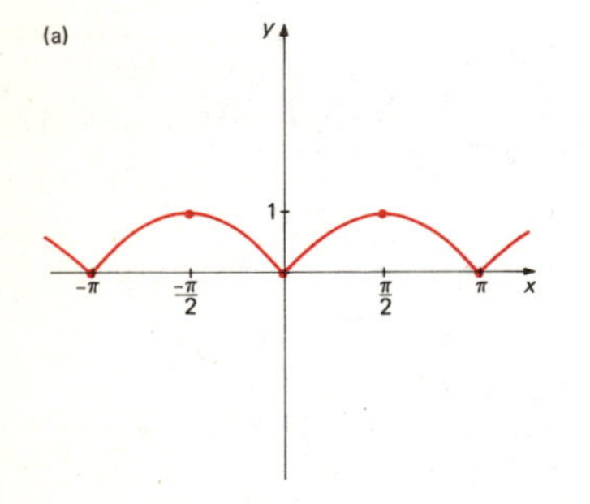

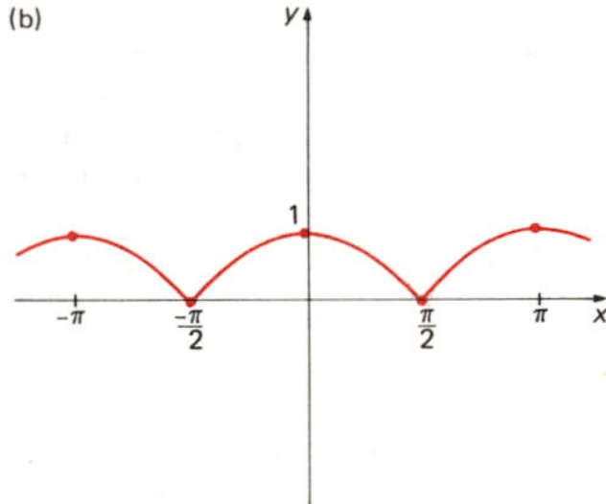

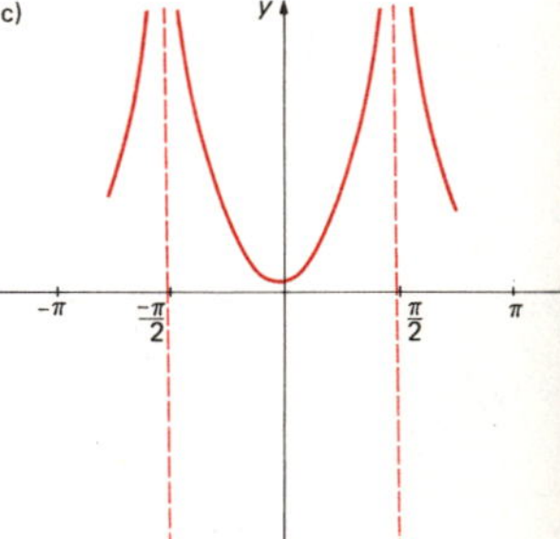

9.3.13 (*a*) Since $f(x) = \sin^2 x = (1 - \cos 2x)/2$, it follows that the fundamental period of $f(x)$ is π.

(*b*) Since $f(x) = \cos^2 x = (1 + \cos 2x)/2$, it follows that the fundamental period of $f(x)$ is π.

9.3.15

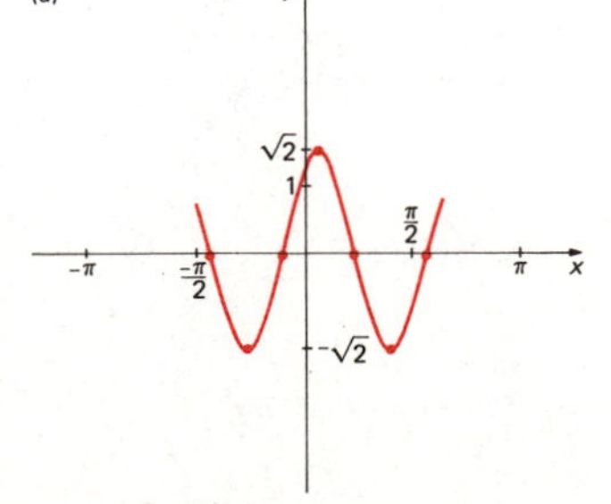

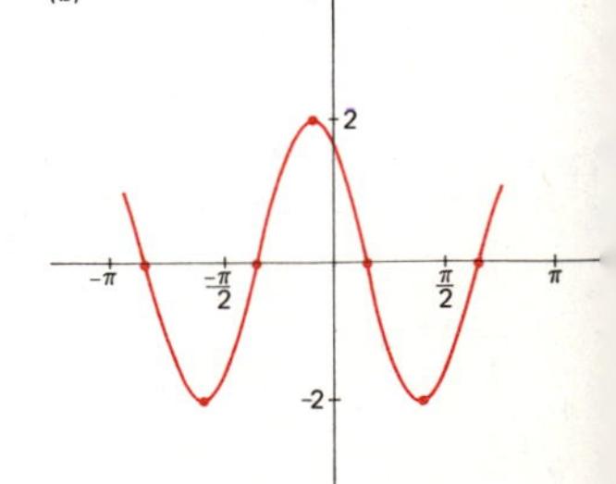

Chapter 10

10.1.1 (*a*) $a=\sqrt{3},\ b=2,\ \beta=30°$
(*b*) $a=\sqrt{3},\ b=1,\ \alpha=60°$
(*c*) $b=3.576,\ c=4.668,\ \beta=50°$
(*d*) $a=5.96,\ c=7.78,\ \beta=40°$
(*e*) $c=31.62,\ \alpha=18°,\ \beta=72°$
(*f*) $b=14.14,\ \alpha=19°,\ \beta=71°$

10.1.3 (*a*) 76.6 (*b*) 6.6 (*c*) 21.0 (*d*) 36.9

10.1.5 (*a*) By the law of sines, if such a triangle exists, then $(\sin\alpha)/5=(\sin 40°)/2$, or $\sin\alpha=(5\sin 40°)/2>1$, which is impossible. Hence *no such triangle exists.*
(*b*) $\beta=29°,\ \alpha=51°,\ a=7.9$
(*c*) By the law of sines, if such a triangle exists, then $(\sin\alpha)/4=(\sin 31°)/2$, or $\sin\alpha=2\sin 31°>1$, which is impossible. Hence *no such triangle exists.*

10.1.7 58.78

10.1.9 (*a*)
$$\frac{(b+c+a)(b+c-a)}{4bc}=\frac{b^2+c^2+2bc-a^2}{4bc}$$
$$=\frac{b^2+c^2+2bc-(b^2+c^2-2bc\cos\alpha)}{4bc}\quad\text{(using law of cosines)}$$
$$=\frac{2bc(1+\cos\alpha)}{4bc}$$
$$=\frac{1+\cos\alpha}{2}$$
(*b*) Argue as in part *a*, again using the law of cosines.
(*c*) Recall that $\cos^2(\alpha/2)=(1+\cos\alpha)/2$. Now use the result in part *a*, keeping in mind that $0°<\alpha/2<90°$ (Why?).
(*d*) Recall that $\sin^2(\alpha/2)=(1-\cos\alpha)/2$. Now use the result in part *b*, keeping in mind that $0°<\alpha/2<90°$.

10.1.11 $500\tan 40°=419.55$

10.1.13 $100\tan 40°=83.91$

10.1.15 By solving the equations $h=x\tan 50°$ and $h=(x+100)\tan 35°$, we obtain $h=170$ approximately.

10.1.17 $30+50\tan 35°=65.01$

10.1.19 Suppose that the height of the hill is h ft. It is easily seen that $(h+100)/h=\tan 50°/\tan 40°$, and hence
$$h=\frac{100\tan 40°}{\tan 50°-\tan 40°}=238\text{ approximately}$$

10.2.1 (*a*) $\sqrt{2}\left(\cos\frac{7\pi}{4}+i\sin\frac{7\pi}{4}\right)$
(*b*) $\sqrt{2}\left(\cos\frac{3\pi}{4}+i\sin\frac{3\pi}{4}\right)$
(*c*) $2\left(\cos\frac{11\pi}{6}+i\sin\frac{11\pi}{6}\right)$
(*d*) $2\left(\cos\frac{\pi}{6}+i\sin\frac{\pi}{6}\right)$
(*e*) $2\left(\cos\frac{7\pi}{6}+i\sin\frac{7\pi}{6}\right)$
(*f*) $2\left(\cos\frac{5\pi}{3}+i\sin\frac{5\pi}{3}\right)$
(*g*) $2\left(\cos\frac{\pi}{3}+i\sin\frac{\pi}{3}\right)$
(*h*) $2\left(\cos\frac{2\pi}{3}+i\sin\frac{2\pi}{3}\right)$
(*i*) $2\left(\cos\frac{4\pi}{3}+i\sin\frac{4\pi}{3}\right)$

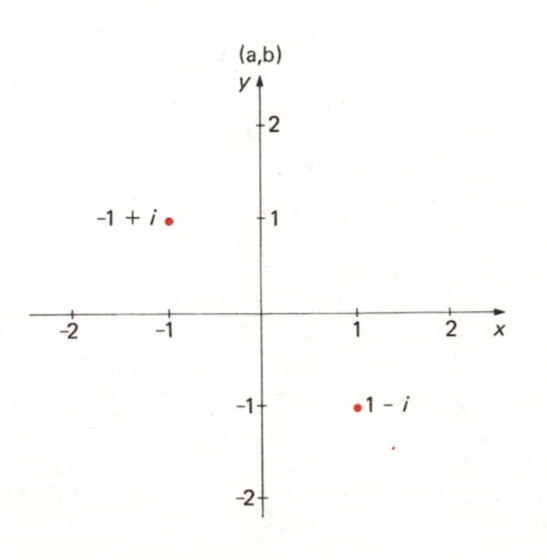

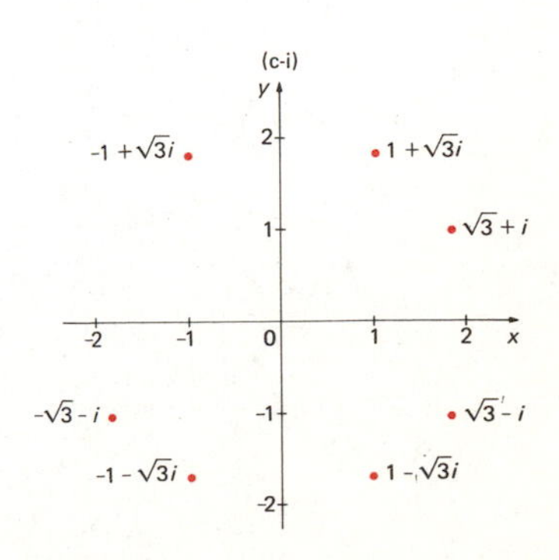

10.2.1 (j) $1\left(\cos\frac{5\pi}{3} + i\sin\frac{5\pi}{3}\right)$ (n) $1\left(\cos\frac{\pi}{4} + i\sin\frac{\pi}{4}\right)$

(k) $1\left(\cos\frac{\pi}{3} + i\sin\frac{\pi}{3}\right)$ (o) $1\left(\cos\frac{7\pi}{4} + i\sin\frac{7\pi}{4}\right)$

(l) $1\left(\cos\frac{2\pi}{3} + i\sin\frac{2\pi}{3}\right)$ (p) $1\left(\cos\frac{3\pi}{4} + i\sin\frac{3\pi}{4}\right)$

(m) $1\left(\cos\frac{4\pi}{3} + i\sin\frac{4\pi}{3}\right)$ (q) $1\left(\cos\frac{5\pi}{4} + i\sin\frac{5\pi}{4}\right)$

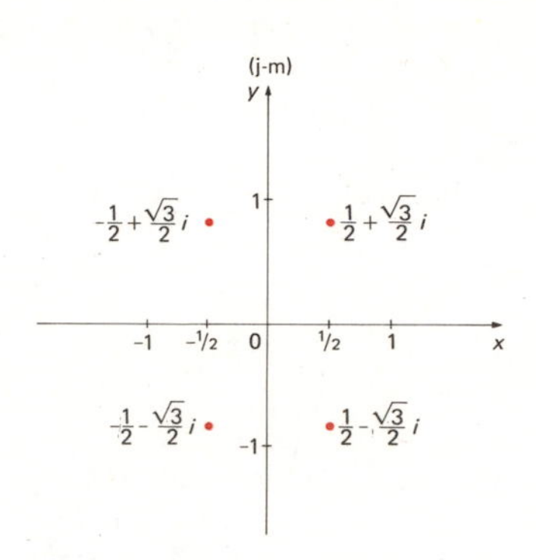

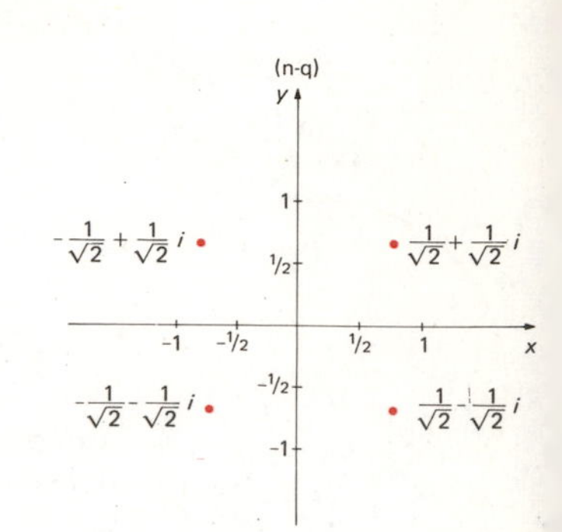

10.2.3 (a) $\cos\frac{10\pi}{4} + i\sin\frac{10\pi}{4} = i$

(b) $\cos\left(\frac{-10\pi}{4}\right) + i\sin\left(\frac{-10\pi}{4}\right) = -i$

(c) $\cos\frac{100\pi}{3} + i\sin\frac{100\pi}{3} = -\frac{1}{2} - \frac{\sqrt{3}}{2}i$

(d) $\cos\frac{250\pi}{6} + i\sin\frac{250\pi}{6} = \frac{1}{2} - \frac{\sqrt{3}}{2}i$

(e) $2^{10}\left(\cos\frac{70\pi}{4} + i\sin\frac{70\pi}{4}\right) = -1024i$

(f) $2^{10}\left[\cos\left(\frac{-70\pi}{4}\right) + i\sin\left(\frac{-70\pi}{4}\right)\right] = 1024i$

10.2.5

n	*nth roots of unity*
4	$1, -1, i, -i$
5	$\cos\frac{2\pi}{5} + i\sin\frac{2\pi}{5}$ $\left(\cos\frac{2\pi}{5} + i\sin\frac{2\pi}{5}\right)^2$ $\left(\cos\frac{2\pi}{5} + i\sin\frac{2\pi}{5}\right)^3$ $\left(\cos\frac{2\pi}{5} + i\sin\frac{2\pi}{5}\right)^4$ 1
7	$\cos\frac{2\pi}{7} + i\sin\frac{2\pi}{7}$ $\left(\cos\frac{2\pi}{7} + i\sin\frac{2\pi}{7}\right)^2$ $\left(\cos\frac{2\pi}{7} + i\sin\frac{2\pi}{7}\right)^3$ . . . $\left(\cos\frac{2\pi}{7} + i\sin\frac{2\pi}{7}\right)^6$ 1
8	$\frac{1+i}{\sqrt{2}}$ $\left(\frac{1+i}{\sqrt{2}}\right)^2$ $\left(\frac{1+i}{\sqrt{2}}\right)^3$ . . . $\left(\frac{1+i}{\sqrt{2}}\right)^7$ 1
9	$\cos\frac{2\pi}{9} + i\sin\frac{2\pi}{9}$ $\left(\cos\frac{2\pi}{9} + i\sin\frac{2\pi}{9}\right)^2$ $\left(\cos\frac{2\pi}{9} + i\sin\frac{2\pi}{9}\right)^3$ . . . $\left(\cos\frac{2\pi}{9} + i\sin\frac{2\pi}{9}\right)^8$ 1
10	$\cos\frac{2\pi}{10} + i\sin\frac{2\pi}{10}$ $\left(\cos\frac{2\pi}{10} + i\sin\frac{2\pi}{10}\right)^2$ $\left(\cos\frac{2\pi}{10} + i\sin\frac{2\pi}{10}\right)^3$ . . . $\left(\cos\frac{2\pi}{10} + i\sin\frac{2\pi}{10}\right)^9$ 1

10.2.5

(a)

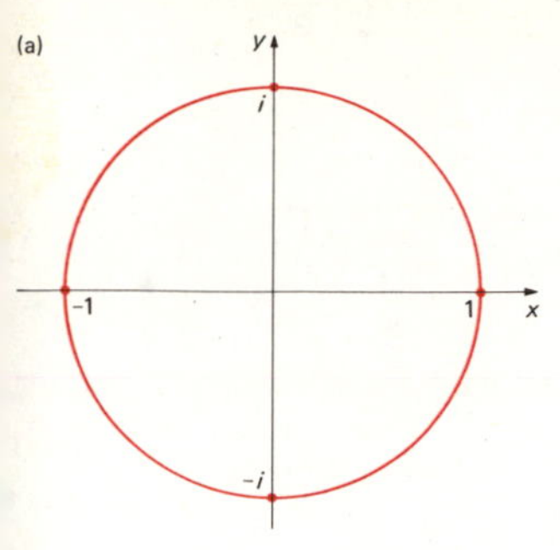

(b)

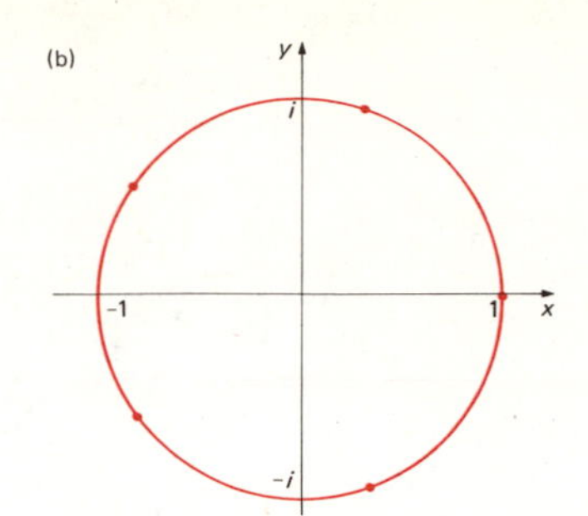

(c)

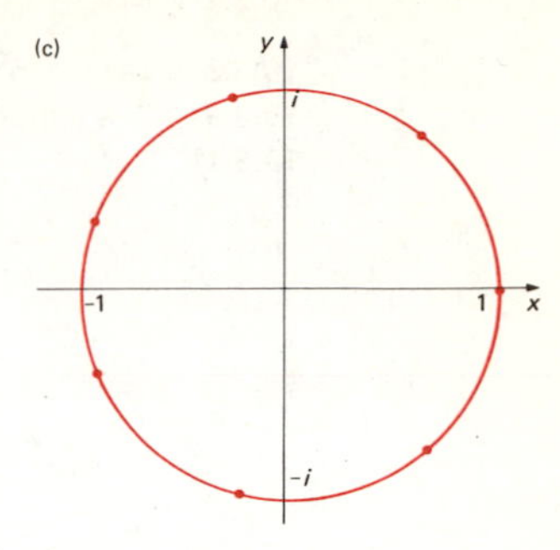

(d)

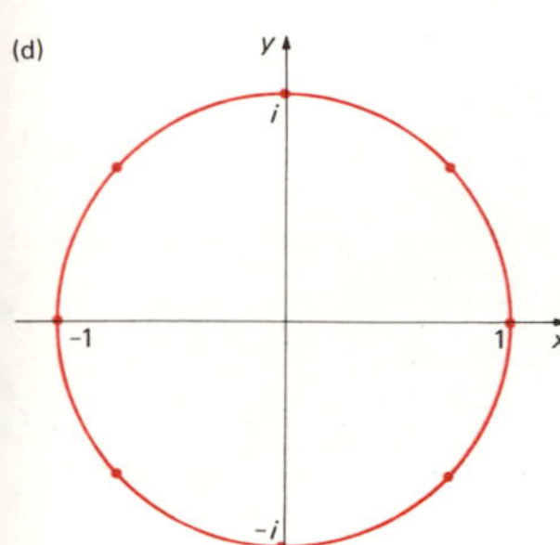

(e)

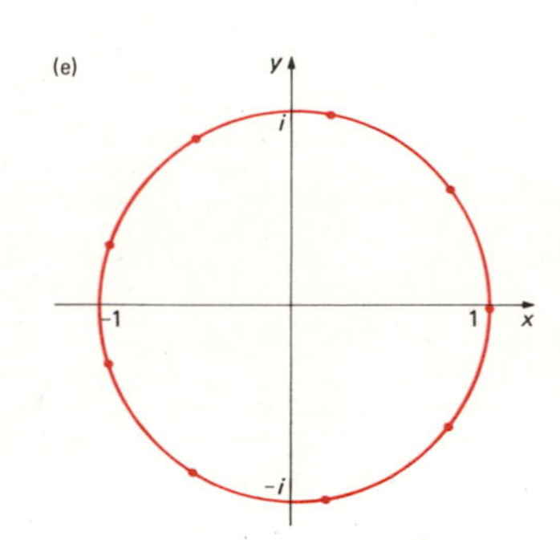

(f)

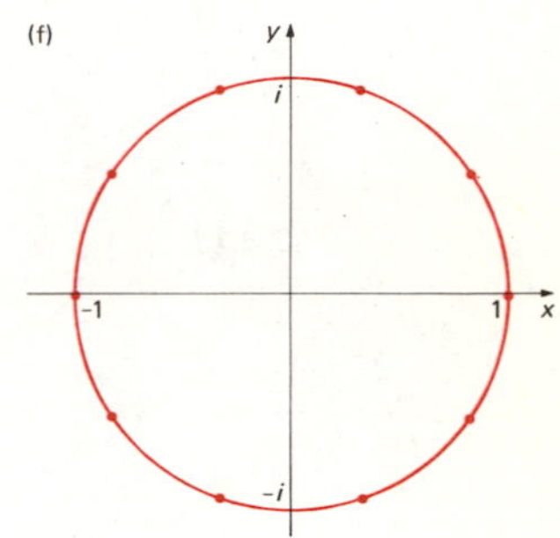

10.2.7 Observe that

$$x_0{}^n = \left(\cos\frac{2\pi}{n} + i\sin\frac{2\pi}{n}\right)^n = \cos 2\pi + i\sin 2\pi = 1$$

and hence x_0 is an nth root of unity. Hence x_0, $x_0{}^2$, $x_0{}^3 \ldots$, $x_0{}^n$ are also nth roots of unity, and are, in fact, *distinct*. Therefore they represent *all* the nth roots of unity (Why?).

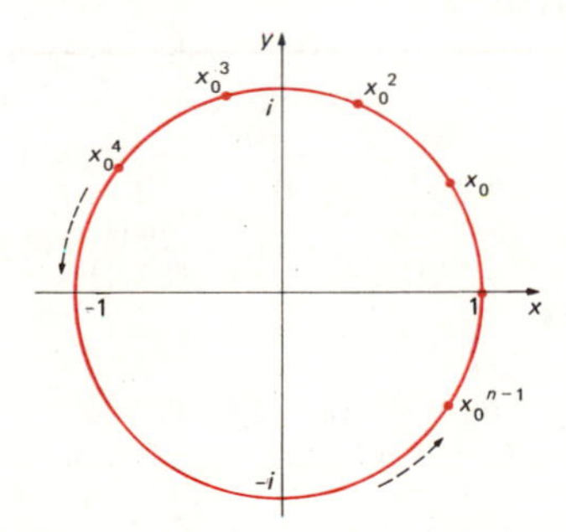

10.2.9 Suppose x is an nth root of unity. Then $x^n = 1$, and hence $1/x^n = 1/1 = 1$. Thus $(1/x)^n = 1$, and hence $1/x$ is also an nth root of unity.

10.2.11 Suppose that $n = -m$ is any negative integer, and thus m is a positive integer. Suppose, further, that $\cos\theta + i\sin\theta \neq 0$. Now, by Prob. 10.2.10, $(\cos\theta + i\sin\theta)^{-1} = \cos(-\theta) + i\sin(-\theta)$. Hence, using De Moivre's theorem for the *positive* integer m, we obtain

$$[(\cos\theta + i\sin\theta)^{-1}]^m = [\cos(-\theta) + i\sin(-\theta)]^m$$
$$= \cos[m(-\theta)] + i\sin[m(-\theta)] = \cos[(-m)\theta] + i\sin[(-m)\theta]$$

Therefore, since $-m = n$, the above calculation shows that $(\cos\theta + i\sin\theta)^n = \cos n\theta + i\sin n\theta$, for all *negative* integers n.

10.3.1 (*a*) $\frac{\pi}{3}$ (*b*) $\frac{-\pi}{3}$ (*c*) $\frac{\pi}{3}$ (*d*) $\frac{-\pi}{3}$ (*e*) $\frac{\pi}{3}$ (*f*) $\frac{2\pi}{3}$

10.3.3 (*a*) $\sqrt{51}/10$ (*b*) $8/10$ (*c*) $120/169$

10.3.9 Yes, since, for all real numbers y, $-1 \le \sin y \le 1$ and $-1 \le \cos y \le 1$.

10.3.11

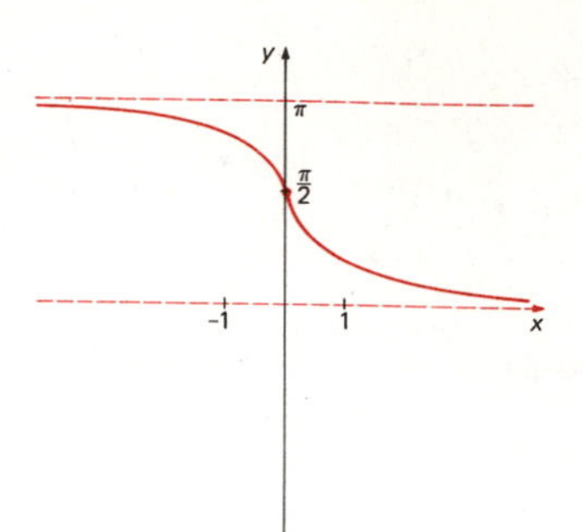

10.3.13

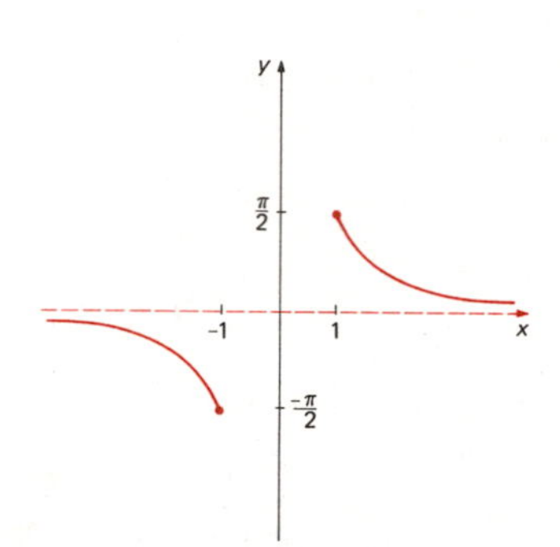

Chapter 11

11.1.1 (*a*) 0 (*b*) 0 (*c*) does not exist (*d*) 2 (*e*) $1/2$ (*f*) $3/4$ (*g*) $3/2$ (*h*) does not exist

11.1.3 (*a*) $y-3=(-1)(x+5)$ (*b*) $y=-5x$ (*c*) $y=-4$ (*d*) $y=5$ (*e*) $x=-1$

11.1.5 (*a*) $y+1={}^{-5}/_{3}(x-1)$ (*b*) $y={}^{3}/_{2}x$ (*c*) $x=5$ (*d*) $y=-2$

11.1.7 $x/3+y/3=1$

11.1.9 $y+4={}^{-4}/_{3}(x-1)$

11.1.11 $(3x-2y-6)+{}^{23}/_{39}(x+4y+8)=0$

11.1.13 $(2x-y-1)-{}^{3}/_{4}(x-2y+1)=0$

11.1.15

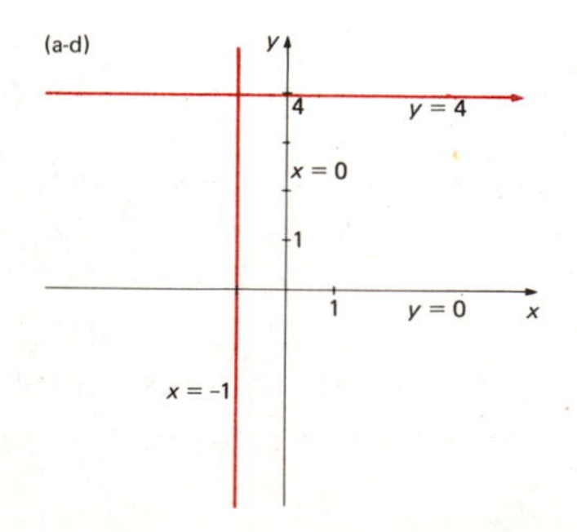

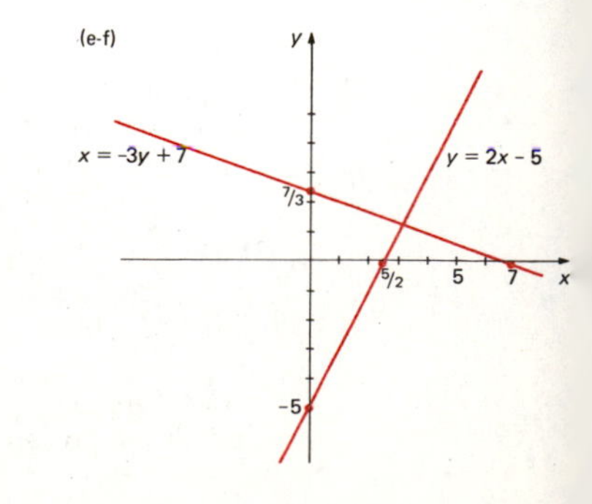

11.1.15 (g,h)

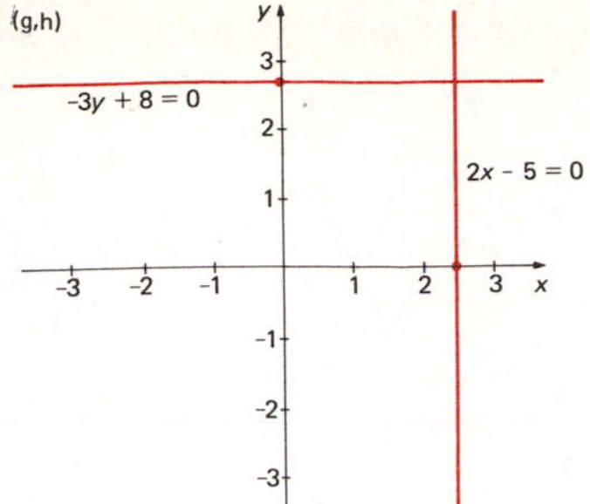

(i,j)

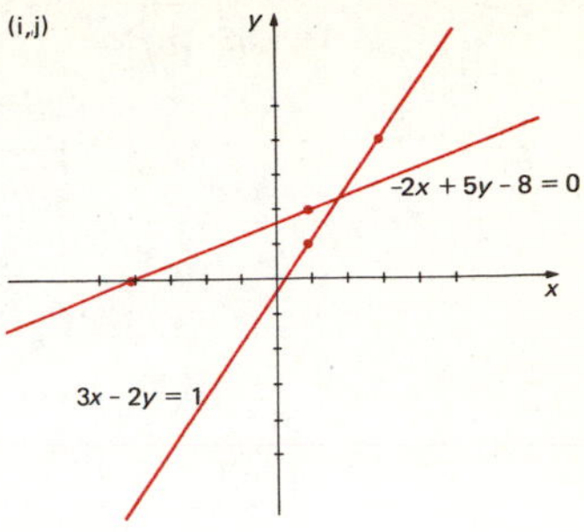

11.1.17 $y - 9/2 = 1(x + 3/2)$, or $y = x + 6$. The x intercept point is $(-6,0)$, and the y intercept point is $(0,6)$.

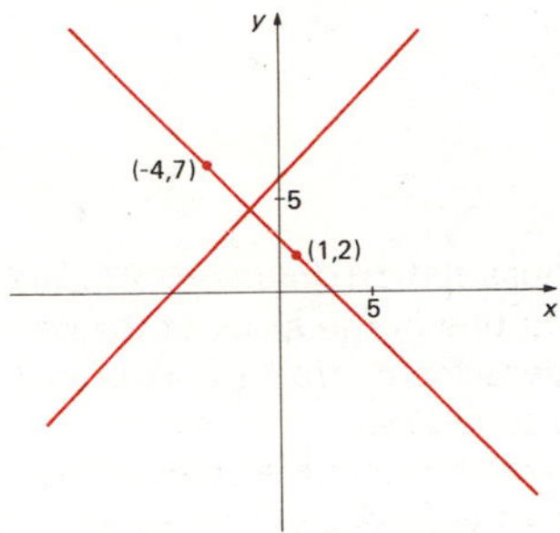

11.1.19 *Hint*: Compute the slopes and lengths of the four sides.
11.1.21 *Hint*: Compute the slopes of the four sides.
11.1.23 $a = -1/10$, $b = -1/7$
11.1.25 $y - 4 = -4/9(x + 3)$ $y - 5 = 7/6(x - 2)$ $y + 1 = -11/3(x - 1)$
11.1.27 Solving simultaneously the equations $2x - y = 1$ and $2x + y = 3$, we obtain $x = 1$, $y = 1$. Hence one of the vertices is $(1,1)$. Similarly we verify that $(2,3)$ and $(-2,7)$ are the other vertices.

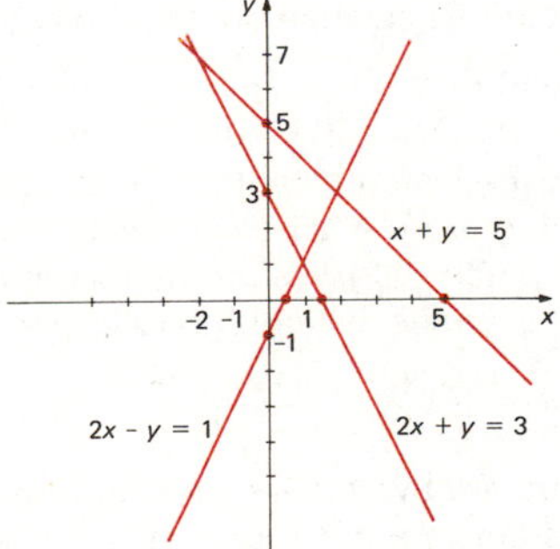

11.1.29 2
11.1.31 (*a*) $\sqrt{4^2 + 3^2} = 5$ (*b*) $(-1,-11/2)$
11.1.33 (*a*) Verify that the slope of the line joining $(2,-1)$ and $(5,1)$ is the same as the slope of the line segment joining $(5,1)$ and $(8,3)$. We may also verify that the length of the line segment joining $(2,-1)$ and $(5,1)$ plus the length of the line segment joining $(5,1)$ and $(8,3)$ is equal to the length of the line segment joining $(2,-1)$ and $(8,3)$.
(*b*) Proceed as described in part *a*.

11.1.35 (*a*)

Line	*x intercept point*	*y intercept point*	*slope*
$-2x + 5y = 8$	$(-4,0)$	$(0,8/5)$	$2/5$
$3x - 2y = 7$	$(7/3,0)$	$(0,-7/2)$	$3/2$
$-x + 8y = 23$	$(-23,0)$	$(0,23/8)$	$1/8$

11.1.35 (*b*)

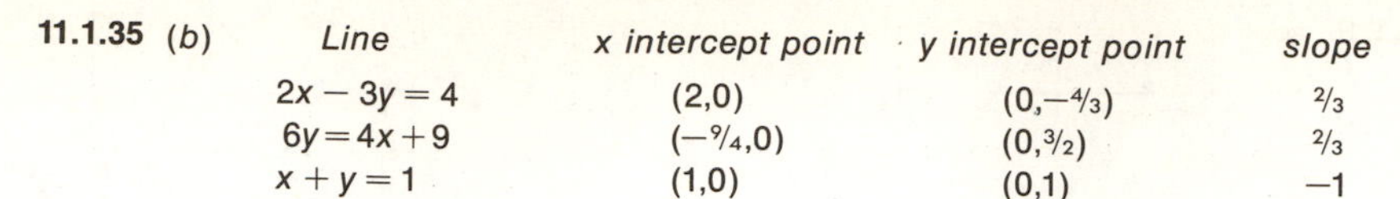

Line	*x intercept point*	*y intercept point*	*slope*
$2x-3y=4$	$(2,0)$	$(0,-4/3)$	$2/3$
$6y=4x+9$	$(-9/4,0)$	$(0,3/2)$	$2/3$
$x+y=1$	$(1,0)$	$(0,1)$	-1

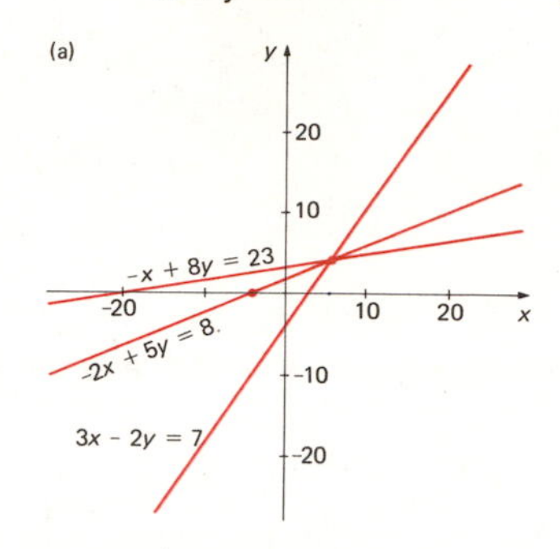

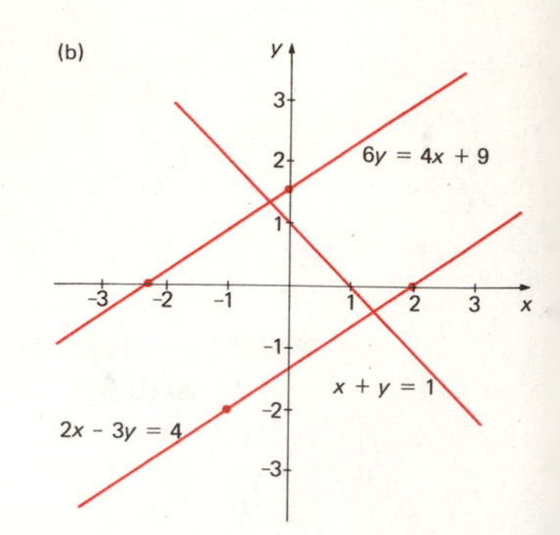

11.1.37 *Hint*: Choose the origin as one of the vertices of the parallelogram, and choose the x axis to lie along one of the sides of the parallelogram. Then use the midpoint formula.

11.1.39 *Hint*: Choose the coordinate axes as described in Prob. 11.1.37. Then use the midpoint and slope formulas.

11.1.41 *Hint*: Choose the origin as one of the vertices of the triangle, and choose the x axis to lie along one of the sides of the triangle. Then find the equations of the three altitudes. Next solve each pair of these equations simultaneously, and verify that you always get the *same* solution.

11.1.43 Suppose that $m>0$. If $x_1<x_2$, then $mx_1<mx_2$ (since $m>0$), and hence $mx_1+b<mx_2+b$. Thus, if $m>0$, then $x_1<x_2$ implies that $f(x_1)<f(x_2)$. Hence f is strictly increasing if $m>0$. The argument is similar if $m<0$ (recall that the direction of an inequality is *reversed* when the inequality is multiplied by a *negative* number).

11.1.45 One way to show this is to imitate the method described in Prob. 11.1.44. This leads to a rather laborious calculation. More elegant proofs appear in the reference given in the text.

11.2.1 (*a*) $x^2+y^2=16$ (*c*) $(x+1)^2+(y+5)^2=9$
(*b*) $(x-1)^2+(y-3)^2=4$ (*d*) $(x-1)^2+(y+2)^2=49$

11.2.3 *Hint*: Let (h,k) be the center of such a circle. Then an equation of our circle is $(x-h)^2+(y-k)^2=25$. Now, since $(1,0)$ lies on the circle, we have $(1-h)^2+(0-k)^2=25$. Similarly, since $(0,2)$ lies on the circle, we also have $(0-h)^2+(2-k)^2=25$. Solving the last two equations, we get the following *two* solutions:

$$h=\frac{1+\sqrt{76}}{2},\ k=\frac{4+\sqrt{76}}{4} \qquad h=\frac{1-\sqrt{76}}{2},\ k=\frac{4-\sqrt{76}}{4}$$

Thus, there are two circles passing through $(1,0)$ and $(0,2)$, each of radius 5. Their equations are $(x-h)^2+(y-k)^2=25$, where h and k are as given above.

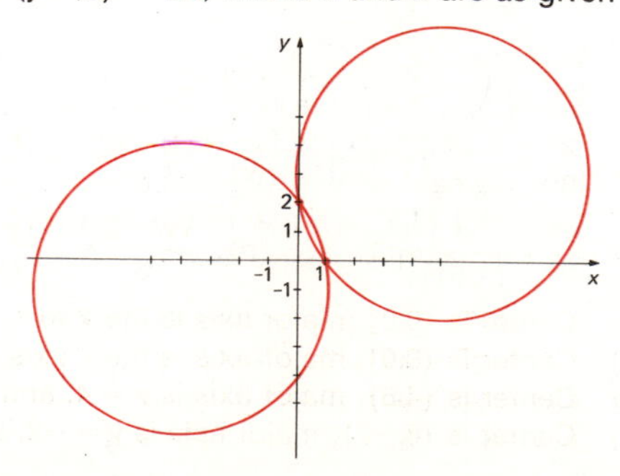

11.2.5 Let $(h,0)$ be the center of such a circle, and let r be its radius. Then an equation of this circle is $(x-h)^2+y^2=r^2$. Now, since $(3,2)$ lies on the circle, we have $(3-h)^2+2^2=r^2$. Also, since $(6,0)$ lies on the circle, we have $(6-h)^2+0^2=r^2$. Now, solving the last two equations, we obtain $h=23/6$ and $r=13/6$. Thus, the desired equation is $(x-23/6)^2+y^2=(13/6)^2$.

11.3.1 (*a*) $y^2=4x$ (*b*) $x^2=4y$ (*c*) $(y-2)^2=4x$ (*d*) $(x-2)^2=4y$ (*e*) $(y+5)^2=4(x-2)$ (*f*) $(x-3)^2=-8(y+3)$ (*g*) $(y+5)^2=-4(x-4)$ (*h*) $(x-3)^2=4(y+6)$

11.3.3 (*a*) $y^2=4(x-1)$ (*b*) $x^2=4(y-1)$ (*c*) $y^2=-8(x-2)$ (*d*) $y^2=8(x+2)$ (*e*) $x^2=-8(y-2)$ (*f*) $x^2=8(y+2)$ (*g*) $(x+1)^2=4(y+4)$ (*h*) $(y+3)^2=4(x+2)$

11.3.5 (*a*) $x=0$ (*b*) $y=0$ (*c*) $x=4$ (*d*) $x=-4$ (*e*) $y=4$ (*f*) $y=-4$ (*g*) $y=-5$ (*h*) $x=-3$

11.3.7 It can be shown that the equations of the two parabolas in Prob. 11.3.6 are: (1) $(y+3/2)^2=6(x+25/24)$ and (2) $(x-3/2)^2=-2(y-17/8)$. The parabola in (1) has vertex $(-25/24,-3/2)$, focus $(-25/24+3/2,-3/2)$, and directrix $x=-25/24-3/2$. The parabola in (2) has vertex $(3/2,17/8)$, focus $(3/2,17/8-1/2)$, and directrix $y=17/8+1/2$.

11.3.9 Solving the equations $y=8-x^2$ and $y=-10+x^2$ simultaneously, we obtain $x=3$, $y=-1$, or $x=-3$, $y=-1$. Thus, the points of intersection of the two given parabolas are $(3,-1)$ and $(-3,-1)$.

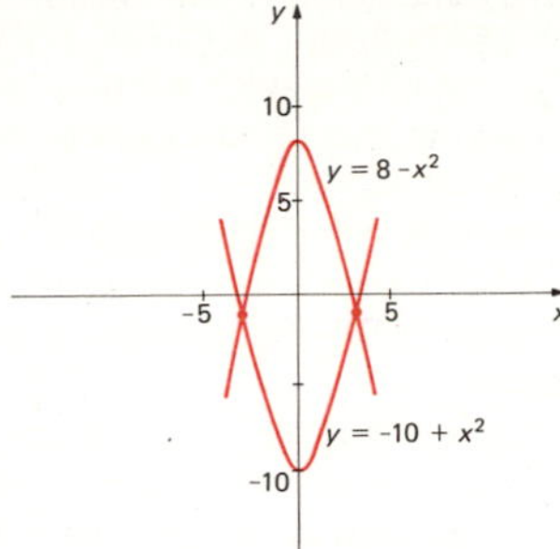

11.3.11 $x^2=-16/5\,y$

11.3.13 $y^2=4x$. This equation represents a parabola with focus $(1,0)$, vertex $(0,0)$, and directrix $x=-1$.

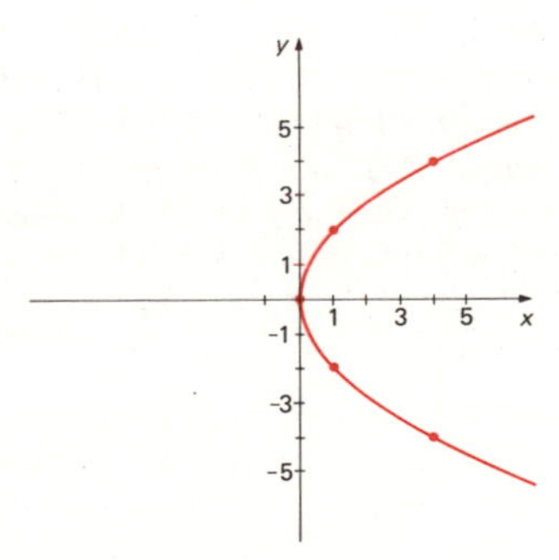

11.4.1 (*a*) $\dfrac{x^2}{3^2}+\dfrac{x^2}{(\sqrt{5})^2}=1$ (*b*) $\dfrac{y^2}{3^2}+\dfrac{x^2}{(\sqrt{5})^2}=1$ (*c*) $\dfrac{(y-6)^2}{6^2}+\dfrac{(x-4)^2}{(\sqrt{20})^2}=1$ (*d*) $\dfrac{(x-8)^2}{3^2}+\dfrac{(y+2)^2}{(\sqrt{5})^2}=1$

11.4.3 (*a*) Center is $(0,0)$, major axis is the x axis, and minor axis is the y axis.
(*b*) Center is $(0,0)$, major axis is the y axis, and minor axis is the x axis.
(*c*) Center is $(4,6)$, major axis is $x=4$, and minor axis is $y=6$.
(*d*) Center is $(8,-2)$, major axis is $y=-2$, and minor axis is $x=8$.

11.4.5 $\frac{y^2}{5^2}+\frac{x^2}{(15/4)^2}=1$

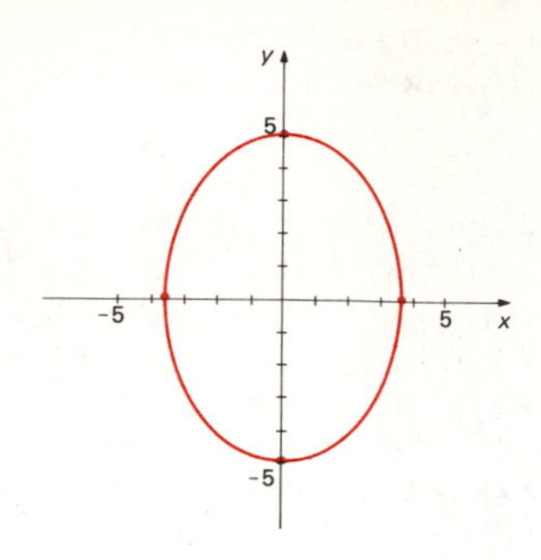

11.4.7 The vertices are (0,5) and (0,−5). The foci are $(0, \sqrt{175/16})$ and $(0, -\sqrt{175/16})$ [or $(0, 5\sqrt{7}/4)$ and $(0, -5\sqrt{7}/4)$]. The major axis is the y axis, and the minor axis is the x axis.

11.4.9 $x^2/3^2 + y^2/2^2 = 1$. This equation represents an ellipse with foci $(\sqrt{5},0)$ and $(-\sqrt{5},0)$, vertices (3,0) and (−3,0), and center (0,0); the major axis is the x axis and minor axis is y axis.

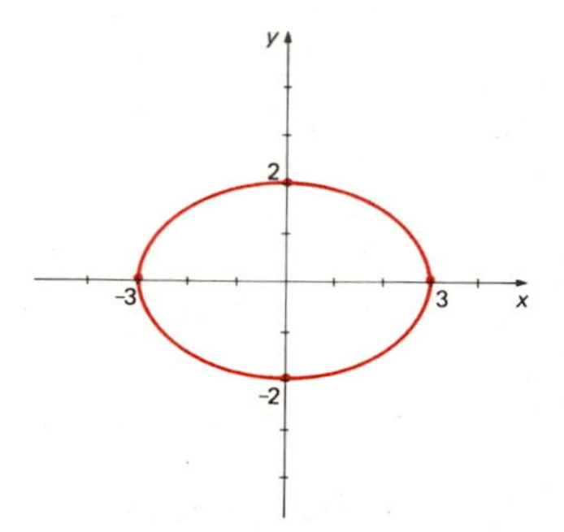

11.5.1 (*a*) $x^2/2^2 - y^2/(\sqrt{5})^2 = 1$ (*b*) $y^2/2^2 - x^2/(\sqrt{5})^2 = 1$

11.5.3 (*a*) Center is (0,0), transverse axis is the x axis, conjugate axis is the y axis, and the asymptotes are $y = (\sqrt{5}/2)x$ and $y = -(\sqrt{5}/2)x$.

(*b*) Center is (0,0), the transverse axis is the y axis, the conjugate axis is the x axis, and the asymptotes are $x = (\sqrt{5}/2)y$ and $x = -(\sqrt{5}/2)y$.

11.5.5 $y^2/4^2 - x^2/(\sqrt{27})^2 = 1$

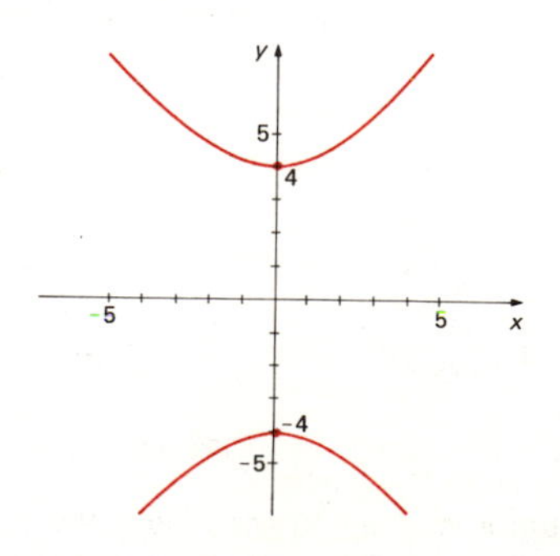

11.5.7 Vertices are (0,4) and (0,−4). Foci are $(0, \sqrt{43})$ and $(0, -\sqrt{43})$. Transverse axis is the y axis, and conjugate axis is the x axis.

11.5.9 Solving the two given equations simultaneously, we obtain $x = 1$, $y = 3$. Thus there is exactly one point of intersection, namely (1,3).

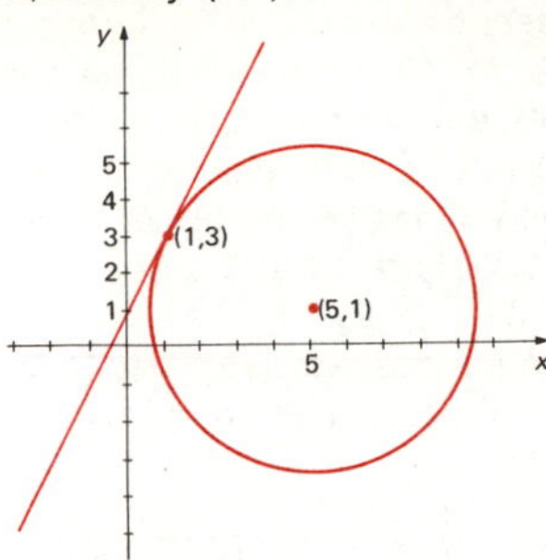

11.5.11 Arguing as in Prob. 11.5.9, we find that there are exactly two points of intersection, namely $(\sqrt{7/2}, 1 - \sqrt{7/2})$ and $(-\sqrt{7/2}, 1 + \sqrt{7/2})$.

11.5.13 (*a*) 2.1. (*b*) 1.9. The values of these slopes are close to 2.

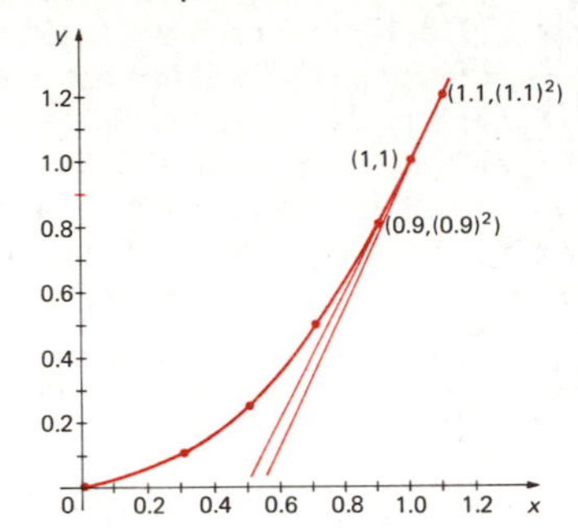

11.5.15

(*a*) 0	(*d*) 2	(*g*) $3c$
(*b*) 1	(*e*) $2c$	(*h*) a
(*c*) c	(*f*) 3	(*i*) $2a + b$

When h is very close to zero, but $h \neq 0$, the slope of each of the secant lines (corresponding to this value of h) is very close to the answers given above. For example, the secant line in part *d* is very close (in position) to the line with slope 2 and passing through the point (1,1).

Chapter 12

12.1.17 Statement A is false. In fact, if $n = 1$, then A becomes "$a + b > 0$," which is *not* necessarily true, since we may have $b \leq -a$. Thus, the first requirement in the principle of mathematical induction is *not* met.

12.2.1 $n(n+1) = n^2 + n$

12.2.3 $n(3n+5)/2 = (3n^2 + 5n)/2$

12.2.5 $n(4n-2)/2 = 2n^2 - n$

12.2.7 $n(4n-14)/2 = 2n^2 - 7n$

12.2.9 $(3^n - 1)/(3-1) = (3^n - 1)/2$

12.2.11 $5(5^n - 1)/(5-1) = (5^{n+1} - 5)/4$

12.2.13 $20/2\,[2(1) + 19(2)] = 400$

12.2.15 $1000(1000+1)/2 = 500{,}500$

12.2.17 Let r be the common ratio. Then, $1 \cdot r^{12-1} = 2048$. Hence $r = 2$. Therefore, the sum of this geometric progression is $1(2^{12} - 1)/(2-1) = 4095$.

12.2.21 The numbers x, $2x + 3$, $3x + 22$ form a geometric progression if and only if $(2x+3)/x$

$= (3x+22)/(2x+3)$. Solving this equation, we get $x=1$ or $x=9$. Thus, the numbers x, $2x+3$, $3x+22$ form a geometric progression if and only if $x=1$ or $x=9$. If $x=1$, the numbers become 1, 5, 25. In this case, the common ratio is 5. If $x=9$, the numbers become 9, 21, 49. In this case, the common ratio is $21/9 = 7/3$. Finally, the given numbers *never* form an arithmetic progression, since the *equation* $(2x+3) - x = (3x+22) - (2x+3)$ *has no solutions.*

12.2.25 Let a, b be the desired numbers. Then $(a+b)/2 = 15$ and $\sqrt{ab} = 9$. Solving these equations, we obtain $a=3$, $b=27$, or $a=27$, $b=3$. Thus, the numbers are 3 and 27, or 27 and 3.

12.2.27 Let a, b be the desired numbers. Then $2ab/(a+b) = 8$, and $\sqrt{ab} = 10$. Solving these equations, we get $a=20$, $b=5$, or $a=5$, $b=20$. Thus, the numbers are 20 and 5, or 5 and 20.

12.2.29 (*a*) $\dfrac{1}{1-1/3} = \dfrac{3}{2}$ (*b*) $\dfrac{1}{1-(-1/2)} = \dfrac{2}{3}$ (*c*) $\dfrac{5}{1-1/5} = \dfrac{25}{4}$ (*d*) $\dfrac{1}{1-(-1/7)} = \dfrac{7}{8}$

12.2.31 Let a be the first term of the given infinite geometric series. Then $a/(1-1/4) = 8$, and hence $a = 6$. Therefore the first six terms are 6, $3/2$, $3/8$, $3/32$, $3/128$, $3/512$.

12.3.1 (*a*) $(x-y)^8 = x^8 - 8x^7y + 28x^6y^2 - 56x^5y^3 + 70x^4y^4 - 56x^3y^5 + 28x^2y^6 - 8xy^7 + y^8$

(*b*) $(3x+y)^7 = 3^7x^7 + 7(3^6)x^6y + 21(3^5)x^5y^2 + 35(3^4)x^4y^3 + 35(3^3)x^3y^4 + 21(3^2)x^2y^5 + 7(3)xy^6 + y^7$

(*c*) $(2x-3y)^5 = 2^5x^5 - 5(2^4x^4)(3y) + 10(2^3x^3)(3y)^2 - 10(2^2x^2)(3y)^3 + 5(2x)(3y)^4 - (3y)^5$

(*d*) $(1-x)^{10} = 1 - 10x + 45x^2 - 120x^3 + 210x^4 - 252x^5 + 210x^6 - 120x^7 + 45x^8 - 10x^9 + x^{10}$

(*e*) $(x-1)^9 = x^9 - 9x^8 + 36x^7 - 84x^6 + 126x^5 - 126x^4 + 84x^3 - 36x^2 + 9x - 1$

(*f*) $(3-y)^6 = 3^6 - 6(3^5)y + 15(3^4)y^2 - 20(3^3)y^3 + 15(3^2)y^4 - 6(3)y^5 + y^6$

12.3.3 $120(3)^3 = 3240$

12.3.7 (*a*) 0 (*c*) $(1+1)^5 = 32$ (*e*) 15

(*b*) 0 (*d*) $(1-1)^5 = 0$ (*f*) 0

12.3.9
$$\binom{n}{r} + \binom{n}{r-1} = \frac{n!}{r!(n-r)!} + \frac{n!}{(r-1)!(n-r+1)!} = \frac{n!}{r!(n-r+1)!}[(n-r+1)+r] = \frac{(n+1)!}{r!(n-r+1)!} = \binom{n+1}{r}$$

12.3.11
$$\frac{f(x+h)-f(x)}{h} = \frac{[3(x+h)^2 - 4(x+h) + 7] - (3x^2 - 4x + 7)}{h} = \frac{h(6x - 4 + 3h)}{h} = 6x - 4 + 3h \quad \text{if } h \neq 0$$

Thus, $[f(x+h) - f(x)]/h = 6x - 4 + 3h$, if $h \neq 0$. Hence, $[f(x+h) - f(x)]/h$ is very close to $6x - 4$ when h is very close to zero, but $h \neq 0$.

12.3.13

h	*Slope of secant line* $= \dfrac{f(1+h) - f(1)}{h}$
0.1	$6 + 6(0.1) + 2(0.1)^2 = 6.62$
−0.1	$6 - 6(0.1) + 2(0.1)^2 = 5.42$

The slopes of the above secant lines are close to 6. Indeed, if h is very close to zero, then the slope of the secant line (corresponding to this value of h) is very close to 6. In such a case, the secant line is very close (in position) to the line with slope 6 and passing through the point (1,2).

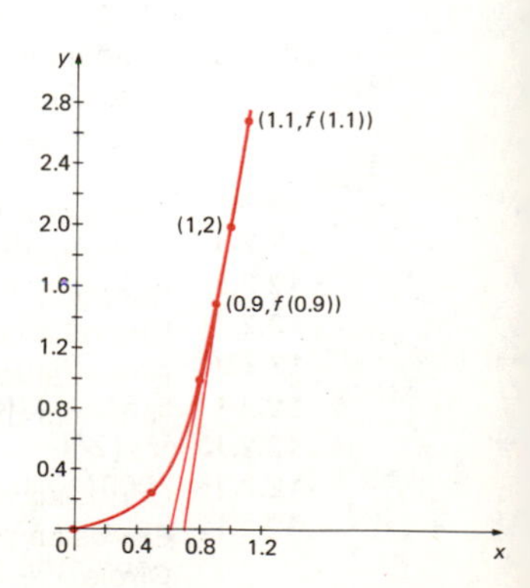

Index